AF805894

Subcellular Biochemistry

Series Editor
Robin Harris

For further volumes:
http://www.springer.com/series/6515

Tony Harris
Editor

Adherens Junctions: From Molecular Mechanisms to Tissue Development and Disease

Springer

Editor
Tony Harris
Department of Cell and Systems Biology
University of Toronto
Toronto, ON
Canada

ISSN 0306-0225
ISBN 978-94-007-4185-0 ISBN 978-94-007-4186-7 (eBook)
DOI 10.1007/978-94-007-4186-7
Springer Dordrecht Heidelberg London New York

Library of Congress Control Number: 2012938445

Springer Science+Business Media Dordrecht 2012
No part of this work may be reproduced, stored in a retrieval system, or transmitted in any form or by any means, electronic, mechanical, photocopying, microfilming, recording or otherwise, without written permission from the Publisher, with the exception of any material supplied specifically for the purpose of being entered and executed on a computer system, for exclusive use by the purchaser of the work.

Printed on acid-free paper

Springer is part of Springer Science+Business Media (www.springer.com)

Contents

Chapter 1
An Introduction to Adherens Junctions: From Molecular Mechanisms to Tissue Development and Disease

Tony J. C. Harris

Abstract Adherens junctions (AJs) are fundamental for the development of animal tissues and organs. The core complex is formed from transmembrane cell–cell adhesion molecules, cadherins, and adaptor molecules, the catenins, that link to cytoskeletal and regulatory networks within the cell. This complex can be considered over a wide range of biological organization, from atoms to molecules, protein complexes, molecular networks, cells, tissues, and overall animal development. AJs have also been an integral part of animal evolution, and play central roles in cancer development and pathogen infection. This book addresses major questions encompassing these aspects of AJ biology. How did AJs evolve? How do the cadherins and catenins interact to assemble AJs and mediate adhesion? How do AJs interface with other cellular machinery to couple adhesion with the whole cell? How do AJs affect cell behaviour and multicellular development? How can abnormal AJ activity lead to disease?

The evolution of animals has been coupled with the evolution of AJs. Animals are made up of multicellular tissues which require cell–cell adhesion. AJs and their components are one of the most pervasive mechanisms used for this adhesion. In Chap. 2, Hiroki Oda discusses how AJs from diverse species can have very similar ultrastructures when observed by electron microscopy. In contrast, the molecules making up the AJs have undergone striking changes. Perhaps most remarkably, the cellular slime mould *Dictyostelium discoideum* contains catenins but not cadherins, and nonetheless forms catenin-associated adhesion complexes resembling AJs at the ultrastructural level. In this case, the adhesion receptor remains unknown. In other primitive species, cadherins are present, contain cytoplasmic binding sites for the catenins, but have much larger extracellular domains compared to the cadherins found in AJs of vertebrate tissues (Oda 2012). The very earliest cadherins may have played roles in capturing extracellular prey, but over evolutionary time, the extracellular domains have been paired down to mediate adhesion in tissues of greater and greater complexity.

T. J. C. Harris (✉)
Department of Cell & Systems Biology, University of Toronto, Toronto, ON M5S 3G5, Canada
e-mail: tony.harris@utoronto.ca

T. Harris (ed.), *Adherens Junctions: From Molecular Mechanisms to Tissue Development and Disease,* Subcellular Biochemistry 60,
DOI 10.1007/978-94-007-4186-7_1, © Springer Science+Business Media Dordrecht 2012

Structural studies have revealed the architecture of the cadherin–catenin complex in vertebrates. In Chap. 3, Noboru Ishiyama and Mitsuhiko Ikura review the structure of this adhesive device. In vertebrate AJs, the cadherins contain five extracellular domains with calcium-binding sites between them. Straightened by calcium-binding, cadherins from one cell extend outward to engage cadherins from neighbouring cells. Structural studies indicate that the cadherin domain furthest from the membrane inserts a tryptophan residue into a hydrophobic pocket of the same domain of the cadherin emanating from the opposite cell. This exchange is reciprocal with the opposite cadherin also inserting a tryptophan residue. This strand-swap dimer formation is a major way cadherins interact in *trans*, but the calcium-binding sites have also been found to form distinct *trans* dimers (X-dimers), and other sites form *cis* dimers between cadherins along the same membrane. On the cytoplasmic side, the catenins wrap around distinct portions of the cadherin tail to form the cadherin–catenin complex. Through competitive interactions and phosphorylation, these binding sites impact the stability and cytoskeletal interactions of the complex (Ishiyama and Ikura 2012).

For AJs to function in adhesion, cadherin–cadherin interactions must withstand forces between cells. In Chap. 4, Deborah Leckband and Sanjeevi Sivasankar compare and contrast biophysical studies of cadherin interactions in light of the known structural information. Interaction strengths have been quantified both in solution and with forces applied to immobilized proteins. These studies have implicated both the strand-swap dimer and the X-dimer in cadherin adhesive interactions. Intriguingly, the X-dimer is weaker in solution compared to when forces are applied to the interaction. Leckband and Sivasankar speculate that this interaction may act as a catch bond, one that increases its strength when pulled. Evidence for additional interaction sites along the length of the molecule is also discussed (Leckband and Sivasankar 2012). In Chap. 5, Sergey Troyanovski discusses how cadherins interact between cells. Mutagenesis studies have shown that the strand-swap dimer plays a key role in adhesion. These dimers are supported by *cis*-dimerization, and together these interactions form small cadherin clusters. Dr. Troyanovski discusses how the X-dimer may be an intermediate state involved in forming the strand-swapped dimer, and that the X-dimer is also important for reversing cadherin interactions as the clusters dissolve. The clustering and reorganization of cadherins is important for building the larger adhesion complexes specifically called 'AJs'. Notably, AJs appear to be comprised of many loosely packed cadherin clusters and can take on a range of higher-order morphologies dependent on associated proteins and on intercellular forces (Troyanovski 2012).

With AJ assembly and remodeling, the adhesion complex interfaces with other complexes within the cell. In Chap. 6, Siew Ping Han and Alpha Yap address cytoskeletal connections, focusing mainly on actin. AJs can form direct links with the actin cytoskeleton through several mechanisms, which may be cell type specific. AJs also impact actin more indirectly by altering the activity of actin regulators. These connections remodel the cytoskeleton and simultaneously reconfigure the AJs to build dynamic protein networks running through and between cells to physically control tissue structure (Han and Yap 2012). In Chap. 7, Yohei Shimono,

Yoshiyuki Rikitake, Kenji Mandai, Masahiro Mori and Yoshimi Takai discuss how another adhesion system based on the immunoglobulin superfamily receptor nectin is coupled with AJs. Similar to cadherins, nectins assemble adhesion complexes through receptor clustering and actin associations. This assembly can be closely coupled to AJ assembly by direct interactions with AJ proteins. Expanding the complexity at the plasma membrane, nectins also interface with other membrane proteins, such as integrins. Cell-type specific expression of different types of nectins is responsible for guiding the assembly of complex epithelial tissues in which multiple cell types are assembled in a precise architecture, such as auditory epithelial cells (Shimono et al. 2012). In Chap. 8, Abbye McEwen, David Escobar and Cara Gottardi explore how adherens junctions can influence signaling through the cytosol to the nucleus. Through cell adhesion, AJs can generally impact signaling dependent on cell contacts or polarized epithelial structure. Additionally, the catenins are known to localize to the nucleus and interact with transcriptional machinery, and the balance between junctional and nuclear localization can impact gene expression. Moreover, AJ proteins interact directly with signaling pathway components to regulate cell growth (McEwen et al. 2012). In Chap. 9, Benjamin Nanes and Andrew Kowalczyk review how endocytic machinery engages cadherin–catenin complexes to remove AJs from the plasma membrane. This engagement is normally inhibited by p120-catenin binding to the juxtamembrane region of the cadherin cytoplasmic tail. When the p120-catenin interaction is lost, endocytic machinery binds the region and drives cadherin endocytosis. Thus, the amount of p120 in a cell can dictate cadherin levels at the plasma membrane. Cadherin endocytosis is additionally regulated by ubiquitination and growth factor signaling pathways, and cadherins can also be removed from the plasma membrane by proteolytic cleavage. The balance of adherens junction assembly and disassembly is important for tissue development and maintenance, and imbalances are linked to disease (Nanes and Kowalczyk 2012). In Chap. 10, Sandrine Etienne-Manneville synthesizes the known connections between AJs and other cellular machinery in a review of AJs during cell migration. Depending on the connections in place, AJs can facilitate the migration of single cells, maintain cells together as a migratory group or prevent cell migration altogether. Understanding how these decisions are made is critical for understanding tissue development and disease (Etienne-Manneville 2012).

AJs are critical for orchestrating multicellular development. In Chap. 11, Annalisa Letizia and Marta Llimargas survey the regulation and roles of AJs through the *Drosophila* life cycle. From the formation of the oocyte to the development of the larvae and its organs, AJs are differentially regulated through altered cytoskeletal associations and endocytosis to shape specific tissues. These models highlight the remarkable pliability of AJ-based adhesion during different types of morphogenesis. *Drosophila* has also provided *in vivo* examples of AJ functions in cell sorting and oriented cell division (Letizia and Llimargas 2012). In Chap. 12, Stephen Armenti and Jeremy Nance further highlight the importance of AJs in morphogenesis, specifically in *C. elegans*. In addition to providing important models of morphogenesis, *C. elegans* offers a twist on AJ biology. Here, core AJ proteins associate with a larger set of proteins to form a multifaceted apical junction which mediates both adhesive and barrier functions. Moreover, core AJ proteins are not essential for basic cell adhesion and epithelial structure, due to redundancies with other members of the larger junctional complex. However, the role of AJs becomes apparent under the mechanical strains of morphogenetic movements (Armenti and Nance 2012). In Chap. 13, Rudolf Winklbauer takes us into the vertebrate embryo, reviewing Xenopus gastrulation. Here, the developing gastrula is sculpted by the remodeling of a multilayered epithelium, in which internal cells make cadherin-based contacts in all directions over their entire plasma membrane. These adhesive contacts give the multilayered tissue fluid-like properties with different portions of the embryo displaying distinct surface tensions. Dissociated Xenopus cells have long been a model of cell sorting based on differential adhesion, but Dr. Winklbauer discusses recent data arguing against a major role for differential cadherin-based adhesion in the developing embryo. Rather, directed cell migration and active cell repulsion may shape the embryo while cadherins flexibly hold tissues together (Winklbauer 2012). In Chap. 14, Barbara Boggetti and Carien Niessen review the roles of AJs in the mouse. Here, AJs are critical for forming the first epithelium of the developing embryo, and conditional knock-outs have revealed essential roles and redundancies between AJ components in a variety of tissues. These findings highlight the importance of AJs for tissue morphogenesis and homeostasis and the mouse provides numerous models for understanding the roles of AJs in human disease (Boggetti and Niessen 2012).

Finally, AJs have major influences on tissue maintenance, repair and disease. In Chap. 15, Terry Lechler reviews the roles of AJs in stem cell biology. AJs mediate interactions among stem cells and with their niches. They often function to physically anchor stem cells within the niche, and additionally direct asymmetric stem cell divisions and the control of cell proliferation and differentiation. Dr. Lechler discusses the main models of basic stem cell biology involving AJs, and then probes how AJs influence applied stem cell biology, including the propagation of embryonic and induced pluripotent stem cells (Lechler 2012). In Chap. 16, Valeri Vasioukhin dissects the connections between AJs and cancer. Association studies have linked the loss of AJ proteins with increased cancer invasiveness and worsened patient outcomes. At the cellular level, the loss of AJ proteins is linked to epithelial-to-mesenchymal transition and the metastasis of cancer cells. AJ proteins can be lost in a number of ways including gene mutation or down-regulation, either transcriptionally or post-transcriptionally. Importantly, experimental studies have verified causal links between AJ loss and cancer progression. AJs appear to suppress cancer through their adhesive activity and their cross-talk with cellular signaling networks (Vasioukhin 2012). In Chap. 17, Georgios Nikitas and Pascale Cossart review how pathogens engage or modify AJs to gain access to cells and body compartments. AJs are at the core of a remarkable range of strategies used by pathogens to invade the host. These include *Listeria monocytogenes*' engagement of cadherins for internalization into the cell, and *Heliobacter pylori*'s destabilization of AJs or *Bacteroides fragilis*' cleavage of cadherins to weaken epithelial barriers (Nikitas and Cossart 2012).

Overall, these reviews provide a comprehensive and current view of how AJs arose; how they assemble; how they integrate with other cellular networks; how they help orchestrate development; and how they impact disease. They also highlight how exciting AJ research is, and will hopefully inspire future discoveries by those established and new in the field.

References

Armenti S, Nance J (2012) Adherens junctions in C. elegans embryonic morphogenesis. Subcell Biochem 60:279–299

Boggetti B, Niessen C (2012) Adherens junctions in mammalian development, homeostasis and disease: lessons from mice. Subcell Biochem 60:321–355

Etienne-Manneville S (2012) Adherens junctions during cell migration. Subcell Biochem 60:225–249

Han SP, Yap A (2012) The cytoskeleton and classical cadherin adhesions. Subcell Biochem 60:111–135

Ishiyama N, Ikura M (2012) The three-dimensional structure of the cadherin-catenin complex. Subcell Biochem 60:39–62

Lechler T (2012) Adherens junctions and stem cells. Subcell Biochem 60:359–377

Leckband D, Sivasankar S (2012) Biophysics of cadherin adhesion. Subcell Biochem 60:63–88

Letizia A, Llimargas M (2012) Adherens junctions and cadherins in Drosophila development. Subcell Biochem 60:251–277

McEwen A, Escobar D, Gottardi C (2012) Signaling from the adherens junction. Subcell Biochem 60:171–196

Nanes B, Kowalczyk A (2012) Adherens junction turnover: regulating adhesion through cadherin endocytosis, degradation, and recycling. Subcell Biochem 60:197–222

Nikitas G, Cossart P (2012) Adherens junctions and pathogens entry. Subcell Biochem 60:415–425

Oda H (2012) Evolution of the cadherin-catenin complex. Subcell Biochem 60:9–35

Shimono Y, Rikitake Y, Mandai K, Mori M, Takai Y (2012) Immunoglobulin superfamily receptors and adherens junctions. Subcell Biochem 60:137–170

Troyanovski S (2012) Adherens junction assembly. Subcell Biochem 60:89–108

Vasioukhin V (2012) Adherens junctions and cancer. Subcell Biochem 60:379–414

Winklbauer R (2012) Cadherin function during Xenopus gastrulation. Subcell Biochem 60:301–320

Part I
How AJs Evolved

Chapter 2
Evolution of the Cadherin–Catenin Complex

Hiroki Oda

Abstract Adherens junctions are the most common junction type found in animal epithelia. Their core components are classical cadherins and catenins, which form membrane-spanning complexes that mediate intercellular binding on the extracellular side and associate with the actin cytoskeleton on the intracellular side. Junctional cadherin–catenin complexes are key elements involved in driving animal morphogenesis. Despite their ubiquity and importance, comparative studies of classical cadherins, catenins and their related molecules suggest that the cadherin/catenin-based adherens junctions have undergone structural and compositional transitions during the diversification of animal lineages. This chapter describes the molecular diversities related to the cadherin–catenin complex, based on accumulated molecular and genomic information. Understanding when and how the junctional cadherin–catenin complex originated, and its subsequent diversification in animals, promotes a comprehensive understanding of the mechanisms of animal morphological diversification.

2.1 Introduction

All multicellular animals composed of differentiated tissues, such as epithelia, are referred to as metazoans. The shaping of their bodies relies on cell–cell adhesion and its regulation. An understanding of how the mechanisms of cell–cell adhesion originated and evolved in animals is required to understand the mechanisms that regulate the morphological diversification of animals.

Metazoan cells adhere to each other using specialized membrane structures termed intercellular junctions. Cells in the differentiated epithelia of vertebrates typically have a junctional complex consisting of a tight junction, an adherens junction and a desmosome (Fig. 2.1a). However, this junction organization is not universal, even within the phylum Chordata. Desmosomes are unique to vertebrates, and tight junctions are unique to vertebrates and urochordates (e.g., ascidians). In the epithelia of cephalochordates (e.g., amphioxus), adherens junctions are the

H. Oda (✉)
JT Biohistory Research Hall, 1-1 Murasaki-cho, Takatsuki, Osaka 569-1125, Japan
e-mail: hoda@brh.co.jp

T. Harris (ed.), *Adherens Junctions: From Molecular Mechanisms to Tissue Development and Disease,* Subcellular Biochemistry 60,
DOI 10.1007/978-94-007-4186-7_2, © Springer Science+Business Media Dordrecht 2012

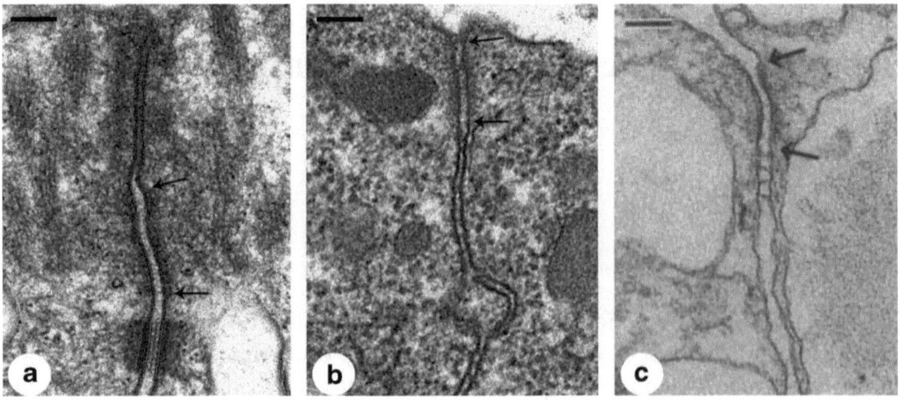

Fig. 2.1 Transmission electron micrographs of intercellular junctions in metazoan epithelia. **a** A junctional complex consisting of a tight junction (*top*), an adherens junction (*middle*) and a desmosome (*bottom*) in the epithelium of a mouse small intestine (Courtesy of Dr. Tomohiro Haruta, JEOL Ltd.). **b** A junctional complex consisting of an adherens junction and a septate junction in the epithelium of a *Drosophila* salivary gland (Courtesy of Dr. Tomohiro Haruta, JEOL Ltd.). **c** A junctional complex consisting of an adherens junction and a septate junction in the epithelium of an *Asterina* (starfish) midgastrula (Copyright 1995 Wiley-Liss Inc. Used with permission from Dan-Sohkawa et al. (1995)). In all panels, the apical end of the lateral cell–cell contact is at the top, and the arrows sandwich an adherens junction

only junction type expressed (Lane et al. 1987). Junctional complexes consisting of an adherens junction and septate junction are widely observed in the epithelia of non-chordate metazoan animals including arthropods, echinoderms and cnidarians (Fig. 2.1b, c). The various junction types show distinct phylogenetic distributions; the most common junction type is the adherens junction (Oda and Takeichi 2011), which is found in all metazoan phyla including the Porifera.

Molecular and genetic studies of vertebrates, *Drosophila melanogaster* and *Caenorhabditis elegans* have revealed that the adherens junctions of these animals share common molecular compositions and organizations, providing strong support for homology of this junction type across the Bilateria (Knust and Bossinger 2002). The core components of adherens junctions are cadherins and catenins, which form complexes that mediate cell–cell adhesion and the association of adherens junctions with the actin cytoskeleton. The cadherin–catenin complex is a basic molecular machinery involved in various morphogenetic processes including cell migration, cell rearrangement, epithelial folding and epithelial-to-mesenchymal transitions.

Despite the ubiquity of cadherin/catenin-based adherens junctions and their importance in shaping animal bodies, comparative studies of cadherins, catenins and related molecules suggest that the adherens junctions have undergone structural and compositional transitions during the diversification of animal lineages. This chapter does not cover the functional and mechanistic details of the cadherin–catenin complex, but instead focuses on describing the molecular diversities related to its components.

2.2 Cadherin and Catenins at Adherens Junctions

2.2.1 Classical Cadherins in Vertebrates

The first molecule to be termed "cadherin" was *Mus musculus* (mouse) E-cadherin (Yoshida-Noro et al. 1984), although it is also referred to as uvomorulin and CDH1 (Hyafil et al. 1981). This cadherin was identified using antibodies that were capable of inhibiting calcium-dependent cell–cell adhesion and cell compaction (Kemler et al. 1977; Hyafil et al. 1981; Yoshida and Takeichi 1982; Nagafuchi et al. 1987; Ringwald et al. 1987). L-CAM, a cell adhesion molecule independently identified in chickens, is an ortholog of mouse E-cadherin (Brackenbury et al. 1981; Gallin et al. 1987). E-cadherin is enriched in adherens junctions in a wide range of epithelial tissues (Boller et al. 1985; Takeichi 1988). N-cadherin (also referred to as CDH2) is abundant in mesodermal and neural tissues and serves as a major adhesion molecule at the adherens junctions in these tissues in the place of E-cadherin (Volk and Geiger 1984; Hatta and Takeichi 1986). VE-cadherin (also referred to as CDH5) is another representative cadherin, which is specifically expressed by endothelial tissues (Heimark et al. 1990) and functions at endothelial adherens junctions. E-, N-, VE- and other cadherins share a common structure and form a molecular family; they consist of five extracellular cadherin domains (ECs) , a transmembrane domain and a cytoplasmic domain (Fig. 2.2a). Their cytoplasmic domains have highly conserved sequences and bind p120- and β-catenins at the juxtamembrane and C-terminal regions, respectively (Fig. 2.2b, c). Each cadherin subtype preferentially binds to the same subtype in a homotypic fashion (Nagafuchi et al. 1987; Nose et al. 1988). The most membrane-distal EC (EC1) contains amino acid sites that are critical in determining the binding specificities of the cadherins (Nose et al. 1990). Another common feature of the cadherins is the presence of a precursor domain that is removed during the maturation process to activate the cadherin (Shirayoshi et al. 1986; Ozawa and Kemler 1990).

In later studies, an increasing number of genes containing ECs has been found in vertebrate and non-vertebrate animals, but many of these ECs are structurally distinct from the original cadherins. Therefore, although the general term "cadherin" refers to any member of the cadherin superfamily (molecules having ECs), members of the first identified cadherin family are referred to as "classical cadherins." Typical ECs contain conserved amino acid sequence motifs, such as "DxD," "DRE" and "DxNDN," which are involved in the Ca^{2+} binding that is necessary for protease resistance and interdomain rigidification (Ozawa et al. 1990; Overduin et al. 1995; Nagar et al. 1996).

2.2.2 **Drosophila** *Homologs of Classical Cadherins*

The first non-vertebrate member of the cadherin superfamily identified was the product of the *Drosophila fat* gene; mutations of this gene cause tumor-like overgrowth

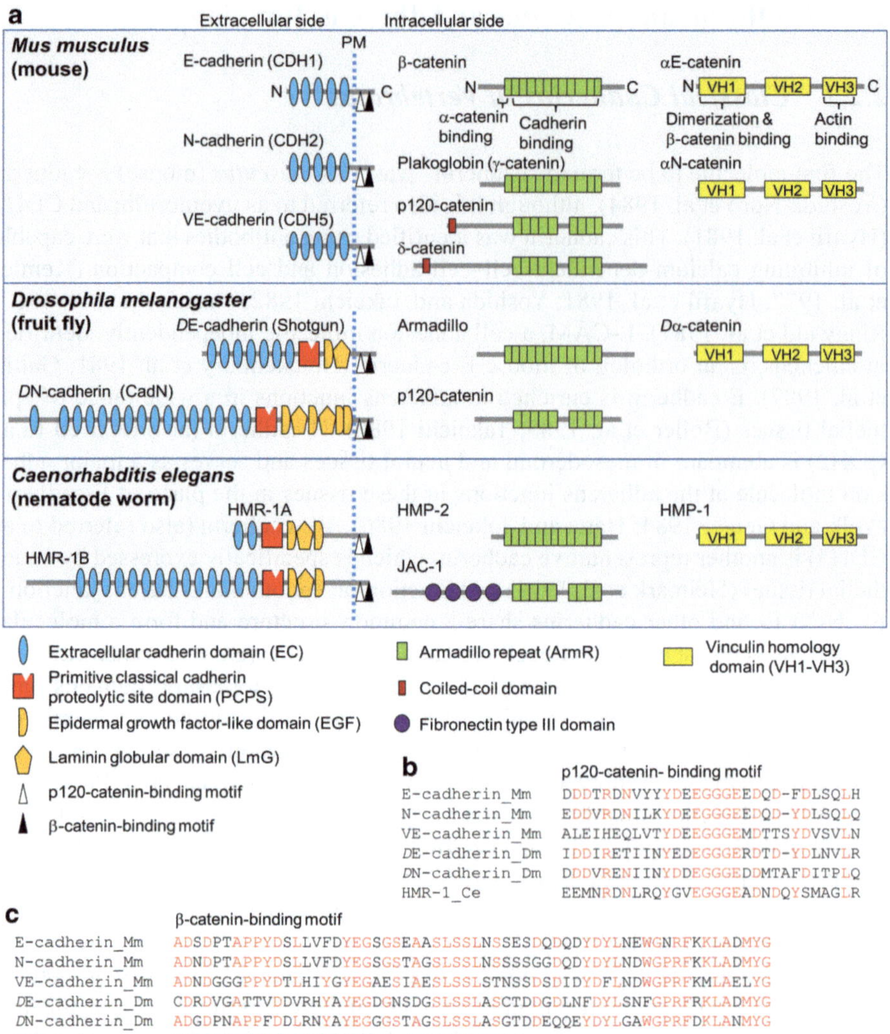

Fig. 2.2 Classical cadherins and catenins in vertebrate and non-vertebrate model species. **a** Schematic illustrations showing the domain structures of classical cadherins and catenins in *Mus musculus* (mouse) and those of their homologues in *Drosophila melanogaster* (fruit fly) and *Caenorhabditis elegans* (nematode worm). The regions that are responsible for interactions between the classical cadherin, β-catenin, αE-catenin and actin are also shown. PM, plasma membrane. **b, c** Amino acid sequence alignments for the p120-catenin- **(b)** and β-catenin- **(c)** binding motifs of the cadherins shown in **(a)** Residues that are identical among four or more of the proteins are colored. HMR-1 is truncated at the C-terminus

of the larval imaginal discs without disrupting the multicellular and epithelial organization (Mahoney et al. 1991). Fat is a single-pass transmembrane protein that contains 34 ECs, four epidermal growth factor-like repeat domains (EGFs) and two laminin A globular domains (LmGs) in its extracellular region (Fig. 2.3a). The amino acid sequence of the cytoplasmic domain of Fat largely differs from those of the classical cadherins.

a

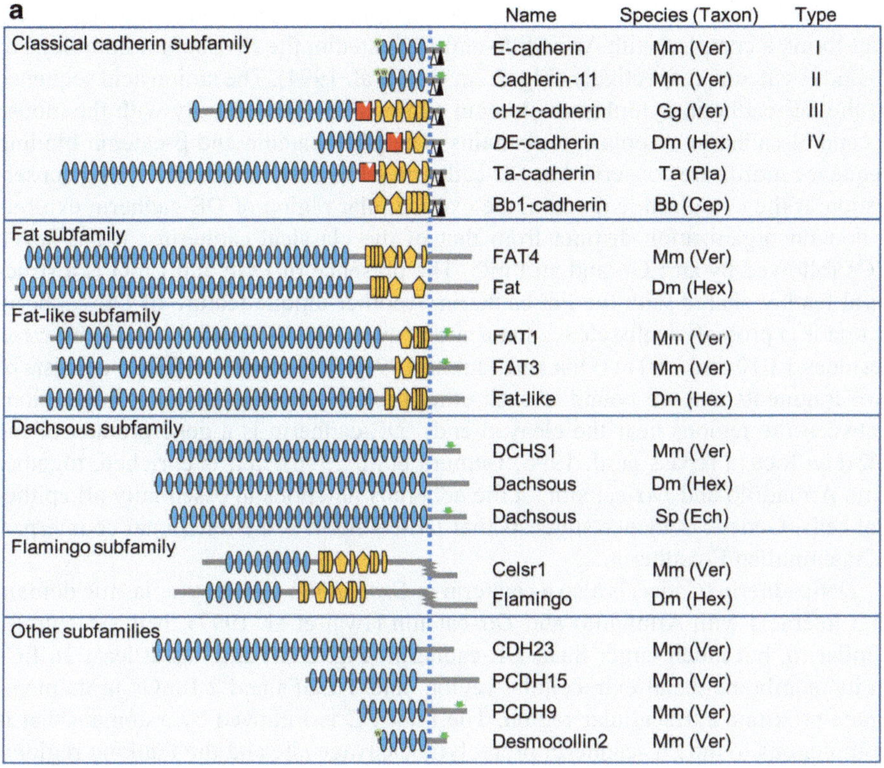

b

```
E-cadherin_Mm     <GSGSEAASLSSLNSSESDQDQDY>
FAT1_Mm         SARESLSEVQSLSSFQS-ESCDDNGYHWDTSDWMPSVPLPDIQEFPNYE
FAT3_Mm         SNKGSNSEVQSLSSFQS-DSGDDNAYHWDTSDWMPGARLSDIEEMPNYE
PCDH9_Mm                                   <HSDN-CWMP>
```

c

```
DCHS1_Mm        YGFPADGKPCVAGALTAI--VAGEEELRGSYNWDYLLSWCPQFQPLASVFT
Dachsous_Sp     FGFTEDGHPSMAGSLSSI--VNSDEELSGSYNWDYLLDWGPQFQPLAHVFA
Cadherin-11_Mm  YGY--EGRGSVAGSLSSLESATTDSDL----DYDYLQNWGPRFKKLADLYG
```

d

```
E-cadherin_Mm    ADSDPTAPPYDSLLVFDYEGSGSEAASLSSLNSSESDQDQDY-DYLNEWGNRFKKLADMYG
Cadherin-11_Mm   ADNDPTAPPYDSIQIYGYEGRGSVAGSLSSLESATTDSDLDY-DYLQNWGPRFKKLADLYG
Desmocollin2_Mm  DDNQKLAQDY--VLTYNYEGKGSAAGSVGCCSDLQEEDGLEFLDHLE---PKFRTLAEVCA
```

Fig. 2.3 Major subfamilies of the cadherin superfamily in metazoans. **a** Schematic illustrations showing the varied domain structures of selected cadherin superfamily members. Domains and motifs are indicated as in Fig. 2.2. Conserved tryptophan (W) residues at the N-terminal regions of type I, type II and desmosomal cadherins are also shown. Abbreviations of species are as follows: Mm, *Mus musculus* (mouse); Gg, *Gallus gallus* (chicken); Dm, *Drosophila melanogaster* (fruit fly); Ta, *Trichoplax adhaerens*; Bb, *Branchiostoma belcheri* (amphioxus); and Sp, *Strongylocentrotus purpuratus* (sea urchin). Abbreviations of taxa in parentheses are as follows: Ver, Subphylum Vertebrata (Phylum Chordata); Hex, Superclass Hexapoda (Phylum Arthropoda); Pla, Phylum Placozoa; Cep, Subphylum Cephalochordata (Phylum Chordata); and Ech, Phylum Echinodermata. **b** Partial sequence similarities in the cytoplasmic domains of classical cadherin (E-cadherin), FAT1, FAT3 and PCDH9. **c** Partial sequence similarities in the cytoplasmic domains of the classical cadherin (cadherin-11), DCHS1 and sea urchin Dachsous. **d** Sequence similarities in the cytoplasmic domains of classical and desmosomal cadherins. Amino acid sequences derived from the regions indicated by green arrows in **a** are aligned in **b** to **d**. Residues that are identical among two or more of the proteins are colored

*D*E-cadherin, a second cadherin in *Drosophila*, was identified as a glycoprotein that forms a complex with Armadillo and *D*α-catenin, the *Drosophila* homologs of β- and α-catenin, respectively (Fig. 2.2a; Oda et al. 1994). The amino acid sequence of the *D*E-cadherin cytoplasmic domain exhibits 33–37% identity with the mouse E- and N-cadherin cytoplasmic domains. The p120-catenin and β-catenin binding sequence motifs are conserved in *D*E-cadherin. However, despite the strong conservation of the cytoplasmic domain, the extracellular region of *D*E-cadherin exhibits a domain organization distinct from that of the classical cadherins; it has seven ECs followed by an EGF and an LmG. The presence of EGF and LmG is a structural feature shared with the Fat cadherin. Another unique feature of *D*E-cadherin is that it is proteolytically cleaved at a site between the EC7 and the EGF (between residues 1,010 and 1,011) (Oda and Tsukita 1999). Mature *D*E-cadherin consists of two fragments that are bound to each other probably by non-covalent interactions between the regions near the cleaved ends. *D*E-cadherin is a gene product of the *shotgun* locus (Tepass et al. 1996; Uemura et al. 1996), and is enriched, together with Armadillo and *D*α-catenin, at the adherens junctions in essentially all epithelial cells. Genetic evidence suggests that *D*E-cadherin is the functional counterpart of mammalian E-cadherin.

*D*N-cadherin (CadN) is also a cadherin in *Drosophila* with a cytoplasmic domain that interacts with Armadillo and *D*α-catenin (Iwai et al. 1997). It is structurally similar to, but much larger than, *D*E-cadherin (Fig. 2.2a) and has at least 16 ECs in its membrane-distal extracellular region, and 4 EGFs and 2 LmGs in its membrane-proximal extracellular region. The final EC is followed by a domain that is homologous to the *D*E-cadherin proteolytic cleavage site and the flanking regions. Since immunochemical data suggest that *D*N-cadherin consists of two fragments as well (Iwai et al. 1997), the proteolytic cleavage is likely to be conserved in *D*N-cadherin. Like vertebrate N-cadherin, *D*N-cadherin is expressed in mesodermal and neural tissues. The functions of *D*N-cadherin are also similar to the functions of the vertebrate N-cadherin (Takeichi 2007). Thus, the relationship between *D*E- and *D*N-cadherin is analogous to the relationship between E and N-cadherin, despite the structural differences existing between *Drosophila* and vertebrate cadherins.

A *D*N-cadherin-like gene (*CadN2*) exists next to the *D*N-cadherin gene in the *Drosophila* genome. However, this cadherin exhibits no detectable adhesion activity, and *CadN2*-null mutants are viable, although subtle functions for CadN2 are detectable (Prakash et al. 2005; Yonekura et al. 2007).

2.2.3 C. elegans *Homologs of Classical Cadherins*

A study of *C. elegans* identified three genes, *hmp-1*, *hmp-2* and *hmr-1*, that are related to α-catenin, β-catenin/Armadillo and classical cadherin, respectively (Fig. 2.2a; Costa et al. 1998). In the *C. elegans* genome, *hmr-1* is the sole cadherin gene related to classical cadherins. The products of *hmp-1*, *hmp-2* and *hmr-1* localize to hypodermal (or epidermal) adherens junctions and their activities are required for hypodermal ventral closure during mid-embryogenesis; however, none of these products are essential for cell–cell adhesion and cell shape regulation before and during gastrulation. Even at early stages, HMR-1 functions in blastoderm compaction and gastrulation, but these functions are redundant with those of an immunoglobulin domain adhesion molecule, SAX-7 (Grana et al. 2010).

hmr-1 encodes two isoforms, HMR-1A and HMR-1B, which have 2 and 13 ECs, respectively, followed by *D*N-cadherin-like membrane-proximal extracellular domains (Broadbent and Pettitt 2002). Whereas the HMR-1A transcript is expressed to play a role in hypodermal morphogenesis, the HMR-1B transcript is transcribed by an alternative, neuron-specific promoter, and subjected to alternative splicing. HMR-1B and *D*N-cadherin resemble each other in their domain organizations and *in vivo* functions, and the relationship between HMR-1A and HMR-1B is analogous to the relationships between *D*E- and *D*N-cadherin and between E- and N-cadherin.

2.2.4 β-Catenin/Armadillo

β-catenin/Armadillo functions as a part of the cadherin–catenin complex in cell–cell adhesion (McCrea et al. 1991; Peifer et al. 1992), and as a signal transducer in the canonical Wnt/Wingless signaling pathway. Armadillo was originally identified as a product of one of the *Drosophila* segment polarity class genes (Riggleman et al. 1989). Vertebrate β-catenin and *Drosophila* Armadillo exhibit essentially the same overall structure; they are divided into three domains, the N-terminal domain, the central domain and the C-terminal domain (Fig. 2.2a). The central domain consists of 12 repeats of ~42 amino acid residues, referred to as Armadillo repeats (ArmR1–ArmR12). These ArmRs each form three α-helices, tightly packed against one another to form a superhelical structure that serves as a scaffold for the binding of the classical cadherin cytoplasmic domain (Huber and Weis 2001). The α-catenin binding site is a 29-amino-acid region of β-catenin that encompasses the junction of the N-terminal domain and ArmR1 (Aberle et al. 1996).

The signaling function of β-catenin depends on the regulation of its stability in the cytoplasm (Peifer and Polakis 2000; Tolwinski and Wieschaus 2004; Brembeck et al. 2006). Binding of Wnt ligands induces the stabilization of the cytoplasmic pool of β-catenin, allowing β-catenin to translocate to the nucleus and to there act as a transcriptional activator in conjunction with DNA-binding proteins, T cell factor (TCF), lymphoid enhancer factor-1 (LEF-1) and Pangolin (Pan). In the absence of Wnt signal input, cytoplasmic β-catenin is efficiently degraded by a destruction complex consisting of the tumor suppressor gene product adenomatous polyposis coli (APC), axin, glycogen synthase kinase 3-beta (GSK-3β) and casein kinase (CKI). The ArmR domain in β-catenin, when free from cadherin and α-catenin, interacts with either components of the destruction complex or TCF/LEF-1/Pan. The N-terminal and C-terminal domains in β-catenin have essential roles in its signaling function. Unlike vertebrates and *Drosophila*, *C. elegans* has three diverged β-catenin genes with separate roles (Eisenmann 2005): *hmp-2*, which is involved in cadherin-mediated adhesion, and *wrm-1* and *bar-1*, which are involved in Wnt signaling.

Plakoglobin (or γ-catenin), a component of adherens junctions and desmosomes, is closely related to β-catenin and only found in vertebrates (Fig. 2.2a). Although plakoglobin and β-catenin exhibit less conservation in their C-terminal domain, their ArmR domains share high sequence identity (approximately 80%), which accounts for the ability of plakoglobin to bind to the classical cadherin cytoplasmic domain. However, plakoglobin also binds to the cytoplasmic domains of other cadherin types, desmoglein and desmocollin, that are responsible for desmosome assembly. Compared to plakoglobin, β-catenin exhibits weaker binding to desmoglein-1, which partly accounts for the specific participation of β-catenin in adherens junction assembly (Choi et al. 2009).

2.2.5 α-Catenin

α-Catenin mediates regulatory interactions between the cadherin-β-catenin complex and the cytoskeleton. Vertebrates have two subtypes of α-catenin, αE- and αN-catenins, which are expressed in epithelial and neural tissues (Nagafuchi et al. 1991; Herrenknecht et al. 1991; Hirano et al. 1992), whereas *Drosophila* and *C. elegans* have a single α-catenin homolog (*D*α-catenin and HMP-1, respectively) (Fig. 2.2a). These α-catenins share essentially the same structural features, including three vinculin-homology domains, VH1, VH2 and VH3, in their N-terminal, middle and C-terminal regions, respectively. αE-catenin binds β-catenin through its N-terminal region (Huber et al. 1997; Koslov et al. 1997; Obama and Ozawa 1997; Nieset et al. 1997) and without β-catenin, it forms homodimers using the same N-terminal region (Koslov et al. 1997; Pokutta and Weis 2000; Drees et al. 2005). The αE-catenin homodimer can bind and bundle F-actin using the C-terminal regions (Rimm et al. 1995). The middle region of αE -catenin binds to other actin-binding proteins, such as vinculin and α-actinin (Kobielak and Fuchs 2004). Unlike mammalian αE-catenin, recombinant full-length HMP-1 is a monomer. The actin-binding ability of the C-terminal region of HMP-1 is usually suppressed by its other regions (Kwiatkowski et al. 2010).

2.2.6 p120-Catenin

p120-Catenin was originally identified as a tyrosine kinase substrate for which tyrosine phosphorylation was induced by transformation with Src in mammalian cells (Reynolds et al. 1989). It was later found to directly bind to the juxtamembrane region of classical cadherins (Fig. 2.2a, b; Reynolds et al. 1994; Daniel and Reynolds 1995; Lampugnani et al. 1997; Yap et al. 1998). p120-catenin has been functionally characterized as a key regulator of classical cadherin stability (Ireton et al. 2002;

Davis et al. 2003; Xiao et al. 2003). p120-catenin is an ArmR-containing protein, like β-catenin and plakoglobin, but belongs to a distinct subgroup referred to as the p120-catenin family, whose members share a conserved central domain comprised of 9 ArmRs. Four members of the vertebrate p120-catenin family, p120-catenin, ARVCF (*A*rmadillo *r*epeat gene deleted in *v*elo-*c*ardio-*f*acial syndrome), δ-catenin and p0071, can bind classical cadherin via their ArmR domains in a mutually exclusive manner (Daniel and Reynolds 1995; Hatzfeld and Nachtsheim 1996; Mariner et al. 2000; Paulson et al. 2000). The vertebrate p120-catenin family also includes components of desmosomes known as plakophilins. In contrast to vertebrates, *Drosophila* and *C. elegans* each possess only one member of the p120-catenin family (p120ctn and JAC-1, respectively). Ablation or depletion of p120-catenin in vertebrate embryos causes severe morphogenetic defects (Fang et al. 2004; Davis and Reynolds 2006); however, the sole p120-catenin of *Drosophila* and *C. elegans* is not essential for the viability and morphogenesis of the animals, although these molecules modulate cadherin-mediated adhesion (Pacquelet et al. 2003; Myster et al. 2003; Pettitt et al. 2003).

2.3 Cadherin Superfamily

2.3.1 Major Subfamilies in Metazoans

A single mammalian genome encodes more than one hundred members of the cadherin superfamily. The *Drosophila* and *C. elegans* genomes include 17 and 12 cadherin genes, respectively (Fung et al. 2008; Pettitt 2005). The sea urchin *Strongylocentrotus purpuratus* genome has fewer cadherin genes than the *Drosophila* and *C. elegans* genomes (Whittaker et al. 2006). Recent advances in genome sequencing have made genomic information available on cadherins from many other metazoan species, including *Ciona intestinalis* (urochordate ascidian), *Branchiostoma floridae* (cephalochordate amphioxus), *Nematostella vectensis* (cnidarian sea anemone) and *Trichoplax adhaerens* (placozoan). The accumulated information indicates that there are at least eight major subfamilies distributed across two or more metazoan phyla (Nollet et al. 2000; Whittaker et al. 2006; Hulpiau and van Roy 2009, 2010), including the classical cadherin, Fat, Fat-like, Dachsous, Flamingo, protocadherin, PCDH15 and CDH23 subfamilies (Fig. 2.3a).

2.3.2 Classical Cadherin

Cadherins with cytoplasmic domains that are closely homologous to those of vertebrate classical cadherins are distributed widely in Metazoans, although a great diversity is seen in their extracellular structures. Irrespective of the species and the extracellular domain, classical cadherin has been re-defined as a molecule having a conserved cytoplasmic domain that interacts (or is expected to interact) with p120-catenin and β-catenin (Oda and Takeichi 2011).

The classical cadherin 5-EC organization has only been observed in the vertebrate and urochordate subphyla, while association of ECs with EGF and LmG is a common feature of all classical cadherins identified in non-chordate metazoans. Another feature shared by most nonchordate classical cadherins is the presence of an extracellular region that is homologous to the *D*E-cadherin extracellular proteolytic cleavage site and its flanking regions, which are referred to as the primitive classical cadherin proteolytic site domain (PCPS; Oda and Tsukita 1999; Oda and Takeichi 2011). Because some PCPSs show weakly detectable partial similarities to ECs, this domain type might have diverged from an EC. However, to avoid potential confusion, the PCPS will hereafter not be considered an EC. The classical cadherins of bilaterian species contain between 2 and 17 ECs, with the exception of the unique molecules described below. In contrast, 25 or more ECs are encoded by the classical cadherin genes found in the genomes of cnidarian and placozoan species (Chapman et al. 2010; Hulpiau and Van Roy 2010; Fahey and Degnan 2010). Expressions and functions of these huge classical cadherins have not been studied.

Bb1- and Bb2-cadherins are a pair of exceptional "cadherins" that have been reported in the amphioxus *Branchiostoma belcheri*, a chordate belonging to the cephalochordate subphylum (Oda et al. 2002, 2004). These molecules each possess a well-conserved classical cadherin cytoplasmic domain, but their extracellular regions consist of no ECs, only LmGs and EGFs. Despite the lack of ECs, Bb1- and Bb2-cadherins can mediate homophilic cell–cell adhesion and cell sorting *in vitro*, although their activities are independent of calcium ions. Importantly, Bb1- and Bb2-cadherins, together with an amphioxus β-catenin homolog, are enriched at adherens junctions in various epithelial tissues (Fig. 2.4e). Bb1- and Bb2-cadherins are not formally included in the cadherin superfamily, but appear to be derivatives of a classical cadherin subfamily member. The sequenced genome of another amphioxus species, *Branchiostoma floridae*, contains orthologs of the Bb1- and Bb2-cadherin genes, as well as an additional gene encoding AmphiCDH, a *D*N-cadherin-like nonchordate-type classical cadherin (Hulpiau and Van Roy 2010).

2.3.3 Fat, Fat-Like and Dachsous

The extracellular regions of members of the Fat and Fat-like subfamilies typically have an array of ~34 ECs, which is followed by EGFs and LmGs. *Drosophila* Fat appears to be orthologous to mammalian FAT4, whereas *Drosophila* Fat-like has a closer relationship to mammalian FAT1, FAT2 and FAT3 (Castillejo-López et al. 2004; Hulpiau and van Roy 2009). The *Drosophila* and mammalian Dachsous cadherins, which have 27 ECs with no EGF and LmG, heterophilically bind the corresponding Fat cadherins *in vitro* (Matakatsu and Blair 2004; Ishiuchi et al. 2009), and the *Drosophila* Dachsous-Fat pair and possibly mouse FAT4 function in a signaling pathway that regulates tissue growth, planar cell polarity and tissue patterning (Reddy and Irvine 2008; Saburi et al. 2008). In *Drosophila* larval epithelial

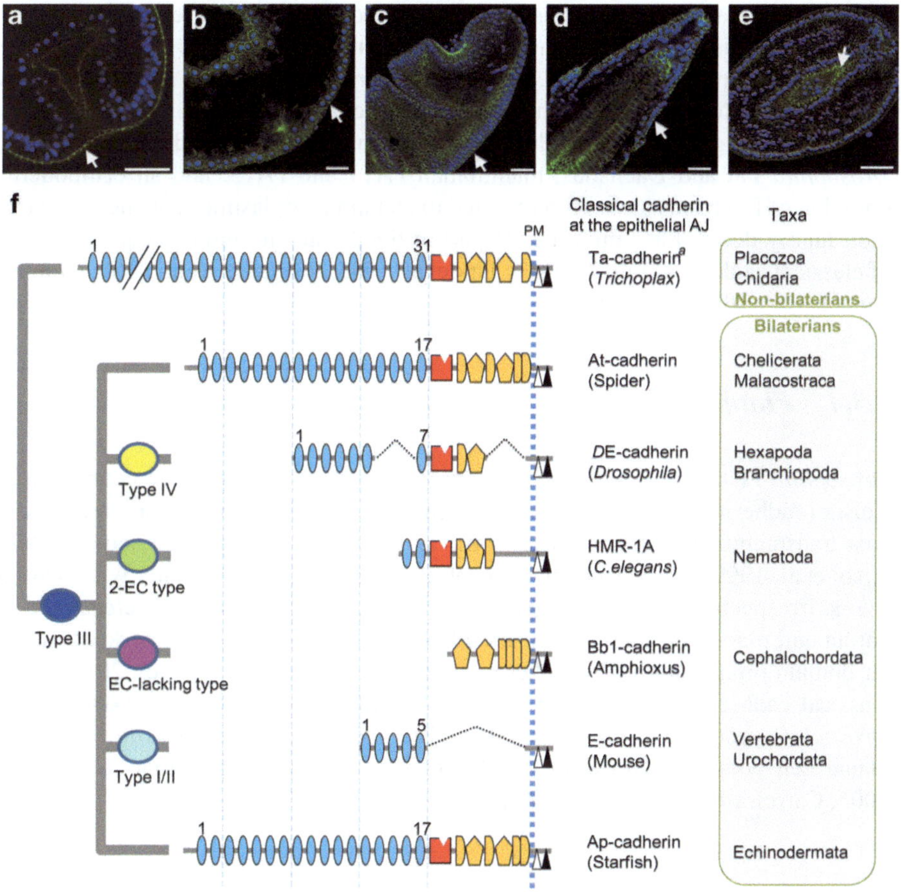

Fig. 2.4 A "lineage-specific domain loss" model for classical cadherin extracellular structure diversification. **a–e** Localization of various classical cadherins at epithelial adherens junctions (indicated by *arrows*). In each panel, immunostaining for classical cadherin is shown in *green*, and staining for DNA is in *blue*. **a** Ap-cadherin (type III) expression in a midgastrula of the starfish *Asterina pectinifera*; **b** At-cadherin (type III) expression in a germ-disc stage embryo of the spider *Achaearanea tepidariorum*; **c** *D*E-cadherin (type IV) expression in a *Drosopohila* gastrula; **d** Af1-cadherin (type IV) expression in a nauplius larva of the branchiopod *Artemia franciscana*; **e** Bb1-cadherin (EC-lacking type) expression in a neurula of the amphioxus *B. belcheri*. Bars, 50 μm. **f** Reconstruction of evolutionary transitions (indicated by circles in *various colors*) that diversified the extracellular domain structures of classical cadherins at the epithelial adherens junctions, based on comparative studies (Oda et al. 2002, 2005; Hulpiau and Van Roy 2010). Gaps are introduced to highlight homologous regions between distinct classical cadherins. Taxa in which the same or similar conditions have been observed are shown in the right column. In this model, type III cadherin represents the ancestral form of classical cadherin for bilaterians. Distinct N-terminal truncations and internal deletions (indicated by *broken back lines*) in type III cadherin occurred in different bilaterian lineages. The epithelial adherens junctions in "ancestral" animals, such as starfish and spider, use type III cadherin, and the epithelial adherens junctions in "derived" animals use structurally reduced forms of classical cadherin such as types I/II and IV. The establishment of type III cadherin may have been preceded by N-terminal truncations. *a*, No expression data is available for Ta-cadherin and other non-bilaterian classical cadherins. PM, plasma membrane; AJ, adherens junction

cells, Fat and Dachsous cadherins are concentrated in a subapical region of cell–cell contact, which is more apical than the adherens junction (Ma et al. 2003). In mouse embryonic neuroepithelial cells, a similar pattern of subcellular localization has been observed for Fat4 and Dachsous1 (Ishiuchi et al. 2009).

Despite being phylogenetically separated from the classical cadherin subfamily, *Drosophila* Fat and Dachsous, mammalian FAT1 and FAT3, and an echinoderm homolog of Dachsous have been reported to contain cytoplasmic sequences that exhibit faintly detectable similarities to part of the β-catenin-binding sequence motif of classical cadherins (Fig. 2.3a–c; Clark et al. 1995; Whittaker et al. 2006; Hulpiau and van Roy 2009).

2.3.4 Flamingo

Drosophila Flamingo (also known as Starry night) and its vertebrate counterparts, Celsrs (*c*adherin, *E*GF-like, *l*aminin A globular, *s*even-pass *r*eceptor), are seven-pass transmembrane proteins categorized as adhesion G protein-coupled receptors (Usui et al. 1999). They each have 8 or 9 ECs and two LmGs together with several EGFs. Irrespective of the phylogenetic distances, the echinoderm, *C. elegans*, cnidarian and placozoan members of the Flamingo subfamily exhibit markedly similar domain organization (Whittaker et al. 2006; Hulpiau and Van Roy 2010). Like classical cadherins, Flamingo exhibits homophilic binding *in vitro*. Notably, this *Drosophila* protein and the vertebrate Celsrs have similar functions in regulating planar cell polarity (Usui et al. 1999; Curtin et al. 2003; Formstone and Mason 2005; Carreira-Barbosa et al. 2008).

2.3.5 Protocadherin

The term “protocadherin” is often confusing, since it is used to refer to many non-classical cadherins without phylogenetic considerations. In vertebrates, cadherins with six or seven ECs are considered to constitute a phylogenetic group, which is termed the protocadherin subfamily (Morishita and Yagi 2007). This subfamily is divided, based on genomic organization, into two subgroups, i.e., the clustered and non-clustered protocadherins. In the mouse, clustered protocadherins, each of which contains six ECs, are encoded by three tandemly aligned gene clusters (α, β and γ) and are predominantly expressed in the nervous system; their *in vivo* functions are poorly understood. Non-clustered protocadherins (e.g., PCDH9) have two unique amino acid sequence motifs in their cytoplasmic regions and are also referred to as δ-protocadherins. Notably, this type of protocadherin is found in a wider range of metazoans including *Nematostella* (Whittaker et al. 2006; Hulpiau and Van Roy 2010) but is missing in *Drosophila* and *C. elegans*. Three vertebrate members of the δ-protocadherin subfamily, paraxial protocadherin, OL-protocadherin (PCDH10) and PCDH19, are known to cooperate with classical cadherins to promote cell sorting and/or cell movements (Chen and Gumbiner 2006; Nakao et al. 2008; Biswas et al. 2010).

2.3.6 PCDH15 and CDH23

Mammalian PCDH15 (protocadherin 15) and CDH23 (cadherin 23) have 11 and 27 ECs, respectively, and interact heterophilically to facilitate mechanosensing in the stereocilia of the mammalian inner ear (Kazmierczak et al. 2007). Mutations in these genes cause hearing loss, termed Usher syndrome. A *Drosophila* homolog of PCDH15, termed Cad99C, regulates microvilli length (D'Alterio et al. 2005). However, *Drosophila* has no CDH23 counterpart. Nonetheless, in the cnidarian sea anemone tentacle, a CDH23-related polypeptide has been detected between the mechanosensory stereocilia (Watson et al. 2008).

2.3.7 Desmosomal Cadherin

The desmosomal cadherins have 4 or 5 ECs and constitute a vertebrate-specific subfamily of the cadherin superfamily and are divided into two types, desmocollins and desmogleins, which bind heterophilically, probably via their EC1 domains. The cytoplasmic binding partners of desmosomal cadherins include plakophilins, plakoglobin and desmoplakin, which play roles in desmosome assembly and intermediate filament anchorage. The cytoplasmic domains of desmosomal cadherins have amino acid sequences that are related to, but significantly divergent from, the β-catenin-binding sequence motif of classical cadherins (Fig. 2.3a, d; Troyanovsky et al. 1994; Hulpiau and van Roy 2009); these sequences appear to be bound to plakoglobin, a close relative of β-catenin, and are required for the anchoring of intermediate filaments by the desmosomal plaque. Amino acid sequences that are partially similar to the p120-catenin-binding sequence motif of classical cadherins are detectable in the juxtamembrane regions of desmogleins and desmocollins (Hulpiau and van Roy 2009) and an association has been shown between p120-catenin and desmoglein 3 (Kanno et al. 2008).

2.4 Evolution of Classical Cadherins

2.4.1 Type I and Type II Cadherins

The major subfamilies of the cadherin superfamily are varied in the number of their ECs, the polypeptide length and the domain composition and organization. However, within each subfamily, domain organization tends to be conserved across the metazoans. In this respect, the classical cadherin subfamily is exceptional.

Members of this subfamily exhibit a great structural diversity in their extracellular regions. Reconstruction of the processes responsible for the generation of the structural diversity of classical cadherins may facilitate an understanding of the evolution of adherens junctions.

Classical cadherins that have been identified in the vertebrate subphylum are classified into type I (e.g., E- and N-cadherins), type II (e.g., VE-cadherin and cadherin-11) and type III (e.g., cHz-cadherin) (Fig. 2.3a). The details of type III cadherins are described below. A shared recent origin of type I, type II and desmosomal cadherins is strongly supported by the presence of the prodomain, which is processed for activation, the extracellular 5-EC organization and the exon-intron structures (Greenwood et al. 1997; Nollet et al. 2000). The differences between type I and type II cadherins are apparent at the amino acid sequence level. The HAV sequence, which is conserved in the EC1 domains of type I cadherins, is missing in type II cadherins. In contrast to type I cadherins, which have a single conserved tryptophan residue at the N-terminal region, most type II cadherins have two tryptophan residues at their N-terminal regions. A crystallographic study has proposed distinct structural mechanisms for adhesion mediated by type I and type II cadherins (Patel et al. 2006). Whereas type I cadherins exhibit broad tissue distribution, type II cadherins tend to be expressed in more restricted cell populations and types. For example, the type I N-cadherin is expressed broadly in the mesodermal and neural tissues, including endothelial cells, but the type II VE-cadherin is expressed only in the endothelial cell populations of the mesoderm (Salomon et al. 1992; Navarro et al. 1998).

In the human genome, at least four genes encode type I classical cadherins, and at least 14 genes encode type II classical cadherins. Many of these genes form clusters. The largest cluster, which is located on the long arm of chromosome 16, comprises two type I cadherin genes, including *E-cadherin*, three type II cadherin genes, including *VE-cadherin*, and a non-classical cadherin (Ksp-cadherin) gene (Kremmidiotis et al. 1998). The urochordate *Ciona intestinalis* genome has only two classical cadherin genes; one is related to the type I cadherins, and the second is related to the type II cadherins (Sasakura et al. 2003). Neither type I nor type II cadherins have been discovered outside of the vertebrate and urochordate subphyla. The complexity of type I and type II cadherins increased due to gene duplications in the vertebrate lineage after it diverged from the urochordate lineage. In addition, vertebrates, but not non-vertebrate animals, have desmosomal cadherins and other cadherins that have ECs closely related to the ECs of type I/type II cadherins but lack the classical cadherin cytoplasmic domain. These cadherins include T-cadherin (CDH13 and H-cadherin), Ksp-cadherin (CDH16) and LI-cadherin (CDH17) (Vestal and Ranscht 1992; Wendeler et al. 2006). Type I and/or type II cadherin genes may have acted as a source of such vertebrate-specific non-classical cadherins during evolution.

2.4.2 Type III Cadherin

Chicken cHz-cadherin was first regarded as a type III cadherin (Tanabe et al. 2004). Despite its vertebrate source, cHz-cadherin is markedly similar to *Drosophila*

*D*N-cadherin with respect to its domain organization. Genes encoding classical cadherins with similar domain organization are also found in other arthropods, echinoderms, amphioxus and fish. These classical cadherins have 14–17 ECs, a PCPS, multiple EGFs and two LmGs arranged specifically in their extracellular regions. Importantly, their mutually similar domain organization has been suggested to result from conservation, not convergence (Oda et al. 2005; Hulpiau and Van Roy 2010). This domain conservation defines the type III cadherins. The only classical cadherin gene in the genome of the sea urchin *Strongylocentrotus purpuratus* encodes a type III cadherin (Whittaker et al. 2006). It is likely that the last common ancestor of all bilaterian animals possessed a type III cadherin. However, the type III cadherin gene is absent from the ascidian and placental mammalian genomes (Tanabe et al. 2004; Hulpiau and Van Roy 2010), suggesting that this cadherin type was lost secondarily at multiple separate points of bilaterian evolution.

The tissue distributions of type III cadherins vary depending on species. cHz-cadherin is expressed in horizontal cells, one of the basic cell types of the chicken retina (Tanabe et al. 2004). Type III cadherins of hexapods (e.g., cricket) and branchiopods (e.g., brine shrimp) are widely expressed in embryonic mesodermal and neural tissues (Oda et al. 2005; Hsu et al. 2009), similar to *D*N-cadherin. In contrast, the type III cadherins of malacostracan crustaceans (e.g., shrimp), chelicerates (e.g., spider) and echinoderms (e.g., starfish) are localized at the adherens junctions in the embryonic epithelia (Fig. 2.4a, b), although the arthropod cadherins are also found in the neural tissues. These observations suggest that the roles of type III cadherins have been altered in a lineage-specific way during evolution.

2.4.3 Type IV Cadherin

*D*E-cadherin and its orthologs are grouped as type IV cadherins, and have 7 ECs, a PCPS, an EGF and an LmG in their extracellular regions. Type IV cadherins have only been found within the branchiopods and hexapods. These cadherins exhibit conserved expression at the adherens junctions in the embryonic epithelia (Fig. 2.4c, d; Oda et al. 2005), and this finding is potentially correlated with the absence of type III cadherin from these tissues in the branchiopods and hexapods. It appears that the domain structure and role of type IV cadherin in adherens junction assembly in the epithelia have been stably maintained during branchiopod and hexapod evolution, indicating the rarity of evolutionary transitions in the extracellular architecture of adherens junctions.

2.4.4 Lineage-Specific Domain Loss

The assumption that type III cadherin represents the bilaterian ancestral form of classical cadherin may facilitate the understanding of not only the wide phylogenetic distribution of this cadherin type, but also the processes that contributed to the structural diversification of classical cadherins. BLAST-based comparisons between individual domains of type IV and type III cadherins and between those of type I/II and type III cadherins have identified homologous regions between the different cadherin types (Oda et al. 2005; Hulpiau and Van Roy 2010; Oda and Takeichi 2011). For example, the membrane-distal 6-EC region of type IV cadherins appears to be homologous to the internal 6-EC region of type III cadherins that is separated from the PCPS by four ECs, and the entire extracellular region of type I/II cadherins appears to be homologous to the membrane-proximal 5-EC region of type III cadherins (Fig. 2.4f). In addition, the region covering the last EC and the single LmG in type IV cadherins is homologous to the region covering the last EC and the first LmG in type III cadherins. Thus, losses of distinct combinations of domains from the type III cadherin may account for the establishment of the type I/II and type IV domain organizations in the branchiopod/hexapod and urochordate/vertebrate lineages.

The "lineage-specific domain loss" model is potentially also applicable to understanding the variously reduced forms of classical cadherin that are observed in other bilaterian animal lineages, including the short form in *C. elegans* and the EC-lacking form in amphioxus. In the echinoderm lineage, sea urchin LvG-cadherin lacks an EC that corresponds to EC2 of starfish Ap-cadherin (Oda et al. 2005). In the hemichordate lineage, which is considered to be a sister lineage to the echinoderms, *Ptychodera flava* Pf1-cadherin has a reduced number of ECs (eight) and a small membrane-proximal deletion (~40 amino acid residues) in its extracellular region (Oda et al. 2002). The validity of the "lineage-specific domain loss" model needs to be tested by collecting more extensive information about the structures of classical cadherin genes from various species. Such effort will also contribute to a better general understanding of the deep phylogenies of animal lineages.

The genomes of cnidarian and placozoan (non-bilaterian eumetazoan) species encode putative classical cadherins that resemble type III cadherins, although they all have many more ECs (Chapman et al. 2010; Hulpiau and Van Roy 2010). Surprisingly, there is a detectable collinear homology between the entire EC region of type III cadherins and the membrane-proximal EC region of the very large classical cadherins of *Trichoplax* and *Nematostella*, implying that size reduction by loss of membrane-distal ECs preceded the establishment of the type III cadherin (Hulpiau and Van Roy 2010). The length and domain composition of the non-bilaterian eumetazoan classical cadherins resemble those of Fat and Fat-like cadherins. The entire extracellular region of δ-protocadherin appears to be derived from the membrane-distal 7-EC region of Fat cadherin (Hulpiau and Van Roy 2010). Size reduction through domain loss is a common strategy in the structural diversification of the cadherin superfamily.

2.4.5 Functional Relevance of Structural Transitions at the Adherens Junction

A remarkable feature of classical cadherin diversification is that structurally reduced-derived forms of classical cadherin tend to be used in the epithelial adherens junctions in "derived" animal lineages (Fig. 2.4f). Assuming that the embryonic surface epithelia are homologous across the eumetazoans, adherens junctions in this tissue type must have undergone distinct transitions in their extracellular architecture in the respective derived lineages. It is yet to be resolved how the classical cadherins with their highly varied sizes are accommodated in the similar intercellular spaces (15–25 nm) of the adherens junctions of different metazoan species (Fig. 2.1). The large non-classical cadherins, PCDH15 and CDH23, form helical filaments bridging the large spaces between stereocilia (150–300 nm) (Kazmierczak et al. 2007; Elledge et al. 2010), a structural mechanism that appears to be advantageous for interciliary force transduction, but that would not be suitable for the large classical cadherins at the adherens junctions.

Particularly, in the vertebrate/urochordate lineage, the PCPS-LmG region was entirely lost in the cadherin responsible for epithelial adherens junction formation. An experimental study using *D*E-cadherin suggests that the EC7/PCPS-LmG region is not essential for type IV cadherins to exhibit strong cell–cell adhesion activity (Haruta et al. 2010). This study also showed that this region, which covers about half of the entire extracellular region of *D*E-cadherin, is unlikely to be the major factor that determines the intercellular space of the adherens junction. Interestingly, early *Drosophila* embryos expressing *D*E-cadherin that lack the EC7/PCPS-LmG region form the normal blastoderm epithelium but exhibit defects in the apical constriction of presumptive mesodermal cells. An important implication of this work is that an abrupt loss of all non-EC domains in the extracellular region of a functional cadherin at the non-chordate adherens junction can occur without disrupting the ability of the animal to form epithelia, although this domain loss may affect morphogenetic processes.

2.5 Ancestry of the Cadherin–Catenin Complex

2.5.1 Poriferans

Poriferans (sea sponges) are the key phylum for exploring the evolutionary origins of intercellular junctions. This phylum comprises four lineages, the Calcispongiae, Demospongiae, Hexactinellida and Homoscleromorpha. Of these four lineages, the Homoscleromorpha lineage is the only lineage in which intercellular junctions resembling the bilaterian adherens junctions have been observed (Ereskovsky et al. 2009). However, molecular information is scarce for the Homoscleromorpha at present. On the other hand, the genome of *Amphimedon queenslandica*, a species of the Demospongiae, has been completely sequenced, revealing the presence of a classical cadherin-like gene, *AmqCadherin1*, and β-, α-, and p120-catenin gene homologs (Fig. 2.5a; Sakarya et al. 2007; Abedin and King 2008; Fahey and Degnan 2010). *AmqCadherin1* encodes a single-pass transmembrane protein with 14 ECs followed by 13 EGFs and 2 LmGs (Fig. 2.5b). However, sequence similarities between the cytoplasmic domains of AmqCadherin1 and classical cadherins are lim-

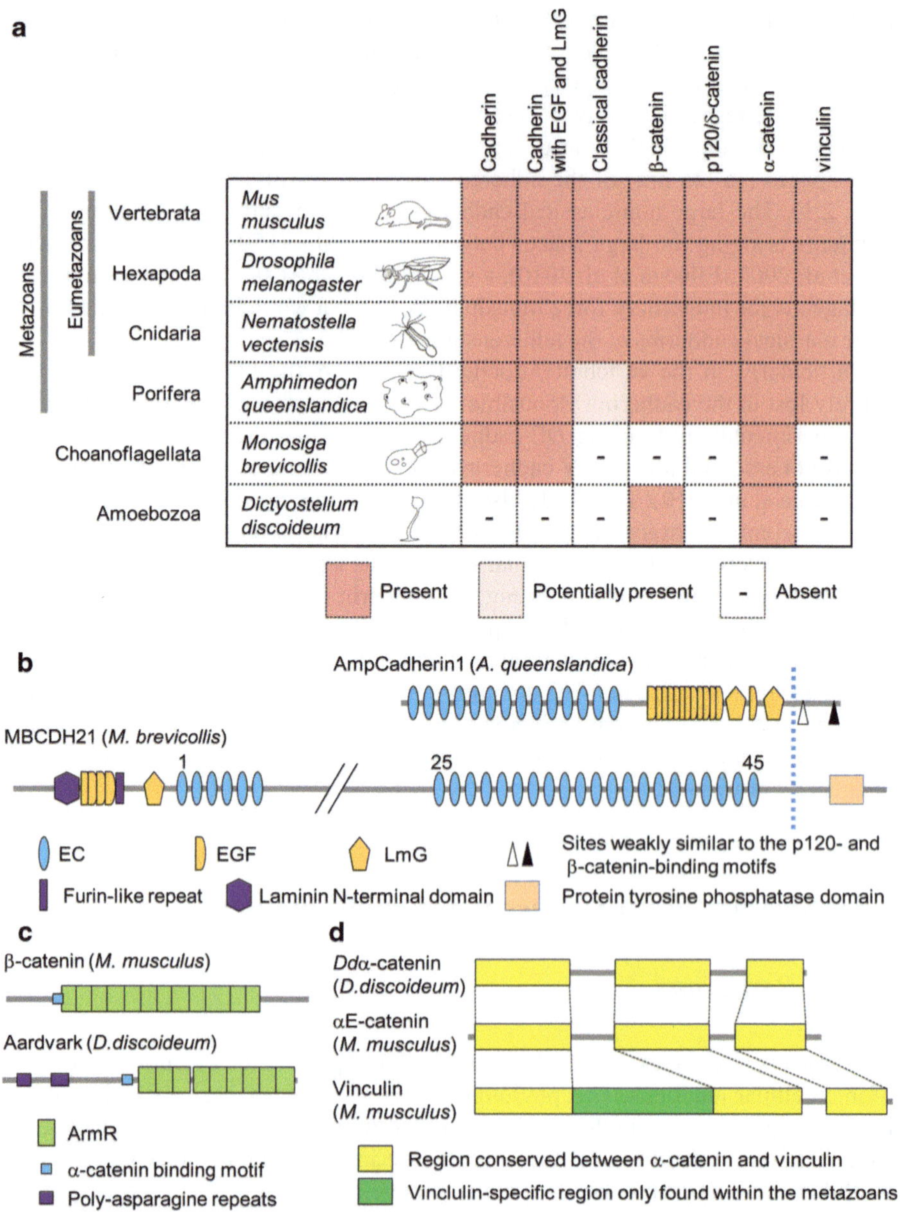

Fig. 2.5 Cadherins and catenins in primitive metazoans and non-metazoans. **a** The presence/absence of cadherin, catenin and vinculin genes in the sequenced genomes as reported by Abedin and King 2008; Fahey and Degnan 2010 and Dickinson et al. 2011. **b** The domain structures of AmqCadherin1, a classical cadherin-like protein from *A. queenslandica*, and MBCDH21, a cadherin with EGF and LmG in *M. breviollis*. **c** A comparison of the domain structures of β-catenin in the mouse and a β-catenin homolog in *D. discoideum*, Aardvark (adapted from Dickinson et al. 2011). **d** A comparison of the domain structures of αE-catenin and vinculin in the mouse and an α-catenin homolog in *D. discoideum*. (adapted from Dickinson et al. 2011)

ited. It has not yet been tested whether AmqCadherin1 physically interacts with the catenin homologs, and it is curious why the extracellular domain structure of AmqCadherin1 is dissimilar from those of the non-bilaterian eumetazoan classical cadherins.

2.5.2 Choanoflagellates

Choanoflagellates are the only non-metazoan organisms in which the presence of ECs has been described (Fig. 2.5a). The genome of the choanoflagellate *Monosiga brevicollis*, a unicellular animal close to the metazoans, contains up to 23 cadherin genes (Abedin and King 2008). However, there are no sequences related to the classical cadherin cytoplasmic domain in the *Monosiga brevicollis* genome. In accordance with this finding, the *Monosiga* genome has no β-catenin gene homolog, although an α-catenin gene homolog is present (Fig. 2.5a; Dickinson et al. 2011). MBCDH21 is the only choanoflagellate gene that represents the combination of EC, EGF and LmG (Fig. 2.5a, b); however, this cadherin is highly diverged from any metazoan cadherins that contain EGF and LmG. MBCDH21 has 45 ECs that are preceded by an LmG and EGFs in the extracellular region, and a protein tyrosine phosphatase domain in the cytoplasmic region. No significant precursor genes for metazoan classical cadherins have been discovered outside of the metazoans.

2.5.3 Slime Molds

Although β-catenin-binding sequence motifs characteristic of classical cadherin cytoplasmic domains have been only found within the metazoans, genes related to β-catenin, as well as those related to α-catenin, show a wider phylogenetic distribution (Fig. 2.5a; Coates 2003; Dickinson et al. 2011). Studies of the cellular slime mold *Dictyostelium discoideum* have provided insights into the evolutionary origins of adherens junctions. This organism grows as a unicellular amoeba, and when starved, *D. discoideum* develops into a multicellular structure termed the fruiting body. Tip cells surrounding the stalk of the fruiting body organize into a simple epithelium (Grimson et al. 2000; Dickinson et al. 2011). Between these tip cells, actin-enriched intercellular junctions resembling the metazoan adherens junctions have been observed by electron microscopy (Grimson et al. 2000). *D. discoideum* has a β-catenin homolog, referred to as Aardvark, which has fewer ArmRs and a truncated C-terminus compared to the metazoan β-catenins (Fig. 2.5c; Grimson et al. 2000). Importantly, however, it retains an α-catenin-binding sequence motif and is localized at the epithelial cell junctions. Consistent with these facts, an α-catenin homolog, referred to as *Dd*α-catenin, exists in *D. discoideum* (Fig. 2.5d; Dickinson et al. 2011). Although α-catenins and vinculins form a molecular family, sequences specific to vinculins have only been found within the metazoans. The ancestral form for this family appears to be α-catenin-like.

Ddα-catenin binds to Aardvark and mouse β-catenin *in vitro* and localizes to regions of cell–cell contact in an Aardvark-dependent manner *in vivo* (Dickinson et al. 2011). Purified *Ddα*-catenin, unlike mammalian αE-catenin, does not form homodimers but it is capable of bundling F-actin. Both *Ddα*-catenin and Aardvark are required to organize and polarize the tip epithelium. However, they are not essential for the formation of the *D. discoideum* tip cell junctions, and *Ddα*-catenin is not concentrated at the junctional regions. It is still unclear whether the ArmR domain of Aardvark interacts with an adhesion molecule. Commonalities and differences in the molecular compositions and interactions for metazoan adherens junctions and the *D. discoideum* cell junctions require further investigation.

2.6 Summary and Future Perspectives

Functional interactions between β-catenin, α-catenin and the actin cytoskeleton predate the origin of cadherin. Because of the early diversification of the cadherin superfamily, a cadherin containing EGFs and LmGs achieved the ability to interact with β-catenin and p120/δ-catenin. This origination of cadherin–catenin interactions was followed by diversification of the extracellular domain organization of classical cadherins. Classical cadherins prior to the origin of bilaterians must have been very large in size, like the current Fat and Fat-like cadherins. Comparative studies suggest that step-by-step size reductions through lineage-specific domain losses resulted in variations in the forms of classical cadherins among metazoans, implying that the extracellular architecture of adherens junctions in the epithelia underwent distinct alterations depending on the lineage. For example, in the vertebrate/urochordate lineage, the 5-EC domain organization for classical cadherin was established, and this was followed by a further diversification of classical and nonclassical cadherins and an increase in the repertoire of the catenins. This relatively recent diversification of the cadherin–catenin system and its derivatives may have contributed to vertebrate-specific morphological complications.

Many questions regarding the evolution of cadherins and catenins are emerging and remain to be answered. Unicellular lineages exist between the metazoans and slime molds, but it is difficult to reconstruct evolutionary transitions between unicellular and multicellular life; non-metazoan β- and α-catenins are rare clues to this issue. The phylogenetic distribution of classical cadherins appears to be restricted to metazoans, whereas cadherins predate the last common ancestor of metazoans and non-metazoan choanoflagellates. Biochemical and cell biological studies of non-bilaterian metazoans, poriferans in particular, are increasingly important for exploring the origin of cadherin-based intercellular junctions. One exciting goal of these studies is to determine the functions of ancient cadherins prior to their being co-opted for junction formation. After being co-opted for junction formation, the cadherin–catenin complex, the extracellular structure of classical cadherin in particular, experienced distinct changes in different metazoan lineages. Did such changes contribute to innovations of morphogenetic mechanisms in the respective lineages? More specifically, what happened to the junctional systems in the earliest chordates? It remains unclear whether the unique conditions of the classical cadherin forms and junction organization in the amphioxus represent the ancestral state for all extant chordates. Analyses of the functional and mechanistic aspects of structural and compositional transitions at the adherens junctions is required to determine the relationships between the diversification of the junctional cadherin–catenin complex and the morphological diversification in animals.

Acknowledgments I would like to thank Shicui Zhang for having supported my access to amphioxus embryos; Tomohiro Haruta for his permission to use unpublished electron micrographs; and Yasuko Akiyama-Oda for her comments on the manuscript.

References

Abedin M, King N (2008) The premetazoan ancestry of cadherins. Science 319:946–948. doi:10.1126/science1151084

Aberle H, Schwartz H, Hoschuetzky H, Kemler R (1996) Single amino acid substitutions in proteins of the armadillo gene family abolish their binding to α-catenin. J Biol Chem 271:1520–1526. doi:10.1074/jbc.271.3.1520

Biswas S, Emond MR, Jontes JD (2010) Protocadherin-19 and N-cadherin interact to control cell movements during anterior neurulation. J Cell Biol 191:1029–1041. doi:10.1083/jcb.201007008

Boller K, Vestweber D, Kemler R (1985) Cell-adhesion molecule uvomorulin is localized in the intermediate junctions of adult intestinal epithelial cells. J Cell Biol 100:327–332

Brackenbury R, Rutishauser U, Edelman GM (1981) Distinct calcium-independent and calcium-dependent adhesion systems of chicken embryo cells. Proc Natl Acad Sci U S A 78:387–391

Brembeck FH, Rosário M, Birchmeier W (2006) Balancing cell adhesion and Wnt signaling, the key role of β-catenin. Curr Opin Genet Dev 16:51–59. doi:10.1016/j.gde.2005.12.007

Broadbent ID, Pettitt J (2002) The *C. elegans hmr-1* gene can encode a neuronal classic cadherin involved in the regulation of axon fasciculation. Curr Biol 12:59–63. doi:10.1016/S0960-9822(01)00624-8

Carreira-Barbosa F, Kajita M, Morel V, Wada H, Okamoto H, Martinez Arias A, Fujita Y, Wilson SW, Tada M (2008) Flamingo regulates epiboly and convergence/extension movements through cell cohesive and signalling functions during zebrafish gastrulation. Development 136:383–392. doi:10.1242/dev.026542

Castillejo-López C, Arias WM, Baumgartner S (2004) The *fat-like* gene of *Drosophila* is the true orthologue of vertebrate Fat cadherins and is involved in the formation of tubular organs. J Biol Chem 279:24034–24043. doi:10.1074/jbc.M313878200

Chapman JA, Kirkness EF, Simakov O, Hampson SE, Mitros T, Weinmaier T, Rattei T, Balasubramanian PG, Borman J, Busam D, Disbennett K, Pfannkoch C, Sumin N, Sutton GG, Viswanathan LD, Walenz B, Goodstein DM, Hellsten U, Kawashima T, Prochnik SE, Putnam NH, Shu S, Blumberg B, Dana CE, Gee L, Kibler DF, Law L, Lindgens D, Martinez DE, Peng J, Wigge PA, Bertulat B, Guder C, Nakamura Y, Ozbek S, Watanabe H, Khalturin K, Hemmrich G, Franke A, Augustin R, Fraune S, Hayakawa E, Hayakawa S, Hirose M, Hwang JS, Ikeo K, Nishimiya-Fujisawa C, Ogura A, Takahashi T, Steinmetz PR, Zhang X, Aufschnaiter R, Eder MK, Gorny AK, Salvenmoser W, Heimberg AM, Wheeler BM, Peterson KJ, Böttger A, Tischler P, Wolf A, Gojobori T, Remington KA, Strausberg RL, Venter JC, Technau U, Hobmayer B, Bosch TC, Holstein TW, Fujisawa T, Bode HR, David CN, Rokhsar DS, Steele RE (2010) The dynamic genome of Hydra. Nature 464:592–596. doi:10.1038/nature08830

Chen X, Gumbiner BM (2006) Paraxial protocadherin mediates cell sorting and tissue morphogenesis by regulating C-cadherin adhesion activity. J Cell Biol 174:301–313. doi:10.1083/jcb.200602062

Choi H J, Gross JC, Pokutta S, Weis WI (2009) Interactions of plakoglobin and β-catenin with desmosomal cadherins: basis of selective exclusion of α- and β-catenin from desmosomes. J Biol Chem 284:31776–31788. doi:10.1074/jbc.M109.047928

Clark HF, Brentrup D, Schneitz K, Bieber A, Goodman C, Noll M (1995) *Dachsous* encodes a member of the cadherin superfamily that controls imaginal disc morphogenesis in *Drosophila.* Genes Dev 9:1530–1542. doi:10.1101/gad.9.12.1530

Coates JC (2003) Armadillo repeat proteins: beyond the animal kingdom. Trends Cell Biol 13:463–471. doi:10.1016/S0962-8924(03)00167-3

Costa M, Raich W, Agbunag C, Leung B, Hardin J, Priess JR (1998) A putative catenin-cadherin system mediates morphogenesis of the *Caenorhabditis elegans* embryo. J Cell Biol 141:297–308. doi:10.1083/jcb.141.1.297

Curtin JA, Quint E, Tsipouri V, Arkell RM, Cattanach B, Copp AJ, Henderson DJ, Spurr N, Stanier P, Fisher EM, Nolan PM, Steel KP, Brown SD, Gray IC, Murdoch JN (2003) Mutation of *Celsr1* disrupts planar polarity of inner ear hair cells and causes severe neural tube defects in the mouse. Curr Biol 13:1129–1133. doi:10.1016/S0960-9822(03)00374-9

D'Alterio C, Tran DD, Yeung MW, Hwang MS, Li MA, Arana CJ, Mulligan VK, Kubesh M, Sharma P, Chase M, Tepass U, Godt D (2005) *Drosophila melanogaster* Cad99C, the orthologue of human Usher cadherin PCDH15, regulates the length of microvilli. J Cell Biol 171:549–558. doi:10.1083/jcb.200507072

Dan-Sohkawa MDS, Kaneko HK, Noda KN (1995) Paracellular, transepithelial permeation of macromolecules in the body wall epithelium of starfish embryos. J Exp Zool 271:264–272. doi:10.1002/jez.1402710404

Daniel JM, Reynolds AB (1995) The tyrosine kinase substrate p120^{cas} binds directly to E-cadherin but not to the adenomatous polyposis coli protein or α-catenin. Mol Cell Biol 15:4819–4824

Davis MA, Reynolds AB (2006) Blocked acinar development, E-cadherin reduction, and intraepithelial neoplasia upon ablation of p120-catenin in the mouse salivary gland. Dev. Cell 10:21–31. doi:10.1016/j.devcel.2005.12.004

Davis MA, Ireton RC, Reynolds AB (2003) A core function for p120-catenin in cadherin turnover. J Cell Biol 163:525–534. doi:10.1083/jcb.200307111

Dickinson DJ, Nelson WJ, Weis WI (2011) A polarized epithelium organized by b- and α-catenin predates cadherin and metazoan origins. Science 331:1336–1339. doi:0.1126/science.1199633

Drees F, Pokutta S, Yamada S, Nelson WJ, Weis WI (2005) α-catenin is a molecular switch that binds E-cadherin-β-catenin and regulates actin-filament assembly. Cell 123:903–915. doi:10.1016/j.cell.2005.09.021

Eisenmann DM (2005) Wnt signaling. WormBook 25:1–17

Elledge HM, Kazmierczak P, Clark P, Joseph JS, Kolatkar A, Kuhn P, Müller U (2010) Structure of the N terminus of cadherin 23 reveals a new adhesion mechanism for a subset of cadherin superfamily members. Proc Natl Acad Sci U S A 107:10708–10712. doi:10.1073/pnas.1006284107

Ereskovsky AV, Borchiellini C, Gazave E, Ivanisevic J, Lapébie P, Perez T, Renard E, Vacelet J (2009) The Homoscleromorph sponge *Oscarella lobularis*, a promising sponge model in evolutionary and developmental biology: model sponge *Oscarella lobularis*. Bioessays 31:89–97. doi:10.1002/bies.080058

Fahey B, Degnan BM (2010) Origin of animal epithelia: insights from the sponge genome. Evol Dev 12:601–617. doi:10.1111/j.1525-142X.2010.00445.x

Fang X, Ji H, Kim SW, Park JI, Vaught TG, Anastasiadis PZ, Ciesiolka M, McCrea PD (2004) Vertebrate development requires ARVCF and p120 catenins and their interplay with RhoA and Rac. J Cell Biol 165:87–98. doi:10.1083/jcb.200307109

Formstone CJ, Mason I (2005) Combinatorial activity of Flamingo proteins directs convergence and extension within the early zebrafish embryo via the planar cell polarity pathway. Dev Biol 282:320–335. doi:10.1016/j.ydbio.2005.03.026

Fung S, Wang F, Chase M, Godt D, Hartenstein V (2008) Expression profile of the cadherin family in the developing *Drosophila* brain. J Comp Neurol 506:469–488. doi:10.1002/cne.21539

Gallin WJ, Sorkin BC, Edelman GM, Cunningham BA (1987) Sequence analysis of a cDNA clone encoding the liver cell adhesion molecule, L-CAM. Proc Natl Acad Sci U S A 84:2808–2812

Grana TM, Cox EA, Lynch AM, Hardin J (2010) SAX-7/L1CAM and HMR-1/cadherin function redundantly in blastomere compaction and non-muscle myosin accumulation during *C. elegans* gastrulation. Dev Biol 344:731–744. doi:10.1016/j.ydbio.2010.05.507

Greenwood MD, Marsden MD, Cowley CM, Sahota VK, Buxton RS (1997) Exon-intron organization of the human type 2 desmocollin gene (DSC2): desmocollin gene structure is closer to "classical" cadherins than to desmogleins. Genomics 44:330–335. doi:10.1006/geno.1997.4894

Grimson MJ, Coates JC, Reynolds JP, Shipman M, Blanton RL, Harwood AJ (2000) Adherens junctions and β-catenin-mediated cell signalling in a non-metazoan organism. Nature 408:727–731 doi:10.1038/35047099

Haruta T, Warrior R, Yonemura S, Oda H. (2010) The proximal half of the *Drosophila* E-cadherin extracellular region is dispensable for many cadherin-dependent events but required for ventral furrow formation. Genes Cells 15:193–208. doi:10.1111/j.1365-2443.2010.01389.x

Hatta K, Takeichi M (1986) Expression of N-cadherin adhesion molecules associated with early morphogenetic events in chick development. Nature 320:447–449. doi:10.1038/320447a0

Hatzfeld M, Nachtsheim C (1996) Cloning and characterization of a new *armadillo* family member, p0071, associated with the junctional plaque: evidence for a subfamily of closely related proteins. J Cell Sci 109:2767–2778

Heimark RL, Degner M, Schwartz SM (1990) Identification of a Ca^{2+}-dependent cell-cell adhesion molecule in endothelial cells. J Cell Biol 110:1745–1756

Herrenknecht K, Ozawa M, Eckerskorn C, Lottspeich F, Lenter M, Kemler R (1991) The uvomorulin-anchorage protein α catenin is a vinculin homologue. Proc Natl Acad Sci U S A 88:9156–9160

Hirano S, Kimoto N, Shimoyama Y, Hirohashi S, Takeichi M (1992) Identification of a neural α-catenin as a key regulator of cadherin function and multicellular organization. Cell 70:293–301. doi:10.1016/0092-8674(92)90103-J

Hsu SN, Yonekura S, Ting CY, Robertson HM, Iwai Y, Uemura T, Lee CH, Chiba A (2009) Conserved alternative splicing and expression patterns of arthropod N-cadherin. PLoS Genet 5(4):e1000441. doi:10.1371/journal.pgen.1000441

Huber AH, Weis WI (2001) The structure of the β-catenin/E-cadherin complex and the molecular basis of diverse ligand recognition by β-catenin. Cell 105:391–402. doi:10.1016/S0092-8674(01)00330–0

Huber O, Krohn M, Kemler R (1997) A specific domain in α-catenin mediates binding to β-catenin or plakoglobin. J Cell Sci 110:1759–1765

Hulpiau P, van Roy F (2009) Molecular evolution of the cadherin superfamily. Int J Biochem Cell Biol 41:349–369. doi:10.1016/j.biocel.2008.09.027

Hulpiau P, van Roy F (2010) New Insights into the Evolution of Metazoan Cadherins. Mol Biol Evol 28:647–657. doi:10.1093/molbev/msq233

Hyafil F, Babinet C, Jacob F (1981) Cell-cell interactions in early embryogenesis: a molecular approach to the role of calcium. Cell 26:447–454. doi:10.1016/0092-8674(81)90214-2

Ireton RC, Davis MA, van Hengel J, Mariner DJ, Barnes K, Thoreson MA, Anastasiadis PZ, Matrisian L, Bundy LM, Sealy L, Gilbert B, van Roy F, Reynolds AB (2002) A novel role for p120 catenin in E-cadherin function. J Cell Biol 159:465–476. doi:10.1083/jcb.200205115

Ishiuchi T, Misaki K, Yonemura S, Takeichi M, Tanoue T (2009) Mammalian Fat and Dachsous cadherins regulate apical membrane organization in the embryonic cerebral cortex. J Cell Biol 185:959–967. doi:10.1083/jcb.200811030

Iwai Y, Usui T, Hirano S, Steward R, Takeichi M, Uemura T (1997) Axon patterning requires *D*N-cadherin, a novel neuronal adhesion receptor, in the Drosophila embryonic CNS. Neuron 19:77–89. doi:10.1016/S0896-6273(00)80349-9

Kanno M, Isa Y, Aoyama Y, Yamamoto Y, Nagai M, Ozawa M, Kitajima Y (2008) p120-catenin is a novel desmoglein 3 interacting partner: identification of the p120-catenin association site of desmoglein 3. Exp Cell Res 314:1683–1692. doi:10.1016/j.yexcr.2008.01.031

Kazmierczak P, Sakaguchi H, Tokita J, Wilson-Kubalek EM, Milligan RA, Müller U, Kachar B (2007) Cadherin 23 and protocadherin 15 interact to form tip-link filaments in sensory hair cells. Nature 449:87–91. doi:10.1038/nature06091

Kemler R, Babinet C, Eisen H, Jacob F (1977) Surface antigen in early differentiation. Proc Natl Acad Sci U S A 74:4449–4452

Knust E, Bossinger O (2002) Composition and formation of intercellular junctions in epithelial cells. Science 298:1955–1959. doi:10.1126/science.1072161

Kobielak A, Fuchs E (2004) Alpha-catenin: at the junction of intercellular adhesion and actin dynamics. Nat Rev Mol Cell Biol 5:614–625. doi:10.1038/nrm1433

Koslov ER, Maupin P, Pradhan D, Morrow JS, Rimm DL (1997) α-catenin can form asymmetric homodimeric complexes and/or heterodimeric complexes with β-catenin. J Biol Chem 272:27301–27306. doi:10.1074/jbc.272.43.27301

Kremmidiotis G, Baker E, Crawford J, Eyre HJ, Nahmias J, Callen DF (1998) Localization of human cadherin genes to chromosome regions exhibiting cancer-related loss of heterozygosity. Genomics 49:467–471. doi:10.1006/geno.1998.5281

Kwiatkowski AV, Maiden SL, Pokutta S, Choi HJ, Benjamin JM, Lynch AM, Nelson WJ, Weis WI, Hardin J (2010) In vitro and in vivo reconstitution of the cadherin-catenin-actin complex from *Caenorhabditis elegans*. Proc Natl Acad Sci U S A 107:14591–14596. doi:10.1073/pnas.1007349107

Lampugnani MG, Corada M, Andriopoulou P, Esser S, Risau W, Dejana E (1997) Cell confluence regulates tyrosine phosphorylation of adherens junction components in endothelial cells. J Cell Sci 110:2065–2077

Lane NJ, Dallai R, Martinucci GB, Burighel P (1987) Cell junctions in amphioxus (Cephalochordata): a thin section and freeze-fracture study. Tissue Cell 19:399–411. doi:10.1016/0040-8166(87)90035-8

Ma D, Yang CH, McNeill H, Simon MA, Axelrod JD (2003) Fidelity in planar cell polarity signalling. Nature 421:543–547. doi:10.1038/nature01366

Mahoney PA, Weber U, Onofrechuk P, Biessmann H, Bryant PJ, Goodman CS (1991) The *fat* tumor suppressor gene in Drosophila encodes a novel member of the cadherin gene superfamily. Cell 67:853–868. doi:10.1016/0092-8674(91)90359-7

Mariner DJ, Wang J, Reynolds AB (2000) ARVCF localizes to the nucleus and adherens junction and is mutually exclusive with $p120^{ctn}$ in E-cadherin complexes. J Cell Sci 113:1481–1490

Matakatsu H, Blair SS (2004) Interactions between Fat and Dachsous and the regulation of planar cell polarity in the *Drosophila* wing. Development 131:3785–3794. doi:10.1242/dev.01254

McCrea PD, Turck CW, Gumbiner B (1991) A homolog of the *armadillo* protein in *Drosophila* (plakoglobin) associated with E-cadherin. Science 254:1359–1361. doi:10.1126/science.1962194

Morishita H, Yagi T (2007) Protocadherin family: diversity, structure, and function. Curr Opin Cell Biol 19:584–592. doi:10.1016/j.ceb.2007.09.006

Myster SH, Cavallo R, Anderson CT, Fox DT, Peifer M (2003) *Drosophila* p120catenin plays a supporting role in cell adhesion but is not an essential adherens junction component. J Cell Biol 160:433–449. doi:10.1083/jcb.200211083

Nagafuchi A, Shirayoshi Y, Okazaki K, Yasuda K, Takeichi M (1987) Transformation of cell adhesion properties by exogenously introduced E-cadherin cDNA. Nature 329:341–343

Nagafuchi A, Takeichi M, Tsukita S (1991) The 102 kd cadherin-associated protein: similarity to vinculin and posttranscriptional regulation of expression. Cell 65:849–857. doi:10.1016/0092-8674(91)90392-C

Nagar B, Overduin M, Ikura M, Rini JM (1996) Structural basis of calcium-induced E-cadherin rigidification and dimerization. Nature 380:360–364. doi:10.1038/380360a0

Nakao S, Platek A, Hirano S, Takeichi M (2008) Contact-dependent promotion of cell migration by the OL-protocadherin-Nap1 interaction. J Cell Biol 182:395–410. doi:10.1083/jcb.200802069

Navarro P, Ruco L, Dejana E (1998) Differential localization of VE- and N-cadherins in human endothelial cells: VE-cadherin competes with N-cadherin for junctional localization. J Cell Biol 140:1475–1484. doi:10.1083/jcb.140.6.1475

Nieset JE, Redfield AR, Jin F, Knudsen KA, Johnson KR, Wheelock MJ (1997) Characterization of the interactions of α-catenin with a-actinin and β-catenin/plakoglobin. J Cell Sci 110:1013–1022

Nollet F, Kools P, van Roy F (2000) Phylogenetic analysis of the cadherin superfamily allows identification of six major subfamilies besides several solitary members. J Mol Biol 299:551–572. doi:10.1006/jmbi.2000.3777

Nose A, Nagafuchi A, Takeichi M (1988) Expressed recombinant cadherins mediate cell sorting in model systems. Cell 54:993–1001. doi:10.1016/0092–8674(88)90114–6

Nose A, Tsuji K, Takeichi M (1990) Localization of specificity determining sites in cadherin cell adhesion molecules. Cell 61:147–155. doi:10.1016/0092–8674(90)90222-Z

Obama H, Ozawa M (1997) Identification of the domain of α-catenin involved in its association with β-catenin and plakoglobin (γ-catenin). J Biol Chem 272:11017–11020

Oda H, Takeichi M (2011) Evolution: Structural and functional diversity of cadherin at the adherens junction. J Cell Biol 193:1137–1146. doi:10.1083/jcb.201008173

Oda H, Tsukita S (1999) Nonchordate classic cadherins have a structurally and functionally unique domain that is absent from chordate classic cadherins. Dev Biol 216:406–422. doi:10.1006/dbio.1999.9494

Oda H, Uemura T, Harada Y, Iwai Y, Takeichi M (1994) A *Drosophila* homolog of cadherin associated with Armadillo and essential for embryonic cell-cell adhesion. Dev. Biol 165:716–726. doi:10.1006/dbio.1994.1287

Oda H, Wada H, Tagawa K, Akiyama-Oda Y, Satoh N, Humphreys T, Zhang S, Tsukita S (2002) A novel amphioxus cadherin that localizes to epithelial adherens junctions has an unusual domain organization with implications for chordate phylogeny. Evol Dev 4:426-434. doi:10.1046/j.1525-142X.2002.02031.x

Oda H, Akiyama-Oda Y, Zhang S (2004) Two classic cadherin-related molecules with no cadherin extracellular repeats in the cephalochordate amphioxus: distinct adhesive specificities and possible involvement in the development of multicell-layered structures. J Cell Sci 117:2757-2767. doi:10.1242/jcs.01045

Oda H, Tagawa K, Akiyama-Oda Y (2005) Diversification of epithelial adherens junctions with independent reductive changes in cadherin form: identification of potential molecular synapomorphies among bilaterians. Evol Dev 7:376–389. doi:10.1111/j.1525-142X.2005.05043.x

Overduin M, Harvey TS, Bagby S, Tong KI, Yau P, Takeichi M, Ikura M (1995) Solution structure of the epithelial cadherin domain responsible for selective cell adhesion. Science 267:386–389. doi: 10.1126/science.7824937

Ozawa M, Kemler R (1990) Correct proteolytic cleavage is required for the cell adhesive function of uvomorulin. J Cell Biol 111:1645–1650

Ozawa M, Engel J, Kemler R (1990) Single amino acid substitutions in one Ca^{2+} binding site of uvomorulin abolish the adhesive function. Cell 63:1033–1038. doi:10.1016/0092–8674(90)90506-A

Pacquelet A, Lin L, Rorth P (2003) Binding site for p120/δ-catenin is not required for *Drosophila* E-cadherin function in vivo. J Cell Biol 160:313–319. doi:10.1083/jcb.200207160

Patel SD, Ciatto C, Chen CP, Bahna F, Rajebhosale M, Arkus N, Schieren I, Jessell TM, Honig B, Price SR, Shapiro L (2006) Type II cadherin ectodomain structures: implications for classical cadherin specificity. Cell 124:1255–1268. doi:10.1016/j.cell.2005.12.046

Paulson AF, Mooney E, Fang X, Ji H, McCrea PD (2000) Xarvcf, *Xenopus* member of the p120 catenin subfamily associating with cadherin juxtamembrane region. J Biol Chem 275:30124–30131. doi:10.1074/jbc.M003048200

Peifer M, Polakis P (2000) Wnt signaling in oncogenesis and embryogenesis-a look outside the nucleus. Science 287:1606–1609. doi:10.1126/science.287.5458.1606

Peifer M, McCrea PD, Green KJ, Wieschaus E, Gumbiner BM (1992) The vertebrate adhesive junction β-catenin and plakoglobin and the *Drosophila* segment polarity gene armadillo form a multigene family with similar properties. J Cell Biol 118:681–691

Pettitt J (2005) The cadherin superfamily. WormBook 29:1–9

Pettitt J, Cox EA, Broadbent ID, Flett A, Hardin J (2003) The *Caenorhabditis elegans* p120 catenin homologue, JAC-1, modulates cadherin-catenin function during epidermal morphogenesis. J Cell Biol 162:15–22. doi:10.1083/jcb.200212136

Pokutta S, Weis WI (2000) Structure of the dimerization and β-catenin-binding region of α-catenin. Mol Cell 5:533–543. doi:10.1016/S1097-2765(00)80447–5

Prakash S, Caldwell JC, Eberl DF, Clandinin TR (2005) *Drosophila* N-cadherin mediates an attractive interaction between photoreceptor axons and their targets. Nat Neurosci 8:443–450. doi:10.1038/nn1415

Reddy B V, Irvine KD (2008) The Fat and Warts signaling pathways: new insights into their regulation, mechanism and conservation. Development 135:2827–2838. doi:10.1242/dev.020974

Reynolds AB, Roesel DJ, Kanner SB, Parsons JT (1989) Transformation-specific tyrosine phosphorylation of a novel cellular protein in chicken cells expressing oncogenic variants of the avian cellular *src* gene. Mol Cell Biol 9:629–638

Reynolds AB, Daniel J, McCrea PD, Wheelock MJ, Wu J, Zhang Z (1994) Identification of a new catenin: the tyrosine kinase substrate $p120^{cas}$ associates with E-cadherin complexes. Mol Cell Biol 14:8333–8342

Riggleman B, Wieschaus E, Schedl P (1989) Molecular analysis of the *armadillo* locus: uniformly distributed transcripts and a protein with novel internal repeats are associated with a *Drosophila* segment polarity gene. Genes Dev 3:96–113. doi:10.1101/gad.3.1.96

Rimm DL, Koslov ER, Kebriaei P, Cianci CD, Morrow JS (1995) α1(E)-catenin is an actin-binding and—bundling protein mediating the attachment of F-actin to the membrane adhesion complex. Proc Natl Acad Sci U S A 92:8813–8817

Ringwald M, Schuh R, Vestweber D, Eistetter H, Lottspeich F, Engel J, Dölz R, Jähnig F, Epplen J, Mayer S (1987) The structure of cell adhesion molecule uvomorulin. Insights into the molecular mechanism of Ca^{2+}-dependent cell adhesion. EMBO J 6:3647–3653

Saburi S, Hester I, Fischer E, Pontoglio M, Eremina V, Gessler M, Quaggin SE, Harrison R, Mount R, McNeill H (2008) Loss of Fat4 disrupts PCP signaling and oriented cell division and leads to cystic kidney disease. Nat Genet 40:1010–1015. doi:10.1016/j.ceb.2005.08.011

Sakarya O, Armstrong KA, Adamska M, Adamski M, Wang IF, Tidor B, Degnan BM, Oakley TH, Kosik KS (2007) A post-synaptic scaffold at the origin of the animal kingdom. PLoS ONE 2(6):e506. doi:10.1371/journal.pone.0000506

Salomon D, Ayalon O, Patel-King R, Hynes RO, Geiger B (1992) Extrajunctional distribution of N-cadherin in cultured human endothelial cells. J Cell Sci 102:7–17

Sasakura Y, Shoguchi E, Takatori N, Wada S, Meinertzhagen IA, Satou Y, Satoh N (2003) A genomewide survey of developmentally relevant genes in *Ciona intestinalis*. X. Genes for cell junctions and extracellular matrix. Dev Genes Evol 213:303–313. doi:1007/s00427-003-0320-1

Shirayoshi Y, Hatta K, Hosoda M, Tsunasawa S, Sakiyama F, Takeichi M (1986) Cadherin cell adhesion molecules with distinct binding specificities share a common structure. EMBO J 5:2485–2488

Takeichi M (1988) The cadherins: cell-cell adhesion molecules controlling animal morphogenesis. Development 102:639–655

Takeichi M (2007) The cadherin superfamily in neuronal connections and interactions. Nat Rev Neurosci 8:11–20. doi:10.1038/nrn2043

Tanabe K, Takeichi M, Nakagawa S (2004) Identification of a nonchordate-type classic cadherin in vertebrates: chicken Hz-cadherin is expressed in horizontal cells of the neural retina and contains a nonchordate-specific domain complex. Dev Dyn 229:899–906. doi:10.1002/dvdy.10493

Tepass U, Gruszynski-DeFeo E, Haag TA, Omatyar L, Török T, Hartenstein V (1996) *shotgun* encodes *Drosophila* E-cadherin and is preferentially required during cell rearrangement in the neurectoderm and other morphogenetically active epithelia. Genes Dev 10:672–685. doi:10.1101/gad.10.6.672

Tolwinski NS, Wieschaus E (2004) Rethinking WNT signaling. Trends Genet 20:177–181. doi:10.1016/j.tig.2004.02.003

Troyanovsky SM, Troyanovsky RB, Eshkind LG, Krutovskikh VA, Leube RE, Franke WW (1994) Identification of the plakoglobin-binding domain in desmoglein and its role in plaque assembly and intermediate filament anchorage. J Cell Biol 127:151–160

Uemura T, Oda H, Kraut R, Hayashi S, Kotaoka Y, Takeichi M (1996) Zygotic *Drosophila* E-cadherin expression is required for processes of dynamic epithelial cell rearrangement in the *Drosophila* embryo. Genes Dev 10:659–671. doi:10.1101/gad.10.6.659

Usui T, Shima Y, Shimada Y, Hirano S, Burgess RW, Schwarz TL, Takeichi M, Uemura T (1999) Flamingo, a seven-pass transmembrane cadherin, regulates planar cell polarity under the control of Frizzled. Cell 98:585–595. doi:10.1016/S0092-8674(00)80046-X

Vestal DJ, Ranscht B (1992) Glycosyl phosphatidylinositol–anchored T-cadherin mediates calcium-dependent, homophilic cell adhesion. J Cell Biol 119:451–461

Volk T, Geiger B (1984) A 135-kd membrane protein of intercellular adherens junctions. EMBO J 3:2249–2260

Watson GM, Pham L, Graugnard EM, Mire P (2008) Cadherin 23-like polypeptide in hair bundle mechanoreceptors of sea anemones. J Comp. Physiol. A Neuroethol. Sens. Neural Behav Physiol 194:811–820. doi:10.1007/s00359-008-0352-0

Wendeler MW, Jung R, Himmelbauer H, Gessner R (2006) Unique gene structure and paralogy define the 7D-cadherin family. Cell Mol Life Sci 63:1564–1573. doi:10.1007/s00018-006-6014-x

Whittaker CA, Bergeron KF, Whittle J, Brandhorst BP, Burke RD, Hynes RO (2006) The echinoderm adhesome. Dev Biol 300:252–266. doi:10.1016/j.ydbio.2006.07.044

Xiao K, Allison DF, Buckley KM, Kottke MD, Vincent PA, Faundez V, Kowalczyk AP (2003) Cellular levels of p120 catenin function as a set point for cadherin expression levels in microvascular endothelial cells. J Cell Biol 163:535–545. doi:10.1083/jcb.200306001

Yap AS, Niessen CM, Gumbiner BM (1998) The juxtamembrane region of the cadherin cytoplasmic tail supports lateral clustering, adhesive strengthening, and interaction with p120ctn. J Cell Biol 141:779–789. doi:10.1083/jcb.141.3.779

Yonekura S, Xu L, Ting CY, Lee CH (2007) Adhesive but not signaling activity of *Drosophila* N-cadherin is essential for target selection of photoreceptor afferents. Dev Biol 304:759–770. doi:10.1016/j.ydbio.2007.01.030

Yoshida C, Takeichi M (1982) Teratocarcinoma cell adhesion: identification of a cell-surface protein involved in calcium-dependent cell aggregation. Cell 28:217–224. doi:10.1016/0092-8674(82)90339-7

Yoshida-Noro C, Suzuki N, Takeichi M (1984) Molecular nature of the calcium-dependent cell-cell adhesion system in mouse teratocarcinoma and embryonic cells studied with a monoclonal antibody. Dev Biol 101:19–27. doi:10.1016/0012-1606(84)90112-X

Part II
How Cadherins and Catenins Interact to Assemble AJs

Chapter 3
The Three-Dimensional Structure of the Cadherin–Catenin Complex

Noboru Ishiyama and Mitsuhiko Ikura

Abstract The cadherin–catenin complex is the major building block of the adherens junction. It is responsible for coupling Ca^{2+}-dependent intercellular junctions with various intracellular events, including actin dynamics and signaling pathways. Determination of three-dimensional structures of cadherins, p120 catenin, β-catenin and α-catenin at atomic-level resolution has allowed us to examine how the structure and function of cell adhesion molecules are further modulated by protein–protein interactions. Structural studies of cadherins revealed the strand-swap-dependent and -independent *trans*-dimerization mechanisms, as well as a potential mechanism for lateral clustering of cadherin *trans*-dimers. Crystallographic and NMR analyses of p120 catenin revealed that it regulates the stability of cadherin-mediated cell–cell adhesion by associating with the majority of the E-cadherin juxtamembrane domain, including residues implicated in clathrin-mediated endocytosis and Hakai-dependent ubiquitination. Crystal structures of the β-catenin/E-cadherin complex and the β-/α-catenin chimera revealed extensive interactions necessary to form the cadherin/β-catenin/α-catenin ternary complex. Structural characterization of α-catenin has revealed conformational changes within the N-terminal and modulatory domains that are crucial for its role as a mechanosensor of cell–cell adhesion. Further insights into the connection between the cadherin–catenin complex and the actin cytoskeleton are integral to better understand how adjoining cells communicate through cell–cell adhesion.

3.1 Introduction

The multi-protein complex consisting of cadherin, a cell adhesion receptor, and its cytosolic binding partners, the catenins, is the major building block of intercellular junctions, such as the adherens junction (AJ). Adjoining cells can be physically

M. Ikura (✉)
Department of Medical Biophysics, University of Toronto, Ontario Cancer Institute, MaRS Toronto Medical Discovery Tower, Room 4-804, 101 College Street, Toronto, ON M5G 1L7, Canada
e-mail: mikura@uhnres.utoronto.ca

N. Ishiyama
Ontario Cancer Institute, MaRS Toronto Medical Discovery Tower, Room 4-902Q, 101 College Street, Toronto, ON M5G 1L7, Canada

T. Harris (ed.), *Adherens Junctions: From Molecular Mechanisms to Tissue Development and Disease,* Subcellular Biochemistry 60,
DOI 10.1007/978-94-007-4186-7_3, © Springer Science+Business Media Dordrecht 2012

connected when the extracellular regions of cadherins on adjacent cell surfaces form Ca^{2+}-dependent homophilic interactions (Hirano et al. 1987; Yoshida and Takeichi 1982). However, the adhesion of the extracellular regions of cadherin alone is insufficient to establish a stable, mature cell–cell contact; the intracellular region of cadherin establishes functional linkages to the actin cytoskeleton and various signaling pathways through catenins (Meng and Takeichi 2009). The resulting architecture of cadherin-catenin-mediated cell–cell junctions is stable enough to support tissue structure and integrity, yet remains sufficiently dynamic to quickly dissolve obsolete connections and foster new connections among neighbouring cells during embryogenesis and wound healing (Takeichi 1995). In contrast, the loss of cadherin expression or dissociation between cadherin and catenins can be induced by a number of factors, including transcriptional regulation, mutation and aberrant cadherin internalization (Mosesson et al. 2008), and has been associated with tumour invasiveness and metastasis (Hanahan and Weinberg 2000; Takeichi 1993).

To better understand how cadherins and catenins regulate cell–cell adhesion mechanisms, cell adhesion molecules found in AJs, namely classical cadherins, p120 catenin, β-catenin and α-catenin, have been subjected to extensive structural characterization for over 15 years. Biophysical techniques, such as X-ray crystallography, nuclear magnetic resonance (NMR) spectroscopy and electron microscopy (EM), have been successfully employed to yield invaluable atomic-level views of their three-dimensional structures, and more importantly the details regarding the protein–protein interactions that are indispensable for the structure and function of the cell adhesion complex. In this chapter, the roles of cadherins and catenins in the regulation of cell–cell adhesion will be discussed based on recently obtained structural information regarding how cadherins and catenins assemble into an entire cell adhesion complex.

3.2 Cadherin

3.2.1 The Overall Structure of Cadherin

Classical (Type I) cadherins, such as E- and N-cadherins, engage in Ca^{2+}-dependent homophilic interactions important for embryogenesis and development (Harris and Tepass 2010; Nishimura and Takeichi 2009). These cadherins comprise the most well characterized subfamily of the cadherin superfamily, which consists of a large number of cell surface receptors recognized by the presence of multiple extracellular cadherin (EC) domains (or ectodomains), ranging from 2 EC domains in the worm cadherin HMR-1 to 34 EC domains in Fat cadherins found in flies to mammals (Nollet et al. 2000). Classical cadherins are single-pass transmembrane proteins that contain five EC domains (EC1-5) on the extracellular side and highly conserved catenin-associating domains on the intracellular side (Fig. 3.1a). Cadherins

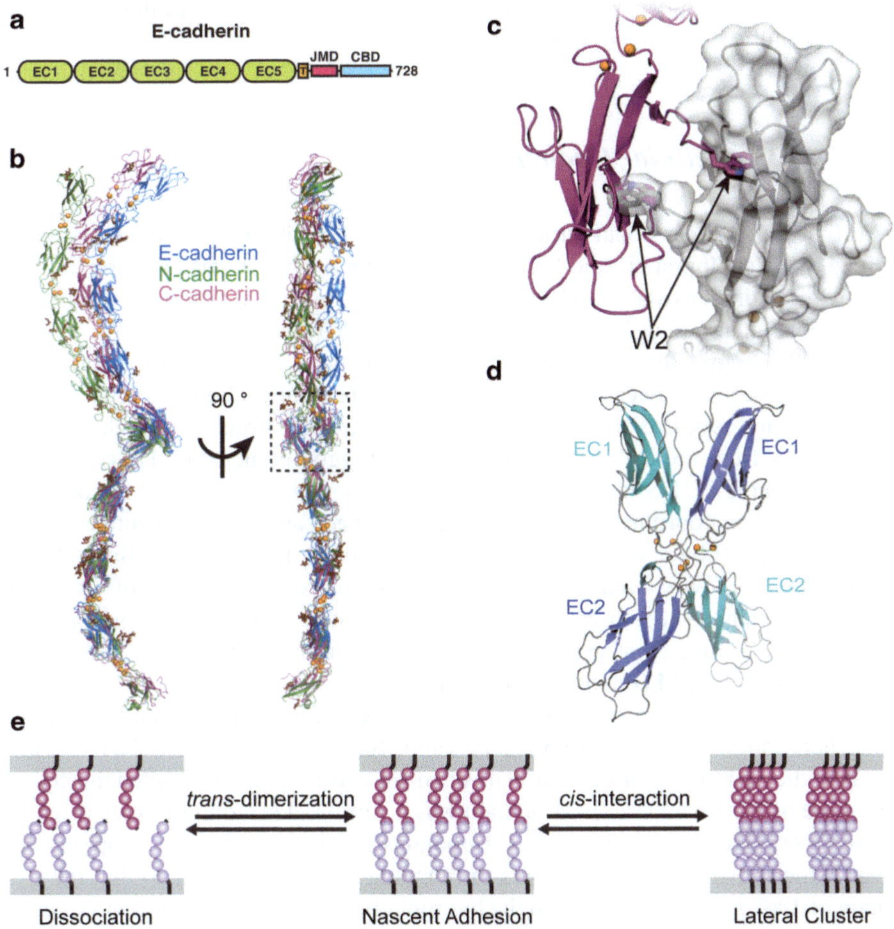

Fig. 3.1 Three-dimensional structures of cadherin ectodomains. **a** Scheme of E-cadherin structure. E-cadherin consists of the extracellular cadherin domains 1–5 (*EC1-5*), the transmembrane region (*T*) and the cytoplasmic tail, which contains the juxtamembrane domain (*JMD*) and the catenin-binding domain (*CBD*). Amino acid residue numbers of cadherin are based on the mature form unless indicated otherwise. **b** Overlay of crystal structures of EC1-5 *trans*-dimers of E-cadherin (*blue*; PDB code 3Q2V) (Harrison et al. 2011), N-cadherin (*green*; PDB code 3Q2W) (Harrison et al. 2011) and C-cadherin (*magenta*; PDB code 1L3W) (Boggon et al. 2002). Structures of EC1-5 monomers from all three cadherins are virtually identical (see the superposed bottom chains), however their *trans*-dimer arrangements vary slightly due to small differences in the strand-swap EC1-EC1 dimer formation. **c** Close-up view of the strand-swap dimer interface of C-cadherin. The surface of one of two EC1 protomers is shown (*gray*) to highlight the hydrophobic pocket where the side chain of W2 from the adjacent protomer is buried. **d** The X-dimer interface of T-cadherin EC1-2 (PDB code 3K5S) (Ciatto et al. 2010) involves the Ca^{2+}-binding sites. **e** Model of cadherin *trans*-dimerization and *cis*-interaction of cadherins during cell–cell adhesion

are essential for connecting intercellular junctions, such as AJs, to the actin cytoskeleton and various signaling pathways.

3.2.2 Extracellular Cadherin Domains

The first three-dimensional structures of N- and E-cadherin EC1 domains determined by crystallographic and NMR studies, respectively, revealed that the individual EC domain consists of ~110 residues forming seven anti-parallel β-strands arranged into an immunoglobulin-like β-sandwich fold (Overduin et al. 1995; Shapiro et al. 1995). More recently, high-resolution structures of the entire extracellular region (EC1-5) have been determined for three classical cadherins: E-, N- and C-cadherins, and revealed a conserved arch-shaped overall structure, which stretches out ~190 Å from the cell surface (Fig. 3.1b) (Boggon et al. 2002; Harrison et al. 2011). All three EC1-5 structures are in the Ca^{2+}-bound state and five EC domains are rigidified by coordinating three Ca^{2+} ions between any two consecutive EC domains connected by a short linker region (Nagar et al. 1996). The binding of Ca^{2+} to cadherin is known to make its extracellular region resistant to proteolytic degradation (Takeichi 1988). In contrast, the Ca^{2+}-free state of EC1-5 is more prone to proteolytic cleavage by trypsin, and has been shown to lose its rigidity and adopt a globular shape by EM (Pokutta et al. 1994).

The first EC (EC1) domain of cadherin is crucial for the affinity and specificity of its homophilic interaction (Nose et al. 1990; Tomschy et al. 1996). The molecular basis of the underlying dimerization mechanism was first unveiled when the crystal structure of the N-cadherin EC1 domain was determined in a dimeric state (Shapiro et al. 1995). The dimerization of EC1 domains involves the 'strand-swap' mechanism which involves two protomers exchanging the first β-strand so that the side chain of Trp2 is firmly buried into the hydrophobic pocket of the adjacent molecule (Fig. 3.1c) (Shapiro and Weis 2009). The presence of Trp as the second residue on the N terminus is critical for this process and is attained by proteolytic cleavage of the cadherin prodomain (discussed later) (Häussinger et al. 2004). The EC1 domains of type II-subfamily of cadherins, such as cadherin-8, cadherin-11 and VE-cadherin, have been also shown to engage in strand-swap dimerization similar to the ones observed for classical cadherins, but their interactions involve two N-terminal Trp residues (Trp2 and Trp4) being inserted into a larger hydrophobic pocket (Brasch et al. 2011; Patel et al. 2006). While the critical role of Trp2 in cadherin-mediated cell–cell adhesion was confirmed by mutagenesis/cell aggregation experiments (Tamura et al. 1998), the crystallographic (Shapiro et al. 1995) and cross-linking (Brieher et al. 1996; Troyanovsky et al. 2003) studies suggested the possibility of strand-swap interaction involved in *cis*-dimerization of cadherins on the same cell surface. Surprisingly weak affinity displayed by the strand-swap interaction of cadherins (K_D=80–720 μM) (Häussinger et al. 2004; Koch et al. 1997) also promoted the idea of *cis*-dimerization or lateral clustering playing a role in activating cadherins to form *trans*-dimerization between two adjoining cells (Leckband

and Prakasam 2006; Stemmler 2008). However, single-molecule studies employing fluorescence resonance energy transfer and atomic force microscopy have shown that *trans*-dimerization of cadherin does not require prior formation of *cis*-dimers (Zhang et al. 2009). This is consistent with most currently available strand-swap dimer structures depicting cadherins as *trans*-dimers: the arch-shaped EC domains place the dimerized EC1 domains to be oriented in the similar direction while the C termini of protomers aim toward the opposite direction (Boggon et al. 2002; Brasch et al. 2011; Harrison et al. 2011; Patel et al. 2006).

3.2.3 Juxtamembrane and Catenin-Binding Domains of Cadherin

The cytoplasmic region of classical cadherin comprises of ~150 residues and contains several highly conserved sequence motifs (Fig. 3.1a) (Nollet et al. 2000). It can be further divided into two major domains: the juxtramembrane domain (JMD) and the catenin-binding domain (CBD). The JMD consists of ~50 residues immediately after the transmembrane domain and provides a specific binding site for p120 catenin and p120-related proteins, such as δ-catenin, ARVCF and p0071 (see below for further discussion) (Ishiyama et al. 2010; Thoreson et al. 2000). On the other hand, the CBD consists of the C-terminal ~100 residues and specifically binds to β-catenin and plakoglobin (Huber and Weis 2001) (see below for further discussion). Circular dichroism and proton NMR spectroscopic studies revealed that the recombinant form of the entire cadherin cytoplasmic region is largely unstructured in solution (Huber et al. 2001).

3.2.4 Other Dimerization Mechanisms of Cadherin

The strand-swap dimer mechanism alone cannot explain the observations from numerous biophysical studies of cadherin *trans*-dimerization suggesting that EC domains other than EC1 also contribute to adhesive forces produced by cadherins at different intermolecular bond distances (Shi et al. 2010; Sivasankar et al. 2001). One such example was recently revealed when the EC1-2 domains of T-cadherin were shown to dimerize through their Ca^{2+}-binding sites depicting a character 'X' (Fig. 3.1d) (Ciatto et al. 2010). T-cadherin is a divergent member of classical cadherin that is glycosylphosphatidylinositol-anchored, and lacks the N-terminal Trp required for the strand-swap dimer. Nonetheless, T-cadherin EC1-2 domains form X-dimers with higher affinity (K_D=~40 μM) than the strand-swap dimer of E-cadherin EC1-2 domains (Ciatto et al. 2010). Interestingly, the X-dimer formation was previously observed with E-cadherin EC1-2 domains when they were crystallized with an N-terminal extension (Nagar et al. 1996). These observations suggest that other classical cadherins may be capable of facilitating *trans*-interaction via

X-dimerization in addition to the strand-swap mechanism. Since there are multiple Ca^{2+}-binding sites along the EC1-5 domains, it raises the possibility of other EC domain pairs, e.g., EC2-3, participating in *trans*-interactions of cadherins at intercellular junctions.

3.2.5 *Clustering Through* Cis-*Interaction*

The transformation of nascent cell–cell adhesion complexes into a stable intercellular junction is likely to involve lateral clustering of *trans*-dimers of cadherin (Fig. 3.1e). While the cytoplasmic region of cadherin has been implicated in cadherin clustering (Ishiyama et al. 2010; Yap et al. 1998), this process likely involves the *cis*-interaction of the EC domains as well (Yap et al. 1997). Visualization of the human epidermis by Cryo-EM tomography revealed a well organized architecture of the desmosome with the cell–cell junction mainly consisting of *trans*-dimers of desmosomal cadherins laterally packed at ~70 Å intervals (Al-Amoudi et al. 2007). The crystal packing contacts observed in the cadherin EC1-5 domain crystals also offer additional clues to how *trans*-dimers of cadherins would participate in *cis*-interaction. In all three independently crystallized EC1-5 domains of E-, N- and C-cadherins, it was observed that an EC1 surface opposite from the strand-swap interface interacts with the EC2 domain of an adjacent molecule, depicting *cis*-interaction of cadherin *trans*-dimer (Boggon et al. 2002; Harrison et al. 2011). Since the occurrence of a common crystal packing interface in three different crystals is extremely rare, it may be indicative of this *cis*-interface playing a physiological role in lateral clustering of cadherin *trans*-dimers. Indeed, mutations within the *cis*-interface (V81D/V175D) of E-cadheirn appears to interfere with proper formation of intercellular junctions (Harrison et al. 2011).

3.2.6 *Post-Translational Modifications*

Post-translational modifications of classical cadherins are a critical part of modulating the structure and function of cadherin adhesion receptors (Takeichi 1988). First, cadherins are *N*-glycosylated at numerous sites in the EC domains as part of the quality control process in the endoplasmic reticulum and Golgi network to ensure proper folding and stability (Boggon et al. 2002). Comparison of non-glycosylated and glycosylated EC1-5 domains of VE-cadherin revealed that the glycosylation of the extracellular region affects the oligomeric state of VE-cadherin (Brasch et al. 2011). It was previously reported that the bacterially expressed recombinant protein of VE-cadherin EC1-4 domains (no posttranslational glycosylation) forms a hexamer made of two *cis*-trimers in *trans*-association (Legrand et al. 2001; Taveau et al. 2008). However, when VE-cadherin EC1-5 domains were produced using mammalian cell expression system, *N*-glycosylated recombinant proteins did not hexamerize, but instead formed strand-swap dimers in solution (Brasch et al. 2011), suggesting that the glycosyl moieties block the surface patches involved in the trimer/hexamer formation observed *in vitro*.

Second, a non-adhesive nascent cadherin molecule contains an N-terminal prodomain (residues 1–156 in mouse E-cadherin) (Takeichi 1988), which is cleaved by furin and other proprotein convertases in the *trans*-Golgi network to present the Asp-Trp pair as the first two residues of the EC1 domain (Ozawa and Kemler 1990; Stemmler 2008). The availability of these residues at the N terminus is critical for homophilic interaction of classical cadherins, as addition of one or two residues have been demonstrated to interfere with strand-swap dimerization (Ciatto et al. 2010; Häussinger et al. 2004; Nagar et al. 1996). Structure determination of the N-cadherin prodomain revealed that this region also has an EC-like fold despite a lack of sequence similarity with the rest of the EC domains (Koch et al. 2004).

Third, the adhesion function can be positively or negatively modulated by post-translational modifications of the cytoplasmic region of cadherin as well. Phosphorylation of Ser686 and Ser692 in the CBD promotes tighter binding with β-catenin, while tyrosine phosphorylation within the JMD residues Tyr599-Tyr600 recruits Hakai E3-ubiquitin ligase and induces ubiquitin-dependent internalization of cadherin (Fujita et al. 2002).

3.3 p120 Catenin

3.3.1 The Overall Structure of p120 Catenin

p120 catenin (p120) was first discovered as a prominent Src tyrosine kinase substrate (Reynolds et al. 1989), and subsequently recognized as an armadillo (arm) repeat containing protein which interacts with the cytoplasmic region of cadherin (Peifer et al. 1994; Reynolds et al. 1994; Reynolds et al. 1992). p120 specifically interacts with the juxtamembrane domain (JMD) that is located between the transmembrane domain and the β-catenin-binding domain of cadherin (Ishiyama et al. 2010; Thoreson et al. 2000). This interaction is critical for regulating the stability of cadherin–catenin cell–cell adhesion complexes at the cell surface, as downregulation of p120 results in aberrant internalization of cadherins (Davis et al. 2003; Xiao et al. 2003). Consistently, the loss, downregulation or mislocalization of p120 in tumors has been linked to poor prognoses (Thoreson and Reynolds 2002).

The p120 arm domain is flanked by an N-terminal regulatory region (NTR) and a C-terminal tail region (CTR), where the size of these regions depends on various isoforms resulting from multiple start codons (residues 1, 55, 102 and 324) and three alternatively spliced exons (A, B and C) (Fig. 3.2a) (Anastasiadis and Reynolds 2000). Numerous phosphorylation sites have been identified within both NTR and CTR (Mariner et al. 2001; Xia et al. 2003), but the functional consequence of these modifications remains unclear (Bauer et al. 1998). In addition, p120 binds to

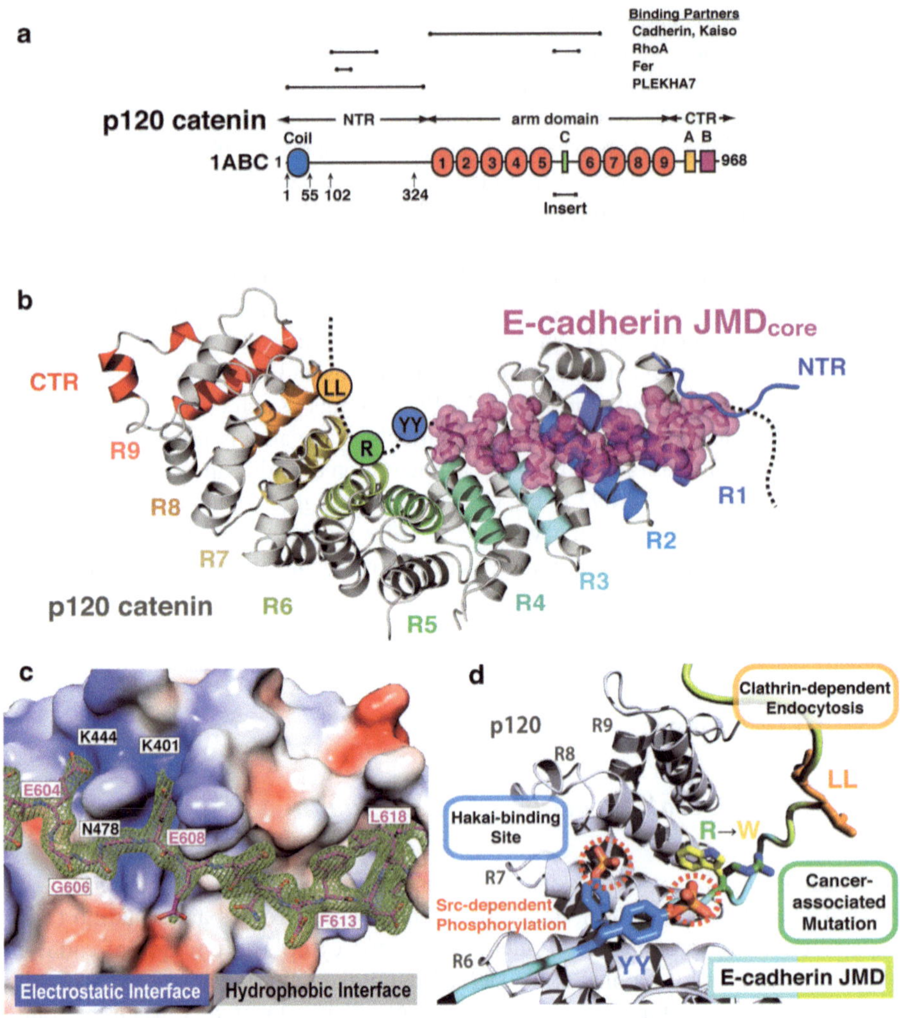

Fig. 3.2 The p120 catenin/E-cadherin juxtamembrane domain (JMD) complex. **a** Scheme of p120 catenin structure. The N-terminal region (*NTR*) contains a coiled-coil region (*Coil*) and four alternate start sites (residues 1, 55, 102 and 324). The arm domain contains nine arm repeats with exon C and a long insert between repeats 5 and 6. The C-terminal region (*CTR*) contains exon A and B. Binding sites for various p120-binding partners are indicated. **b** Crystal structure of the p120 catenin/E-cadherin JMD complex. p120 contains nine arm repeats (*R1–9*) with most repeats consisting of three α-helices (H1, H2 and H3), except for *R1* and *R8* (H3 shown in *different colours*; PDB codes 3L6X & 3L6Y) (Ishiyama et al. 2010). The JMD_{core} is shown in *magenta* (stick and space filling representation). **c** The surface electrostatic potential of the JMD_{core}-binding site of p120 with positively and negatively charged regions in *blue* and *red*, respectively (Ishiyama et al. 2010). $2F_o$-F_c electron density map (*green mesh*; contoured at σ=1.5) of the JMD_{core} (*magenta*) is shown. In the electrostatic interface, basic residues of p120 (e.g., K401 & K444) pair up with acidic residues of JMD_{core} (e.g., E604 & E608) to form several salt bridges. In the hydrophobic interface, the side chain of L618 is buried in the hydrophobic pocket of *R1*. **d** Model of dynamic and static interactions between p120 and the E-cadherin JMD. NMR studies revealed that JMD residues 580–590 (containing the endocytic LL motif (*orange*)) constitute the dynamic p120-binding site (*yellow*). JMD residues 591–625 (containing the JMD_{core}, the R593W cancer-associated mutation site in human E-cadherin (*green*) and the YY phosphorylation/Hakai-binding site (*blue*)) constitute the static p120-binding site. (*cyan*) (Ishiyama et al. 2010)

the Kaiso transcription repressor, which modulates non-canonical Wnt signaling (Daniel and Reynolds 1999; Kim et al. 2004; Park et al. 2006), through its arm domain. The NTR and a large insert within the arm domain of p120 have been shown to interact with Rho-GTPases, such as RhoA and Rac1 (Anastasiadis 2007; Yanagisawa et al. 2008). The recently determined crystal structure of the p120-E-cadherin complex has provided a first look at how this catenin specifically recognizes the cadherin JMD and regulates the internalization of cadherin–catenin complexes via endocytosis (Ishiyama et al. 2010).

3.3.2 Armadillo Domain of p120

p120 was initially thought to contain as many as 10 arm repeats based on its amino acid sequence (Anastasiadis and Reynolds 2000; Reynolds and Roczniak-Ferguson 2004). A typical arm repeat consists of a ~40-residue motif forming three helices (H1, H2 and H3) arranged into a triangular shape (Huber et al. 1997). The crystal structure of a modified form of human p120 isoform 4A (containing a deletion in the arm insert region) in complex with the mouse E-cadherin JMD core fragment (JMD_{core}; residues 602–619) revealed that p120 contains a central arc-shaped arm domain (residues 368–825) with 9 arm repeats accompanied by a mostly disordered NTR (residues 324–367) and CTR (residues 826–933) (Fig. 3.2b) (Ishiyama et al. 2010). The p120 arm domain is similar to that of a closely related desmosomal molecule plakophilin-1 (PKP1) (Choi and Weis 2005) as both arm domains contain a long unstructured insert region between arm repeats 5–6, and the two structures can be superposed with a root mean square distance of 1.2 Å over 324 C_{α} atoms (Ishiyama et al. 2010). In addition, the CTR of p120 forms two α-helices that fold over the hydrophobic surface of arm repeat 9 (Fig. 3.2b). Though other p120 isoforms contain longer NTR, CTR and insert region than p120-4 A, the structure of the 9-repeat arm domain is likely conserved in all p120 isoforms, as well as other members of the p120-subfamily (p120, ARVCF, δ-catenin and p0071) and the PKP-subfamily (PKP-1,-2 and -3) of arm repeat proteins (McCrea and Gu 2010; McCrea and Park 2007).

3.3.3 p120-E-Cadherin Interfaces

Previous deletion and mutagenesis studies have determined that the 18-residue JMD_{core} region is critical for the binding of E-cadherin with p120, and more importantly for epithelial compaction (Ireton et al. 2002; Thoreson et al. 2000). In the crystal structure of the p120-JMD complex, the bound JMD_{core} peptide is stretched over the N-terminal half of the p120 arm domain in the opposite orientation (Fig. 3.2b) (Ishiyama et al. 2010). The JMD_{core} occupies part of the basic groove of the p120 arm domain formed by H3 helices of arm repeats 1–7 (Fig. 3.2c). The interaction between p120 and the JMD_{core} involves approximately 2400 Å^2 of occluded solvent accessible surface area, and this large interface can be further divided into two different types of intermolecular interactions. The N-terminal acidic region (residues 602–610) of the JMD_{core} and p120 arm repeats 1–5 are involved in extensive electrostatic interactions, including the formation of five salt bridges formed between acidic residues (e.g., Glu604 and Glu608) from the cadherin JMD_{core} and basic residues (e.g., Lys401 and Lys444) from p120 (Fig. 3.2c) (Ishiyama et al. 2010). In the middle, the triple Gly motif (residues 605–607) of the JMD_{core} forms a turn that fits into a trough formed by p120, where the backbone of the JMD_{core} forms critical hydrogen bonds with p120 residues, including Asn478 (Fig. 3.2c). In comparison, the C-terminal half of the JMD_{core} (residues 611–619) and the N terminus of p120 are mainly involved in hydrophobic interactions (Fig. 3.2c). This region of the JMD_{core} is wedged between arm repeat 1 and the NTR of p120, resulting in the insertion or anchoring of the Leu618 side chain into an N-terminal hydrophobic pocket of the p120 arm domain. In VE-cadherin, Src-induced phosphorylation of Tyr611 (F613 in E-cadherin) in this region has been shown to prevent the binding of p120 (Potter et al. 2005). The importance of both electrostatic and hydrophobic interactions between p120 and the cadherin JMD is underscored by the strict conservation of the triple Gly motif with two flanking Glu residues (EGGGE) and the anchoring Leu in the JMD_{core} sequence from fly DE-cadherin to human E-cadherin (Ishiyama et al. 2010; Nollet et al. 2000).

The significance of the crystallographically determined JMD binding site of p120 was confirmed when single-residue mutations of p120, K401M and N478A, were shown to completely abolish the interactions of p120 with E- and N-cadherins by *in vitro* pull-down assays (Ishiyama et al. 2010). Compared to control p120, expression of these p120 mutants in p120-downregulated Madin-Darby canine kidney and MCF-7 cells resulted in complete loss of the p120-E-cadherin interaction and significantly reduced expression of E-cadherin at the cell perimeter. Furthermore, NMR and isothermal titration calorimetry studies revealed that p120 tightly binds to the majority of the JMD (residues 591–625), including the core, residues associated with a hereditary diffuse gastric cancer mutation of human E-cadherin (R593W) (Kaurah et al. 2007) and residues associated with Hakai-dependent ubiquitination/internalization of E-cadherin (Fujita et al. 2002), with a sub-micromolar affinity (Ishiyama et al. 2010). At the same time, it participates in a weak, dynamic interaction with the N-terminal clathrin-dependent endocytic motif (L587–L588), protecting it from endocytic proteins, such as the AP2 clathrin adaptor complex (Fig. 3.2d) (Kelly et al. 2008; Miyashita and Ozawa 2007a; Miyashita and Ozawa 2007b). These observations strongly suggest that these interfaces between p120 and E-cadherin JMD_{core} are crucial for p120 to colocalize with E-cadherin at the cell surface and to regulate the stability of cadherin–catenin complexes by countering cadherin internalization mechanisms at AJs (Ishiyama et al. 2010).

3.4 β-Catenin

3.4.1 The Overall Structure of β-Catenin

β-Catenin is an archetypal member of the armadillo repeat protein family, and plays an integral role in establishing adherens junctions by directly interacting with cadherin (McCrea and Gumbiner 1991; McCrea et al. 1991). It is also a critical transcriptional coactivator in the canonical Wnt signaling pathway that controls cell fate and proliferation when it forms a complex with members of Tcf/LEF-1 transcription factors in the nucleus (Graham et al. 2000). As β-catenin is involved in various protein–protein interactions that are crucial for embryogenesis, development and tumorigenesis, its population is tightly regulated by a dedicated degradation mechanism (Angers and Moon 2009). In the cadherin–catenin cell adhesion complex, β-catenin can simultaneously interact with cadherin, a cell adhesion molecule, and α-catenin, an actin-binding protein. The formation of a cadherin/β-catenin/α-catenin ternary complex is essential for linking cadherin-mediated cell–cell adhesion with actin dynamics (Meng and Takeichi 2009). The primary sequence of β-catenin is highly conserved from insects to humans, and its critical biological role in vertebrates is especially highlighted by strict conservation (>95% sequence identity) of a 781-residue sequence from frogs to humans. The overall structure of β-catenin consists of three distinct domains, an N-terminal tail containing the α-catenin-binding site, a central arm domain that binds to the cytoplasmic region of cadherin and a C-terminal tail (Fig. 3.3a) (Shapiro and Weis 2009).

3.4.2 Armadillo Domain of β-Catenin

The arm domain (residues 146–662) of β-catenin is comprised of 12 arm repeats, which are sequentially packed together through hydrophobic interfaces into a superhelical architecture (Fig. 3.3a) (Huber et al. 1997). Structural determination of this domain was facilitated by crystallizing a protease-resistant, structurally stable fragment of β-catenin, containing residues 134–671, determined by limited trypsin digestion (Huber et al. 1997). Most of the arm repeats are comprised of ~40 residues forming three α-helices, H1, H2 and H3, that are arranged into a triangular shape (Fig. 3.3b). The β-catenin arm domain also contains atypical repeats: repeats 1 and 7 are missing H1 and repeat 10 contains a 15-residue insert between H2 and H3. The arm domain is slightly twisted and this results in consecutively ordered H3 helices forming a concave groove that is 95 Å long and 20 Å wide (Huber et al. 1997). Multiple basic residues from H3 and the first turn of H1 in arm repeats 1–10 give this groove a large positively charged surface critical for interacting with various ligands, including E-cadherin (Huber and Weis 2001). Besides the basic arm groove, the arm domain has an exposed hydrophobic pocket on repeat 1, which is also involved in ligand binding. More recently, structure determination of a full-length β-catenin from zebrafish has revealed that the C-terminal tail forms an additional α-helix that packs against the hydrophobic patch of repeat 12 (Xing et al. 2008).

3.4.3 β-Catenin-E-Cadherin Interfaces

Crystal structures of the β-catenin arm domain bound to either an unphosphorylated or phosphorylated E-cadherin cytoplasmic tail revealed the molecular basis

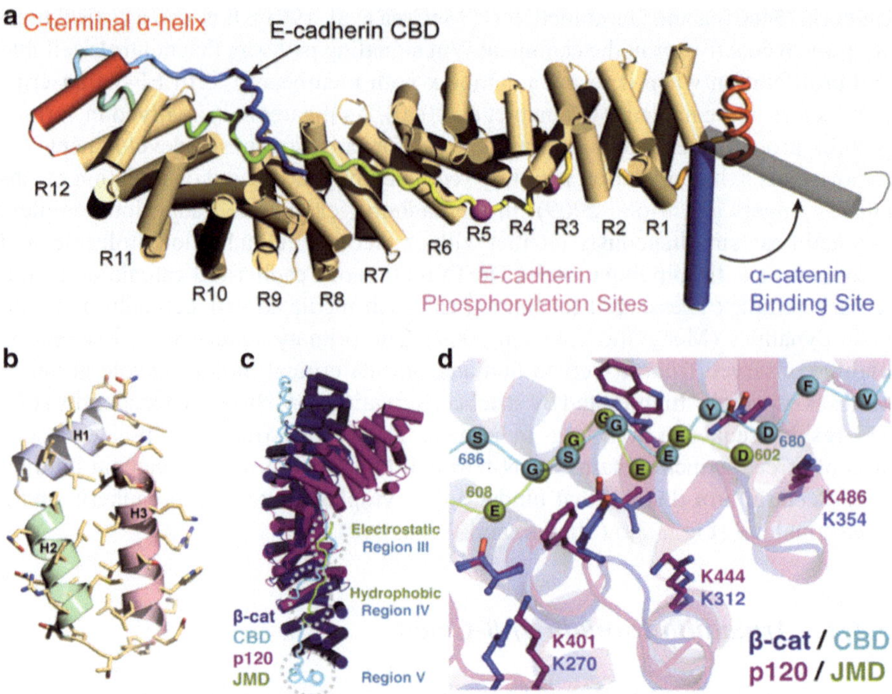

Fig. 3.3 The β-catenin/E-cadherin catenin-binding domain (CBD) complex. **a** The full-length structure of β-catenin (PDB codes 1I7W, 1I7X & 2Z6G) consists of the central arm-repeat domain with 12 repeats (*R1–12*; *orange*) flanked by the N-terminal tail containing the α-catenin-binding site (*blue*) and the C-terminal tail that forms an additional α-helix (Huber et al. 1997; Huber and Weis 2001; Xing et al. 2008). The crystal structure of β-catenin/E-cadherin complex revealed that the entire arm domain is involved in associating with the CBD (*rainbow coloured tube*). The N-terminal tail helix is expected to change its conformation to accommodate the binding of α-catenin. **b** The arm repeat 5 of β-catenin contains three α-helices (*H1–3*) with a hydrophobic core. **c** Superposition of the p120/juxtamembrane domain $(JMD)_{core}$ and β-catenin/CBD complexes. p120 (*purple*) and β-catenin (*navy*) arm domains are shown as cylinders. JMD (*green*) and CBD (*cyan*) are shown as tubes. **d** Comparison of the basic arm grooves of p120 and β-catenin. p120 arm repeats 2–4 and β-catenin arm repeats 4–6 (PDB codes 1I7X & 3L6X) are superposed (Ishiyama et al. 2010). The Cα atoms of the JMD_{core} (*green*) and CBD (*cyan*) are shown as spheres

of the intimate interaction between β-catenin and E-cadherin (Huber and Weis 2001). Although the crystallized β-catenin-E-cadherin complex contained the entire E-cadherin cytoplasmic region (residues 577–728), the structure of β-catenin-bound cadherin was limited to the CBD (residues 628–728) with the JMD (residues 577–627) remaining unbound and disordered (Huber et al. 2001). This is consistent with previous observation that β-catenin specifically interacts with the CDB of E-cadherin (Aberle et al. 1994; Yap et al. 1998). The interaction between β-catenin and E-cadherin involves all 12 armadillo repeats of β-catenin and the majority of the E-cadherin CBD. Similar to the p120-JMD complex (Ishiyama et al. 2010), the bound CBD peptide generally follows the concave groove of the arm domain in the opposite orientation so that the C terminus of CBD is bound to the N-terminal hydrophobic patch of β-catenin (Fig. 3.3a) (Huber and Weis 2001). The extensive binding interfaces can be further divided into five different regions (regions I–V) (Huber and Weis 2001). Most notably, region III forms the central interface involving residues 667–684 of E-cadherin forming critical electrostatic and hydrogen bonding interactions with arm repeats 4–9 of β-catenin (Fig. 3.3c). The transcription factor Tcf3 also utilizes the same binding interface to interact with β-catenin (Graham et al. 2000). Single-residue mutation of β-catenin residues, K312 and K435, involved in intermolecular salt bridge formation at this interface has been shown to abolish the interaction with the E-cadherin CBD (Graham et al. 2000). Interestingly, similar salt bridges are essential for the interaction between p120 and the E-cadherin JMD (Fig. 3.3d) (Ishiyama et al. 2010). On the other hand, region IV involves β-catenin-E-cadherin interactions that depend on the phosphorylation state of CBD residues 684–699. While the unphosphorylated CBD displayed disordered structure in region IV, phosphorylation of Ser684, Ser686 and Ser692 resulted in a stable interface, with phosphorylated Ser686 and Ser692 involved in additional ionic and hydrogen bonding interactions (Huber and Weis 2001). β-catenin has been shown to bind unphosphorylated E-cadherin cytoplasmic tail with a K_D of 36 nM, but phosphorylation of CBD by casein kinase II increases the affinity to a K_D of 52 pM (Choi et al. 2006). In contrast, phosphorylation of β-catenin residue Tyr654 located in arm repeat 12 by Src kinase has been shown to disrupt the binding of β-catenin to E-cadherin (Roura et al. 1999). The binding state of region IV also affects region V, which involves hydrophobic interaction between two anti-parallel α-helices formed at the C terminus of the CBD and the hydrophobic patch of β-catenin arm repeat 1 (Huber and Weis 2001).

3.4.4 α-Catenin-Binding Site

The α-catenin-binding site of β-catenin is located in the N-terminal tail (residues 118–149) immediately adjacent to region V of the β-catenin-cadherin interface (Aberle et al. 1994). The structure of the α-catenin-binding site in β-catenin in the absence of α-catenin has been observed as disordered (Huber et al. 1997) or as a long α-helix that further extends H2 of arm repeat 1 (Xing et al. 2008). When the α-catenin-binding site of β-catenin binds to α-catenin, this region forms two helices: a long helix (residues 120–141) and a short helix (residues 145–149) connected by a 3-residue linker (Pokutta and Weis 2000). These observations suggest that the cadherin-bound β-catenin could bind to α-catenin without any steric hindrance by forming a discrete α-helix within the α-catenin binding site (Huber and Weis 2001). A closely related plakoglobin (γ-catenin) also associates with E-cadherin and α-catenin at AJs, but does not recruit α-catenin to desmosome where it associates with desmosomal cadherins (Witcher et al. 1996). The crystal structure of a plakoglobin-E-cadherin CBD complex showed that observed interactions are virtually identical to the interactions between β-catenin and E-cadherin CBD

(Choi et al. 2009). However, further biochemical studies reveal that the α-catenin-binding site is part of the desmosomal cadherin binding site, explaining the mutually exclusive nature of plakoglobin localization at the desmosome and its association with α-catenin (Choi et al. 2009).

3.5 α-Catenin

3.5.1 The Overall Structure of a-Catenin

α-catenin is a 102 kDa cytosolic protein implicated in anchoring the cadherin–catenin cell adhesion complex to the actin cytoskeleton at adherens junctions (Kobielak and Fuchs 2004). Unlike β-catenin and p120 catenin, it does not contain any armadillo repeat motifs and does not directly bind to the cadherin cytoplasmic region (Nagafuchi et al. 1991; Ozawa and Kemler 1992). Instead it is closely related to an actin-binding protein vinculin and it indirectly associates with cadherin by binding to the N-terminal segment of β-catenin bound to cadherin (Aberle et al. 1994; Ozawa et al. 1990). In addition, an αE-catenin homodimer has been shown to cross-link actin filaments (Rimm et al. 1995) as well as interfere with Arp2/3-dependent actin polymerization/branching (Drees et al. 2005).

There are three known α-catenin subtypes in mammals, E (epithelial), N (neuronal), and T (prevalent in heart and testis), but invertebrates only express one homolog of α-catenin (Costa et al. 1998; Oda et al. 1993). Previous studies have revealed that α-catenin contains three major domains: an N-terminal (N) domain involved in β-catenin-binding and homodimerization (Aberle et al. 1994; Pokutta and Weis 2000); a modulatory (M) domain involved in binding to vinculin (Yang et al. 2001; Yonemura et al. 2010); and a C-terminal (C) domain involved in binding and bundling of actin filaments (Rimm et al. 1995) (Fig. 3.4a). All three domains contain vinculin homology regions (VH1, VH2 and VH3 in N, M and C domains, respectively) where α-catenin and vinculin share 25–35% sequence identity (Herrenknecht et al. 1991; Nagafuchi et al. 1991). As the N and C domains of α-catenin contain discrete binding sites for β-catenin and actin filaments, respectively, it was assumed that α-catenin would act as a stable linker between the cadherin–catenin complex and actin filaments (Gates and Peifer 2005; Weis and Nelson 2006). However, this 'traditional' model was called into question when αE-catenin was shown to interact with actin filaments only as a homodimer and not while being part of the cadherin–catenin complex by binding to β-catenin (Drees et al. 2005; Yamada et al. 2005). The *Caenorhabditis elegans* α-catenin homolog HMP-1, on the other hand, does not homodimerize and is auto-inhibited for actin binding as part of the complex or as a monomer (Kwiatkowski et al. 2010). These observations led to three models with different conformational states of α-catenin regulating its ability to interact with actin filaments (Fig. 3.4b) (Drees et al. 2005). The first model involves α-catenin dissociating from the cadherin–catenin complex to form homodimers to interact with actin filaments. The second model involves α-catenin bound to the cadherin–catenin complex adopting an active conformation

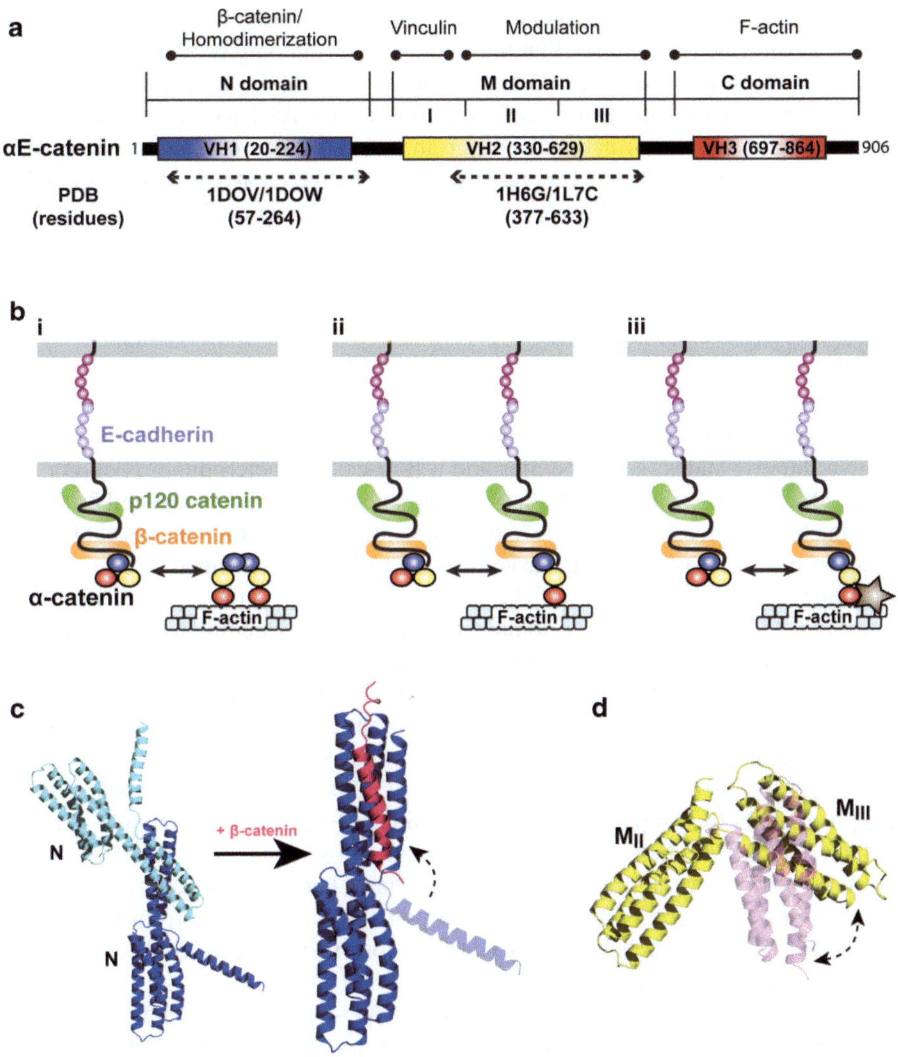

Fig. 3.4 The molecular architecture of α-catenin. **a** Scheme of the αE-catenin structure. It consists of N, M and C domains, which contain vinculin homology regions 1, 2 and 3, respectively. The N domain (*blue*) facilitates its homodimerization as well heterodimerization with β-catenin. The M domain (*yellow*) includes the vinculin binding site (Yonemura et al. 2010) and the adhesion modulation region (Yang et al. 2001). The C domain (*red*) is responsible for interacting with actin filaments (Rimm et al. 1995). **b** Models of α-catenin-mediated connections between the cadherin–catenin complex and actin filaments (F-actin). Model 1 shows auto-inhibited α-catenin dissociating from the cadherin–catenin complex resulting in its homodimerization, which induces the conformational change necessary to interact with F-actin (i). Model 2 shows auto-inhibited α-catenin bound to the cadherin–catenin complex changing its conformation to facilitate the direct connection to F-actin (ii). Model 3 shows auto-inhibited α-catenin bound to the cadherin–catenin complex changing its conformation to recruit other actin-binding proteins (*brown*) to the complex, allowing indirect connection to F-actin (iii). **c** Crystal structures of the α-catenin N domain in the homodimer arrangement (PDB code 1DOV) and the β-α-catenin chimeric protein depicting the heterodimer arrangement (PDB code 1DOW) (Pokutta and Weis 2000). **d** Crystal structures of the M fragments have been determined in the open and closed conformation. (PDB codes 1H6G & 1L7C)

to directly bind to actin filaments. The third model involves α-catenin bound to the cadherin–catenin complex binding to other actin-binding proteins to indirectly interact with actin filaments. Consistent with these models, several recent studies have reported that α-catenin acts as a mechanosensor at AJs: actomyosin-dependent forces trigger the conformational change in auto-inhibited α-catenin, which then recruits vinculin to cell–cell contact sites and links the cadherin–catenin complex with actin filaments both directly and indirectly (le Duc et al. 2010; Smutny et al. 2010; Yonemura et al. 2010).

3.5.2 N-Terminal Dimerization Domain

The structure of the N domain (residues 1–264) of αE-catenin has been determined in two dimeric states: a homodimer and a β-/α-catenin heterodimer (Pokutta and Weis 2000). The K_D values for both αE-catenin homodimer formation and β-catenin binding have been estimated to be in the single micromolar range (Drees et al. 2005; Shapiro and Weis 2009). The heterodimer structure was determined by crystallizing a chimeric protein consisting of the α-catenin binding region of β-catenin (residues 118–151) fused to the N domain fragment of αE-catenin starting at residue 57. It consists of two sets of four-helix bundles connected by a long central helix. The N-terminal helical bundle contains an α-helix formed by β-catenin residues 120–141 (Fig. 3.4c). Interestingly, this chimeric structure of the α-catenin N domain highly resembles the N-terminal D1 domain structure of vinculin (its VH1 region shares 27% sequence identity) (Bakolitsa et al. 2004; Borgon et al. 2004), except the β-catenin helix is replaced by vinculin residues 9–33 forming its first N-terminal helix. Structural studies also suggest α-catenin and vinculin employ different heterodimerization mechanisms, as the vinculin/talin heterodimer complex structure resulted in a five-helix N-terminal bundle (with a talin fragment forming the fifth α-helix) instead of a mixed four-helix bundle observed in the β-/α-catenin chimera structure (Izard et al. 2004).

In comparison, the homodimer structure of the α-catenin N domain was determined by crystallizing a proteolysis-resistant fragment (residue 82–279) containing a region (residues 96–226) necessary for homodimerization (Koslov et al. 1997). N domain residues 82–258 in the homodimer state virtually adopt the same structure as the heterodimer, except for two α-helices in the N termini (residues 86–142) of two protomers which form an intermolecular four-helix bundle (Fig. 3.4c) (Pokutta and Weis 2000). Differences in the homodimer and heterodimer structures of the N domain suggests that the first α-helix (residues 57–83) of α-catenin observed in the β/α-catenin heterodimer is likely to pivot between open and closed conformations, making the homo- and hetero-dimerization of α-catenin mutually exclusive events (Fig. 3.4c) (Pokutta and Weis 2000). This is consistent with observations from other studies suggesting that αE-catenin bound to the cadherin-β-catenin complex does not directly associate with the actin cytoskeleton (Drees et al. 2005; Yamada et al. 2005).

3.5.3 Modulatory Domain

The modulatory (M) domain of αE-catenin consists of residues 277–631, which can be further divided into three subdomains: I, II and III (Fig. 3.4a). This region contains the VH2 region (residues 377–585) and shares 31% sequence identity with residues 582–796 of vinculin (Herrenknecht et al. 1991). Previously determined crystal structures of M domain fragments mostly consist of subdomains II (M_{II}, residues 391–506) and III (M_{III}, residues 507–631) (Pokutta et al. 2002; Yang et al. 2001), and form a pair of four-helix bundles connected by a hinge region (Fig. 3.4d). Comparison of five independent crystal structures of the M fragment reveals that the hinge region connecting M_{II} and M_{III} appears to be quite flexible as the angle between two bundles ranges from 57–100°. Recently, αE-catenin subdomain I (M_I) was determined to contain the vinculin-binding site (residues 325–360), but the association of αE-catenin with vinculin is normally inhibited by M_{III} (Yonemura et al. 2010). These observations have led to a proposal that actomyosin-dependent conformational change within the M domain of αE-catenin attenuates the M_{III}-inhibition, resulting in the recruitment of vinculin to AJs (Yonemura et al. 2010).

3.5.4 C-Terminal Actin-Binding Domain

The C-terminal (C) domain of α-catenin is responsible for actin filament binding, and shares considerable sequence similarity with the D5 actin-binding domain of vinculin in the VH3 region (34% sequence identity) (Fig. 3.4a) (Herrenknecht et al. 1991). Previous studies have estimated the αE-catenin-F-actin interaction to have a K_D value of 0.3 ± 0.4 μM, which is in the same affinity range between vinculin and F-actin (Johnson and Craig 1995), with the stoichiometry of one α-catenin dimer to 14 actin monomers (equivalent to an actin filament helical repeating unit) (Rimm et al. 1995). Nevertheless, α-catenin appears to have a distinct actin-binding mechanism involving an additional 42-residue tail (residues 865–906) that is not present in the C terminus of vinculin (Pokutta et al. 2002). A larger isoform of αN-catenin present during development has been shown to contain a 48-residue insertion after Gly810 in the C domain (Uchida et al. 1994). Although a high-resolution structure of the C domain of α-catenin remains elusive, VH3 region is expected have a similar fold as the five-helix bundle found in the D5 of vinculin (Bakolitsa et al. 1999). Determination of full-length vinculin structures revealed that one of two critical actin-binding interfaces within the D5 domain is occluded when vinculin is in its inactive closed conformation (Bakolitsa et al. 2004; Borgon et al. 2004). As vinculin has been shown to adopt an open conformation upon binding to various ligands, e.g., talin and phosphatidylinositol-4,5-bisphosphate (Bakolitsa et al. 2004; Winkler et al. 1996), it is tempting to speculate that activation of α-catenin could also involve modulation of inter-domain interactions (Fig. 3.4b).

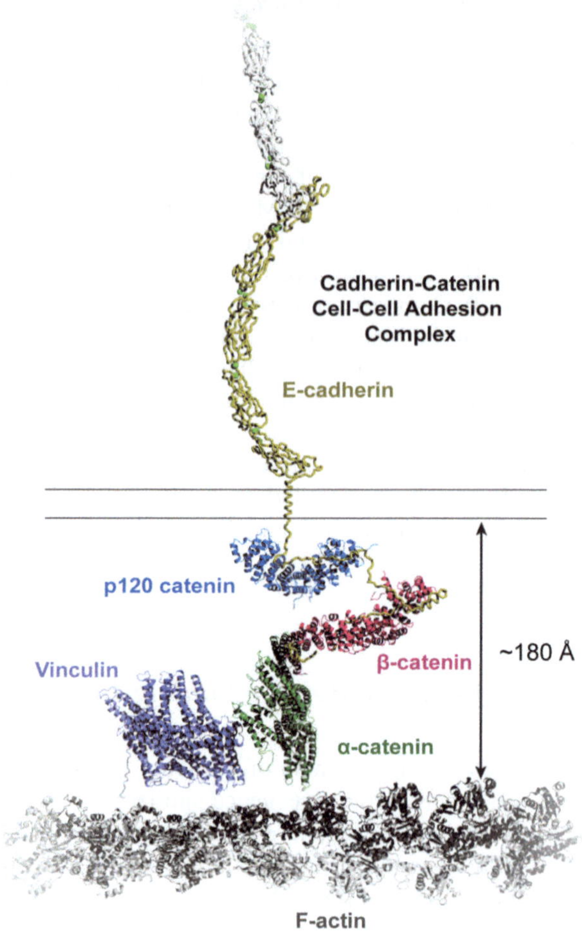

Fig. 3.5 Hypothetical model of the cadherin–catenin cell adhesion complex. The cadherin–catenin cell–cell adhesion complex consists of E-cadherin (PDB code 3Q2V), p120-catenin (PDB code 3L6X), β-catenin (PDB code 1I7W) and α-catenin (PDB codes 1DOW & 1H6G). α-catenin could either directly interact with F-actin (PDB code 3B63) or indirectly via vinculin (PDB code 1ST6) or other actin-binding molecules

3.6 The Cadherin–Catenin Cell Adhesion Complex

3.6.1 Hypothetical Model of the Cadherin–Catenin Complex

Since the determination of first high-resolution structures of E- and N-cadherin EC1 domains by NMR and X-ray crystallography over 15 years ago (Overduin et al. 1995; Shapiro et al. 1995), a nearly complete collection of three-dimensional structures of cadherins, catenins and their complexes have been determined, and more importantly, have provided invaluable atomic-level details about cadherin-catenin-dependent cell–cell adhesion mechanisms. To gain further insights into the multimeric arrangement of the cadherin–catenin complex in its entirety, a hypothetical model of the cadherin–catenin cell adhesion complex was constructed (Fig. 3.5).

The core cell–cell adhesion complex consists of the E-cadherin ectodomain (PDB code 3Q2V; Harrison et al. 2011), the p120/JMD complex (PDB code 3L6X; Ishiyama et al. 2010), the β-catenin/CBD complex (PDB codes 1I7W and 1I7X; Huber and Weis 2001), and α-catenin fragments including the β-/α-catenin complex and the M domain (PDB codes 1DOW and 1H6G; Pokutta and Weis 2000; Yang et al. 2001). The model indicates that a single cadherin–catenin complex could take up an intracellular space with the dimensions of ~140 Å x ~140 Å x ~180 Å (length × width × height). However, cadherin–catenin complexes found in AJs are likely to occupy less space per complex by facilitating lateral clustering of both extracellular and intracellular components (Fig. 3.1e). In a mature intercellular junction, the presence of cadherin-bound p120, β-catenin and α-catenin in a tight space between the plasma membrane and the actin filament would restrict endocytic machineries and kinases from gaining access to the cytoplasmic tail of E-cadherin. The model also indicates the close proximity of the N terminus of p120 to the arm domain of β-catenin. This is consistent with the role of p120 in recruiting Fer kinase through its NTR to modulate the cadherin-β-catenin interaction (Lee et al. 2008; Xu et al. 2004). Additional structural studies are still pending to decipher whether α-catenin could interact with F-actin (PDB code 3B63; Cong et al. 2008) directly and/or indirectly via vinculin (PDB code 1ST6; Bakolitsa et al. 2004) and other actin-binding proteins (Fig. 3.4b).

3.7 Conclusion

Cadherin-mediated cell–cell adhesion requires intimate and intricate interactions between cadherins, catenins, and the actin cytoskeleton network. The structure and function of individual cell adhesion molecules are further modulated by protein–protein interactions, sometimes involving only a few amino acid residues. Three-dimensional structures of multiple classical cadherins, p120 catenin, β-catenin and parts of α-catenin have now been determined at atomic-level resolution, bringing considerable advantages to researchers in the field to further explore the relationships between the cadherin–catenin complex and various intracellular networks, including the actin cytoskeleton and numerous signaling pathways. With the recent recognition of α-catenin as a mechanosensor of cell–cell adhesion, precise structural information regarding the intermolecular relationships among cadherins, catenins and the actin cytoskeleton is indispensable to understand how adjoining cells communicate through cell–cell adhesion.

References

Aberle H, Butz S, Stappert J, Weissig H, Kemler R, Hoschuetzky H (1994) Assembly of the cadherin-catenin complex in vitro with recombinant proteins. J Cell Sci 107:3655–3663

Al-Amoudi A, Díez D, Betts M, Frangakis A (2007) The molecular architecture of cadherins in native epidermal desmosomes. Nature 450:832–837

Anastasiadis PZ (2007) p120-ctn: A nexus for contextual signaling via Rho GTPases. Biochim Biophys Acta 1773:34–46

Anastasiadis PZ, Reynolds AB (2000) The p120 catenin family: complex roles in adhesion, signaling and cancer. J Cell Sci 113:1319–1334

Angers S, Moon RT (2009) Proximal events in Wnt signal transduction. Nat Rev Mol Cell Biol 10:468–477

Bakolitsa C, Cohen DM, Bankston LA, Bobkov AA, Cadwell GW, Jennings L, Critchley DR, Craig SW, Liddington RC (2004) Structural basis for vinculin activation at sites of cell adhesion. Nature 430:583–586

Bakolitsa C, de Pereda JM, Bagshaw CR, Critchley DR, Liddington RC (1999) Crystal structure of the vinculin tail suggests a pathway for activation. Cell 99:603–613

Bauer A, Lickert H, Kemler R, Stappert J (1998) Modification of the E-cadherin-catenin-complex in mitotic Madin-Darby Canine Kidney epithelial cells. J Biol Chem 273:28314–28321

Boggon TJ, Murray J, Chappuis-Flament S, Wong E, Gumbiner BM, Shapiro L (2002) C-cadherin ectodomain structure and implications for cell adhesion mechanisms. Science 296:1308–1313

Borgon RA, Vonrhein C, Bricogne G, Bois PRJ, Izard T (2004) Crystal structure of human vinculin. Structure 12:1189–1197

Brasch J, Harrison OJ, Ahlsen G, Carnally SM, Henderson RM, Honig B, Shapiro L (2011) Structure and binding mechanism of vascular endothelial cadherin: a divergent classical cadherin. J Mol Biol 408:57–73

Brieher WM, Yap AS, Gumbiner BM (1996) Lateral dimerization is required for the homophilic binding activity of C-cadherin. J Cell Biol 135:487–496

Choi HJ, Weis WI (2005) Structure of the armadillo repeat domain of plakophilin 1. J Mol Biol 346:367–376

Choi HJ, Huber AH, Weis WI (2006) Thermodynamics of β-catenin-ligand interactions: the roles of the N- and C-terminal tails in modulating binding affinity. J Biol Chem 281:1027–1038

Choi HJ, Gross JC, Pokutta S, Weis WI (2009) Interactions of plakoglobin and β-catenin with desmosomal cadherins: basis of selective exclusion of α- and β-catenin from desmosomes. J Biol Chem 284:31776–31788

Ciatto C, Bahna F, Zampieri N, Vansteenhouse HC, Katsamba PS, Ahlsen G, Harrison OJ, Brasch J, Jin X, Posy S, Vendome J, Ranscht B, Jessell TM, Honig B, Shapiro L (2010) T-cadherin structures reveal a novel adhesive binding mechanism. Nat Struct Mol Biol 17:339–348

Cong Y, Topf M, Sali A, Matsudaira P, Dougherty M, Chiu W, Schmid MF (2008) Crystallographic conformers of actin in a biologically active bundle of filaments. J Mol Biol 375:331–336

Costa M, Raich W, Agbunag C, Leung B, Hardin J, Priess JR (1998) A putative catenin-cadherin system mediates morphogenesis of the *Caenorhabditis elegans* embryo. J Cell Biol 141:297–308

Daniel JM, Reynolds AB (1999) The catenin p120(ctn) interacts with Kaiso, a novel BTB/POZ domain zinc finger transcription factor. Mol Cell Biol 19:3614–3623

Davis MA, Ireton RC, Reynolds AB (2003) A core function for p120-catenin in cadherin turnover. J Cell Biol 163:525–534

Drees F, Pokutta S, Yamada S, Nelson WJ, Weis WI (2005) α-catenin is a molecular switch that binds E-cadherin-β-catenin and regulates actin-filament assembly. Cell 123:903–915

Fujita Y, Krause G, Scheffner M, Zechner D, Leddy HE, Behrens J, Sommer T, Birchmeier W (2002) Hakai, a c-Cbl-like protein, ubiquitinates and induces endocytosis of the E-cadherin complex. Nat Cell Biol 4:222–231

Gates J, Peifer M (2005) Can 1000 reviews be wrong? Actin, α-Catenin, and adherens junctions. Cell 123:769–772

Graham TA, Weaver C, Mao F, Kimelman D, Xu W (2000) Crystal structure of a β-catenin/Tcf complex. Cell 103:885–896

Hanahan D, Weinberg RA (2000) The hallmarks of cancer. Cell 100:57–70

Harris TJC, Tepass U (2010) Adherens junctions: from molecules to morphogenesis. Nat Rev Mol Cell Biol 11:502–514

Harrison OJ, Jin X, Hong S, Bahna F, Ahlsen G, Brasch J, Wu Y, Vendome J, Felsovalyi K, Hampton CM, Troyanovsky RB, Ben-Shaul A, Frank J, Troyanovsky SM, Shapiro L, Honig, B

(2011) The extracellular architecture of adherens junctions revealed by crystal structures of type I cadherins. Structure 19:244–256

Häussinger D, Ahrens T, Aberle T, Engel J, Stetefeld J, Grzesiek S (2004) Proteolytic E-cadherin activation followed by solution NMR and X-ray crystallography. EMBO J 23:1699–1708

Herrenknecht K, Ozawa M, Eckerskorn C, Lottspeich F, Lenter M, Kemler R (1991) The uvomorulin-anchorage protein α catenin is a vinculin homologue. Proc Natl Acad Sci USA 88:9156–9160

Hirano S, Nose A, Hatta K, Kawakami A, Takeichi M (1987) Calcium-dependent cell-cell adhesion molecules (cadherins): subclass specificities and possible involvement of actin bundles. J Cell Biol 105:2501–2510

Huber AH, Nelson WJ, Weis WI (1997) Three-dimensional structure of the armadillo repeat region of β-catenin. Cell 90:871–882

Huber AH, Stewart DB, Laurents DV, Nelson WJ, Weis WI (2001) The cadherin cytoplasmic domain is unstructured in the absence of β-catenin. A possible mechanism for regulating cadherin turnover. J Biol Chem 276:12301–12309

Huber AH, Weis WI (2001) The structure of the β-catenin/E-cadherin complex and the molecular basis of diverse ligand recognition by β-catenin. Cell 105:391–402

Ireton RC, Davis MA, van Hengel J, Mariner DJ, Barnes K, Thoreson MA, Anastasiadis PZ, Matrisian L, Bundy LM, Sealy L, Gilbert B, van Roy F, Reynolds AB (2002) A novel role for p120 catenin in E-cadherin function. J Cell Biol 159:465–476

Ishiyama N, Lee S-H, Liu S, Li G-Y, Smith MJ, Reichardt LF, Ikura M (2010) Dynamic and static interactions between p120 catenin and E-cadherin regulate the stability of cell-cell adhesion. Cell 141:117–128

Izard T, Evans G, Borgon RA, Rush CL, Bricogne G, Bois PRJ (2004) Vinculin activation by talin through helical bundle conversion. Nature 427:171–175

Johnson RP, Craig SW (1995) F-actin binding site masked by the intramolecular association of vinculin head and tail domains. Nature 373:261–264

Kaurah P, MacMillan A, Boyd N, Senz J, De Luca A, Chun N, Suriano G, Zaor S, Van Manen L, Gilpin C, Nikkel S, Connolly-Wilson M, Weissman S, Rubinstein WS, Sebold C, Greenstein R, Stroop J, Yim D, Panzini B, McKinnon W, Greenblatt M, Wirtzfeld D, Fontaine D, Coit D, Yoon S, Chung D, Lauwers G, Pizzuti A, Vaccaro C, Redal MA, Oliveira C, Tischkowitz M, Olschwang S, Gallinger S, Lynch H, Green J, Ford J, Pharoah P, Fernandez B, Huntsman D (2007) Founder and recurrent *CDH1* mutations in families with hereditary diffuse gastric cancer. JAMA 297:2360–2372

Kelly BT, McCoy AJ, Späte K, Miller SE, Evans PR, Höning S, Owen DJ (2008) A structural explanation for the binding of endocytic dileucine motifs by the AP2 complex. Nature 456:976–979

Kim SW, Park J-I, Spring CM, Sater AK, Ji H, Otchere AA, Daniel JM, McCrea PD (2004) Non-canonical Wnt signals are modulated by the Kaiso transcriptional repressor and p120-catenin. Nat Cell Biol 6:1212–1220

Kobielak A, Fuchs E (2004) α-catenin: at the junction of intercellular adhesion and actin dynamics. Nat Rev Mol Cell Biol 5:614–625

Koch AW, Pokutta S, Lustig A, Engel J (1997) Calcium binding and homoassociation of E-cadherin domains. Biochemistry 36:7697–7705

Koch AW, Farooq A, Shan W, Zeng L, Colman DR, Zhou M-M (2004) Structure of the neural (N-) cadherin prodomain reveals a cadherin extracellular domain-like fold without adhesive characteristics. Structure 12:793–805

Koslov ER, Maupin P, Pradhan D, Morrow JS, Rimm DL (1997) α-catenin can form asymmetric homodimeric complexes and/or heterodimeric complexes with β-catenin. J Biol Chem 272:27301–27306

Kwiatkowski AV, Maiden SL, Pokutta S, Choi H-J, Benjamin JM, Lynch AM, Nelson WJ, Weis WI, Hardin J (2010) In vitro and in vivo reconstitution of the cadherin-catenin-actin complex from *Caenorhabditis elegans*. Proc Natl Acad Sci USA 107:14591–14596

le Duc Q, Shi Q, Blonk I, Sonnenberg A, Wang N, Leckband D, de Rooij J (2010) Vinculin potentiates E-cadherin mechanosensing and is recruited to actin-anchored sites within adherens junctions in a myosin II-dependent manner. J Cell Biol 189:1107–1115

Leckband D, Prakasam A (2006) Mechanism and dynamics of cadherin adhesion. Annu Rev Biomed Eng 8:259–287

Lee S-H, Peng I-F, Ng YG, Yanagisawa M, Bamji SX, Elia LP, Balsamo J, Lilien J, Anastasiadis PZ, Ullian EM, Reichardt LF (2008) Synapses are regulated by the cytoplasmic tyrosine kinase Fer in a pathway mediated by p120catenin, Fer, SHP-2, and β-catenin. J Cell Biol 183:893–908

Legrand P, Bibert S, Jaquinod M, Ebel C, Hewat E, Vincent F, Vanbelle C, Concord E, Vernet T, Gulino D (2001) Self-assembly of the vascular endothelial cadherin ectodomain in a Ca^{2+}-dependent hexameric structure. J Biol Chem 276:3581–3588

Mariner DJ, Anastasiadis P, Keilhack H, Böhmer FD, Wang J, Reynolds AB (2001) Identification of Src phosphorylation sites in the catenin $p120^{ctn}$. J Biol Chem 276:28006–28013

McCrea PD, Gu D (2010) The catenin family at a glance. J Cell Sci 123:637–642

McCrea PD, Gumbiner BM (1991) Purification of a 92-kDa cytoplasmic protein tightly associated with the cell-cell adhesion molecule E-cadherin (uvomorulin). Characterization and extractability of the protein complex from the cell cytostructure. J Biol Chem 266:4514–4520

McCrea PD, Park JI (2007) Developmental functions of the P120-catenin sub-family. Biochim Biophys Acta 1773:17–33

McCrea PD, Turck CW, Gumbiner B (1991) A homolog of the armadillo protein in Drosophila (plakoglobin) associated with E-cadherin. Science 254:1359–1361

Meng W, Takeichi M (2009) Adherens junction: molecular architecture and regulation. Cold Spring Harbor Perspect. Biol 1:a002899

Miyashita Y, Ozawa M (2007a) A dileucine motif in its cytoplasmic domain directs β-catenin-uncoupled E-cadherin to the lysosome. J Cell Sci 120:4395–4406

Miyashita Y, Ozawa M (2007b) Increased internalization of p120-uncoupled E-cadherin and a requirement for a dileucine motif in the cytoplasmic domain for endocytosis of the protein. J Biol Chem 282:11540–11548

Mosesson Y, Mills GB, Yarden Y (2008) Derailed endocytosis: an emerging feature of cancer. Nat Rev Cancer 8:835–850

Nagafuchi A, Takeichi M, Tsukita S (1991) The 102 kd cadherin-associated protein: similarity to vinculin and posttranscriptional regulation of expression. Cell 65:849–857

Nagar B, Overduin M, Ikura M, Rini JM (1996) Structural basis of calcium-induced E-cadherin rigidification and dimerization. Nature 380:360–364

Nishimura T, Takeichi M (2009) Remodeling of the adherens junctions during morphogenesis. Curr Top Dev Biol 89:33–54

Nollet F, Kools P, van Roy F (2000) Phylogenetic analysis of the cadherin superfamily allows identification of six major subfamilies besides several solitary members. J Mol Biol 299:551–572

Nose A, Tsuji K, Takeichi M (1990) Localization of specificity determining sites in cadherin cell adhesion molecules. Cell 61:147–155

Oda H, Uemura T, Shiomi K, Nagafuchi A, Tsukita S, Takeichi M (1993) Identification of a *Drosophila* homologue of α-catenin and its association with the *armadillo* protein. J Cell Biol 121:1133

Overduin M, Harvey TS, Bagby S, Tong KI, Yau P, Takeichi M, Ikura M (1995) Solution structure of the epithelial cadherin domain responsible for selective cell adhesion. Science 267:386–389

Ozawa M, Kemler R (1990) Correct proteolytic cleavage is required for the cell adhesive function of uvomorulin. J Cell Biol 111:1645–1650

Ozawa M, Kemler R (1992) Molecular organization of the uvomorulin-catenin complex. J Cell Biol 116:989–996

Ozawa M, Ringwald M, Kemler R (1990) Uvomorulin-catenin complex formation is regulated by a specific domain in the cytoplasmic region of the cell adhesion molecule. Proc Natl Acad Sci USA 87:4246–4250

Park J-I, Ji H, Jun S, Gu D, Hikasa H, Li L, Sokol SY, McCrea PD (2006) Frodo links Dishevelled to the p120-catenin/Kaiso pathway: distinct catenin subfamilies promote Wnt signals. Dev Cell 11:683–695

Patel SD, Ciatto C, Chen CP, Bahna F, Rajebhosale M, Arkus N, Schieren I, Jessell TM, Honig B, Price SR, Shapiro L (2006) Type II cadherin ectodomain structures: implications for classical cadherin specificity. Cell 124:1255–1268

Peifer M, Berg S, Reynolds AB (1994) A repeating amino acid motif shared by proteins with diverse cellular roles. Cell 76:789–791

Pokutta S, Weis WI (2000) Structure of the dimerization and β-catenin-binding region of α-catenin. Mol Cell 5:533–543

Pokutta S, Herrenknecht K, Kemler R, Engel J (1994) Conformational changes of the recombinant extracellular domain of E-cadherin upon calcium binding. Eur J Biochem 223:1019–1026

Pokutta S, Drees F, Takai Y, Nelson WJ, Weis WI (2002) Biochemical and structural definition of the l-afadin- and actin-binding sites of α-catenin. J Biol Chem 277:18868–18874

Potter MD, Barbero S, Cheresh DA (2005) Tyrosine phosphorylation of VE-cadherin prevents binding of p120- and β-catenin and maintains the cellular mesenchymal state. J Biol Chem 280:31906–31912

Reynolds AB, Roczniak-Ferguson A (2004) Emerging roles for p120-catenin in cell adhesion and cancer. Oncogene 23:7947–7956

Reynolds AB, Roesel DJ, Kanner SB, Parsons JT (1989) Transformation-specific tyrosine phosphorylation of a novel cellular protein in chicken cells expressing oncogenic variants of the avian cellular src gene. Mol Cell Biol 9:629–638

Reynolds AB, Herbert L, Cleveland JL, Berg ST, Gaut JR (1992) p120, a novel substrate of protein tyrosine kinase receptors and of $p60^{v\text{-}src}$, is related to cadherin-binding factors β-catenin, plakoglobin and armadillo. Oncogene 7:2439–2445

Reynolds AB, Daniel J, Mccrea PD, Wheelock MJ, Wu J, Zhang Z (1994) Identification of a new catenin: the tyrosine kinase substrate $p120^{cas}$ associates with E-cadherin complexes. Mol Cell Biol 14:8333–8342

Rimm DL, Koslov ER, Kebriaei P, Cianci CD, Morrow JS (1995) α_1(E)-catenin is an actin-binding and -bundling protein mediating the attachment of F-actin to the membrane adhesion complex. Proc Natl Acad Sci USA 92:8813–8817

Roura S, Miravet S, Piedra J, García de Herreros A, Duñach M (1999) Regulation of E-cadherin/Catenin association by tyrosine phosphorylation. J Biol Chem 274:36734–36740

Shapiro L, Fannon AM, Kwong PD, Thompson A, Lehmann MS, Grübel G, Legrand JF, Als-Nielsen J, Colman DR, Hendrickson WA (1995) Structural basis of cell-cell adhesion by cadherins. Nature 374:327–337

Shapiro L, Weis WI (2009) Structure and biochemistry of cadherins and catenins. Cold Spring Harbor Perspect. Biol 1:a003053

Shi Q, Maruthamuthu V, Li F, Leckband D (2010) Allosteric cross talk between cadherin extracellular domains. Biophys J 99:95–104

Sivasankar S, Gumbiner B, Leckband D (2001) Direct measurements of multiple adhesive alignments and unbinding trajectories between cadherin extracellular domains. Biophys J 80:1758–1768

Smutny M, Cox HL, Leerberg JM, Kovacs EM, Conti MA, Ferguson C, Hamilton NA, Parton RG, Adelstein RS, Yap AS (2010) Myosin II isoforms identify distinct functional modules that support integrity of the epithelial zonula adherens. Nat Cell Biol 12:696–702

Stemmler MP (2008) Cadherins in development and cancer. Mol Biosyst 4:835–850

Takeichi M (1988) The cadherins: cell-cell adhesion molecules controlling animal morphogenesis. Development 102:639–655

Takeichi M (1993) Cadherins in cancer: implications for invasion and metastasis. Curr Opin Cell Biol 5:806–811

Takeichi M (1995) Morphogenetic roles of classic cadherins. Curr Opin Cell Biol 7:619–627

Tamura K, Shan WS, Hendrickson WA, Colman DR, Shapiro L (1998) Structure-function analysis of cell adhesion by neural (N-) cadherin. Neuron 20:1153–1163

Taveau J-C, Dubois M, Le Bihan O, Trépout S, Almagro S, Hewat E, Durmort C, Heyraud S, Gulino-Debrac D, Lambert O (2008) Structure of artificial and natural VE-cadherin-based adherens junctions. Biochem Soc Trans 36:189–193

Thoreson MA, Anastasiadis PZ, Daniel JM, Ireton RC, Wheelock MJ, Johnson KR, Hummingbird DK, Reynolds AB (2000) Selective uncoupling of p120^{ctn} from E-cadherin disrupts strong adhesion. J Cell Biol 148:189–202

Thoreson MA, Reynolds AB (2002) Altered expression of the catenin p120 in human cancer: implications for tumor progression. Differentiation 70:583–589

Tomschy A, Fauser C, Landwehr R, Engel J (1996) Homophilic adhesion of E-cadherin occurs by a co-operative two-step interaction of N-terminal domains. EMBO J 15:3507–3514

Troyanovsky RB, Sokolov E, Troyanovsky SM (2003) Adhesive and lateral E-cadherin dimers are mediated by the same interface. Mol Cell Biol 23:7965–7972

Uchida N, Shimamura K, Miyatani S, Copeland NG, Gilbert DJ, Jenkins NA, Takeichi M (1994) Mouse αN-catenin: two isoforms, specific expression in the nervous system, and chromosomal localization of the gene. Dev Biol 163:75–85

Weis WI, Nelson WJ (2006) Re-solving the cadherin-catenin-actin conundrum. J Biol Chem 281:35593–35597

Winkler J, Lünsdorf H, Jockusch BM (1996) The ultrastructure of chicken gizzard vinculin as visualized by high-resolution electron microscopy. J Struct Biol 116:270–277

Witcher LL, Collins R, Puttagunta S, Mechanic SE, Munson M, Gumbiner B, Cowin P (1996) Desmosomal cadherin binding domains of plakoglobin. J Biol Chem 271:10904–10909

Xia X, Mariner DJ, Reynolds AB (2003) Adhesion-associated and PKC-modulated changes in serine/threonine phosphorylation of p120-catenin. Biochemistry 42:9195–9204

Xiao K, Allison DF, Buckley KM, Kottke MD, Vincent PA, Faundez V, Kowalczyk AP (2003) Cellular levels of p120 catenin function as a set point for cadherin expression levels in microvascular endothelial cells. J Cell Biol 163:535–545

Xing Y, Takemaru K, Liu J, Berndt JD, Zheng JJ, Moon RT, Xu W (2008) Crystal structure of a full-length β-catenin. Structure 16:478–487

Xu G, Craig AW, Greer P, Miller M, Anastasiadis PZ, Lilien J, Balsamo J (2004) Continuous association of cadherin with β-catenin requires the non-receptor tyrosine-kinase Fer. J Cell Sci 117:3207–3219

Yamada S, Pokutta S, Drees F, Weis WI, Nelson WJ (2005) Deconstructing the cadherin-catenin-actin complex. Cell 123:889–901

Yanagisawa M, Huveldt D, Kreinest P, Lohse CM, Cheville JC, Parker AS, Copland JA, Anastasiadis PZ (2008) A p120 catenin isoform switch affects Rho activity, induces tumor cell invasion, and predicts metastatic disease. J Biol Chem 283:18344–18354

Yang J, Dokurno P, Tonks NK, Barford D (2001) Crystal structure of the M-fragment of α-catenin: implications for modulation of cell adhesion. EMBO J 20:3645–3656

Yap AS, Brieher WM, Pruschy M, Gumbiner BM (1997) Lateral clustering of the adhesive ectodomain: a fundamental determinant of cadherin function. Curr Biol 7:308–315

Yap AS, Niessen CM, Gumbiner BM (1998) The juxtamembrane region of the cadherin cytoplasmic tail supports lateral clustering, adhesive strengthening, and interaction with p120ctn. J Cell Biol 141:779–789

Yonemura S, Wada Y, Watanabe T, Nagafuchi A, Shibata M (2010) α-Catenin as a tension transducer that induces adherens junction development. Nat Cell Biol 12:533–542

Yoshida C, Takeichi M (1982) Teratocarcinoma cell adhesion: Identification of a cell-surface protein involved in calcium-dependent cell aggregation. Cell 28:217–224

Zhang Y, Sivasankar S, Nelson WJ, Chu S (2009) Resolving cadherin interactions and binding cooperativity at the single-molecule level. Proc Natl Acad Sci USA 106:109–114

Chapter 4
Biophysics of Cadherin Adhesion

Deborah Leckband and Sanjeevi Sivasankar

Abstract Since the identification of cadherins and the publication of the first crystal structures, the mechanism of cadherin adhesion, and the underlying structural basis have been studied with a number of different experimental techniques, different classical cadherin subtypes, and cadherin fragments. Earlier studies based on biophysical measurements and structure determinations resulted in seemingly contradictory findings regarding cadherin adhesion. However, recent experimental data increasingly reveal parallels between structures, solution binding data, and adhesion-based biophysical measurements that are beginning to both reconcile apparent differences and generate a more comprehensive model of cadherin-mediated cell adhesion. This chapter summarizes the functional, structural, and biophysical findings relevant to cadherin junction assembly and adhesion. We emphasize emerging parallels between findings obtained with different experimental approaches. Although none of the current models accounts for all of the available experimental and structural data, this chapter discusses possible origins of apparent discrepancies, highlights remaining gaps in current knowledge, and proposes challenges for further study.

4.1 Introduction

The assembly and maintenance of intercellular junctions is central to the role of cadherins in morphogenesis and disease. A challenge is to determine how classical cadherins assemble junctions, and how sequence differences, mutations, and post-translational modifications alter this function. Classical cadherins are transmembrane proteins. The extracellular segment, which embeds the adhesive function, folds into five extracellular (EC) domains, numbered 1–5 from the N-terminal domain (EC1-5) (Fig. 4.1a).

D. Leckband (✉)
Department of Chemistry, Department of Chemical and Biomolecular Engineering, Center for Biophysics and Computational Biology, University of Illinois at Urbana-Champaign, 600 South Mathews Ave, Urbana, IL, 61801, USA
e-mail: leckband@illinois.edu

S. Sivasankar
Department of Physics and Astronomy, Iowa State University, A319 Zaffarano Hall, Aimes, IA 50011, USA
e-mail: sivsank@iastate.edu

T. Harris (ed.), *Adherens Junctions: From Molecular Mechanisms to Tissue Development and Disease,* Subcellular Biochemistry 60,
DOI 10.1007/978-94-007-4186-7_4, © Springer Science+Business Media Dordrecht 2012

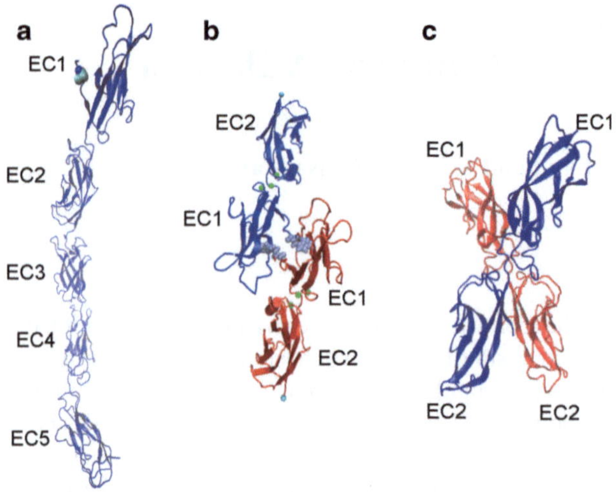

Fig. 4.1 Cadherin structures. **a** Crystal structure of the extracellular region of *Xenopus* C-Cadherin showing the W2 residue (cyan van der Waals structure) (Boggon and Eck 2004). **b** Strand-swapped dimer between E-cadherin EC1-2 fragments. Here the W2 residues (gray van der Waals structures) bridge the apposing EC1 domains, and the calcium ions are shown as green Van der Waals structures. **c** X-dimer of the W2A mutant of E-cadherin EC1-2 fragments. The adjacent domains form a tetrahedral structure with extensive contacts at the inter-domain junction. (Harrison et al. 2010)

The several approaches used to investigate the mechanism of cadherin binding probe different aspects of cadherin function. Structure determinations contributed to our present understanding of binding interfaces and global protein organization at inter-membrane junctions. Static images do not, however, reveal crucial biophysical properties such as kinetic rates, affinities, and adhesion energies that determine the assembly dynamics and mechanical integrity of intercellular junctions. Thus, solution binding and mechanical measurements generate complementary mechanistic insights into cadherin functions, but different experimental methods can also produce seemingly contradictory results. A goal of this review is to summarize investigations of cadherin adhesion, in the context of functional data that must be accounted for by cadherin binding models. We emphasize, in particular, emerging parallels between diverse experimental findings and the evolving picture of the mechanism of cadherin adhesion. We also discuss seemingly disparate findings and their possible origins, and define future challenges towards developing a comprehensive model of cadherin adhesion that accounts for all of the experimental data.

4.2 Characteristics of Cadherin Binding

4.2.1 Trans *Cadherin Bonds*

We first consider functional signatures, which models of cadherin adhesion must capture, in order to account for the wide range of experimental data. First, cadherins form *trans*, adhesive bonds with both similar and dissimilar classical cadherins on apposing cells. This is evidenced by classical cadherins' ability to promote the aggregation of nonadhesive cells when those cells are transfected with cadherins (Nose et al. 1988). Recombinant fragments of the ectodomains of *Xenopus* cleavage stage C-cadherin, epithelial E-cadherin, and neural N-cadherin all support cell adhesion to cadherin-coated substrata (Bixby and Zhang 1990; Brieher et al. 1996; Gavard et al. 2004; Pokutta et al. 1994). This capacity of isolated ectodomains to support the adhesion of cadherin-expressing cells or to aggregate beads demonstrates the adhesive function and localizes that function to the extracellular domain.

4.2.2 Cis*-Interactions and Cadherin Adhesion*

Several lines of evidence indicate that cadherins' adhesive function is also affected by their lateral organization on cell membranes. This organization occurs at two levels: namely, *cis* dimerization and the assembly of cadherins into larger scale clusters. Studies of soluble recombinant, ectodomains of C-cadherin as well as C-cadherin expressed on cells provided biochemical evidence for the existence of lateral dimers (Geng et al. 2004; Kim et al. 2005; Klingelhofer et al. 2002; Takeda et al. 1999; Troyanovsky et al. 2003). The functional significance of *cis* dimers was first demonstrated by biochemical studies in which dimers of C-cadherin ectodomains resulted in greater cell adhesion than did immobilized monomers (Brieher et al. 1996). Atomic force microscopy (AFM) and single molecule fluorescence imaging also identified dimers and larger 10–250 nm clusters on cell surfaces (Chtcheglova et al. 2010; Iino et al. 2001). These results suggested that lateral dimerization is one mechanism to enhance adhesion. It is unclear whether this is due to increased avidity, which results from an increase in the number of bonds formed, or to the allosteric enhancement of the intrinsic affinity of individual cadherin bonds.

The ability of C-cadherin fragments to form lateral dimers suggests that the ectodomain embeds a *cis* binding interface(s) (Brieher et al. 1996). However, a distinct interface that could mediate lateral-dimerization and account for different experimental data has yet to be identified. Potential contacts were identified in some structures (Sect. 4.3.2), but *cis*-dimers were not detected in biophysical studies of soluble extracellular domains. (Haussinger et al. 2002; Pokutta et al. 1994; Zhang et al. 2009).

Chemical cross-linking and immunoprecipitation results suggested that lateral and adhesive bonds may share the same interface (Troyanovsky et al. 2003). Inherent cadherin flexibility could enable cadherins to use the same binding interface for either *cis* or *trans* binding. The ectodomains are often portrayed as rigidly curved structures, but molecular dynamics simulations (Sotomayor and Schulten 2008) and electron microscopy images (He et al. 2003; Koch et al. 1999; Pokutta et al. 1994) indicate that, in the presence of calcium, the ectodomains can adopt other configurations than seen in the crystal lattice (Boggon et al. 2002).

Whether lateral dimerization is an intrinsic property of all classical cadherins remains to be established. The evidence so far suggests that some cadherins can form *cis*-dimers, but the distribution of cadherin monomers, dimers, and higher-order aggregates is likely to be dynamic, and may depend on cell–cell adhesion. Cadherins also organize into much larger clusters on cell surfaces. Because extensive clustering requires Myosin II, Ena/VASP, and PIP3 (Gavard et al. 2004; Scott et al. 2006; Smutney et al. 2010), it is unlikely to be an intrinsic property of the cadherin structure, and will not be considered here.

4.2.3 Role of EC1 in Cadherin-Dependent Cell Adhesion

The importance of EC1 for cadherin adhesion was first demonstrated by studies in which exchanging EC1 domains of different cadherin subtypes altered cadherin-dependent segregation of cells that expressed different cadherins. In a cell-sorting assay, cells that expressed different cadherins at similar levels were shown to segregate away from one another in agitated cell suspensions, but the cells intermixed when they expressed the same cadherin (Nose et al. 1988). Specifically, cells expressing a chimeric protein, in which the EC1 domain of P-cadherin was replaced by the EC1 domain from E-cadherin, only formed aggregates with cells that expressed full-length P-cadherin (Nose et al. 1990). This localized the cell binding specificity to the N-terminal EC1 domain, and suggested that the identity of the EC1 domain was sufficient to specify cell aggregation patterns.

A variety of domain deletion analyses also supported the essential role of EC1 in cell adhesion (Chappuis-Flament et al. 2001; Shan et al. 2004). For example, substrata coated with domain deletion fragments that retained EC1-2 also support the adhesion of cells expressing C-cadherin (Chappuis-Flament et al. 2001). Additionally, cells expressing an N-cadherin mutant that only contained EC1-2 formed cell–cell aggregates (Shan et al. 2004).

4.2.4 Functional Evidence for Contributions from Other Regions of Cadherin Ectodomains

Experimental evidence suggests, however, that *trans*-binding between EC1 domains is not sufficient to account for the range of observed cadherin adhesive behavior. First, EC1 is necessary, but not sufficient, for cadherin-based cell adhesion. The EC1-2 region of N-cadherin appears to be the minimum fragment necessary for homophilic adhesion (Shan et al. 2004).

Second, genetic analyses of E-cadherin mutations associated with inherited gastric cancers identified clusters of mutations that are distributed along the entire extracellular domain, both within EC1 and outside of this domain in EC2-5 (Becker et al. 1999; Berx et al. 1998; Handschuh et al. 1999, 2001; Luber et al. 2000). Several of these mutants are expressed on the cell surface, but they impair cadherin's adhesive function to different extents (Becker et al. 1999; Berx et al. 1998). Intriguingly, several of the most deleterious mutations are within EC2 and EC3

(Fuchs et al. 2004; Handschuh et al. 1999, 2001; Luber et al. 2000). Mutations at the EC3-EC4 and EC4-EC5 junctions have a milder affect on adhesion (Handschuh et al. 1999, 2001; Prakasam et al. 2006a).

Third, aberrant glycosylation alters cadherin-specific cell functions including cell adhesion, barrier integrity, signaling, and interactions with the cytoskeleton (Geng et al. 2004; Guo et al. 2009; Jamal et al. 2009; Liwosz et al. 2006; Nita-Lazar et al. 2010; Pinho et al. 2009, 2011; Zhao et al. 2008a, b). In some cancers, abnormally high N-glycosylation of the membrane-proximal EC4 and EC5 domains of E-cadherin is associated with impaired intercellular adhesion and signaling. Mutating the eight N-glycosylation sites in the N-cadherin ectodomain increased the prevalence of dimers on cell membranes and enhanced ERK signaling (Guo et al. 2009). More limited mutagenesis localized N-glycosylation sites having the greatest impact on N-cadherin functions to EC2-3 (Jamal et al. 2009; Liwosz et al. 2006; Pinho et al. 2011; Zhao et al. 2008a).

Finally, cadherin blocking and activating antibodies have been identified that recognize membrane proximal EC domains. The E-cadherin blocking antibody DECMA-1 recognizes membrane proximal domains (Ozawa et al. 1990). An antibody that binds the EC5 domain of C-cadherin also activates strong *Xenopus* blastomere adhesion, and reverses the inhibitory effect of activin (Zhong et al. 1999).

These several lines of experimental evidence suggest that the entire ectodomain may contribute to cadherin's adhesive function in as yet incompletely understood ways. The following sections discuss structural and biophysical evidence for different cadherin interactions and possible mechanisms accounting for these experimental observations.

4.3 Structural Evidence for Cadherin Interactions

4.3.1 EC1-Domain Interactions

Structural studies of the extracellular domains (EC1-5) and of smaller fragments identified protein interactions responsible for many of the functional signatures discussed above. A consistently observed interface between EC1 domains in nearly all crystal structures of classical cadherins is termed the "strand swapped dimer" (Fig. 4.1b) (Katsamba et al. 2009). Essentially all biophysical and cell adhesion assays support the view that this is a *trans* adhesive bond. At this interface, tryptophan at position 2 (W2) inserts into a complementary hydrophobic pocket on EC1 of the apposing protein (Fig. 4.1b). Consistent with this being the central adhesive interface, mutating the conserved W2 residue to alanine (W2A) substantially reduces cell adhesion in a variety of assays (Pertz et al. 1999; Prakasam et al. 2006a; Shan et al. 2004; Tamura et al. 1998), however, W2A mutants localize to cell–cell junctions (Kitagawa et al. 2000; Tamura et al. 1998) and W2A fragments weakly aggregate beads (Prakasam et al. 2006a). Rotary shadowing electron micrographs of recombinant E-cadherin ectodomains also showed apparent association at the

N-terminal tips of the proteins (Pertz et al. 1999; Tomschy et al. 1996). Together, these experimental findings support the view that this EC1 interface mediates *trans* adhesion.

An additional contact seen in crystal structures of the EC1-2 fragment of E-cadherin, T-cadherin, and W2A mutants—termed the "X-dimer" (Fig. 4.1c)—is at the EC1-2 junction (Ciatto et al. 2010; Harrison et al. 2010; Nagar et al. 1996), where adjacent proteins interact through extensive nonpolar contacts between EC1 and EC2 domains in a tetrahedral configuration. An interface in the structure of E-cadherin EC1-2 (Nagar et al. 1996) was at first postulated to be a cloning artifact (Haussinger et al. 2004), but similar contacts in the structures of W2A mutants and of T-cadherin (Ciatto et al. 2010; Harrison et al. 2010) altered this view. The ability of mutations at this X-dimer interface to impede the rate of *trans*-dimerization led to the hypothesis that this complex is a transient intermediate in the kinetic pathway to strand dimerization (Harrison et al. 2010). T-cadherin also supports cell–cell adhesion (Ciatto et al. 2010), albeit more weakly than N-cadherin, indicating that this interface can also resist force.

4.3.2 Structures of Possible Cis *Binding Interfaces*

Evidence for *cis*-dimers has also been inferred from crystal packing interfaces and from electron microscopy images of ectodomains. A potential candidate for a *cis*-binding interface was observed in the structures of C-, E-, and N-cadherins (Boggon et al. 2002; Harrison et al. 2011). At this contact, EC1 contacts the EC2 domain of an adjacent protein in the crystal lattice (Boggon et al. 2002; Harrison et al. 2011). Mutations at this interface disrupt the organization of cadherin junctions (Harrison et al. 2011), thus supporting a role for this contact in cadherin ordering at junctions. Interactions between amino acids within this interface were not, however, confirmed by NMR measurements of E-cadherin EC1-2 (Haussinger et al. 2002), by electron micrographs of cadherin ectodomains (Pokutta et al. 1994), or by single-molecule fluorescence measurements (Zhang et al. 2009), indicating that the dissociation constant (K_d) exceeds 1 mM (Harrison et al. 2011). Other contacts in crystal lattices that were originally attributed to *cis* interactions include calcium bridging at the interdomain junction between parallel E-cadherin EC1-2 fragments (Nagar et al. 1996). In hindsight, this structure was found to be the X-dimer (Fig. 4.1c). Mutating acidic calcium-binding residues at this junction disrupts adhesion (Prakasam et al. 2006b), although this could be due to perturbations of the X-dimer intermediate or to allosteric perturbation of W2 docking (Harrison et al. 2005; Haussinger et al. 2002; Sotomayor and Schulten 2008; Vunnam and Pedigo 2011b). Another potential *cis* bond involves EC4, which is required for the oligomerization of soluble VE-cadherin ectodomains (Bibert et al. 2002; Hewat et al. 2007; Lambert et al. 2005; Taveau et al. 2008). Intriguingly, only non-glycosylated VE-cadherin appears to form hexamers (Brasch et al. 2011).

In addition to structure determinations, several other biophysical approaches were used to interrogate cadherin function and to test models for cadherin-based cell adhesion. These can be divided generally into solution-binding and adhesion measurements. In solution, freely diffusing cadherins associate under force-independent conditions, but in adhesion measurements, cadherins are confined to surfaces and subject to force, as they would be at cell–cell junctions. Adhesion-based approaches provide complementary information about the number, dynamics, and strength of cadherin bonds. Sections 4.4 and 4.5 describe experimental results obtained with different biophysical methods, and discuss results in the context of structures and of cell adhesion.

4.4 Solution Studies of Cadherin Ectodomain Interactions

4.4.1 Analytical Ultracentrifugation (AUC) Measurements of Binding Affinities

Affinity differences that are at the heart of type I classical cadherin interactions have been characterized using *Sedimentation Equilibrium* and *Sedimentation Velocity* experiments by Analytical Ultra Centrifugation (AUC). AUC can be used to characterize the hydrodynamic and thermodynamic properties of macromolecules in solution, by monitoring their sedimentation in a centrifugal field (Lebowitz et al. 2002). In *sedimentation equilibrium* experiments, at small centrifugal forces, an equilibrium is established where sedimentation is balanced by diffusional transport. Analysis of this sedimentation equilibrium yields information on the molar mass of the proteins, their states of association, and the free energies of binding (Lebowitz et al. 2002). Alternatively, in a *sedimentation velocity* experiment, a larger centrifugal force causes rapid protein sedimentation. Although sedimentation velocity experiments cannot determine binding affinities, analysis of the evolving concentration gradients can be used to determine whether the kinetics is fast or slow relative to the time-scale of the experiment (Lebowitz et al. 2002). In contrast to surface plasmon resonance (SPR) and other dynamic approaches described below, sedimentation velocity AUC cannot yield quantitative rate constants, but instead qualitatively assesses whether the molecules exchange rapidly or slowly within a ~45 min period.

Sedimentation equilibrium AUC measurements of the dissociation constant (K_d) of wild type and mutant classical cadherin homodimerization are summarized in Table 4.1. These measurements focused on two-domain protein constructs (EC1-2), except for the full-length ectodomain of C-cadherin (EC1-5), for which the determined K_d for homodimerization was 64 μM (Table 4.1) (Chappuis-Flament et al. 2001). Wild type E-cadherin EC1-2 expressed in mammalian and in bacterial cells have similar affinities with K_d values of 97 μM (Katsamba et al. 2009) and 80 μM (Koch et al. 1997), respectively. Sedimentation equilibrium AUC experiments also show that the solution binding affinity for N-cadherin EC1-2 is fourfold higher than

Table 4.1 AUC measurements of the dissociation constants for cadherin homodimerization

Protein	Description	Mean K_d (μM)	References
C-cadherin EC1-5 construct			
WT	Wild type	64	Chappuis-Flament et al. (2001)
E-cadherin EC1-2 constructs			
WT	Wild type	80±20	Koch et al. (1997)
WT	Wild type	96.5±10	Katsamba et al. (2009)
WT	Wild type	98.6±15	Ciatto et al. (2010)
W2A	Strand-swap mutant	916±47	
Ala-Ala N-terminal extension	Strand-swap mutant	811±97	
E89A	Strand-swap mutant	293±11	
Asp-Trp deletion at N-terminus	Strand-swap mutant	662±28	Harrison et al. (2010)
K14E	X-dimer mutant	117±8	
K14S	X-dimer mutant	96.0±1.0	
Y142R	X-dimer mutant	77.4±1.4	
W2A K14E	Double mutant	Monomer	
W2F	Reduced strain on A*/A strand	246±2	Vendome et al. (2011)
A inserted between 2 and 3	Reduced strain on A*/A strand	1,517±726	
AA inserted between 2 and 3	Reduced strain on A*/A strand	195±8.6	
E11D	Enhanced strand-swapping	71±12	
W2F	Reduced strain on A*/A strand	246±2	
P5A P6A	Alternate interface	3.7±0.1	
P5S P6S	Alternate interface	2.9±0.04	
P5G P6G	Alternate interface	2.7±1.68	
P5A	Alternate interface	2.9±2.1	
P6A	Alternate interface	4.8±1.65	
N-cadherin EC1-2 constructs			
WT	Wild type	25.8±1.5	Katsamba et al. (2009)
P5A P6A	Alternate interface	3.6±0.2	Vendome et al. (2011)

the E-cadherin EC1-2 at 25°C (Table 4.1) (Katsamba et al. 2009). It is noteworthy that a fourfold difference is not large relative to the thermal energy, amounting to only ~0.8 kcal/mole at 37°C. Such differences would, however, be amplified by large numbers of cadherin bonds at junctions.

Sedimentation equilibrium AUC measurements also determined the binding affinities of cadherin mutants that abolish strand swapping. Mutating the conserved W2 to Ala (W2A) significantly increases the K_d to 916 μM (Table 4.1) (Harrison et al. 2010). Similarly, strand swapping is stabilized by the formation of a salt bridge between the side chain of Glu89 and the N-terminus of the swapped strand. Extending the N terminus (Ala-Ala–extension mutant) or replacing Glu89 with an uncharged residue (E89A) eliminates this salt-bridge. Relative to wild type, these mutants increased the K_d values to 811 and 293 μM, respectively (Harrison et al. 2010). Finally, abolishing the strand swapping entirely, by deleting two N-terminal residues ('Asp-Trp deletion' mutant), which removes the entire swapped structural element, increased the K_d to 662 μM (Table 4.1) (Harrison et al. 2010).

Mutations that relieve strain in the swapping strand in cadherin monomers also decrease the dimerization affinity because the short, swapping-strand in the closed-monomer is strained. This is due to its anchorage at one end by the conserved W2 and at the other by a Ca^{2+}-Glu11 ion pair. This conformational strain provides the driving force for strand expulsion and swapping (Vendome et al. 2011). Mutating the conserved W2 to Phe (W2F) decreases strain in the monomer, and increases K_d to 246.5 μM (Table 4.1) (Vendome et al. 2011). Similarly, increasing the length of the swapping strand reduces strain, such that inserting one or two alanines increases the K_d values to 1,517 and 195 μM, respectively (Table 4.1) (Vendome et al. 2011). On the other hand, increasing the strain by shortening the Glu11 side-chain, while preserving the Ca^{2+}-binding site (E11D mutant), reduces the K_d slightly to 71.2 μM (Table 4.1) (Vendome et al. 2011).

A conserved proline-proline motif in the swapped strand ensures that the cadherin pair cannot form a continuous hydrogen-bonded β-sheet (Vendome et al. 2011; Vunnam and Pedigo 2011a), such that mutating Pro5, Pro6, or both results in an unnaturally tight dimer. With such E-cadherin mutants, the *trans* dimerization K_d value decreases by almost two orders of magnitude (Table 4.1) (Vendome et al. 2011), and the P5AP6A double mutant similarly decreases the N-cadherin EC1-2 K_d (Vendome et al. 2011). Notably, despite the fourfold difference in affinity between the wild-type proteins, the K_d values of the double mutants of E- and N-cadherin are essentially identical (Vendome et al. 2011).

In contrast to strand-swapping mutants, AUC experiments show that mutants that cannot form X-dimers have K_d values that are virtually indistinguishable from wild type protein (Table 4.1) (Harrison et al. 2010). This suggests that the thermodynamics of strand dimerization is not substantially affected by X-dimer interface mutations (Harrison et al. 2010), which instead primarily affect the rate of equilibration. Sedimentation velocity AUC experiments showed that the wild-type protein shows sedimentation behavior characteristic of rapidly exchanging monomer-dimer equilibrium, whereas the X-dimer mutants exhibit a slowly exchanging equilibrium, where little inter-conversion between monomers and dimers occurs on the measurement timescale (~45 min) (Harrison et al. 2010).

4.4.2 Surface Plasmon Resonance (SPR) Measurements of Relative Binding Affinities

Surface Plasmon Resonance (SPR) (Homola 2008) measurements compared relative homophilic and heterophilic K_d values of classical cadherins (Katsamba et al. 2009). SPR quantifies the time-dependent change in ligand binding to immobilized receptors, as ligand solution is flown over the receptor-coated sensor chip. From these data, one can obtain association and dissociation rates, as well as the affinities for simple receptor-ligand binding. However, determinations of affinities between molecules that also dimerize in solution are complicated by the two competing equilibria. Obtaining quantitative homophilic cadherin K_d values required more complicated analyses of binding kinetics and equilibria, so that it was only possible to determine relative dissociation constants (Katsamba et al. 2009). In agreement with the AUC measurements (Table 4.1), the homophilic dimerization K_d of N-cadherin EC1-2 was lower than the K_d of E-cadherin EC1-2 (Katsamba et al. 2009). The relative K_d for the heterophilic interaction between N- and E-cadherin is intermediate between the homophilic values (Katsamba et al. 2009).

SPR measurements similarly explored the effect of strand-dimer and X-dimer mutations on cadherin K_d values (Harrison et al. 2010). Although wild type E-cadherin EC1-2 forms homo-dimers, the K14E X-dimer mutant does not interact with either the immobilized K14E mutant or the wild type protein (Harrison et al. 2010). In agreement with sedimentation velocity AUC experiments, this suggests that the association rate for the dimerization of X-dimers mutants is very slow (Harrison et al. 2010).

4.4.3 Single-Molecule Fluorescence Measurements of Classical Cadherin Conformation

Fluorescence-based techniques like Förster Resonance Energy Transfer (also known as Fluorescence Resonance Energy Transfer or FRET) (Roy et al. 2008) and sub nanometer single-molecule localization (Pertsinidis et al. 2010) have been used to measure the conformation of pairs of isolated cadherin molecules. In a FRET experiment, the cadherins are tagged on a specific EC domain with one of two fluorescent dyes, called donor and acceptor dyes respectively (Figs. 4.2a, c). The cadherins are allowed to interact and the distance between the fluorescent probes is measured with nanometer resolution. Dyes separated by more than ~10 nm do not interact, and the donor emits photons upon its excitation by laser. However if the two dyes are closer than ~10 nm, then the donor transfers its energy to the acceptor, which emits photons of a different wavelength. By monitoring the relative intensities of the donor and acceptor fluorescence, nanometer scale distance changes can be resolved (Roy et al. 2008). In single-molecule localization, fluorescent dyes are attached to a specific EC domain, the position of the dyes are localized with sub nanometer resolution, and the distance between these domains in the cadherin complex is determined (Pertsinidis et al. 2010).

Single molecule FRET measurements between soluble cadherins labeled on the N-terminal domain (Fig. 4.2a) showed that, in the presence of Ca^{2+}, a majority of cadherin monomers homodimerize via their EC1 domains (Fig. 4.2b) (Zhang et al. 2009). Since these experiments could not differentiate between cadherins interacting in *cis* or in *trans* orientations, recombinant dimers were engineered to

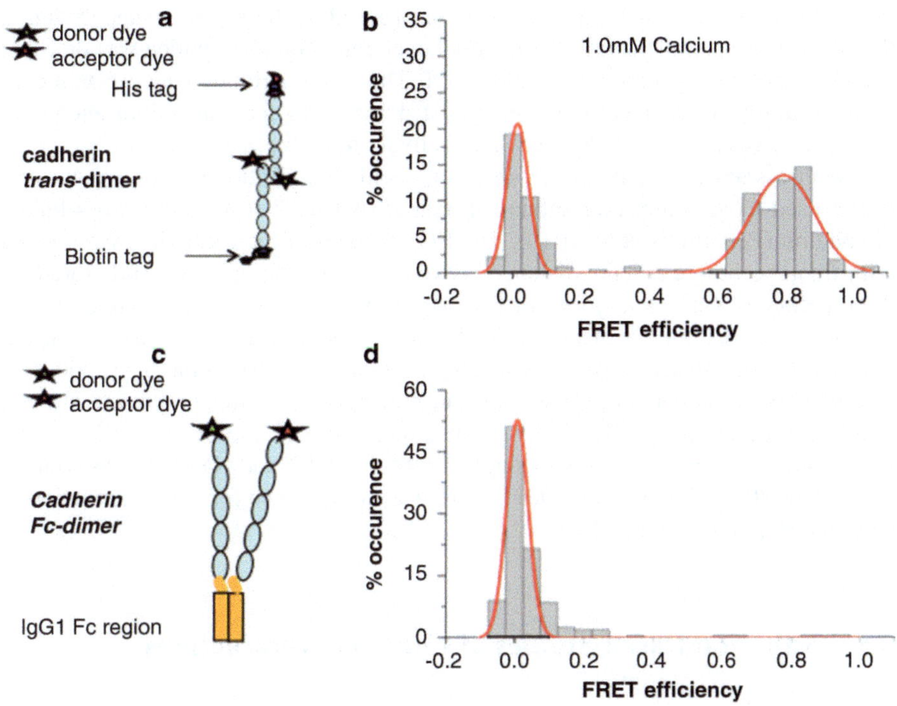

Fig. 4.2 Adapted from (Zhang et al. 2009). **a** *Trans*-dimers formed between E-cadherin monomers fluorescently labeled on the EC1 domains. **b** The majority of *trans*-dimers in 1.0 mM Ca^{2+} exhibit a FRET value of 0.8, which corresponds to a distance of 4 nm between EC1 domains. **c** Fluorescently labeled E-cadherins placed in a close *cis* orientation (cadherin-Fc dimer construct). **d** E-cadherin-Fc dimers in 1.0 mM Ca^{2+} exhibit very few events with a FRET efficiency above 0.5, indicating that these proteins do not form a *cis* bond

force the ectodomains into a close *cis*-orientation (Chappuis-Flament et al. 2001). The outermost domain of the resulting *cis* dimers were then dual-labeled with donor and acceptor fluorophores (Fig. 4.2c), and the distance between them was monitored by FRET. In this case, the FRET signals indicated that, although the cadherins were in close proximity, they did not interact in *cis* (Fig. 4.2d) (Zhang et al. 2009).

Similar studies investigated the reaction pathway for *trans* dimerization (Sivasankar et al. 2009). Two alternative pathways for tryptophan exchange have been proposed (Miloushev et al. 2008). In the induced-fit pathway, cadherin monomers with buried W2 residues, form a W2-independent, initial encounter complex (Fig. 4.1c). Subsequent conformational changes result in W2 strand swapping. In the selected-fit pathway, cadherin monomers adopt an "active" conformation that exposes the W2 residues before binding. Subsequent collisions between "activated" cadherin monomers result in the formation of a stand-swapped dimer. In order to resolve these questions regarding the strand exchange mechanism, the W2A mutant was used to block the selected-fit pathway (Sivasankar et al. 2009). Thus individual W2A mutants would only interact, if the cadherins dimerized via an induced-fit mechanism. Because W2A mutants cannot proceed to the strand-swapped dimer, this strategy kinetically captured the initial encounter complex, which was detected and characterized by single molecule FRET. This result strongly suggests that cadherins initially interact via their outermost domains to form an initial encounter complex (Sivasankar et al. 2009), which is likely to be the X-dimer.

An ultra-stable, sub nanometer, single-molecule localization microscope was recently used to measure inter subunit distances of E-cadherin dimers cross-linked in solution (Pertsinidis et al. 2010). The EC5 domain of the cadherins was labeled with fluorescent dye and the distance between the EC5 domains was measured. In the presence of Ca^{2+}, a majority of E-cadherin dimers adopted an extended *trans*-conformation (EC5-EC5 distance=32.2 nm) (Pertsinidis et al. 2010), consistent with the crystal structure of the C-cadherin ectodomain (Boggon et al. 2002). A smaller population had an EC5-EC5 distance of $\approx$25 nm, which may be an alternative conformation due to flexibility of the dimer complex and/or association of the inner domains. In agreement with single molecule FRET and NMR studies (Haussinger et al. 2002; Zhang et al. 2009), there was no evidence for *cis*-dimerization in solution (Pertsinidis et al. 2010).

4.5 Adhesion Based Studies of Cadherin Interactions

The capacity for cadherin bonds to resist force is a central function of classical cadherins such that adhesion-based measurements provide additional, functionally relevant information about the dynamics and strengths of cadherin bonds. The physics of the force-dependent rupture of noncovalent, bonds enables determinations of different dynamic and physical properties of protein bonds that complement force-independent measurements such as solution binding affinities (Sect. 4.4). Mechanical measurements use force to accelerate bond failure, and the rupture force (bond strength) reflects the activation energy for unbinding, intrinsic dissociation rate, and the shape of the interaction potential (Dudko 2009; Dudko et al. 2006, 2008; Evans and Ritchie 1997). Force measurements can also quantify protein-mediated adhesion energies directly (Leckband and Israelachvili 2001). The principals of typical force measurement approaches and the information they provide are reviewed elsewhere (Evans 1998; Evans and Calderwood 2007; Leckband and Israelachvili 2001).

4.5.1 Surface Force Apparatus Measurements

The surface force apparatus quantifies the interaction energy between two surfaces, as a function of the separation distance, within ±0.1 nm (Israelachvili 1992; Israelachvili and Adams 1978; Leckband and Israelachvili 2001). This approach has been used extensively to study the interactions of several proteins (Johnson et al. 2004, 2005b; Leckband et al. 1995a, b, 2011; Leckband and Prakasam 2006; Menon et al. 2009; Sivasankar et al. 1998, 2001; Yeung et al. 1999; Zhu et al. 2003), including cadherins. In several examples, including CD2, CD58, antibodies, streptavidin, the immune proteins DC-SIGN and DC-SIGNR, cytochrome b5 and cytochrome c, and the protein dimensions measured with this approach agreed quantitatively with crystallographic data (Bayas et al. 2007; Johnson et al. 2004, 2005a; Leckband 2000; Leckband et al. 1994, 1995b, 2011; Yeung et al. 1999; Zhu et al. 2002).

Surface force measurements of the distance-dependence of interactions between opposing cadherin monolayers identified three main features of cadherin binding. First, the measurements identified three distinct cadherin bonds that require different EC domains (Sivasankar et al. 2001; Zhu et al. 2003). Second, they quantified differences between adhesion energies of cadherin subtypes (Prakasam et al. 2006b). Third, they demonstrated that cadherin subtypes cross-react with heterophilic adhesion energies that are intermediate between those of homophilic bonds (Leckband and Prakasam 2006; Prakasam et al. 2006b). Further investigations explored the impact of cancer-associated, calcium-site mutations (Prakasam et al. 2006a) and of *N*-glycosylation (unpublished) on cadherin adhesion.

The distance dependent force between oriented monolayers of C- and N-cadherin ectodomains immobilized on supported lipid membranes (Fig. 4.3a) detected three adhesive interactions that occur at three, distinct membrane separations (Fig. 4.3b). These three cadherin adhesions were separated by ~4 nm—the length of one EC domain (Prakasam et al. 2006b; Zhu et al. 2003). Although the existence of multiple adhesive bonds was initially unexpected, two of the three bonds identified are consistent with current structural data. Adhesion at the membrane separation of 39 nm requires EC1 (Zhu et al. 2003) and W2 (Shi et al. 2010), and is at a distance consistent with a strand-swapped dimer under tension (Sotomayor and Schulten 2008). Adhesion at the separation of 32 nm requires EC1-2, but not W2 (Shi et al. 2010), and is at a distance that is geometrically consistent with the X-dimer complex (Fig. 4.3b) (Hong et al. 2011). The third and strongest adhesion measured under these conditions requires EC3, and is at a membrane distance (26 nm) at which EC3 domains could interact directly.

Studies of the impact of cancer-associated mutations at calcium binding sites in E-cadherin revealed the functional consequences of mutations far from the EC1 domain in the cadherin sequence. Alanine substitutions at D103 and D134 in EC1-2 of E-cadherin ablated EC1-EC1 adhesion and reduced EC3-dependent adhesion (Prakasam et al. 2006a). In comparison with the X-dimer structure and with solution binding studies of N-cadherin D103A and D134A mutants (Prasad and Pedigo 2005; Vunnam and Pedigo 2011b), the D103A mutant likely inhibits X-dimer formation: this mutant retains the interdomain structure but does not form the strand swapped dimer (Vunnam and Pedigo 2011b). The D134A mutant abolishes all calcium binding at the junction and hence the X-dimer intermediate (Prakasam et al. 2006a; Vunnam and Pedigo 2011b), but the effect of these mutations on EC3-dependent adhesion also suggests that these perturbations affect distant sites in the protein. These findings provided evidence for long-ranged

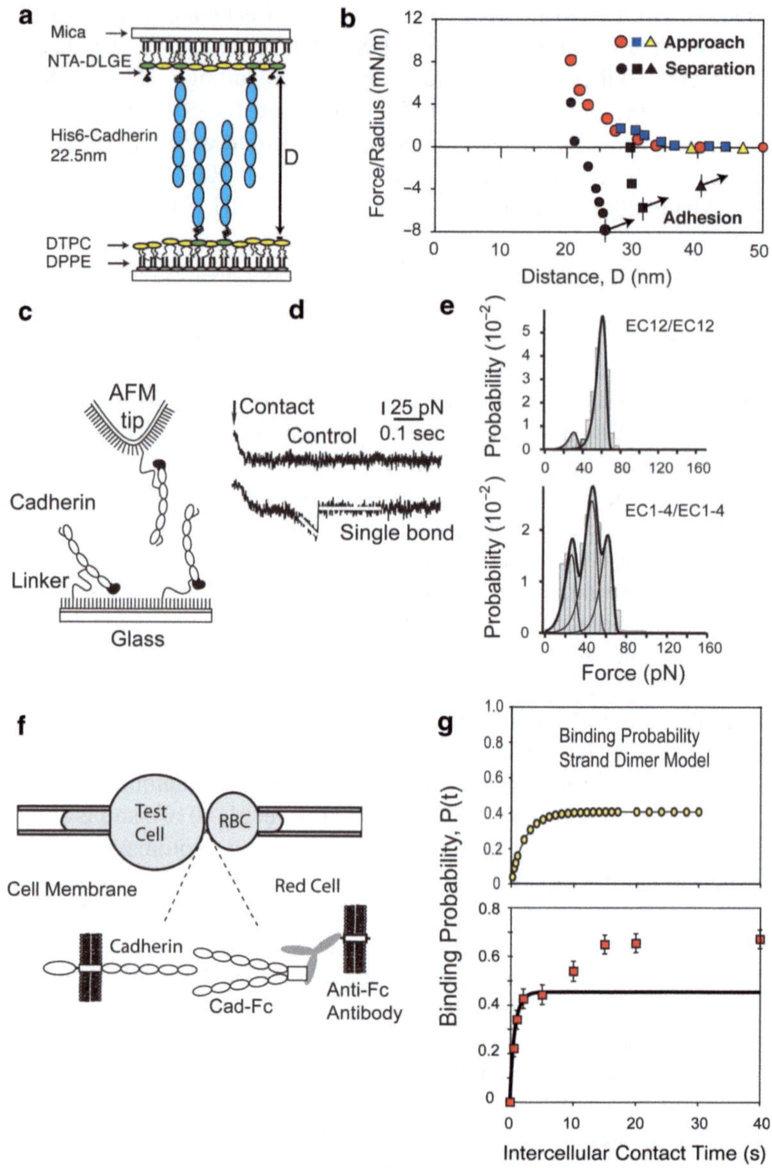

Fig. 4.3 Summary of force-based measurements of cadherin interactions. **a** Sample configuration used in surface force measurements. The distance, D, is the separation between lipid membranes. C-cadherin extracellular domains with C-terminal His tags were immobilized on supported lipid bilayers. The outer membrane leaflet contains di-lauryl glycerol ester (*DLGE*) with nitrile-tri-acetic-acid head groups (*NTA-DLGE*) mixed with di-lauryl-phosphatidyl-choline (*DLPC*). The lipid adjacent to the mica is di-lauryl-phosphatidyl-ethanolamine (*DLPE*). **b** Force normalized by the radius of the curved substrate, F/R versus the distance between the membranes. Forces were measured during approach (*colored symbols*) and during separation (*black symbols*). The outward directed arrows indicate the three distinct positions at which the ectodomains adhere, and the

inter-domain cooperativity, which could play a role in propagating binding information to the cytoplasmic domain.

Consistent with long-range interdomain cooperativity, the D216A mutation at the EC2-3 junction (Handschuh et al. 1999, 2001) is far from all postulated *trans* and *cis* binding interfaces, but it reduces cell–cell adhesion, increases migration, and is associated with aggressive metastatic gastric tumors (Becker et al. 1999; Handschuh et al. 1999, 2001). This mutation did not significantly perturb the E-cadherin ectodomain structure (Prakasam et al. 2006a), but it eliminated EC1-EC1 adhesion and significantly weakened the EC3-dependent bond. This both demonstrated the protein-level impact of these mutations on cell adhesion and demonstrated the long-ranged structural effect of the perturbations.

Taken together, the force-distance measurements identified distinct, adhesive interactions at different membrane distances that require different EC domains (Fig. 4.3b). Except for the EC3-dependent interaction, two of these adhesions are compatible with existing structural and solution binding data. The outermost bond is the strand-swapped dimer: it requires W2 and fails at the membrane separation compatible with the complex under tension (Sotomayor and Schulten 2008). The existence of additional interactions beyond EC1 was initially controversial, but the bond at the intermediate distance is geometrically consistent with the recently identified X-dimer (Hong et al. 2011; Leckband and Prakasam 2006).

Homophilic adhesion energies attributed to the strand-swapped dimer between chicken N-, canine E-, and *Xenopus* C-cadherin ectodomains differed by at most fourfold. Homophilic E-cadherin adhesion exceeded N-cadherin adhesion, and differs from relative solution binding affinities (Katsamba et al. 2009). However, these adhesion data were corroborated by single molecule measurements (Shi et al. 2008) (Sect. 4.5.2) and by measured cadherin affinities at the cell surface (Sect. 4.5.3). The difference in relative magnitudes of E- and N-cadherin adhesion in the context of solution binding data could be due to sequence differences between the cadherin subtypes used.

Heterophilic interactions similarly exhibited multiple adhesive bonds at quantitatively identical spacing as the homophilic bonds (Prakasam et al. 2006b). The quantified EC1-dependent adhesion energies were intermediate between those of

---

pull-off force, indicated by the outward directed arrows, gives the adhesion. **c** AFM set up showing the probe tip and substrate with sparsely immobilized cadherin ectodomains. **d** Force versus tip-surface separation curves in an AFM measurement. In the absence of adhesion, the trace is flat (*top*), but the formation of a single bond causes the force to increase (*dip*) and then snap back to zero at bond rupture. **e** Force histograms measured between EC12 (*top*) and EC1-4 (*bottom*) fragments. The solid lines are Gaussian fits to models for two (*top*) and three bonds (*bottom*). **f** Micropipette manipulation experiment. Two cells are aspirated into apposed pipettes (*top*) and then repetitively brought in and out of contact. The test cell expresses cadherin (*bottom*). A red blood cell is covalently modified with monoclonal anti-Fc antibody, which captures Fc-tagged cadherin dimers (*bottom*). **g** Binding probability time courses. The top panel shows the theoretical binding probability for the strand swapping mechanism. The cadherin binding kinetics exhibit a fast initial rise to $P \sim 0.5$, followed by a 2–5 s lag, and a second rise to a higher binding probability at $P \sim 0.7$. The solid line through the data is the fit of the first binding step to the strand-swap mechanism

the homophilic bonds (Prakasam et al. 2006b), as was later also shown qualitatively by SPR (Katsamba et al. 2009).

4.5.2 Single Molecule Bond Rupture Measurements

Single molecule AFM measurements quantify the forces to rupture single protein-ligand, e.g. cadherin-cadherin bonds as a function of the rate at which the bonds are pulled. In measurements with cadherins, the proteins are attached to the small tip of a cantilever in the AFM and to an opposing surface (Fig. 4.3c). Sparsely immobilized proteins adhere when the cantilever is brought into contact with sparse proteins on the test surface, and retracting the cantilever increases the force on the bond, until it fails (Fig. 4.3d). Bond rupture events are stochastic, so that histograms of rupture forces are generated from hundreds of measurements (Fig. 4.3e). In typical bond rupture measurements, referred to as force spectroscopy, the most probable rupture force depends on the rate of pulling, the bond dissociation rate, and the distance between the ground state and the transition state (Dudko 2009; Dudko et al. 2006; Evans and Ritchie 1997; Suzuki and Dudko 2010). Analyses of these force histograms determine the number of distinct bonds formed, their strengths, and the bond dissociation rates (Dudko 2009; Dudko et al. 2006, 2007, 2008; Evans 2001; Evans and Ritchie 1997, Evans and Calderwood 2007). Alternatively, bond lifetimes determined under constant force ("force-clamp") generate similar, complementary information (Bayas et al. 2006). Four, independent research groups used single molecule force measurements to investigate both homophilic and heterophilic cadherin interactions (Baumgartner et al. 2000; Bayas et al. 2006; du Roure et al. 2006; Perret et al. 2004; Shi et al. 2008, 2010; Tsukasaki et al. 2007).

Single molecule studies mainly compared binding characteristics of EC1-5, EC domain deletions, and W2A mutants of different classical cadherins. Findings with EC1-2 fragments are consistent with solution-based measurements, cadherin structures, and force-distance measurements. E-cadherin forms multiple, independent bonds with a hierarchy of strengths and dissociation rates (Perret et al. 2004). Force histograms measured between EC1-2 fragments of C-cadherin or E-cadherin identified two, weak bonds with fast dissociation rates (Fig. 4.3e, top) (Bayas et al. 2006; Perret et al. 2004; Shi et al. 2010). One of the bonds requires W2 (Shi et al. 2010).

In light of structural data, the W2-independent interaction is likely the X-dimer. The measured strength of the putative X-dimer is ~25% that of the strand swapped dimer, when the rupture forces were quantified at similar pulling rates (Sivasankar et al. 2009). Force spectroscopy measurements (Sect. 4.4) also indicate that the W2-independent bond is weaker, at the pulling rates examined (Shi et al. 2010). However, further studies at different forces and with different proteins, e.g. K14E X-dimer mutant would conclusively define the kinetic and mechanical properties of these two EC1-2 bonds.

In addition to the two, EC1-2 dependent bonds, force histograms measured with full length EC1-5 domains of canine, human, and mouse E-cadherin (Perret et al.

2004; Shi et al. 2008; Tsukasaki et al. 2007), chicken N-cadherin (Shi et al. 2008), and *Xenopus* C-cadherin (Bayas et al. 2006; Shi et al. 2008) form an additional, stronger bond with a slow dissociation rate. *Xenopus* C-cadherin EC1245, which lacks EC3, only exhibited two weak, fast bonds as observed with EC1-2, suggesting that this third interaction requires EC3 in some way (Shi et al. 2010). This result corroborated surface force measurements with the same proteins (Shi et al. 2010; Zhu et al. 2003). AFM measurements of C-cadherin domain deletion mutants EC1-3, EC1-4, and EC1-5 also exhibited three, distinct peaks in force histograms (Fig. 4.3e, bottom) (Shi et al. 2010). The strength of the third, additional interaction increases with the ectodomain length (Shi et al. 2010), indicating that EC4 and EC5 augment this third bond, possibly by stabilizing the binding interface.

4.5.3 Kinetic Measurements of Cadherin-Mediated Cell–Cell Binding

An alternative approach to both force measurements and solution-binding studies uses micropipette manipulation to quantify the kinetics of binding between single cells that are partially aspirated into apposing micropipettes (Fig. 4.3f). The intercellular binding probability is the number of cell–cell binding events divided by the total number of times the cells are repetitively brought into contact, and reflects the number of intercellular bonds (Chesla et al. 1998). The time-dependence of the binding probability depends on the binding mechanism, the kinetic rates and affinities, the contact time, and the cell–cell contact area (Chesla et al. 1998). The two-dimensional affinities and dissociation rates of adhesion proteins on the cell membrane are determined from fits of the data to kinetic rate equations that describe mathematically the postulated binding mechanism. Such measurements determined the two-dimensional affinities and kinetic rates for several proteins, including selectins, T-cell/MHC, integrins, MHC/CD8, and C-cadherin (Chen et al. 2008; Chesla et al. 2000; Chien et al. 2008; Huang et al. 2004, 2007, 2010; Long et al. 2001; Piper et al. 1998; Williams et al. 2001; Zhang et al. 2005).

For simple receptor-ligand interactions such as the strand swapping mechanism, the binding probability is described by a simple exponential that rises smoothly to a limiting plateau (Fig. 4.3g, top) (Chien et al. 2008). However, the binding kinetics of *Xenopus* C-cadherin occurs in two stages: an initial fast step with a low binding probability is followed by a lag or induction phase and then a subsequent rise to a second, higher binding probability (Fig. 4.3g, bottom) (Chien et al. 2008). This kinetic signature was measured with *Xenopus* C-cadherin (Chien et al. 2008); human and canine E-cadherin; chicken, mouse, and human N-cadherin (unpublished). Similar results were obtained for both homophilic and heterophilic binding.

C-cadherin domain deletions identified domains necessary for the two kinetic steps. The C-cadherin EC1-2 and EC1245 fragments only displayed the fast, initial binding step (Chien et al. 2008), which is attributed to strand swapping. The strand exchange model also describes the EC12 and EC1245 kinetic profiles and the

Table 4.2 Two dimensional homodimerization affinities and dissociation rates from cell binding kinetics

Cadherin on Test Cell	Density (#/μm²)	Cadherin-Fc on Red Cell	Density (#/μm²)	2D Affinity (× 10^{-4} μm²)	Dissociation rate (s^{-1})
C-cadherin	18	C-cadherin	10	11 ± 2	0.6 ± 0.2
C-cadherin	7	C-EC1245	10	30 ± 9	0.3 ± 0.1
C-cadherin	7	C-EC12	155	1.4 ± 0.5	0.9 ± 0.2
C-cadherin W2A	24	C-cadherin	452	0.12 ± 0.05	0.10 ± 0.03
N-cadherin	15	N-cadherin	69	1.9 ± 0.3	1.1 ± 0.4
E-cadherin	16	E-cadherin	44	3.3 ± 0.5	1.0 ± 0.3
C-cadherin	14	N-cadherin	38	3.5 ± 0.2	1.3 ± 0.3
C-cadherin	18	E-cadherin	33	3.3 ± 0.9	1.3 ± 0.4
N-cadherin	16	E-cadherin	33	2.6 ± 0.4	1.2 ± 0.5

first binding step measured with EC1-5 (Fig. 4.3g, bottom). Model fits to the data thus determine the two-dimensional (2D) binding affinity and dissociation rates for EC1-EC1 bonds (Table 4.2). The EC12-dependent affinities for EC1245 and EC1-5 were identical, but the EC1-2 affinity was ~20 fold lower. The latter difference may be partly due to protein length differences, which affect two-dimensional affinities (Huang et al. 2004). The W2A mutation altered the kinetic profile relative to the wild type protein, and the residual binding affinity, which is presumably due to X-dimerization, was ~100 fold lower (Chien et al. 2008) (Table 4.2).

Studies of C-cadherin domain deletion mutants showed that EC3 is necessary for the second rise to the high probability state observed with EC1-5. This kinetic profile cannot be described by a proposed transient intermediate (Harrison et al. 2010), suggesting that a different mechanism, possibly involving *cis* interactions, underlies this behavior. Consistent with this notion, a glycosylation mutant that alters the prevalence of N-cadherin dimers on the cell surface (Guo et al. 2009) also changes the kinetics in a manner suggesting that the second step involves lateral dimerization (unpublished). Kinetic analyses also suggest that initial cell–cell binding nucleates the second step. These findings are qualitatively consistent with recent simulations, which suggest that initial *trans* binding facilitates *cis* dimerization in intermembrane gaps (Wu et al. 2010, 2011).

4.6 Conclusions and Future Directions

Accumulating experimental data are revealing several parallels between structures, solution-binding data, and adhesion measurements that reconcile in part what previously appeared to be contradictory findings (Fig. 4.4). The EC1-2 domains are the most extensively studied fragments, and studies provide the greatest qualitative agreement among different experimental measurements. All approaches identified a W2-dependent interaction between EC1 domains that is consistent with the strand-swapped dimer. Single molecule AFM, surface force measurements, FRET, and

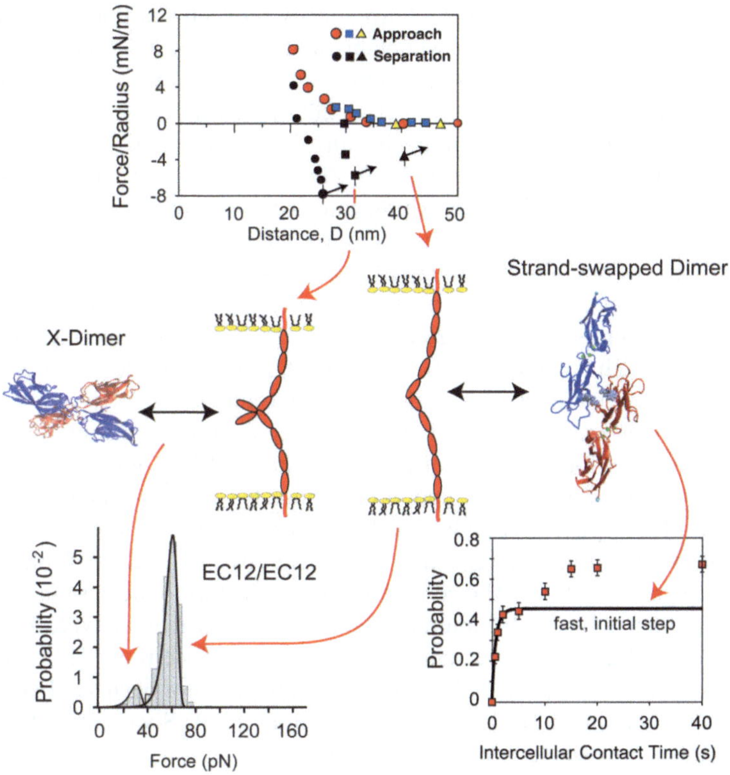

Fig. 4.4 Comparison of current biophysical and structural data. The structures of the *trans* and X-dimers (*center*) are compared with cadherin binding signatures from three different biophysical measurements discussed in the text. The *red arrows* highlight the correspondence between solution binding and/or geometrical/structural characteristics of the *trans* or X-dimer (*center*) and the different features in surface force measurements (*top*), AFM data (*bottom left*), and cell binding kinetics (*bottom right*)

intercellular binding kinetics also identified a W2-independent interaction between EC1-2 fragments (Bayas et al. 2006; Chien et al. 2008; Harrison et al. 2010; Perret et al. 2004; Prakasam et al. 2006a; Shi et al. 2010; Sivasankar et al. 2009), which is consistent with the X-dimer interface (Ciatto et al. 2010; Harrison et al. 2010). The W2-independent middle bond detected in force-distance measurements is geometrically consistent with the X-dimer (Hong et al. 2011; Shi et al. 2010; Sivasankar et al. 2001; Zhu et al. 2003). As summarized below, some differences remain. We discuss studies that may further reconcile the diverse observations discussed in this chapter, and generate a comprehensive functional and structural understanding of classical cadherins.

First, one apparent difference between experimental results is the low affinity and short lifetime of the X-dimer measured under force-independent conditions (Harrison et al. 2010) compared with surface force measurements (Bayas et al. 2006; Leckband and Sivasankar 2000; Perret et al. 2004; Shi et al. 2010). There are, however, increasing examples of differences between equilibrium binding and adhesion-based measurements, most notably in the case of catch bonds, which have no apparent strength in the absence of force but strengthen when pulled (Marshall et al. 2003; Thomas 2008, 2009; Zhu et al. 2008). The adhesive behavior of catch bonds, rather than the solution binding affinity, is the functionally relevant property (Marshall et al. 2003; Thomas 2008, 2009; Zhu et al. 2008). It will be interesting to determine whether cadherins also exhibit catch bond behavior.

A second issue concerns the existence of a unique *cis* dimer interface and the possible role(s) of domains other than EC1-2. A *cis* interface that was postulated on the basis of several crystal structures (Harrison et al. 2011), was not confirmed by solution NMR (Haussinger et al. 2002), electron microscopy (Pokutta et al. 1994), or fluorescence (Zhang et al. 2009). By contrast, a possible role for EC3-5 is supported by biophysical data, and could explain the impact of *N*-glycosylation mutants on *cis* dimerization (Guo et al. 2009). A functional interface involving EC3-5 could also account for the effect of D216A at the EC2/EC3 junction on adhesion (Handschuh et al. 1999), the disruption of various E-cadherin-dependent functions by EC4 and EC5 hyper-glycosylation (Jamal et al. 2009; Pinho et al. 2011; Zhao et al. 2008a), or inhibition of E-cadherin adhesion by the DECMA-1 blocking antibody (Ozawa et al. 1990). The conundrum is due in part to the current absence of structural evidence for EC3-5 interactions, despite experimental evidence that this region affects adhesion. At the same time, the *cis* interface proposed on the basis of structures (Harrison et al. 2011; Wu et al. 2010) has not been verified by other approaches (Haussinger et al. 2002; Pokutta et al. 1994; Sivasankar et al. 2009). Further studies are needed to resolve these findings.

Third, physical chemical differences between molecular interactions in solution (3D) versus inter-membrane gaps (2D) also likely affect experimental outcomes. The effect of molecular confinement on protein folding is well known (Cheung et al. 2005; Cheung and Thirumalai 2007; Dhar et al. 2010), but the impact on protein functions at cell–cell junctions is only recently attracting attention. Functionally significant differences between 3D and 2D affinities are not explained solely by simple geometric corrections. Molecular length, cell topology, clustering, and lateral diffusivity also affect 2D affinities (Chen et al. 2008; Huang et al. 2004; Williams et al. 2001). Computer simulations suggest that molecular confinement could facilitate weak lateral cadherin interactions. Although this could enhance the putative *cis* interaction seen in structures (Wu et al. 2011), it could also promote other EC domain interactions for which biophysical evidence exists. For example, the two-stage kinetic signature (Fig. 4.3g), which may involve *cis* interactions, requires EC3-5 (Chien et al. 2008). As yet, there is no comprehensive, experimentally testable theoretical model for cadherin binding. Consequently, there is currently no method for testing models that might reconcile apparent differences between experimentally measured adhesion and solution binding data.

So far, a complete picture has yet to emerge that reconciles all of the available structural and functional data. However, new structures, additional biophysical studies, and now computer simulations continue to generate new insights into cadherin binding mechanisms. Some differences have yet to be resolved, but recent results demonstrate the increasing consistencies between experimental findings and highlight physical chemical bases for experimental outcomes.

Acknowledgments DEL was supported by NSF CBET 0853705 and by NIH R21 HD059002. SS was supported by the Basil O'Connor Starter Scholar Award from the March of Dimes Foundation (#5-FY10-51).

References

Baumgartner W, Hinterdorfer P, Ness W, Raab A, Vestweber D, Schindler H, Drenckhahn D (2000) Cadherin interaction probed by atomic force microscopy. Proc Natl Acad Sci USA 97:4005–4010

Bayas MV, Leung A, Evans E, Leckband D (2006) Lifetime measurements reveal kinetic differences between homophilic cadherin bonds. Biophys J 90:1385–1395

Bayas MV, Kearney A, Avramovic A, van der Merwe PA, Leckband DE (2007) Impact of salt bridges on the equilibrium binding and adhesion of human CD2 and CD58. J Biol Chem 282:5589–5596

Becker KF, Kremmer E, Eulitz M, Becker I, Handschuh G, Schuhmacher C, Muller W, Gabbert HE, Ochiai A, Hirohashi S, Hofler H (1999) Analysis of E-cadherin in diffuse-type gastric cancer using a mutation-specific monoclonal antibody. Am J Pathol 155:1803–1809

Berx G, Nollet F, van Roy F (1998) Dysregulation of the E-cadherin/catenin complex by irreversible mutations in human carcinomas. Cell Adhes Commun 6:171–184

Bibert S, Jaquinod M, Concord E, Ebel C, Hewat E, Vanbelle C, Legrand P, Weidenhaupt M, Vernet T, Gulino-Debrac D (2002) Synergy between extracellular modules of vascular endothelial cadherin promotes homotypic hexameric interactions. J Biol Chem 277:12790–12801

Bixby JL, Zhang R (1990) Purified N-cadherin is a potent substrate for the rapid induction of neurite outgrowth. J Cell Biol 110:1253–1260

Boggon TJ, Eck MJ (2004) Structure and regulation of Src family kinases. Oncogene 23:7918–7927

Boggon TJ, Murray J, Chappuis-Flament S, Wong E, Gumbiner BM, Shapiro L (2002) C-cadherin ectodomain structure and implications for cell adhesion mechanisms. Science 296:1308–1313

Brasch J, Harrison OJ, Ahlsen G, Carnally SM, Henderson RM, Honig B, Shapiro L (2011) Structure and binding mechanism of vascular endothelial cadherin: a divergent classical cadherin. J Mol Biol 408:57–73

Brieher WM, Yap AS, Gumbiner BM (1996) Lateral dimerization is required for the homophilic binding activity of C-cadherin. J Cell Biol 135:487–496

Chappuis-Flament S, Wong E, Hicks LD, Kay CM, Gumbiner BM (2001) Multiple cadherin extracellular repeats mediate homophilic binding and adhesion. J Cell Biol 154:231–243

Chen W, Zarnitsyna VI, Sarangapani KK, Huang J, Zhu C (2008) Measuring receptor-ligand binding kinetics on cell surfaces: from adhesion frequency to thermal fluctuation methods. Cell Mol Bioeng 1:276–288

Chesla SE, Selvaraj P, Zhu C (1998) Measuring two-dimensional receptor-ligand binding kinetics by micropipette. Biophys J 75:1553–1572

Chesla SE, Li P, Nagarajan S, Selvaraj P, Zhu C (2000) The membrane anchor influences ligand binding two-dimensional kinetic rates and three-dimensional affinity of FcgammaRIII (CD16). J Biol Chem 275:10235–10246

Cheung MS, Thirumalai D (2007) Effects of crowding and confinement on the structures of the transition state ensemble in proteins. J Phys Chem B 111:8250–8257

Cheung MS, Klimov D, Thirumalai D (2005) Molecular crowding enhances native state stability and refolding rates of globular proteins. Proc Natl Acad Sci USA 102:4753–4758

Chien YH, Jiang N, Li F, Zhang F, Zhu C, Leckband D (2008) Two stage cadherin kinetics require multiple extracellular domains but not the cytoplasmic region. J Biol Chem 283:1848–1856
Chtcheglova, LA, Wildling L, Waschke J, Drenckhahn D, Hinterdorfer P (2010) AFM functional imaging on vascular endothelial cells. J Mol Recogn 23:589–596
Ciatto C, Bahna F, Zampieri N, VanSteenhouse HC, Katsamba PS, Ahlsen G, Harrison OJ, Brasch J, Jin X, Posy S, Vendome J, Ranscht B, Jessell TM, Honig B, Shapiro L (2010) T-cadherin structures reveal a novel adhesive binding mechanism. Nat Struct Mol Biol 17:339–347
Dhar A, Samiotakis A, Ebbinghaus S, Nienhaus L, Homouz D, Gruebele M, Cheung MS (2010) Structure, function, and folding of phosphoglycerate kinase are strongly perturbed by macromolecular crowding. Proc Natl Acad Sci USA 107:17586–17591
Dudko OK (2009) Single-molecule mechanics: new insights from the escape-over-a-barrier problem. Proc Natl Acad Sci USA 106:8795–8796
Dudko OK, Hummer G, Szabo A (2006) Intrinsic rates and activation free energies from single-molecule pulling experiments. Phys Rev Lett 96:108101
Dudko OK, Mathe J, Szabo A, Meller A, Hummer G (2007) Extracting kinetics from single-molecule force spectroscopy: nanopore unzipping of DNA hairpins. Biophys J 92:4188–4195
Dudko OK, Hummer G, Szabo A (2008) Theory, analysis, and interpretation of single-molecule force spectroscopy experiments. Proc Natl Acad Sci USA 105:15755–15760
du Roure O, Buguin A, Feracci H, Silberzan P (2006) Homophilic interactions between cadherin fragments at the single molecule level: an AFM study. Langmuir 22:4680–4684
Evans E (1998) Energy landscapes of biomolecular adhesion and receptor anchoring at interfaces explored with dynamic force spectroscopy. Faraday Discuss:1–16
Evans E (2001) Probing the relation between force—lifetime—and chemistry in single molecular bonds. Ann Rev Biophys Biomol Struct 30:105–128
Evans E, Ritchie K (1997) Dynamic strength of molecular adhesion bonds. Biophys J 72:1541–1555
Evans EA, Calderwood DA (2007) Forces and bond dynamics in cell adhesion. Science 316:1148–1153
Fuchs M, Hutzler P, Handschuh G, Hermannstadter C, Brunner I, Hofler H, Luber B (2004) Dynamics of cell adhesion and motility in living cells is altered by a single amino acid change in E-cadherin fused to enhanced green fluorescent protein. Cell Motil Cytoskeleton 59:50–61
Gavard J, Lambert M, Grosheva I, Marthiens V, Irinopoulou T, Riou JF, Bershadsky A, Mege RM (2004) Lamellipodium extension and cadherin adhesion: two cell responses to cadherin activation relying on distinct signalling pathways. J Cell Sci 117:257–270
Geng F, Shi BZ, Yuan YF, Wu XZ (2004) The expression of core fucosylated E-cadherin in cancer cells and lung cancer patients: prognostic implications. Cell Res 14:423–433
Guo HB, Johnson H, Randolph M, Pierce M (2009) Regulation of homotypic cell-cell adhesion by branched N-glycosylation of N-cadherin extracellular EC2 and EC3 domains. J Biol Chem 284:34986–34997
Handschuh G, Candidus S, Luber B, Reich U, Schott C, Oswald S, Becke H, Hutzler P, Birchmeier W, Hofler H, Becker KF (1999) Tumour-associated E-cadherin mutations alter cellular morphology, decrease cellular adhesion and increase cellular motility. Oncogene 18:4301–4312
Handschuh G, Luber B, Hutzler P, Hofler H, Becker KF (2001) Single amino acid substitutions in conserved extracellular domains of E-cadherin differ in their functional consequences. J Mol Biol 314:445–454
Harrison OJ, Corps EM, Berge T, Kilshaw PJ (2005) The mechanism of cell adhesion by classical cadherins: the role of domain 1. J Cell Sci 118:711–721
Harrison OJ, Bahna F, Katsamba PS, Jin X, Brasch J, Vendome J, Ahlsen G, Carroll KJ, Price SR, Honig B, Shapiro L (2010) Two-step adhesive binding by classical cadherins. Nat Struct Mol Biol 17:348–357
Harrison OJ, Jin X, Hong S, Bahna F, Ahlsen G, Brasch J, Wu Y, Vendome J, Felsovalyi K, Hampton CM, Troyanovsky RB, Ben-Shaul A, Frank J, Troyanovsky SM, Shapiro L, Honig B (2011) The extracellular architecture of adherens junctions revealed by crystal structures of type I cadherins. Structure 19:244–256

Haussinger D, Ahrens T, Sass HJ, Pertz O, Engel J, Grzesiek S (2002) Calcium-dependent homoassociation of E-cadherin by NMR spectroscopy: changes in mobility, conformation and mapping of contact regions. J Mol Biol 324:823–839

Haussinger D, Ahrens T, Aberle T, Engel J, Stetefeld J, Grzesiek S (2004) Proteolytic E-cadherin activation followed by solution NMR and X-ray crystallography. EMBO J 23:1699–1708

He W, Cowin P, Stokes DL (2003) Untangling desmosomal knots with electron tomography. Science 302:109–113

Hewat EA, Durmort C, Jacquamet L, Concord E, Gulino-Debrac D (2007) Architecture of the VE-cadherin hexamer. J Mol Biol 365:744–751

Homola J (2008) Surface plasmon resonance sensors for detection of chemical and biological species. Chem Rev 108:462–493

Hong S, Troyanovsky RB, Troyanovsky SM (2011) Cadherin exits the junction by switching its adhesive bond. J Cell Biol 192:1073–1083

Huang J, Chen J, Chesla SE, Yago T, Mehta P, McEver RP, Zhu C, Long M (2004) Quantifying the effects of molecular orientation and length on two-dimensional receptor-ligand binding kinetics. J Biol Chem 279:44915–44923

Huang J, Edwards LJ, Evavold BD, Zhu C (2007) Kinetics of MHC-CD8 interaction at the T cell membrane. J Immunol 179:7653–7662

Huang J, Zarnitsyna VI, Liu B, Edwards LJ, Jiang N, Evavold BD, Zhu C (2010) The kinetics of two-dimensional TCR and pMHC interactions determine T-cell responsiveness. Nature 464:932–936

Iino R, Koyama I, Kusumi A (2001) Single molecule imaging of green fluorescent proteins in living cells: E-cadherin forms oligomers on the free cell surface. Biophys J 80:2667–2677

Israelachvili J (1992) Adhesion forces between surfaces in liquids and condensable vapours. Surface Sci Rep 14:110–159

Israelachvili JN, Adams GE (1978) Measurement of forces between two mica surfaces in aqueous electrolyte solutions in the range 0–100 nm. J Chem Soc Faraday Trans I 75:975–1001

Jamal BT, Nita-Lazar M, Gao Z, Amin B, Walker J, Kukuruzinska MA (2009) N-glycosylation status of E-cadherin controls cytoskeletal dynamics through the organization of distinct beta-catenin- and gamma-catenin-containing AJs. Cell Health Cytoskelet 2009:67–80

Johnson CP, Fujimoto I, Perrin-Tricaud C, Rutishauser U, Leckband D (2004) Mechanism of homophilic adhesion by the neural cell adhesion molecule: use of multiple domains and flexibility. Proc Natl Acad Sci USA 101:6963–6968

Johnson CP, Fragneto G, Konovalov O, Dubosclard V, Legrand JF, Leckband DE (2005a) Structural studies of the neural-cell-adhesion molecule by X-ray and neutron reflectivity. Biochemistry 44:546–554

Johnson CP, Fujimoto I, Rutishauser U, Leckband DE (2005b) Direct evidence that neural cell adhesion molecule (NCAM) polysialylation increases intermembrane repulsion and abrogates adhesion. J Biol Chem 280:137–145

Katsamba P, Carroll K, Ahlsen G, Bahna F, Vendome J, Posy S, Rajebhosale M, Price S, Jessell TM, Ben-Shaul A, Shapiro L, Honig BH (2009) Linking molecular affinity and cellular specificity in cadherin-mediated adhesion. Proc Natl Acad Sci USA 106:11594–11599

Kim YJ, Johnson KR, Wheelock MJ (2005) N-cadherin-mediated cell motility requires cis dimers. Cell Commun Adhes 12:23–39

Kitagawa M, Natori M, Murase S, Hirano S, Taketani S, Suzuki ST (2000) Mutation analysis of cadherin-4 reveals amino acid residues of EC1 important for the structure and function. Biochem Biophys Res Commun 271:358–363

Klingelhofer J, Laur OY, Troyanovsky RB, Troyanovsky SM (2002) Dynamic interplay between adhesive and lateral E-cadherin dimers. Mol Cell Biol 22:7449–7458

Koch AW, Pokutta S, Lustig A, Engel J (1997) Calcium binding and homoassociation of E-cadherin domains. Biochemistry 36:7697–7705

Koch AW, Bozic D, Pertz O, Engel J (1999) Homophilic adhesion by cadherins. Curr Opin Struct Biol 9:275–281

Lambert O, Taveau JC, Him JL, Al Kurdi R, Gulino-Debrac D, Brisson A (2005) The basic framework of VE-cadherin junctions revealed by cryo-EM. J Mol Biol 346:1193–1196

Lebowitz J, Lewis MS, Schuck P (2002) Modern analytical ultracentrifugation in protein science: a tutorial review. Protein Sci 11:2067–2079

Leckband D (2000) Measuring the forces that control protein interactions. Annu Rev Biophys Biomol Struct 29:1–26

Leckband D, Israelachvili J (2001) Intermolecular forces in biology. Q Rev Biophys 34:105–267

Leckband D, Prakasam A (2006) Mechanism and dynamics of cadherin adhesion. Annu Rev Biomed Eng 8:259–287

Leckband D, Sivasankar S (2000) Mechanism of homophilic cadherin adhesion. Curr Opin Cell Biol 12:587–592

Leckband DE, Schmitt FJ, Israelachvili JN, Knoll W (1994) Direct force measurements of specific and nonspecific protein interactions. Biochemistry 33:4611–4624

Leckband D, Muller W, Schmitt FJ, Ringsdorf H (1995a) Molecular mechanisms determining the strength of receptor-mediated intermembrane adhesion. Biophys J. 69:1162–1169

Leckband DE, Kuhl T, Wang HK, Herron J, Muller W, Ringsdorf H (1995b). 4-4-20 anti-fluorescyl IgG Fab' recognition of membrane bound hapten: direct evidence for the role of protein and interfacial structure. Biochemistry 34:11467–11478

Leckband DE, Menon S, Rosenberg K, Graham SA, Taylor ME, Drickamer K (2011) Geometry and adhesion of extracellular domains of DC-SIGNR neck length variants analyzed by force-distance measurements. Biochemistry 50:6125–6132

Liwosz A, Lei T, Kukuruzinska MA (2006) N-glycosylation affects the molecular organization and stability of E-cadherin junctions. J Biol Chem 281:23138–23149

Long M, Zhao H, Huang KS, Zhu C (2001) Kinetic measurements of cell surface E-selectin/carbohydrate ligand interactions. Annals Biomed Eng 29:935–946

Luber B, Candidus S, Handschuh G, Mentele E, Hutzler P, Feller S, Voss J, Hofler H, Becker KF (2000) Tumor-derived mutated E-cadherin influences beta-catenin localization and increases susceptibility to actin cytoskeletal changes induced by pervanadate. Cell Adhes Commun. 7:391–408

Marshall BT, Long M, Piper JW, Yago T, McEver RP, Zhu C (2003) Direct observation of catch bonds involving cell-adhesion molecules. Nature 423:190–193

Menon S, Rosenberg K, Graham SA, Ward EM, Taylor ME, Drickamer K, Leckband DE (2009) Binding-site geometry and flexibility in DC-SIGN demonstrated with surface force measurements. Proc Natl Acad Sci USA 106:11524–11529

Miloushev VZ, Bahna F, Ciatto C, Ahlsen G, Honig B, Shapiro L, Palmer AG 3rd (2008) Dynamic properties of a type II cadherin adhesive domain: implications for the mechanism of strand-swapping of classical cadherins. Structure 16:1195–1205

Nagar B, Overduin M, Ikura M, Rini JM (1996) Structural basis of calcium-induced E-cadherin rigidification and dimerization. Nature 380:360–364

Nita-Lazar M, Rebustini I, Walker J, Kukuruzinska MA (2010) Hypoglycosylated E-cadherin promotes the assembly of tight junctions through the recruitment of PP2A to adherens junctions. Exp Cell Res 316:1871–1884

Nose A, Nagafuchi A, Takeichi M (1988) Expressed recombinant cadherins mediate cell sorting in model systems. Cell 54:993–1001

Nose A, Tsuji K, Takeichi M (1990) Localization of specificity determining sites in cadherin cell adhesion molecules. Cell 61:147–155

Ozawa M, Hoschutzky H, Herrenknecht K, Kemler R (1990) A possible new adhesive site in the cell-adhesion molecule uvomorulin. Mech Dev 33:49–56

Perret E, Leung A, Feracci H, Evans E (2004) Trans-bonded pairs of E-cadherin exhibit a remarkable hierarchy of mechanical strengths. Proc Natl Acad Sci USA 101:16472–16477

Pertsinidis A, Zhang Y, Chu S (2010) Subnanometre single-molecule localization, registration and distance measurements. Nature 466:647–651

Pertz O, Bozic D, Koch AW, Fauser C, Brancaccio A, Engel J (1999) A new crystal structure, Ca^{2+} dependence and mutational analysis reveal molecular details of E-cadherin homoassociation. EMBO J 18:1738–1747

Pinho SS, Osorio H, Nita-Lazar M, Gomes J, Lopes C, Gartner F, Reis CA (2009) Role of E-cadherin N-glycosylation profile in a mammary tumor model. Biochim Biophys Res Comm 379:1091–1096
Pinho SS, Seruca R, Gartner F, Yamaguchi Y, Gu J, Taniguchi N, Reis CA (2011) Modulation of E-cadherin function and dysfunction by N-glycosylation. Cell Mol Life Sci 68:1011–1020
Piper JW, Swerlick RA, Zhu C (1998) Determining force dependence of two-dimensional receptor-ligand binding affinity by centrifugation. Biophys J 74:492–513
Pokutta S, Herrenknecht K, Kemler R, Engel J (1994) Conformational changes of the recombinant extracellular domain of E-cadherin upon calcium binding. Eur J Biochem 223:1019–1026
Prakasam A, Chien YH, Maruthamuthu V, Leckband DE (2006a) Calcium site mutations in cadherin: impact on adhesion and evidence of cooperativity. Biochemistry 45:6930–6939
Prakasam AK, Maruthamuthu V, Leckband DE (2006b) Similarities between heterophilic and homophilic cadherin adhesion. Proc Natl Acad Sci USA 103:15434–15439
Prasad A, Pedigo S (2005) Calcium-dependent stability studies of domains 1 and 2 of epithelial cadherin. Biochemistry 44:13692–13701
Roy R, Hohng S, Ha T (2008) A practical guide to single-molecule FRET. Nature Methods 5:507–516
Scott JA, Shewan AM, den Elzen NR, Loureiro JJ, Gertler FB, Yap AS (2006) Ena/VASP proteins can regulate distinct modes of actin organization at cadherin-adhesive contacts. Mol Biol Cell. 17:1085–1095
Shan W, Yagita Y, Wang Z, Koch A, Fex Svenningsen A, Gruzglin E, Pedraza L, Colman DR (2004) The minimal essential unit for cadherin-mediated intercellular adhesion comprises extracellular domains 1 and 2. J Biol Chem 279:55914–55923
Shi Q, Chien YH, Leckband D (2008) Biophysical properties of cadherin bonds do not predict cell sorting. J Biol Chem 283:28454–28463
Shi Q, Maruthamuthu V, Leckband D (2010) Allosteric cross-talk between cadherin ectodomains. Biophys J 99:95–104
Sivasankar S, Subramaniam S, Leckband D (1998) Direct molecular level measurements of the electrostatic properties of a protein surface. Proc Natl Acad Sci USA 95:12961–12966
Sivasankar S, Gumbiner B, Leckband D (2001) Direct measurements of multiple adhesive alignments and unbinding trajectories between cadherin extracellular domains. Biophys J 80:1758–1768
Sivasankar S, Zhang Y, Nelson WJ, Chu S (2009) Characterizing the initial encounter complex in cadherin adhesion. Structure 17:1075–1081
Smutney M, Cox H, Leerberg J, Conti M, Ferguson C, Hamilton N, Parton R, Adeslstein R, Yap A (2010) Myosin II isoforms identify distinct functional modules that support integrity of the epithelial zonula adherens. Nat Cell Biol 12:696–702
Sotomayor M, Schulten K (2008) The allosteric role of the Ca^{2+} switch in adhesion and elasticity of C-cadherin. Biophys J 94:4621–4633
Suzuki Y, Dudko OK (2010) Single-molecule rupture dynamics on multidimensional landscapes. Phys Rev Lett 104:048101
Takeda H, Shimoyama Y, Nagafuchi A, Hirohashi S (1999) E-cadherin functions as a cis-dimer at the cell-cell adhesive interface in vivo. Nat Struct Biol 6:310–312
Tamura K, Shan WS, Hendrickson WA, Colman DR, Shapiro L (1998) Structure-function analysis of cell adhesion by neural (N-) cadherin. Neuron 20:1153–1163
Taveau JC, Dubois M, Le Bihan O, Trepout S, Almagro S, Hewat E, Durmort C, Heyraud S, Gulino-Debrac D, Lambert O (2008) Structure of artificial and natural VE-cadherin-based adherens junctions. Biochem Soc Trans 36:189–193
Thomas W (2008) Catch bonds in adhesion. Annu Rev Biomed Eng 10:39–57
Thomas WE (2009) Mechanochemistry of receptor-ligand bonds. Curr Opin Struct Biol 19:50–55
Tomschy A, Fauser C, Landwehr R, Engel J (1996) Homophilic adhesion of E-cadherin occurs by a co-operative two-step interaction of N-terminal domains. EMBO J 15:3507–3514
Troyanovsky RB, Sokolov E, Troyanovsky SM (2003) Adhesive and lateral E-cadherin dimers are mediated by the same interface. Mol Cell Biol 23:7965–7972

Tsukasaki Y, Kitamura K, Shimizu K, Iwane AH, Takai Y, Yanagida T (2007) Role of multiple bonds between the single cell adhesion molecules, nectin and cadherin, revealed by high sensitive force measurements. J Mol Biol 367:996–1006

Vendome J, Posy S, Jin X, Bahna F, Ahlsen G, Shapiro L, Honig B (2011) Molecular design principles underlying beta-strand swapping in the adhesive dimerization of cadherins. Nat Struct Mol Biol 18:693–700

Vunnam N, Pedigo S (2011a) Prolines in betaA-sheet of neural cadherin act as a switch to control the dynamics of the equilibrium between monomer and dimer. Biochemistry 50:6959–6965

Vunnam N, Pedigo S (2011b) Sequential binding of calcium leads to dimerization in neural cadherin. Biochemistry 50:2973–2982

Williams TE, Nagarajan S, Selvaraj P, Zhu C (2001) Quantifying the impact of membrane microtopology on effective two-dimensional affinity. J Biol Chem 276:13283–13288

Wu Y, Jin X, Harrison O, Shapiro L, Honig BH, Ben-Shaul A (2010) Cooperativity between trans and cis interactions in cadherin-mediated junction formation. Proc Natl Acad Sci USA 107:17592–17597

Wu Y, Vendome J, Shapiro L, Ben-Shaul A, Honig B (2011) Transforming binding affinities from three dimensions to two with application to cadherin clustering. Nature 475:510–513

Yeung C, Purves T, Kloss AA, Kuhl TL, Sligar S, Leckband D (1999) Cytochrome c recognition of immobilized, orientational variants of cytochrome b5: direct force and equilibrium binding measurements. Langmuir 15:6829–6836

Zhang F, Marcus WD, Goyal NH, Selvaraj P, Springer TA, Zhu C (2005) Two-dimensional kinetics regulation of alphaLbeta2-ICAM-1 interaction by conformational changes of the alphaL-inserted domain. J Biol Chem 280:42207–42218

Zhang Y, Sivasankar S, Nelson WJ, Chu S (2009) Resolving cadherin interactions and binding cooperativity at the single-molecule level. Proc Natl Acad Sci USA 106:109–114

Zhao H, Liang Y, Xu Z, Wang L, Zhou F, Li Z, Jin J, Yang Y, Fang Z, Hu Y, Zhang L, Su J, Zha X (2008a) N-glycosylation affects the adhesive function of E-Cadherin through modifying the composition of adherens junctions (AJs) in human breast carcinoma cell line MDA-MB-435. J Cell Biochem 104:162–175

Zhao Y, Sato Y, Isaji T, Fukuda T, Matsumoto A, Miyoshi E, Gu J, Taniguchi N (2008b) Branched N-glycans regulate the biological functions of integrins and cadherins. Febs J 275:1939–1948

Zhong Y, Brieher WM, Gumbiner BM (1999) Analysis of C-cadherin regulation during tissue morphogenesis with an activating antibody. J Cell Biol 144:351–359

Zhu B, Davies EA, van der Merwe PA, Calvert T, Leckband DE (2002) Direct measurements of heterotypic adhesion between the cell surface proteins CD2 and CD48. Biochemistry 41:12163–12170

Zhu B, Chappuis-Flament S, Wong E, Jensen IE, Gumbiner BM, Leckband D (2003) Functional analysis of the structural basis of homophilic cadherin adhesion. Biophys J 84:4033–4042

Zhu C, Yago T, Lou J, Zarnitsyna VI, McEver RP (2008) Mechanisms for flow-enhanced cell adhesion. Ann Biomed Eng 36:604–621

Chapter 5
Adherens Junction Assembly

Sergey Troyanovsky

Abstract Classical cadherins are a family of transmembrane proteins that mediate cell–cell adhesion at adherens junctions. A complex chain of *cis*- and *trans*- interactions between cadherin ectodomains establishes a cadherin adhesive cluster. A principal adhesive interaction in such clusters is an exchange of β strands between the first extracellular cadherin domains (EC1). The structure of cadherin adhesive clusters can be modified by other adherens junction proteins including additional transmembrane proteins, nectins and various intracellular proteins that directly or indirectly interact with the intracellular cadherin region. These interactions determine the dynamics and stability of cadherin adhesive structures.

5.1 Introduction

The assembly of the vast majority of multiprotein structures includes two distinct steps—nucleation and elongation. The latter step is often based on cooperative interactions between the structure's subunits. The assembly process is followed by the reverse process of structure disassembly. The balance between these two opposite processes determines the size of the structure and its dynamics. It is highly likely that adherens junctions are not an exception and that the same principles underlie their homeostasis. Assembly of adherens junctions is likely to be initiated at specific sites of cell–cell contacts by a nucleation reaction, the nature of which remains to be determined. Recent advances in the field have shown that junction assembly is based on a set of cooperative *trans* and *cis* interactions between cadherin ectodomains. These binding reactions produce adhesive clusters in which cadherin molecules are arranged in specific linear arrays. These reactions of cadherin adhesive cluster self-assembly are, perhaps, the most ubiquitous and currently the best understood event in formation of adherens junctions. Importantly, these reactions are specific to vertebrate classic cadherins; invertebrate cadherins employ another, much less studied, set of extracellular interactions (Harrison et al. 2011).

S. Troyanovsky (✉)
Department of Dermatology, Feinberg School of Medicine, Northwestern University, Chicago, IL 60611, USA
e-mail: s-troyanovsky@northwestern.edu

T. Harris (ed.), *Adherens Junctions: From Molecular Mechanisms to Tissue Development and Disease,* Subcellular Biochemistry 60,
DOI 10.1007/978-94-007-4186-7_5, © Springer Science+Business Media Dordrecht 2012

It is not known how these adhesive clusters arrange themselves into mature adherens junctions. The junctions are not static: they continuously loose and gain cadherin. The reorganization of the adhesive clusters into the adherens junctions and their subsequent disassembly are, perhaps, regulated by diverse intracellular signaling pathways and the cytoskeleton. The complexity and redundancy of these intracellular mechanisms are likely key reasons for the morphological and structural pleomorphism of adherens junctions, which can be classified by a number of subtypes (*zonulae adhaerentes, fasciae adhaerentes, puncta adhaerentia* and many others, see Franke 2009). This diversity of adherens junctions reflects the varying requirements for cell–cell junction positioning, their strength and their dynamics in different types of cells.

In this chapter we will discuss some basic principles of cadherin–cadherin interactions resulting in the assembly and disassembly of adherens junctions.

5.2 From Cadherin Monomer to Cadherin Adhesive Clusters

Adherens junctions are formed as a result of two independent but coordinated cellular activities. The first one is cadherin adhesiveness, which, as we discuss below, is based on *cis*- and *trans*-interactions between cadherin molecules. The second one is the activity of actin cytoskeleton controlling protrusion-retracting cycles of plasma membranes of the contacting cells. In cell culture of MDCK epithelial cells, the initial junction contact is established by lamellipodia of two adjacent cells (McNeill et al. 1993; Adams and Nelson 1998). In mouse keratinocytes, the initial contact is made by interdigitating filopodia that form transient point contacts, which then zipper into a continuous mature junction (Vasioukhin and Fuchs 2001). In both cases, formation of adherens junctions coincides with extensive reorganization of the actin cytoskeleton. Inactivation of this reorganization by inhibitors of actin polymerization or actomyosin contractility affects junction formation. How these two activities are coordinated is one of the key unknown aspects of cadherin adhesion. Detailed understanding of cadherin adhesive interactions is essential for unraveling how the actin cytoskeleton regulates cadherin adhesion.

5.2.1 Cadherin Strand-Swapping is at the Core of Cadherin-Based Cell–Cell Adhesion

Different experimental approaches have compellingly shown that the cadherin adhesive site is localized to the EC1 domain. This was first indicated by domain shuffling experiments (Nose at al. 1990). This work showed that cells expressing an E/P-cadherin chimera with a P-cadherin-derived EC1 domain co-aggregate with P-cadherin-expressing cells. Similar experiments, but based on a co-immunoprecipitation assay, confirmed the key role of the EC1 domain in binding specificity

(Klingelhöfer et al. 2000). These biochemical data have been corroborated by an electron microscopy study that showed intercadherin interactions through the EC1 domain (Tomschy et al. 1996). Two independent cryo-electron-tomography studies of desmosomes (He et al. 2003; Al-Amoudi et al. 2007) also documented the aminoterminal location of the adhesive sites. Finally, two recent FRET-based studies, which used elegantly designed cadherin molecules bearing fluorescent tags at different locations of the E-cadherin extracellular region, also showed that cadherin adhesion is established by the EC1 domain (Zhang et al. 2009; Kim et al. 2011).

Point mutagenesis of the EC1 domain in conjunction with co-immunoprecipitation and zonal sedimentation (Chitaev and Troyanovsky 1998; Tamura et al. 1998; Shan et al. 2000; Kitagawa et al. 2000; Laur et al. 2002) provided strong evidence that the cadherin adhesive site corresponds to the strand-swap dimer interface detected first in N-cadherin three-dimensional structure by Shapiro et al. (1995) and then in many other type I cadherins (see Posy et al. 2008). The involvement of this site in adhesion was further indicated by cross-linking experiments performed with engineered cadherin cysteine mutants (Troyanovsky et al. 2003; Harrison et al. 2005) and later documented by two independent FRET studies (Zhang et al. 2009; Kim et al. 2011).

Strand-swap cadherin dimerization is based on the exchange of N-terminal β strands of the EC1 domains (A* strand) between pairing cadherin molecules (Fig. 5.1, see below). Since the amino-terminal amino group stabilizes strand swapping by the salt bridge with Glu89, strand swapping is destroyed by the prodomain present in the unprocessed cadherin or by extra aminoterminal amino acids in recombinant cadherins (Troyanovsky 2005). Proteolytic removal of the prodomain is a key event activating cadherin adhesiveness (Häussinger et al. 2004).

Structural analysis of the full-size cadherin ectodomain shows that the cadherin rod is bent, so that the long axes of the EC1 and EC5 domains are at a nearly right angle (Boggon et al. 2002; Harrison et al. 2011). Such curvature of the cadherin ectodomain presents EC1 domain in a way that it can swap its A* strand with the cadherin from the adjacent cell more efficiently than with cadherin located at the same plasma membrane. The curved structure of the cadherin ectodomain had been shown by EM in 1989, and it was proposed that such a conformation of cell adhesion receptors may represent a general evolutionary solution to the specific problems of cell–cell adhesion: the flexible bent may absorb the stress when adjoining cell surfaces are in motion (Becker et al. 1989).

Biochemical examination of cadherin–cadherin interactions in cell culture is completely consistent with the structural data: it has shown that strand swapping results in the formation of both lateral (or *cis*) and adhesive (or *trans*) cadherin dimers on the cell surface (Chitaev and Troyanovsky 1998; Harrison et al. 2005). Importantly, from structural stand-point, these lateral and adhesive dimers are the same: the only difference between them is that cadherins in the dimers originate either from the same or from opposite cell surfaces. Inactivation of calcium-binding interface or placing cells in low-calcium media attenuates *trans*, but has no effect on *cis* strand-swapping (Klingelhöfer et al. 2002). Apparently, the loss of correct ectodomain curvature impedes *trans* and promotes *cis* dimer formation. As we will

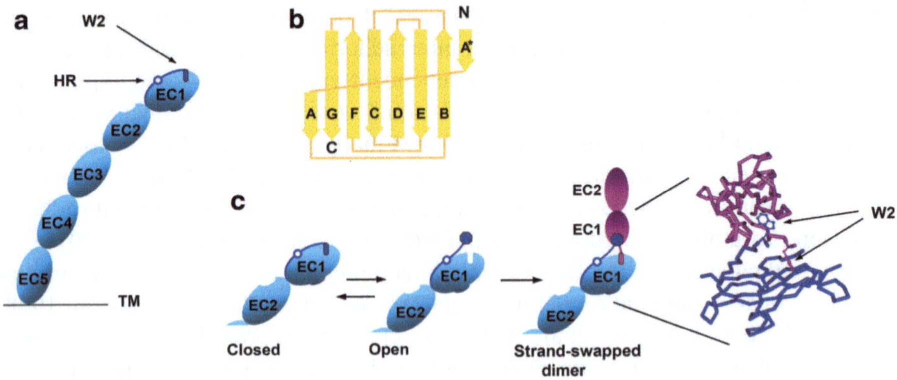

Fig. 5.1 Cadherin dimerization using the strand-swap interface. **a** Schematic representation of the cadherin ectodomain. It consists of five homological cadherin-like domains (EC1-EC5). The A* and A strands of EC1 are dark blue. They are separated by the hinge region (*open circle*, HR). The most important residue of the A* strand is Trp2 (*dark blue rectangle*, W2). The cadherin molecule shown is in the closed conformation—its Trp2 residue is inserted into its own core. **b** Topology diagram of the classical cadherin EC1 domain. Note that the domain consists of seven β strands. The first strand is broken into two parts, strands A* and A. Strand A* forms a contact with strand B. This interaction can be intra- or inter-molecular. **c** Schematic representation of the strand-swapping process. Only EC1 and EC2 domains are shown. In the presence of calcium ions, the closed cadherin conformation is unstable and is in equilibrium with the open conformation in which Trp2 is exposed to solvent. Two cadherin molecules in open conformation produce a strand-swap cadherin dimer. The structural model of the strand-swapped dimer (only EC1 domains of both molecules are shown) is on the right. Note that the W2 residues of both molecules in the dimer are in nearly perpendicular planes

discuss below, it is not quite clear how strand-swapping proceeds at low calcium. It is possible that some amount of *cis* strand-swap dimers can be assembled even at high calcium, but no available data suggests that these strand-swap *cis* dimers play any role in adhesion.

In conclusion, adherens junction homeostasis is a process of strand-swap dimer formation, clustering of these dimers, and stabilization and disintegration of the resulting clusters. These subjects will be discussed below. In addition, the formation of strand-swap dimers and concomitant processes should trigger signaling effects informing the cells about cell–cell contact formation. These outside-in signaling pathways make up an important but weakly explored area of cadherin adhesion that lies outside the focus of our review.

5.2.2 Unique Features of Cadherin Strand Swapping

Crystallographic, biophysical, and computational studies provide a clear understanding of the cadherin strand-swap process (Shapiro et al. 1995; Posy et al. 2008; Vendome et al. 2011; Vunnam and Pedigo 2011a). A key player in this binding reaction is the amino-terminal β strand (the A*/A strand) of the EC1 domain. This strand is followed by the residue Glu11 that anchors this strand to the EC1-EC2 interface through calcium ions. The A*/A strand contacts B and G strands of the seven β strands that constitute the core of the EC1 domain. A conserved N-terminal segment of the A*/A strand (the A* strand), which comprises residues 1–3, including Trp2, forms β-sheet hydrogen bonds with the B strand. In addition, the A* strand is linked to the rest of the molecule by a salt bridge between the N-terminal amino-group and a conserved Glu89. Trp2 is inserted in the core of the EC1 domain where it forms multiple hydrophobic bonds. In monomeric cadherin all of these contacts are intramolecular. Upon strand-swapping, the A* strand is swapped to another EC1 domain with which it forms exactly the same contacts (Fig. 5.1).

The second segment of the A*/A strand (residues 7–10, the A strand) is immobile; it is locked in place by the hydrogen bonds with the G strand. The mobile A* and immobile A strands are separated by a three-residue-long hinge region which, with few exceptions, contains two consecutive Pro residues in positions 5 and 6. This region does not make any hydrogen bonds with either strand (Vendome et al. 2011).

The two Pro residues conformationally strain the A strand between Trp2 and Glu11 (Vendome et al. 2011; Vunnam and Pedigo 2011b). Since in the presence of calcium ions, the A strand is tightly fixed to the rest of the EC1 domain, the strain can be relieved only by releasing Trp2 from its pocket. Therefore, the strain imposed by Pro5/Pro6 residues prevents the stable anchorage of Trp2 to its own EC1. Once the A* strand is relocated to another EC1, the resulting intermolecular contact is much more stable than the intramolecular one because the strain is released. Specific mutations that release the strain in the cadherin monomer, thereby stabilizing the A* strand anchorage to its own protomer, significantly reduce the affinity of strand swap binding. Thus, strand swapping is based on the instability of the A* strand that is imposed by the A strand and Ca^{2+}-binding.

This mechanism of cadherin strand-swapping has two important consequences for the assembly of adherens junctions. First, strand-swapping is a relatively slow binding process and, therefore, depends on the duration of the cadherin–cadherin encounter. Second, extracellular conditions (like temperature or ion concentrations) or interactions with other proteins that increase the A* strand instability can facilitate the strand swapping.

Regardless of the large binding interface, the strand-swap dimers are unstable. For example, the K_D of E-cadherin strand-swap dimerization is about 100 μM (Harrison et al. 2010). It suggests that lifetime of the dimers should be in the millisecond range. It has been originally proposed that the instability of strand swap dimers is based on their competition with intramolecular anchorage of the A* strand (Chen et al. 2005). However, more recent experiments clearly show that the main reason for strand-swap dimer instability is the competition with another type of cadherin dimer, the X-dimer. The inability to produce X-dimers increases the dimer's lifetime almost indefinitely (Harrison et al. 2011; Vunnan et al. 2011). As we discuss below, the X dimer requirement for the disassembly of strand-swap dimers is a very important feature of adherens junctions.

5.2.3 Cadherin X-Dimerization Maintains Strand-Swap Dimer Dynamics

One of the remarkable features of strand-swap dimers is that despite their low affinity, they are detectable by a co-immunoprecipitation assay, which typically requires much stronger interactions. Recent examinations of the strand-swap dimerization kinetics provided a clue in understanding this obvious paradox.

In addition to strand-swap dimers, another type of cadherin dimer has been reported for two-domain (EC1-EC2) E-cadherin fragments (Nagar et al. 1996; Pertz et al. 1999). The paired molecules in this dimer contact each other via interdomain calcium-binding interfaces leading to X-shaped arrangement of two molecules. Initially, this "X" mode of dimerization has been regarded as a crystal-packing artifact (Häussinger et al. 2004) since cadherin forms such dimers only upon blocking its natural amino-terminus by an N-terminal extension. However, recently obtained data unraveled the important functional significance of the cadherin X-dimer.

It was found that this extremely unstable dimer (K_D~ 900 μM) serves as a kinetically important intermediate in strand-swap dimerization (Harrison et al. 2010; Vunnam et al. 2011). Cadherin bearing a compromised X-dimer interface exhibits a slowly exchanging monomer-dimer equilibrium: monomers have very slow kinetics of strand-swap association but, once formed, dimers have extremely slow kinetics of dissociation. These experiments definitively showed that the X-dimer represents an initial encounter complex in a strand-swap binding reaction, the requirements for which had also been proposed based on the results of single molecule tracking experiments (Sivasankar et al. 2009), and also revealed a role in disassembly.

Our examination of X-dimer mutants expressed in A-431 cells suggests, however, that X-dimerization might not be so essential in cadherin strand-swapping in real cell–cell junctions (Hong et al. 2011). We have proposed that two factors enhance the production of strand-swap dimers in living cells thereby lifting the X-dimer requirement. The first factor is cadherin "presentation": in cell–cell junctions two encountering EC1 domains may be presented such that they are set for swapping. The second factor is a slow diffusion of cadherin molecules on the cell surface: each cadherin–cadherin encounter has a long enough duration to allow two A* strands to swap. Importantly, the experiments with X-dimer mutants clearly showed that strand-swap adhesive bond cannot be disassembled without its reconfiguration into X-dimer. This observation suggests that in order to disassemble adherens junctions, cadherin has to change its adhesive bond from a strand-swap to an "X" configuration. How this strand-swap-to-X-dimer transition works and whether cells can regulate this transition is an exciting avenue for future research.

The X-dimer requirement for strand-swap dimer dissociation changes our understanding of adherens junction disassembly in calcium-switch assays, which are widely used in cadherin adhesion studies. Since X-dimer formation requires the calcium-binding interface, the disruption of this interface by EDTA or other calcium chelators locks cadherin into the strand-swap configuration (Harrison et al. 2010; Vunnam et al. 2011). The strand-swap *trans* dimers may still dissociate because cell rounding triggered by calcium switch can physically disrupt the strand-swap *trans* bond. In contrast, strand-swap *cis* bonds are stabilized. This explains the fast accumulation of cadherin *cis* dimers in low calcium conditions. Furthermore, the inability of strand-swap dimers to dissociate explains another inconsistency in the field—the stability of strand-swapped dimers in co-immunoprecipitation assays. Indeed, the lysis buffers used for these experiments (Chitaev and Troyanovsky 1998; Shan et al. 2000; Ozawa 2002; Troyanovsky et al. 2007) typically contain EDTA or other calcium chelators to prevent cadherin proteolysis. The absence of calcium ions would lock cadherin into the strand-swap dimer conformation and allow dimer detection.

5.2.4 A Specific Form of Cis *Interaction Reinforces and Clusters Strand-Swap Dimers*

Theoretical studies show that at a K_D of about 100 μM, cadherin cannot self-assemble adhesive clusters (Kusumi et al. 1999; Wu et al. 2010): at such low affinity, some specific intracellular mechanisms have to assist cadherin recruitment into the adhesive clusters. However, live-imaging experiments with the tailless cadherin mutant clearly showed that the extracellular cadherin region alone can produce cadherin clusters (Hong et al. 2010). In order to do so, the cadherin extracellular region has to participate in some type of *cis* interactions that stabilizes strand-swapping and promotes clustering (Wu et al. 2010).

It had long been proposed that cadherin forms *cis* dimers and that the *cis* dimers are essential for cadherin adhesion. However, neither biochemical approaches—including cross-linking or co-immunoprecipitation assays (Troyanovsky 2005)—nor a FRET study of the recombinant cadherin ectodomains (Zhang et al. 2009), have presented compelling evidence for cadherin *cis* dimerization. The only *cis* cadherin dimers that have been detected are the strand-swap lateral dimers. But these dimers, especially prominent in low calcium conditions, apparently play no role in adhesion (Ozawa 2002).

Current data suggests that precursory cadherin *cis* dimers, if they do form, are very weak and transient. They may be maintained through the cadherin transmembrane domain (Huber et al. 1999) or through unknown intracellular interactions. Such transient *cis* dimers may be important for increasing local cadherin density. Obviously, more work remains to be done to identify such transient forms of cadherin *cis* dimerization and to assess their roles in adhesion.

While the quest for stable, precursory cadherin *cis* dimers has not yet produced any definitive results, structural analysis of the crystal packing interactions in the E-, N-, and C-cadherin crystals has revealed that strand-swapped cadherin *trans* dimers do form *cis* contacts (Boggon et al. 2002; Harrison et al. 2011). This *cis* interface comprises a nonsymmetrical interaction between the concave face of the EC1 domain of one molecule and the convex face of the EC2 domain of the partner cadherin. The EC1 *cis* binding surface is opposite to the *trans* dimer interface. The interaction is stabilized by a small hydrophobic core and several intermolecular

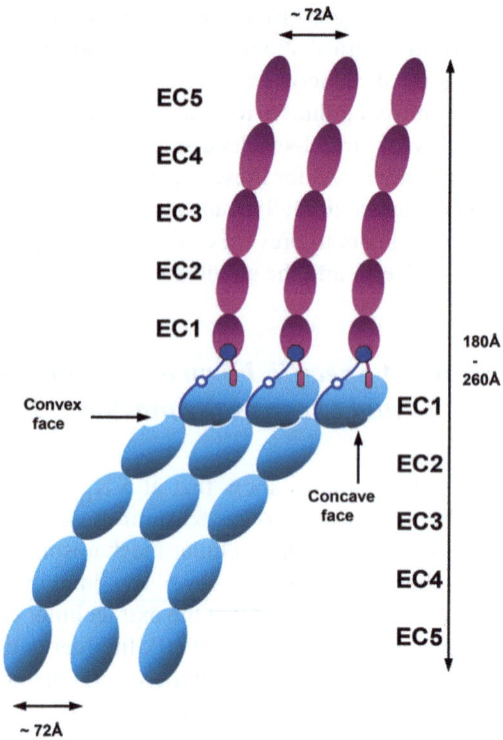

Fig. 5.2 Schematic representation of the cadherin adhesive cluster formed by *cis* and *trans* intercadherin interactions. Blue molecules are organized in a linear array through *cis* interactions. The periodicity of the array is 72 Å. Each molecule in the array is engaged in strand-swapped *trans* interactions with magenta molecules, which belong to the opposite cell. Each of these molecules is part of its own array. Note that the opposing arrays are at right angles

hydrogen bonds. Each cadherin molecule can provide simultaneously both its concave EC1 surface and its convex EC2 surface for two identical *cis* interactions. Thus, the *cis* interface arranges cadherin molecules into linear arrays (Fig. 5.2). Each cadherin in such a *cis* array also has a single *trans* bond with the cadherin located at the opposing plasma membrane. Importantly, since *trans* bonded cadherin molecules are nearly perpendicular to each other, the linear arrays of cadherin molecules on the opposing surfaces crisscross at right angles.

Computational analyses suggest that the formation of such perpendicular *trans* interacting linear arrays could be a driving force for self-assembly of cadherin adhesive clusters (Wu et al. 2011). To demonstrate this experimentally, two point mutations (V81D/L175D) that destroy the hydrophobic core of the *cis* interface were introduced into E-cadherin and the molecular structure of the resulting mutant and its recruitment into adherens junctions were studied (Harrison et al. 2011). This work showed that the *cis* interface is functional—its inactivation completely abolished adherens junction assembly. Importantly, the mutant is still able to produce *trans* dimers, but the resulting dimers are much less stable. Because of this, the cadherin *cis* mutant can be recruited into cell–cell contacts by a "diffusion trap" mechanism (Perez et al. 2008). However, the junctions formed by the *cis* mutant are extremely transient and unstable.

Taken together, these data illuminate the important role of *cis* interactions in junction formation. The *cis* interactions are too weak to be detected in solution and are not expected to produce stable *cis* dimers on an extrajunctional cell surface. However, in cooperation with *trans* interactions, they produce stable and ordered adhesive structures. Importantly, strand-swap *trans* bonds are stable only once they are interconnected by *cis* interactions. For cadherin to exit such structures the strand-swap *trans* bond must be exchanged for the X-dimer *trans* bond. Intercadherin *cis* interactions can also be significantly distorted by cadherin interactions with cytoplasmic proteins, other transmembrane proteins and the cytoskeleton. In theory, these additional components can increase or decrease stability of the adhesive bonds by adding new levels of cadherin–cadherin *cis* interactions or by preventing the ectodomain *cis* bond formation, respectively. The contribution of these elements to adherens junction assembly is discussed in the next section.

5.3 From Adhesive Clusters to Adherens Junctions

5.3.1 Evidence for Reorganization

The data discussed above shows that cadherin adhesive clusters may self-assemble through a combination of *trans* and *cis* interactions. In the resulting clusters cadherin molecules are organized in linear arrays. The intercadherin distance in an array is about 7.2 nm. Indeed, electron microscopy examination of desmosomes, which consist of close relatives of classical cadherins, desmosomal cadherins, revealed periodical organization of intercellular rod-like structures, approximately 7 nm apart (Al-Amoudi et al. 2007). Paradoxically, no signs of such an organized structure have been detected in EM studies of adherens junctions (Hirokawa and Heuser 1981; Miyaguchi 2000).

A study of adherens junctions of chicken retinal pigment epithelium, using quick-freeze, deep-etch electron microscopy (Miyaguchi 2000), revealed no periodicity in adherens junction organization. Instead, intramembrane particles, approximately 7.7 nm in diameter, were irregularly packed within the inner face of the membrane. While it is impossible to exclude that the linear cadherin arrangement was lost in this study during EM sample preparation, strikingly, the average density of the intramembrane particles was only 700 per μM^2. The number of the rod-like intermembrane structures that apparently corresponds to cadherin molecules was even smaller (approximately 100 rods per μM^2). Similar low density of the rod-like intermembrane structures was evident in the study of cell–cell junctions in intestinal epithelium using a quick-freeze, deep-etch, rotary-replication technique (Hirokawa and Heuser 1981). Such densities are much less that the predicted density of cadherin molecules in the cadherin adhesion clusters (~17,000 molecules per μM^2) (Harrison et al. 2011). Even in desmosomes, the dense and ordered cadherin organization was found only in their specific "hyperadhesive" state that is maintained by intracellular signaling (Garrod and Kimura 2008, see below).

To reconcile these EM observations with the structural data described above, one may propose that the formation of the ordered adhesive clusters is a transient process, which is immediately followed by their internal reconfiguration into more loose structures. Another possibility is that other transmembrane or intracellular proteins associated with cadherin distort the assembly of organized cadherin clusters. In any scenario, the reconfiguration of cadherin adhesive clusters into adherens junctions should include multiple cycles of cadherin adhesive dimer assembly and disassembly that, as discussed above, require strand-swap to X-dimer transitions. The requirement of this transition for cluster remodeling may explain why X-dimer interface cadherin mutants induce a dramatic dominant negative effect on cadherin adhesion in epithelial cells (Harrison et al. 2010; Hong et al. 2011). Cadherin cluster remodeling may also explain a very rapid turnover of cadherin molecules in adherens junctions (de Beco et al. 2009; Hong et al. 2010). Another piece of circumstantial evidence of the reconfiguration of cadherin clusters is the high pleomorphism of adherens junctions with respect to their morphology and protein composition (Meng and Takeichi 2009). The most prominent is the difference between apical adherens junctions (*zonulae adhaerentes*) and spot-like adherens junctions (*puncta adhaerentia*) present on the lateral (bounded) cell surfaces. The apical adherens junctions typically associate with a group of cytosolic actin-binding proteins such as vinculin, VASP, and EPLIN (Meng and Takeichi 2009). Their transmembrane adhesive domain also co-associates with another transmembrane adhesion receptor, nectin (Okabe et al. 2004; see also below). In contrast, lateral spot-like junctions do not exhibit association with these proteins, while they are also interconnected to the actin cytoskeleton. This interconnection is important for their basal to apical flow (Kametani and Takeichi 2007). Collectively, this evidence, while circumstantial, suggests that the self-assembly of the cadherin adhesive clusters is only a first step in adherens junction assembly.

The reconfiguration of the adhesive clusters or their assembly modifications could be, in theory, very important to how adherens junctions mature. This process could reconfigure uniformly packed cadherins into cell type-specific clusters. One may propose that cadherin cluster reconfiguration is mediated through additional types of intercadherin interactions, anchorage of cadherin to the cytoskeleton, and, finally, via interactions with other adhesion proteins, such as nectins or JAMs. While currently too little information is available to describe detailed mechanisms of cluster reconfiguration, we briefly outline the main possible driving forces of this process below.

5.3.2 Potential Role of Catenins

A linear array of cadherin molecules, which is formed during cadherin clustering, brings the intracellular cadherin tails into proximity. Such specific arrangement of the cadherin-catenin complexes on the intracellular face of plasma membrane may initiate new binding reactions that are too weak to be detected in solution using regular in vitro binding assays. Moreover, these inter-catenin interactions may induce specific conformational changes in catenin molecules, which, in turn, may open or establish new binding interfaces.

One such potentially important interaction resulting in α-catenin dimerization was observed in the crystal lattice of the α-catenin VH2 domain (Yang et al. 2001; Pokutta et al. 2002). The binding interface of this dimer is localized within the C-terminal four-helix bundle (residues 507–632) of this domain. The dimer is formed by the perpendicular packing of helices E and H against their counterparts. Dimerization of VH2 domain in solution mediated by this interface was also detected using a cross-linking assay. Importantly, the α-catenin region involved in this dimerization exactly corresponds to the adhesion modulation domain, which had been mapped by experiments with cadherin-α-catenin chimeric proteins (Imamura et al. 1999). The authors of this work showed that the chimera consisting of a β-catenin-uncoupled mutant of E-cadherin and an α-catenin VH2 domain mediates aggregation of cadherin-deficient L cells. It has been proposed that this adhesion modulation domain is involved in cadherin clustering. Importantly, since the paired catenin molecules in the dimer are in an antiparallel orientation, this dimerization is unlikely to occur in the parallel arrays of catenins that could be formed in the process of cadherin cluster self-assembly described above. Therefore, for a VH2 domain dimerization interface to be used, the linear cadherin arrays need to be broken and the entire cluster must be reorganized in a particular way.

Another potential α-catenin-dependent mechanism for remodeling the cluster is the binding of α-catenin VH3 domain to the actin filaments. A similar mechanism has been shown to be important in focal adhesions. It was shown that the α-catenin relative, vinculin, forms dimers through its VH3 domain (Bakolitsa et al. 1999; Johnson and Craig 2000; Janssen et al. 2006). Importantly, the dimerization of the VH3 domain of vinculin is proposed to be triggered by its binding to F-actin. The model suggests that actin filaments may be directly involved in molecular organization of vinculin-containing structures. Therefore, α-catenin-mediated reconfiguration of adherens junctions may also involve the actin cytoskeleton.

Finally, very interesting intermolecular interactions were detected in the p120 crystal lattice (Ishiyama et al. 2010). Here, the cadherin-p120 complexes were found to be arranged into linear head-to-tail oligomers with ~6 nm periodicity, which is close to periodicity of cadherin in the self-assembled arrays. Interestingly, residues of both, E-cadherin and p120, are involved in this interaction. Its most crucial feature is the conserved p120 residue W363: it is positioned within the paired p120 molecule, in a hydrophobic cleft between Arm repeats six and seven.

This secondary, catenin-based lateral ligation of cadherin molecules may have two consequences. First, it may reinforce the cadherin cluster if the ligation is compatible with *cis* interactions between cadherin ectodomains. Alternatively, if they are not, such interactions may change the position of cadherins in the cluster, thereby disengaging extracellular *cis* interactions. In both cases, the formation of such intracellular layers of cadherin–cadherin bonds can lift the requirement for extracellular *cis* interactions for cadherin cluster stability: strand-swapped adhesive bonds can be reinforced in the remodeled clusters by catenin-dependent inter-cadherin associations. Therefore, instead of a *cis* interface, cadherin positioning in the remodeled clusters can be determined by catenin conformations and the cytoskeleton.

The advantage of these new *cis* bonds is that they can be directly regulated by a cell signaling network.

5.3.3 Nectins

Another obvious mechanism of cadherin adhesive cluster remodeling is cadherin interaction with other transmembrane proteins that can interfere with *trans* or *cis* intercadherin interactions. Once bound to cadherin, such transmembrane proteins may induce unspecific steric clashes into the process of cadherin cluster assembly. They also may specifically target cadherin *cis*-binding interfaces, thereby weakening strand-swap *trans* bonds. In both cases, cadherin molecules bound to such transmembrane proteins would be excluded from adherens junction assembly. A possible example of this mechanism is a cadherin interaction with a large transmembrane proteolytic enzyme, γ-secretase: the complex consisting of E-cadherin and γ-secretase is mostly present within the extrajunctional lateral surface of epithelial cells (Kiss et al. 2008).

The most interesting example is another group of proteins that, through interactions with cadherin, can mediate specific distortion in the cadherin cluster assembly. The most promising candidates for such a role are the transmembrane immunoglobulin-like cell adhesion receptors, nectins. These proteins form calcium-independent adhesive clusters by their own in cadherin-deficient cells (Takahashi et al. 1999). While the mechanism of nectin clustering and adhesion is far from being clear, similar to cadherin clustering, it is, apparently, based on the self-assembly mechanism. It is suggested by two observations; (i) nectins can form *cis* and *trans* bonds (Momose et al. 2002; Narita et al. 2011), and (ii) nectin binding to the large cytosolic scaffolding protein, afadin, the only known intracellular nectin-binding partner, is not essential for nectin junction formation (Takahashi et al. 1999; Krummenacher et al. 2003).

Importantly, upon co-expression with cadherin, this nectin/afadin complex loses its independence and co-localizes with adherens junctions (Takahashi et al. 1999; Asakura et al. 1999). Whether nectin molecules produce the same *trans* and *cis* contacts in the adherens junctions, as in the cadherin-free adhesive clusters, remains to be determined. Furthermore, it is not known how and at what step these two adhesive systems interact. The function of the association between these two adhesive systems, particularly with respect to the structure of cadherin adhesion, is also unknown.

One of the possible mechanisms of interactions between cadherin and nectin adhesive systems is their intracellular association through α-catenin and afadin. This possibility is suggested by a number of observations: (i) It was shown that the intracellular C-terminal region of nectins forms a stable complex with the PDZ domain of afadin. Afadin-uncoupled mutants of nectin form adhesive clusters, which are not integrated into adherens junctions (Takahashi et al. 1999; Krummenacher et al. 2003). (ii) Afadin binds with a low affinity to the VH2 domain of α-catenin

(Tachibana et al. 2000; Pokutta et al. 2002). (iii) Experiments with cadherin-α-catenin chimeras definitively showed that this VH2 domain of α-catenin is essential for cadherin-nectin co-clustering (Tachibana et al. 2000).

It was suggested that nectins are crucial for the nucleation of cadherin adhesion (Takai et al. 2008; Sato et al. 2006). However, several observations are not consistent with this point of view. While afadin was shown to be required for the general organization of cell–cell contacts in epithelial cells, it is not essential for the assembly of the individual adherens junctions (Zhadanov et al. 1999; Ikeda et al. 1999). Vice versa, α-catenin-deficient cells can recruit cadherin into nectin-deficient, adherens junction-like structures (Tachibana et al. 2000; Troyanovsky et al. 2011). A chimeric protein consisting of a β-catenin-uncoupled cadherin mutant and the α-catenin VH3 domain, which is unable to interact with afadin, still can form junctions upon expression in cadherin-deficient L cells (Imammura et al. 1999). Finally, tailless cadherin mutants rapidly form junctions in the calcium-switch assay (Hong et al. 2010). Taken together, these observations demonstrate that nectin association with cadherin via the intracellular domain is not a key step in the formation of cadherin adhesive clusters.

The data described above show that nectin can be co-recruited with cadherin into adherens junctions and that α-catenin-afadin interactions play a role in this process. A very important and still open question is whether the interaction between these adhesive systems occurs at the level of nectin and cadherin monomers or at the level of independently pre-assembled cadherin and nectin clusters. This question is important because the entire process of cadherin cluster assembly can be distorted if adherens junctions can be assembled from cadherin-nectin *cis* dimers. For example, *cis* interactions between nectin's extracelluar domains can provide an alternative mechanism for the reinforcement of cadherin strand-swap *trans* dimers. The possibility that extrajunctional, free cadherin molecules can interact with nectin is suggested by experiments with dominant negative cadherin mutants: they destroy both cadherin and nectin adhesion (Tanaka et al. 2003). Alternatively, nectins may only be able to recognize and interact with preassembled cadherin clusters. In this scenario, independently formed cadherin and nectin clusters would associate along their periphery. In this case nectins would play a role in organizing small cadherin clusters into mature adherens junctions.

Adding even more complexity to the problem of cadherin-nectin interactions is a recent work that, using a Xenopus developmental model, suggested that cadherin and nectin molecules can interact through their extracellular regions (Morita et al. 2010). Again, whether this interaction is specific to some oligomeric forms of cadherin, remains to be studied.

5.3.4 Intercellular Distance

As discussed above, the cadherin/catenin complex interacts with a number of other molecules and structures. These interactions can significantly change not only the global distribution of cadherin clusters but also their internal organization. The changes can be cell type-specific or can be specific to the type of adherens junctions—for instance, the zonula adherens or the puncta adherens. Through regulation of the lateral alignment of cadherin molecules in the clusters, a junction can change its strength and its signaling potentials. In addition to transmembrane proteins and to intracellular bridging by catenins, the lateral alignment of cadherin in the junctions can also be controlled by the junctional intercellular distance. This variable largely depends on two opposite forces, stretching the junction by actomyosin contraction and compressing the junction by actin polymerization. Indeed, experiments performed in several laboratories demonstrate that junctional tension controls different parameters of adherens junctions including their protein composition (Ladoux et al. 2010; le Duc et al. 2010; Yonemura et al. 2010; Taguchi et al. 2011). It was proposed, therefore, that the cadherin-catenin complex is a mechanosensor that transmits force between F-actin and the cadherin adhesive bond (Yonemura et al. 2010; le Duc et al. 2010). However, it is also possible that a change in cadherin lateral alignment induced by junctional tension controls different properties of adherens junctions.

Electron microscopy shows that the distance between two adjoining plasma membranes in the junctions varies from 15–30 nm (McNutt and Weinstein 1973; Hirokawa and Heuser 1981; Drenckhahn and Franz 1986; Miyaguchi 2000). The distance between the opposite cadherin C-termini of the strand-swap cadherin dimer is 37–38 nm (Harrison et al. 2011). To be accommodated in the narrow intermembrane space, cadherin dimers, therefore, must transverse this space at an angle. Indeed, such a configuration of cadherin *trans* dimers in crystal lattice narrows the distance between the presumptive membranes to 18–25 nm. By changing the angle between cadherin and membrane, cells can potentially change *cis* and *trans* binding interfaces. Such structural changes can be crucial. Indeed, a cadherin inclination that is compatible with *cis* and strand-swap *trans* interactions would ultimately stabilize adherens junctions. In contrast, an angle that is incompatible with these interactions would result in junction disassembly. Therefore, the angle between cadherin and membrane can govern the strength of cadherin adhesion as well as the junction assembly-disassembly process.

Clear evidence for cadherin reorganization within particular adhesive structures, desmosomes, was obtained in the Garrod laboratory (Garrod and Kimura 2008). Desmosomal cadherins and classic cadherins share the same strand-swap *trans* dimerization binding site (Posy et al. 2006). However, desmosomal cadherins lack a classic cadherin-like *cis* interface, suggesting that desmosomal cadherin *trans* dimers have a specific lateral alignment (Harrison et al. 2011). Nevertheless, because of extensive structural similarities, the major principles of adhesion in adherens junctions and desmosomes may be similar. It was shown that desmosomal cadherins in desmosomes have, at least, two types of arrangements. The mature or "hyperadhesive" desmosomes are calcium-independent and exhibit a dense midline. The adjoining membranes in these desmosomes are 30 nm apart. Cryo-electron tomography of rapidly frozen epidermal desmosomes (Al-Amoudi et al. 2007) and computer modeling (Garrod et al. 2005) showed that cadherin molecules in mature desmosomes form arrays with a periodicity of 7.5 nm. Such an arrangement is very similar to that of classic cadherin in crystal lattices. In migrating cells, however, desmosomes become calcium-dependent and lose their midline and cadherin periodicity, and their intercellular space narrows to about 27 nm. This dramatic change in desmosome organization is regulated by PKCα-dependent signaling pathways (Garrod et al. 2005). Therefore, the rearrangement of cadherin molecules within adhesive structures can be a general mechanism regulating junctional dynamics and functions. Future works should address this important issue.

5.4 From Adherens Junctions to Cadherin Monomer

Cadherin-mediated adherens junctions are not static. Live imaging experiments have shown that they are in constant and directional motion (Kametani and Takeichi 2007; Hong et al. 2010). Spot-like adherens junctions are assembled in the basal area of the lateral cell surface and move in the apical direction. Reaching the apical surface, these junctions integrate into the zonula adherens. Such basal-to-apical movement of adherens junctions suggests that the adhesive bonds cementing the junction are strong enough to sustain the stress induced by this motion, which is unlikely to be completely synchronized in two neighboring cells. FRAP (fluorescence recovery after photobleaching) experiments showed that adherens junctions continuously loose and gain cadherin molecules (Yamada et al. 2005; Stehbens et al. 2006; Thoumine et al. 2006; Hong et al. 2010). This exchange of cadherin has been traditionally regarded as a result of dynamic equilibrium between junctional and extrajunctional cadherin (Kusumi et al. 1999). However, experiments performed at both cellular and molecular levels indicate that the mechanism of cadherin exchange in adherens junctions is far more complex: certain active processes continuously remove cadherin molecules from the junction. Active removal of cadherin from junctions has been suggested by the fact that ATP depletion completely stalls cadherin strand-swap dimer dynamics and rapidly blocks dimer disassembly in calcium-switch assay (Troyanovsky et al. 2006). Photoconversion of Dendra2-tagged cadherin in adherens junctions further demonstrated that cadherin molecules are locked in the adherens junctions of ATP-depleted cells (Hong et al. 2010). Importantly, both, live-cell imaging and biochemical approaches, have shown that ATP depletion does not interfere with the recruitment of the plasma membrane exposed cadherin into the junctions (Troyanovsky et al. 2006; Hong et al. 2010). This imbalance between cadherin recruitment and its release rapidly traps nearly all available cadherin in intercellular junctions. This data suggests that adhesive and lateral interactions between cadherin molecules in adherens junctions are strong enough to immobilize cadherin. To unlock cadherin from such a stable immobile state, some specific, energy-consuming processes are required. The active processes disassembling the junctions are far from clear. They can range from ATP-dependent conformational changes that destroy particular catenin-dependent intercadherin *cis* bonds, discussed in the previous section, to a more complex active process that physically removes cadherin from the junctions.

Among the possible mechanisms of the removal of cadherin molecules from the junctions is clathrin-dependent endocytosis. Indeed, broad inhibition of endocytosis by 0.4 M sucrose in A431 cells (Troyanovsky et al. 2006; Hong et al. 2010) as well as the inactivation of clathrin-dependent endocytosis by more specific inhibitors, dynasore or MiTMAB, in MDCK cells (de Beco et al. 2009) were shown to block cadherin exchange in adherens junctions. However, the process that unlocks cadherin and removes it from the junctions is clearly much more complex. For example, our attempts to prevent a release of cadherin from the junctions by clathrin depletion (Troyanovsky unpublished) or by point mutations of cadherin endocytic motifs (Hong et al. 2010) in A431 cells failed: both maneuvers blocked cadherin endocytosis but were ineffective in slowing down cadherin dynamics in the junctions. Similarly, the same inhibitors, dynasore and MiTMAB that blocked cadherin junctional turnover in MDCK cells produce little effect in MCF7 cells (de Beco et al. 2009).

As we showed recently, cadherin undergoes a strand-swap-to-X-dimer transition before exiting the adherens junction (Hong et al. 2011). This suggests that the mechanism unlocking cadherin from the junction includes the reconfiguration of the main adhesive bonds. Such reconfiguration could be the same ATP-consuming process that is detected in the ATP-depletion experiments. Apparently, the intracellular mechanisms that participate in the maturation of adherens junctions after initial cadherin clustering may play the leading role in disengagement of cadherin from the junction. How exactly this strand-swap-to-X-dimer transition is initiated and performed remains to be studied.

5.5 Perspectives and Future Directions

In our review we have highlighted recent progress in understanding the molecular mechanisms of adherens junction assembly. The data we discussed show that the initial formation of adhesive contact is based on cadherin *trans* dimerization via a strand-swapping mechanism. The resulting strand-swap dimers are unstable unless they are clustered through *cis* interactions. Despite some advances, we still have no answers to many outstanding questions. For example, virtually no data suggests whether any specific nucleation process triggers this initial cadherin clustering. Little is also known about how these initial clusters are organized into mature adherens junctions and how structural and morphological diversity of the junctions is achieved. Finally, what is the mechanism of adherens junction dynamics and disassembly? Answering these questions is a critical step in our understanding of various pathologies that are associated with abnormalities in cell–cell adhesion.

References

Adams CL, Nelson WJ (1998) Cytomechanics of cadherin-mediated cell-cell adhesion. Curr Opin Cell Biol 10:572–577

Al-Amoudi A, Díez DC, Betts MJ, Frangakis AS (2007) The molecular architecture of cadherins in native epidermal desmosomes. Nature 450:832–837

Asakura T, Nakanishi H, Sakisaka T, Takahashi K, Mandai K, Nishimura M, Sasaki T, Takai Y (1999) Similar and differential behaviour between the nectin-afadin-ponsin and cadherin-catenin systems during the formation and disruption of the polarized junctional alignment in epithelial cells. Genes Cells 4:573–581

Bakolitsa C, de Pereda JM, Bagshaw CR, Critchley DR, Liddington RC (1999) Crystal structure of the vinculin tail suggests a pathway for activation. Cell 99:603–613

Boggon TJ, Murray J, Chappuis-Flament S, Wong E, Gumbiner BM, Shapiro L (2002) C-cadherin ectodomain structure and implications for cell adhesion mechanisms. Science 296:1308–1313

Becker JW, Erickson HP, Hoffman S, Cunningham BA, Edelman GM (1989) Topology of cell adhesion molecules. Proc Natl Acad Sci U S A 86:1088–1092

Chen CP, Posy S, Ben-Shaul A, Shapiro L, Honig BH (2005) Specificity of cell-cell adhesion by classical cadherins: critical role for low-affinity dimerization through beta-strand swapping. Proc Natl Acad Sci U S A 102:8531–8536

Chitaev NA, Troyanovsky SM (1998) Adhesive but not lateral E-cadherin complexes require calcium and catenins for their formation. J Cell Biol 142:837–846

de Beco S, Gueudry C, Amblard F, Coscoy S (2009) Endocytosis is required for E-cadherin redistribution at mature adherens junctions. Proc Natl Acad Sci U S A 106:7010–7015

Drenckhahn D, Franz H (1986) Identification of actin-, alpha-actinin-, and vinculin-containing plaques at the lateral membrane of epithelial cells. J Cell Biol 102:1843–1852

Franke WW (2009) Discovering the molecular components of intercellular junctions—a historical view. Cold Spring Harb Perspect Biol 1:a003061

Garrod D, Kimura TE (2008) Hyper-adhesion: a new concept in cell-cell adhesion. Biochem Soc Trans 36:195–201

Garrod DR, Berika MY, Bardsley WF, Holmes D, Tabernero L (2005) Hyper-adhesion in desmosomes: its regulation in wound healing and possible relationship to cadherin crystal structure. J Cell Sci 118:5743–5754

Harrison OJ, Corps EM, Kilshaw PJ (2005) Cadherin adhesion depends on a salt bridge at the N-terminus. J Cell Sci 118:4123–4130

Harrison OJ, Bahna F, Katsamba PS, Jin X, Brasch J, Vendome J, Ahlsen G, Carroll KJ, Price SR, Honig B, Shapiro L (2010) Two-step adhesive binding by classical cadherins. Nat Struct Mol Biol 17:348–357

Harrison OJ, Jin X, Hong S, Bahna F, Ahlsen G, Brasch J, Wu Y, Vendome J, Felsovalyi K, Hampton CM, Troyanovsky RB, Ben-Shaul A, Frank J, Troyanovsky SM, Shapiro L, Honig B (2011) The extracellular architecture of adherens junctions revealed by crystal structures of type I cadherins. Structure 19:244–256

Häussinger D, Ahrens T, Aberle T, Engel J, Stetefeld J, Grzesiek S (2004) Proteolytic E-cadherin activation followed by solution NMR and X-ray crystallography. EMBO J 23:1699–1708

He W, Cowin P, Stokes DL (2003) Untangling desmosomal knots with electron tomography. Science 302:109–113

Hirokawa N, Heuser JE (1981) Quick-freeze, deep-etch visualization of the cytoskeleton beneath surface differentiations of intestinal epithelial cells. J Cell Biol 91:399–409

Hong S, Troyanovsky RB, Troyanovsky SM (2010) Spontaneous assembly and active disassembly balance adherens junction homeostasis. Proc Natl Acad Sci U S A 107:3528–3533

Hong S, Troyanovsky RB, Troyanovsky SM (2011) Cadherin exits the junction by switching its adhesive bond. J Cell Biol 192:1073–1083

Huber O, Kemler R, Langosch D (1999) Mutations affecting transmembrane segment interactions impair adhesiveness of E-cadherin. J Cell Sci 112:4415–4423

Ikeda W, Nakanishi H, Miyoshi J, Mandai K, Ishizaki H, Tanaka M, Togawa A, Takahashi K, Nishioka H, Yoshida H, Mizoguchi A, Nishikawa S, Takai Y (1999) Afadin: a key molecule essential for structural organization of cell-cell junctions of polarized epithelia during embryogenesis. J Cell Biol 146:1117–1132

Imamura Y, Itoh M, Maeno Y, Tsukita S, Nagafuchi A (1999) Functional domains of alpha-catenin required for the strong state of cadherin-based cell adhesion. J Cell Biol 144:1311–1322

Ishiyama N, Lee SH, Liu S, Li GY, Smith MJ, Reichardt LF, Ikura M (2010) Dynamic and static interactions between p120 catenin and E-cadherin regulate the stability of cell-cell adhesion. Cell 141:117–128

Janssen ME, Kim E, Liu H, Fujimoto LM, Bobkov A, Volkmann N, Hanein D (2006) Three-dimensional structure of vinculin bound to actin filaments. Mol Cell 21:271–281

Johnson RP, Craig SW (2000) Actin activates a cryptic dimerization potential of the vinculin tail domain. J Biol Chem 275:95–105

Kametani Y, Takeichi M (2007) Basal-to-apical cadherin flow at cell junctions. Nat Cell Biol 9:92–98

Kim SA, Tai CY, Mok LP, Mosser EA, Schuman EM (2011) Calcium-dependent dynamics of cadherin interactions at cell-cell junctions. Proc Natl Acad Sci U S A 108:9857–9862

Kiss A, Troyanovsky RB, Troyanovsky SM (2008) p120-catenin is a key component of the cadherin-gamma-secretase supercomplex. Mol Biol Cell 19:4042–4050

Kitagawa M, Natori M, Murase S, Hirano S, Taketani S, Suzuki ST (2000) Mutation analysis of cadherin-4 reveals amino acid residues of EC1 important for the structure and function. Biochem Biophys Res Commun 271:358–363

Klingelhöfer J, Troyanovsky RB, Laur OY, Troyanovsky S (2000) Amino-terminal domain of classic cadherins determines the specificity of the adhesive interactions. J Cell Sci 113:2829–2836

Klingelhöfer J, Laur OY, Troyanovsky RB, Troyanovsky SM (2002) Dynamic interplay between adhesive and lateral E-cadherin dimers. Mol Cell Biol 22:7449–7458

Krummenacher C, Baribaud I, Eisenberg RJ, Cohen GH (2003) Cellular localization of nectin-1 and glycoprotein D during herpes simplex virus infection. J Virol 77:8985–8999

Kusumi A, Suzuki K, Koyasako K (1999) Mobility and cytoskeletal interactions of cell adhesion receptors. Curr Opin Cell Biol 11:582–590

Ladoux B, Anon E, Lambert M, Rabodzey A, Hersen P, Buguin A, Silberzan P, Mège RM (2010) Strength dependence of cadherin-mediated adhesions. Biophys J 98:534–542

Laur OY, Klingelhöfer J, Troyanovsky RB, Troyanovsky SM (2002) Both the dimerization and immunochemical properties of E-cadherin EC1 domain depend on Trp(156) residue. Arch Biochem Biophys 400:141–147

le Duc Q, Shi Q, Blonk I, Sonnenberg A, Wang N, Leckband D, de Rooij J (2010) Vinculin potentiates E-cadherin mechanosensing and is recruited to actin-anchored sites within adherens junctions in a myosin II-dependent manner. J Cell Biol 189:1107–1115

McNeill H, Ryan TA, Smith SJ, Nelson WJ (1993) Spatial and temporal dissection of immediate and early events following cadherin-mediated epithelial cell adhesion. J Cell Biol 120:1217–1226

McNutt NS, Weinstein RS (1973) Membrane ultrastructure at mammalian intercellular junctions. Prog Biophys Mol Biol 26:45–101

Meng W, Takeichi M (2009) Adherens junction: molecular architecture and regulation. Cold Spring Harb Perspect Biol 1:a002899

Miyaguchi K (2000) Ultrastructure of the zonula adherens revealed by rapid-freeze deep-etching. J Struct Biol 132:169–178

Momose Y, Honda T, Inagaki M, Shimizu K, Irie K, Nakanishi H, Takai Y (2002) Role of the second immunoglobulin-like loop of nectin in cell-cell adhesion. Biochem Biophys Res Commun 293:45–49

Morita H, Nandadasa S, Yamamoto TS, Terasaka-Iioka C, Wylie C, Ueno N (2010) Nectin-2 and N-cadherin interact through extracellular domains and induce apical accumulation of F-actin in apical constriction of Xenopus neural tube morphogenesis. Development 137:1315–1325

Nagar B, Overduin M, Ikura M, Rini JM (1996) Structural basis of calcium-induced E-cadherin rigidification and dimerization. Nature 380:360–364
Narita H, Yamamoto Y, Suzuki M, Miyazaki N, Yoshida A, Kawai K, Iwasaki K, Nakagawa A, Takai Y, Sakisaka T (2011) Crystal Structure of the cis-Dimer of Nectin-1: implications for the architecture of cell-cell junctions. J Biol Chem 286:12659–12669
Nose A, Tsuji K, Takeichi M (1990) Localization of specificity determining sites in cadherin cell adhesion molecules. Cell 1990 61:147–155
Okabe N, Ozaki-Kuroda K, Nakanishi H, Shimizu K, Takai Y (2004) Expression patterns of nectins and afadin during epithelial remodeling in the mouse embryo. Dev Dyn 230:174–186
Ozawa M (2002) Lateral dimerization of the E-cadherin extracellular domain is necessary but not sufficient for adhesive activity. J Biol Chem 277:19600–19608
Perez TD, Tamada M, Sheetz MP, Nelson WJ (2008) Immediate-early signaling induced by E-cadherin engagement and adhesion. J Biol Chem 283:5014–5022
Pertz O, Bozic D, Koch AW, Fauser C, Brancaccio A, Engel J (1999) A new crystal structure, Ca^{2+} dependence and mutational analysis reveal molecular details of E-cadherin homoassociation. EMBO J 18:1738–1747
Pokutta S, Drees F, Takai Y, Nelson WJ, Weis WI (2002) Biochemical and structural definition of the l-afadin- and actin-binding sites of alpha-catenin. J Biol Chem 277:18868–18874
Posy S, Shapiro L, Honig B (2008) Sequence and structural determinants of strand swapping in cadherin domains: do all cadherins bind through the same adhesive interface? J Mol Biol 378:954–968
Shan WS, Tanaka H, Phillips GR, Arndt K, Yoshida M, Colman DR, Shapiro L (2000) Functional cis-heterodimers of N- and R-cadherins. J Cell Biol 148(3):579–590
Sato T, Fujita N, Yamada A, Ooshio T, Okamoto R, Irie K, Takai Y (2006) Regulation of the assembly and adhesion activity of E-cadherin by nectin and afadin for the formation of adherens junctions in Madin-Darby canine kidney cells. J Biol Chem 281:5288–5299
Shapiro L, Fannon AM, Kwong PD, Thompson A, Lehmann MS, Grübel G, Legrand JF, Als-Nielsen J, Colman DR, Hendrickson WA (1995) Structural basis of cell-cell adhesion by cadherins. Nature 374:327–337
Sivasankar S, Zhang Y, Nelson WJ, Chu S (2009) Characterizing the initial encounter complex in cadherin adhesion. Structure 17:1075–1081
Stehbens SJ, Paterson AD, Crampton MS, Shewan AM, Ferguson C, Akhmanova A, Parton RG, Yap AS (2006) Dynamic microtubules regulate the local concentration of E-cadherin at cell-cell contacts. J Cell Sci 119:1801–1811
Tachibana K, Nakanishi H, Mandai K, Ozaki K, Ikeda W, Yamamoto Y, Nagafuchi A, Tsukita S, Takai Y (2000) Two cell adhesion molecules, nectin and cadherin, interact through their cytoplasmic domain-associated proteins. J Cell Biol 150:1161–1176
Taguchi K, Ishiuchi T, Takeichi M (2011) Mechanosensitive EPLIN-dependent remodeling of adherens junctions regulates epithelial reshaping. J Cell Biol 194:643–656
Takahashi K, Nakanishi H, Miyahara M, Mandai K, Satoh K, Satoh A, Nishioka H, Aoki J, Nomoto A, Mizoguchi A, Takai Y (1999) Nectin/PRR: an immunoglobulin-like cell adhesion molecule recruited to cadherin-based adherens junctions through interaction with Afadin, a PDZ domain-containing protein. J Cell Biol 145:539–549
Takai Y, Ikeda W, Ogita H, Rikitake Y (2008) The immunoglobulin-like cell adhesion molecule nectin and its associated protein afadin. Annu Rev Cell Dev Biol 24:309–342
Tamura K, Shan WS, Hendrickson WA, Colman DR, Shapiro L (1998) Structure-function analysis of cell adhesion by neural (N-) cadherin. Neuron 20:1153–1163
Tanaka Y, Nakanishi H, Kakunaga S, Okabe N, Kawakatsu T, Shimizu K, Takai Y (2003) Role of nectin in formation of E-cadherin-based adherens junctions in keratinocytes: analysis with the N-cadherin dominant negative mutant. Mol Biol Cell 14:1597–1609
Thoumine O, Lambert M, Mège RM, Choquet D (2006) Regulation of N-cadherin dynamics at neuronal contacts by ligand binding and cytoskeletal coupling. Mol Biol Cell 17:862–875
Tomschy A, Fauser C, Landwehr R, Engel J (1996) Homophilic adhesion of E-cadherin occurs by a co-operative two-step interaction of N-terminal domains. EMBO J 15:3507–3514

Troyanovsky S (2005) Cadherin dimers in cell-cell adhesion. Eur J Cell Biol 84:225–233
Troyanovsky RB, Sokolov E, Troyanovsky SM (2003) Adhesive and lateral E-cadherin dimers are mediated by the same interface. Mol Cell Biol 23:7965–7972
Troyanovsky RB, Sokolov EP, Troyanovsky SM (2006) Endocytosis of cadherin from intracellular junctions is the driving force for cadherin adhesive dimer disassembly. Mol Biol Cell 17:3484–3493
Troyanovsky RB, Laur O, Troyanovsky SM (2007) Stable and unstable cadherin dimers: mechanisms of formation and roles in cell adhesion. Mol Biol Cell 18:4343–4352
Troyanovsky RB, Klingelhöfer J, Troyanovsky SM (2011) α-Catenin contributes to the strength of E-cadherin-p120 interactions. Mol Biol Cell 22:4247–4255
Vasioukhin V, Fuchs E (2001) Actin dynamics and cell-cell adhesion in epithelia. Curr Opin Cell Biol 13:76–84
Vendome J, Posy S, Jin X, Bahna F, Ahlsen G, Shapiro L, Honig B (2011) Molecular design principles underlying β-strand swapping in the adhesive dimerization of cadherins. Nat Struct Mol Biol 18:693–700
Vunnam N, Flint J, Balbo A, Schuck P, Pedigo S (2011) Dimeric states of neural- and epithelial-cadherins are distinguished by the rate of disassembly. Biochemistry 50:2951–2961
Vunnam N, Pedigo S (2011a) Calcium-induced strain in monomer promotes dimerization in neural-cadherin. Biochemistry 50(39):8437–8444. doi:10.1021/bi200902s
Vunnam N, Pedigo S (2011b) Prolines in βA-sheet of neural cadherin act as a switch to control the dynamics of the equilibrium between monomer and dimer. Biochemistry 50:6959–6965
Wu Y, Jin X, Harrison O, Shapiro L, Honig BH, Ben-Shaul A (2010) Cooperativity between trans and cis interactions in cadherin-mediated junction formation. Proc Natl Acad Sci U S A 107:17592–17597
Yamada S, Pokutta S, Drees F, Weis WI, Nelson WJ (2005) Deconstructing the cadherin-catenin-actin complex. Cell 123:889–901
Yang J, Dokurno P, Tonks NK, Barford D (2001) Crystal structure of the M-fragment of alpha-catenin: implications for modulation of cell adhesion. EMBO J 20:3645–3656
Yonemura S, Wada Y, Watanabe T, Nagafuchi A, Shibata M (2010) Alpha-catenin as a tension transducer that induces adherens junction development. Nat Cell Biol 12:533–542
Zhadanov AB, Provance DW Jr, Speer CA, Coffin JD, Goss D, Blixt JA, Reichert CM, Mercer JA (1999) Absence of the tight junctional protein AF-6 disrupts epithelial cell-cell junctions and cell polarity during mouse development. Curr Biol 9:880–888
Zhang Y, Sivasankar S, Nelson WJ, Chu S (2009) Resolving cadherin interactions and binding cooperativity at the single-molecule level. Proc Natl Acad Sci U S A 106:109–114

Part III
How AJs Interface with Other Cellular Machinery

Chapter 6
The Cytoskeleton and Classical Cadherin Adhesions

Siew Ping Han and Alpha S. Yap

Abstract This chapter discusses the biochemical and functional links between classical cadherin adhesion systems and the cytoskeleton. Cadherins are best understood to cooperate with the actin cytoskeleton, but there is increasing evidence for the role of junctional microtubules in regulating cadherin biology. Cadherin adhesions and the junctional cytoskeleton are both highly dynamic systems that undergo continual assembly, turnover and remodeling, and yet maintain steady state structures necessary for intercellular adhesion. This requires the functional coordination of cadherins and cadherin-binding proteins, actin regulatory proteins, organizers of microtubule assembly and structure, and signaling pathways. These components act in concert to regulate junctional organization in response to extracellular forces and changing cellular contexts, which is essential for intercellular cohesion and tissue integrity.

6.1 Introduction

It has long been appreciated that classical cadherin molecules work in close cooperation with the cytoskeleton. This is best understood for the actin cytoskeleton, and there is emerging evidence that cadherin adhesions also interact with microtubules. Cadherin-based contacts are characteristically enriched in actin filaments (F-actin), ranging from the apical junctional complexes of simple epithelia to synaptic connections between neurons. Indeed, it is often useful to think of cadherin biology as arising from the coordinated action of adhesion and cytoskeletal systems, integrated together into a dynamic ensemble. In this chapter we discuss how this integration may come about. We focus on understanding the physical linkages between the cadherin molecular complex and the actin cytoskeleton; the cellular mechanisms that regulate actin filament homeostasis and organization at junctions; and how adhesion and the cytoskeleton are functionally coordinated. In closing we will also briefly review modes of interaction between cadherins and microtubules.

A. S. Yap (✉) · S. P. Han
Division of Molecular Cell Biology, Institute for Molecular Bioscience,
The University of Queensland, St. Lucia, 4072 Brisbane, Queensland, Australia
e-mail: a.yap@uq.edu.au

T. Harris (ed.), *Adherens Junctions: From Molecular Mechanisms to Tissue Development and Disease,* Subcellular Biochemistry 60,
DOI 10.1007/978-94-007-4186-7_6, © Springer Science+Business Media Dordrecht 2012

6.2 Physical Interactions between Cadherins and the Actin Cytoskeleton

A key concept is the notion that cadherins interact physically with the actin cytoskeleton. In its most simple form, cadherin-associated proteins are postulated to bind directly to actin filaments, thereby anchoring the cadherin molecular complex to the cortical actin cytoskeleton (Rimm et al. 1995). Such direct interactions have been proposed to potentially stabilize the cadherin at the cell surface, promote cadherin clustering and mechanically couple cells together.

6.2.1 Linking Cadherin Junctions and the Actin Cytoskeleton

The concept of physical association between cadherins and actin arose from several lines of evidence. First, the integrity of cadherin-based junctions and that of the actin cytoskeleton are interdependent. Cadherin interactions are perturbed when the actin cytoskeleton is disrupted. This is observed when cells are treated with drugs such as cytochalasins and latrunculins, that perturb actin filament integrity, and also when key actin regulators are disrupted by molecular genetic approaches, such as mutation, or depletion by RNAi or homologous recombination (Angres et al. 1996; Baum and Perrimon 2001; Mangold et al. 2011; Verma et al. 2004). Conversely, the assembly of cadherin-based interactions can reorganize the actin cytoskeleton and, as we shall see, can induce the activation of signals to the cytoskeleton and recruit key cytoskeletal regulators to cadherin adhesions (Gloushankova et al. 1997; Kovacs et al. 2002b; Vasioukhin et al. 2000).

Second, cadherin junctions are typically enriched in F-actin and often display what appear to be cortical interactions of actin filaments. These apparent interactions take different morphological forms. In simple epithelial cells grown in culture, E-cadherin often concentrates in an apical ring that has been interpreted to represent the zonula adherens (Meng et al. 2008; Otani et al. 2006; Smutny et al. 2010, 2011). Here, F-actin also accumulates in a ring-like structure, that likely comprises bundles whose ends terminate in the junctions themselves (Scott et al. 2006; Yonemura et al. 1995). Analysis of fluorescence recovery after photobleaching of GFP-actin at junctions also suggests that newly-formed actin filaments may extend from the junctional membrane to the apical actin ring (Kovacs et al. 2011). In contrast, the early *C. elegans* embryo is covered by an epithelial layer, the hypodermis, which is characterized by prominent actin cables that run around the circumference of the embryo, terminating in cadherin-enriched junctions between the cells (Costa et al. 1998; Kwiatkowski et al. 2010). Analagous connections between actin cables and junctions are seen in epithelia of the early *Drosophila* embryo. In particular, an actomyosin network has been identified in the apical zones of such epithelia as they undergo apical constriction during gastrulation and strands from this actomyosin network extend into cadherin junctions (Martin et al. 2010).

The dynamic behaviours of these cytoskeletal-junction couplings also provide evidence of physical connection. For example, during apical constriction in early *Drosophila* embryos, the actomyosin networks display contractile pulses that lead to the repositioning of cadherin junctions (Martin et al. 2009). Conversely, loss of cadherin or catenins in the *C. elegans* hypodermis causes the circumferential actin cables to lose connection to the cell–cell junctions (Costa et al. 1998; Kwiatkowski et al. 2010). Recent studies in cell culture also demonstrate that cadherins can participate in mechanotransduction (Ladoux et al. 2010; le Duc et al. 2010; Liu et al. 2010), sensing forces exerted on sites where cadherins are engaged in homophilic interactions. Such mechanotransduction requires an intact actin cytoskeleton (le Duc et al. 2010), consistent with functional coupling of cadherin receptors and cytoskeleton. Observations such as these strongly imply some form of physical connection between components of the actin cytoskeleton and cell–cell junctions themselves.

The third line of evidence derives from a range of biochemical and cellular studies that suggest the capacity for cadherin complexes to interact with, or to regulate, the cortical actin cytoskeleton. Early hints included the observation that a pool of cadherin was relatively resistant to extraction in non-ionic buffers (Ayalon et al. 1994) and evidence that cadherin could associate with DNAase1 (Ozawa et al. 1990). Detergent insolubility has often been used as an index of cytoskeletal association, although this is not specific (Parton and Simons 2007); while DNAase1 binds to F-actin, albeit with lower affinity than to monomeric G-actin (Weber et al. 1994). At a cellular level, adhesion of cells to cadherin adhesive ligands immobilized on substrata or beads can induce the accumulation of F-actin at adhesion sites (Hara et al. 2004; Helwani et al. 2004; Kovacs et al. 2002b; Kraemer et al. 2007; Lambert et al. 2002), suggesting some capacity of cadherin to locally reorganize cortical actin. Overall, these several lines of data strongly suggest that some functionally significant populations of cadherins interact physically with actin filaments.

6.2.2 Molecular Linkages Between Cadherins and Actin Filaments

What then is the molecular basis for this physical association? For much of the past two decades, attention has focused on the potential role of α-catenin in mediating this linkage (Fig. 6.1a). Early interest in this notion arose from several lines of data. The detergent insolubility of cadherin-catenin complexes, interpreted as an index of F-actin-association, was decreased in deletion mutants of cadherin that were unable to associate with α-catenin (Nagafuchi and Takeichi 1988, 1989; Ozawa et al. 1990). Moreover, cadherin mutants unable to bind catenins were poorly adhesive and expression of α-catenin conferred adhesiveness on α-catenin-deficient cells (Watabe et al. 1994). Further, fusion of α-catenin could confer adhesive capacity upon cadherin mutants lacking the catenin-binding domain (Nagafuchi et al. 1994). Taken

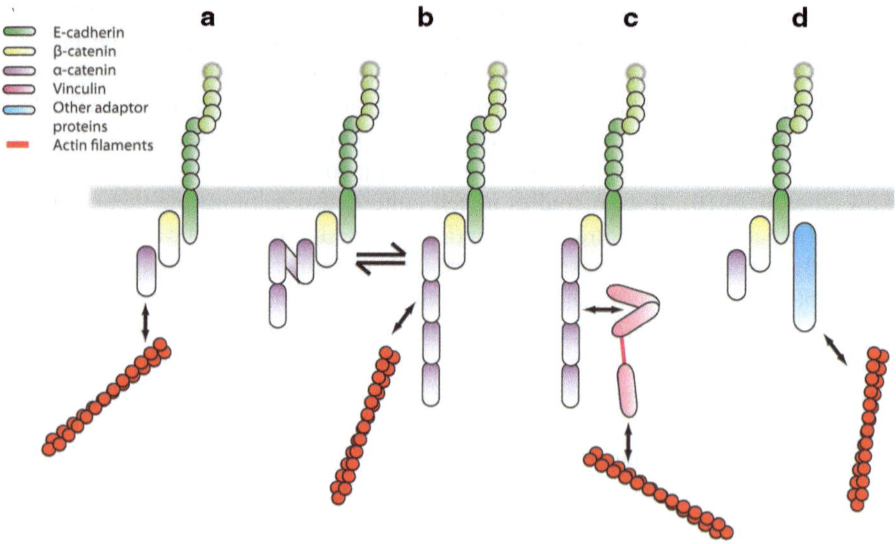

Fig. 6.1 Evolving models of cadherin-actin linkages. **a** Cadherin binds β-catenin, which binds α-catenin, which in turn binds F-actin. **b** Activation of α-catenin and exposure of hidden binding sites is required for F-actin binding. **c** α-catenin recruits adaptor proteins that bind F-actin. **d** Other adaptor proteins mediate linkage between cadherins and F-actin

with evidence that actin integrity also contributes to cadherin adhesion (Matsuzaki et al. 1990), this correlation between detergent-solubility, adhesion and catenin-binding suggested that the catenins might be involved in cytoskeletal association.

However, it is difficult to rigorously test whether proteins bind to actin filaments within intact cells. Instead, more definitive evidence of actin-binding capacity relies on assessing whether purified proteins can interact *in vitro*. Thus a key insight came from the demonstration that recombinant α(E)-catenin could interact with purified F-actin (Rimm et al. 1995). These experiments used differential centrifugation to demonstrate that bacterially-expressed α-catenin, but not β-catenin, cosedimented with purified actin filaments, a classic assay for F-actin binding. Subsequently, the F-actin-binding domain was mapped to the C-terminus of α-catenin (Pokutta and Weis 2002) and the capacity for mammalian α(E)-catenin to bind actin was extended to α-catenin from *Dictyostelium discoideum* (Kwiatkowski et al. 2010), suggesting that it is a conserved property of α-catenin. In contrast, the cytoplasmic tail of E-cadherin did not bind actin filaments *in vitro* (Rimm et al. 1995). However, while β-catenin alone did not bind F-actin, it co-sedimented with actin filaments when mixed with α-catenin. This suggested a simple quaternary model where α-catenin, scaffolded to cadherin through β-catenin, bound directly to F-actin, thereby indirectly linking cadherins to actin filaments (Rimm et al. 1995).

It should be noted that this model was extrapolated from a series of binary protein–protein interactions: cadherin cytoplasmic tail/β-catenin, β-catenin/α-catenin and α-catenin/F-actin. Confidence in the model was shaken when it was reported that the reconstituted cadherin/β-catenin/α-catenin trimer did not bind actin filaments (Yamada et al. 2005). Instead, the capacities of α-catenin to bind to F-actin or to cadherin/β-catenin appeared to be mutually exclusive. Actin filament binding and the regulation of filament dynamics was a property of free α-catenin, functioning as a homodimer (Drees et al. 2005), but this property was lost when α-catenin incorporated into the cadherin -β-catenin complex.

It is not possible to resolve this conundrum at the present time. Failure to identify a quaternary cadherin/catenin/F-actin complex *in vitro* has been interpreted as evidence that cadherins may not interact, directly or indirectly, with actin filaments (Weis and Nelson 2006). However, as noted above, there is strong biological evidence to suggest some form of physical association. Moreover, some of the biology continues to place α-catenin at a key point in this interaction. Of note, genetic studies indicate that both the β-catenin-binding domain of α-catenin and its F-actin-binding domain are necessary for its function in the early *Dictyostelium* embryo (Kwiatkowski et al. 2010). One possibility is that binding of α-catenin to F-actin may require some "activation" step that is not readily recapitulated with recombinant proteins (Fig. 6.1b). Of note, whereas the C-terminal fragment of *Dictyostelium* α-catenin could bind F-actin, this did not occur with the full length recombinant protein (Kwiatkowski et al. 2010), suggesting that actin-binding might sometimes be masked in the full-length α-catenin protein. Indeed, it has been suggested that α-catenin may be tension-sensitive, with cryptic sites that are revealed by contractile tension (Yonemura et al. 2010).

Alternatively, α-catenin may mediate cadherin-actin binding, but through other associated proteins (Fig. 6.1c). These include EPLIN (Abe and Takeichi 2008; Taguchi et al. 2011), α-actinin (Knudsen et al. 1995; Yamada et al. 2005), and vinculin (Bakolitsa et al. 2004; Watabe-Uchida et al. 1998; Weiss et al. 1998; Yamada et al. 2005), all of which can directly bind actin filaments. Thus recombinant EPLIN could confer actin-binding to the cadherin/β-catenin/ α-catenin complex (Abe and Takeichi 2008). Moreover, the recruitment of EPLIN to cell–cell junctions requires both α-catenin (Abe and Takeichi 2008) and mechanical tension (Taguchi et al. 2011). Similarly, the recruitment of vinculin to cadherin junctions requires both α-catenin and contractility, perhaps reflecting mechanosensitivity of α-catenin itself (Yonemura et al. 2010). An emerging theme here is the role that contractile tension plays in the junctional recruitment of these proteins.

Finally, other actin-binding proteins exist at cadherin junctions that have the potential to link cadherins to actin filaments (Fig. 6.1d). These include cortactin (Helwani et al. 2004; Ren et al. 2009b; El Sayegh et al. 2004) and myosin VI (Maddugoda et al. 2007; Mangold et al. 2011), proteins which bear actin-binding domains and that have been found to interact with cadherin by co-immunoprecipitation analysis (El Sayegh et al. 2004; Geisbrecht and Montell 2002; Maddugoda et al. 2007; Mangold et al. 2011). It should be noted that these alternative models are not mutually exclusive. Thus it is possible that cadherin uses different molecular mechanisms to physically bind actin filaments, perhaps in different biological contexts.

6.3 The Regulation of Actin Filament Homeostasis and Organization at Cadherin Adhesions

The close proximity of actin filaments and cadherin junctions begs the question of how this specialized cytoskeleton is generated. Of note, actin filaments are intrinsically dynamic, including those found at cell–cell junctions. This is well-described for the perijunctional actin cytoskeleton seen in polarized epithelia, which is characterized by a ring-like accumulation of F-actin just proximate to the cadherin junctions. However, assessment of molecular turnover by fluorescence recovery after photobleaching of GFP-tagged actin indicates that these stable structures are comprised of filaments that turnover with half-lives of approximately 30–40 s (Kovacs et al. 2011; Mangold et al. 2011; Yamada et al. 2005). Thus the morphologically-stable ring of F-actin seen at established cell–cell junctions is generated from intrinsically dynamic components. Ultimately, understanding the molecular basis for such cytoskeletal homeostasis requires the characterization of mechanisms that mediate filament assembly and turnover; regulate filament organization; and ensure that these events occur at cadherin adhesions with fidelity in space and time. Many molecular regulators that potentially contribute to these processes have been identified at cadherin adhesive junctions, although our understanding is far from comprehensive. It should also be noted that much of what we currently understand has been derived from studies in epithelia, where the predominant cadherin is E-cadherin. It is possible that many of these principles will be conserved in other actin-enriched cadherin junctions, such as N-cadherin-mediated neuronal synapses. However, this remains to be thoroughly experimentally addressed. Accordingly, much of this review will focus on epithelial models, with reference to other systems where appropriate.

6.3.1 Actin Filament Homeostasis at Cadherin Adhesions

Actin exists in two forms: monomeric globular actin (G-actin), which can self-assemble into polymeric filamentous actin (F-actin) (Fig. 6.2a). Actin is an ATPase that cycles between an ATP-bound state, which favours polymerization, and an ADP-bound state, which favours depolymerization. F-actin is also a polar structure, whose ends differ in their propensity to add or lose monomer. These ends can be defined morphologically at the ultrastructural level by the orientation of S1 myosin fragments when they bind to actin filaments, which generates a "pointed" and "barbed" end to each filament (Moore et al. 1970). Filament polymerization is favoured at the barbed end. The initial nucleation of G-actin into dimers or trimers, however, is energetically unfavourable, but once nucleated, elongation of F-actin is rapid and diffusion-limited (Pollard et al. 2000; Drenckhahn and Pollard 1986; Mullins et al. 1998). Actin filament nucleation thus represents a key rate-limiting step in actin assembly and a number of proteins have been identified that catalyse this step,

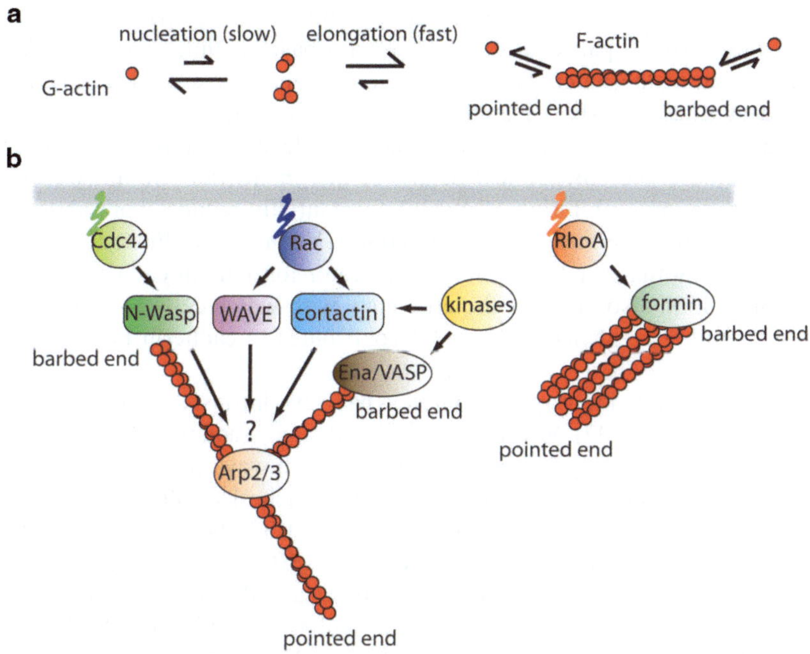

Fig. 6.2 Actin dynamics. **a** G-actin and F-actin exist in a state of dynamic equilibrium, wherein the rate of actin polymerisation is determined by the slow nucleation step. G-actin incorporation is favoured at barbed ends, generating a polar actin filament. **b** Key actin regulators are controlled by multiple and overlapping pathways. Arp2/3 nucleates dendritic networks, formins nucleate and elongate linear filament bundles, and Ena/VASP promotes filament growth

typically in response to cell signaling (Insall and Machesky 2009). This allows actin assembly to be tightly regulated within cells.

The demonstration in epithelia that filaments in the junctional cytoskeleton undergo turnover implies that these junctions are likely to be sites of actin nucleation. Indeed, this has been demonstrated directly (Braga et al. 1997; Kovacs et al. 2002b, 2011; Vasioukhin et al. 2000; Verma et al. 2004; Yamada et al. 2005). Consistent with this, two major actin nucleators have been identified at E-cadherin junctions (Fig. 6.2b).

Arp2/3-mediated Actin Nucleation The Arp2/3 complex consists of seven subunits—ARP2 and ARP3, which together resemble an actin dimer, and accessory proteins ARPC1–5 (Goley and Welch 2006). When the Arp2/3 complex is bound to an existing filament, known as the mother filament, the ARP2 and ARP3 subunits nucleate the formation of a new filament, known as the daughter filament (Amann and Pollard 2001; Mullins et al. 1998). Binding of Arp2/3 to the mother filament requires conformational changes of the ARP2 and ARP3 subunits, and also two actin subunits on the mother filament (Rouiller et al. 2008). The geometry of binding to mother filaments means that Arp2/3 tends to generate branched networks of filaments that can subsequently undergo reorganization (Goley and Welch 2006). Arp2/3 participates in many actin-based cellular processes, including regulation of the junctional cytoskeleton (Bernadskaya et al. 2011). Arp2/3 is found at E-cadherin-based epithelial cell–cell contacts (Kovacs et al. 2011) and, indeed, can interact with the E-cadherin molecular complex (Kovacs et al. 2002b), although the biochemical details of this interaction are not well understood. Furthermore, inhibition of Arp2/3 substantially reduces both actin nucleation and F-actin levels at cadherin junctions (Verma et al. 2004; Kovacs et al. 2011) indicating that Arp2/3 is a major nucleator for actin at cadherin junctions. Arp2/3 inhibition also impedes the efficiency with which cells formed junctions with one another, evidence that Arp2/3 contributes to cadherin biology (Verma et al. 2004).

Of note, though, the Arp2/3 complex has low intrinsic nucleation activity (Higgs and Pollard 1999). Within cells, it is activated—and thereby made responsive to physiological regulation—by signaling pathways that exert their action on Arp2/3 by intermediary proteins (Insall and Machesky 2009; Machesky et al. 1999; Padrick and Rosen 2010; Rottner et al. 2010). The best-understood of these nucleation-promoting factors (NPFs) are the WASP/WAVE protein family (Pollitt and Insall 2009). Each of these proteins bears a multi-domain C-terminus that contains an Arp2/3-binding acidic domain and G-actin-binding WH2 domains (this multi-domain C-terminal region is called the VCA (or WCA) domain). In the current model of Arp2/3 activation, the VCA domain of activated N-WASP binds and causes conformational changes in Arp2/3, while also delivering G-actin to the ARP2 and ARP3 subunits in the primed complex (Campellone and Welch 2010). Ubiquitously-expressed members of this NPF family, WAVE2 and N-WASP, are found at epithelial junctions where they influence both the junctional cytoskeleton and cadherin junctions themselves (Ivanov et al. 2005; Kovacs et al. 2011; Yamazaki et al. 2007). Of note, each of these NPFs is itself regulated by cell signaling, especially through the small GTPases, Rac and Cdc42, which preferentially regulate WAVE2 and N-WASP respectively (Pollitt and Insall 2009). This suggests that Arp2/3 at E-cadherin based adhesions may be regulated by these Rho-family GTPases. However, the role of N-WASP at cell contacts may not be so direct. Although N-WASP is found at cadherin-based contacts, it can be dispensable for Arp2/3-mediated actin nucleation at the epithelial zonula adherens where it regulates actin dynamics at a post-nucleation step (Kovacs et al. 2011). The precise effect of these pathways on Arp2/3 thus remains to be fully elucidated.

Formin-mediated Actin Nucleation In contrast to Arp2/3, formins are thought to mediate the nucleation of unbranched actin filaments (Chesarone et al. 2010; Kovar and Pollard 2004). They are characterised by the presence of formin homology (FH) domains, usually in combination with a G-actin binding domain (GBD) and a Dia-autoregulatory domain (DAD). Formins are autoinhibited by interactions between the GBD and DAD, which is released by Rho-family GTPase signaling, thereby linking their activity to upstream signal regulation (Li and Higgs 2003; Nezami et al. 2006; Rose et al. 2005). In addition, formins also interact with other proteins, including Src family kinases and profilin, thus acting as structural scaffolds that coordinate the recruitment of multiple factors involved in actin remodeling (Wallar and Alberts 2003). In contrast to Arp2/3, which remains at the pointed ends of daughter filaments, formins stay attached to the barbed end as filament elongation proceeds, a phenomenon known as processive barbed end association (Xu et al. 2004). Structural analysis of the yeast formin BniP suggests that a dimer of FH2 domains forms a "doughnut" that encloses the actin filament, with each FH2 domain binding two actin subunits (Otomo et al. 2005). Since only one of these contacts needs to be released to allow the incorporation of G-actin, this arrangement allows both FH2 domains to remain continuously bound while permitting filament growth.

Two members of the formin family, mDia1 and formin-1, have been described at epithelial cell–cell junctions. mDia1 was reported to localize to junctions of simple epithelial cells in a Rho-dependent fashion, whereas formin-1 localized to keratinocyte junctions in an α-catenin-dependent manner. Both contribute to junctional integrity (Carramusa et al. 2007; Kobielak et al. 2004; Sahai and Marshall 2002), however their precise role in cytoskeletal regulation at junctions has yet to be elucidated. Depletion of mDia1 decreased junctional F-actin (Carramusa et al. 2007), but nucleation was not directly measured. Nor do we yet understand what functional relationship there may be between Arp2/3 and formin-mediated actin regulation. It is possible that these nucleators are responsible for different actin populations in different contexts or different types of cells. Also, they may cooperate, for example during filopodial biogenesis when formins reportedly remodel filaments initially nucleated by Arp2/3 (Mellor 2010).

Post-nucleation Regulation of Filament Homeostasis Although nucleation is a rate-limiting step, filament turnover is also regulated at various stages following nucleation. Of note, actin filament elongation is inhibited when barbed ends are capped and thus the regulated extension of nascent filaments likely requires the action of anti-capping mechanisms (Bear 2008). Interestingly, formins themselves can permit filament growth in the presence of capping proteins (Kovar 2006). Other potential anti-capping proteins are the Enabled/Vasodilator-stimulated phosphoprotein (Ena/Vasp) proteins, which support actin assembly although they do not act as nucleators (Bear et al. 2002). Mena and VASP are members of this family that localize to epithelial cadherin junctions and disruption of their function perturbs actin organization and assembly (Scott et al. 2006; Vasioukhin et al. 2000). Ena/VASP proteins may also regulate filament homeostasis by other mechanisms, such as promoting the delivery of actin monomers to barbed ends (Hansen and Mullins 2010). These must also be coordinated with the action of proteins that inhibit assembly, such as the inhibition of Arp2/3 by cytosolic α-catenin (Drees et al. 2005), and proteins that sever filaments for turnover and recycling (Ono 2007). The structural order and dynamic responsiveness of actin cytoskeletons thus are a testament to the complex regulatory mechanisms that ensure spatially localized and temporally coordinated activity of the actin assembly machinery.

6.3.2 Regulating Filament Organization at Cadherin Adhesions

The biological impact of the actin cytoskeleton depends on the manner in which its constituent filaments are organized. Key parameters include the length of filaments, whether they are isolated or cross-linked and, in the latter case, the geometry of cross-linking. Broadly speaking, cross-linked filaments may be found in networks or bundles. Coupled to filament dynamics, these patterns of organization can influence the force generation by and mechanical properties of the cytoskeleton.

As we have already seen, patterns of filament organization may be influenced even at the initial step of nucleation (Achard et al. 2010; Michelot and Drubin 2011). Thus, Arp2/3 is commonly thought to generate branched networks of filaments, whereas formins are thought to work on unbranched filaments (Mellor 2010). The emerging evidence that both sets of nucleators may act at cadherin junctions would suggest that different patterns of filament organization may exist. However, this has been difficult to rigorously assess. The small diameter of microfilaments (approximately 8–10 nm) means that their detailed organization can currently only be captured by electron microscopy (Svitkina et al. 2003). By this means, dense F-actin bundles have been identified just adjacent to the zonula adherens in simple, polarized epithelia (Miyaguchi 2000); this likely corresponds to the apical rings of F-actin that can be seen by light microscopy. However, more detailed understanding of filament organization at cell–cell junctions remains elusive, as it is a major technical challenge to evaluate finer filament organization, notably networks, within intact cells. Filament architecture can be readily perturbed by fixation conditions (Small et al. 1999) and three-dimensional analysis through electron tomography (Urban et al. 2010) is best suited to flat, thin structures such as lamellipodia.

It is unlikely, though, that the pattern of filament organization is set solely by the nucleators involved. Filament re-organization, including de-branching and re-bundling, occurs in dynamic networks (Svitkina et al. 2003). Some of the actin regulators found at cadherin adhesions are likely to exert significant effects by regulating aspects of dynamic organization. For example, the phosphoprotein, cortactin, is a junctional component that associates with E-cadherin or N-cadherin in epithelia and fibroblasts, respectively, and supports the junctional actin cytoskeleton (El Sayegh et al. 2005; El Sayegh et al. 2004; Helwani et al. 2004; Ren et al. 2009b). At the molecular level, cortactin can cooperate with Arp2/3 to influence several steps in actin assembly (Ren et al. 2009a). Although it has only weak intrinsic NPF activity (Uruno et al. 2001), it can bind N-WASP (Weaver et al. 2002), and also acts post-nucleation to stabilize nascent filament branches (Weaver et al. 2001). At another post-nucleation stage, Ena/VASP proteins have been implicated in converting branched actin networks into bundles (Svitkina et al. 2003). Ena/VASP proteins bear F-actin-binding sites as well as an oligomerization motif that promotes tetramerization (Krause et al. 2003). The presence of multiple F-actin-binding sites in Ena/Vasp tetramers allows them to participate in filament bundling in addition to filament elongation, which could influence filament organization at junctions.

6.3.3 Controlling the Spatio-Temporal Fidelity of Actin Regulation at Cadherin Adhesions

Actin filament dynamics and organization are spatially regulated within cells. This general principle applies to cadherin adhesive contacts, as exemplifed by the observation that actin nucleation occurs at the junctions (Kovacs et al. 2002a; Kovacs et al. 2011; Vasioukhin et al. 2000; Yamada et al. 2005) and, more generally, that regulatory proteins implicated in modulating the junctional actin cytoskeleton concentrate at those junctions (Helwani et al. 2004; Kovacs et al. 2011; Scott et al. 2006; Taguchi et al. 2011). Additionally, the junctional recruitment of cytoskeletal regulators can display temporal or contextual specificity (Maddugoda et al. 2007). For example, EPLIN (Taguchi et al. 2011) and myosin VI (Maddugoda et al. 2007) appear to be recruited to junctions as cell–cell contacts mature and are, indeed, necessary for that junctional maturation. Cellular mechanisms must then exist to confer stringent spatial and contextual (or temporal) specificity upon such recruitment.

Protein–protein interactions provide a powerful general mechanism to determine the spatial specificity of cytoskeletal regulators. Indeed, many proteins that regulate the junctional actin cytoskeleton can interact with cadherins themselves, as identified by co-immunoprecipitation studies. Such intermolecular interactions provide an attractive mechanism to ensure spatial specificity of recruitment to adhesive junctions. Given the range of actin regulators that can be found at cadherin adhesive junctions, it is not surprising that multiple mechanisms are implicated in their recruitment. Many such interactions are likely to be indirect. Of note, α-catenin binds via β-catenin and proteins such as EPLIN (Abe and Takeichi 2008) and vinculin (Watabe-Uchida et al. 1998; Weiss et al. 1998) are recruited to cadherin by association with α-catenin, although vinculin can also bind β-catenin (Peng et al. 2010). Proteins such as Arp2/3 (Kovacs et al. 2002b) and cortactin (El Sayegh et al. 2005; El Sayegh et al. 2004) co-immunoprecipitate with cadherin, but the molecular basis for these interactions has yet to be elucidated. Other proteins, such as Ena/VASP (Gates et al. 2009; Scott et al. 2006) and myosin II (Smutny et al. 2010), concentrate at cadherin adhesive junctions, such as the epithelial zonula adherens, but are not known to associate with the cadherin molecular complex. It is possible that their localization occurs through protein–protein interactions that are too weak to be identified by co-immunoprecipitation analysis or by association with other proteins found in the region of, but not directly linked to, cadherin receptors. It is likely that multiple intermolecular interactions will contribute to recruiting different actin regulators to cadherin adhesions.

Importantly, many interactions between actin regulators and cadherins are not constitutive but instead are subject to cellular regulation. Indeed, many key signaling pathways that regulate the cytoskeleton are found at cadherin adhesions. These include small GTPases of the Rho family, lipid kinases such as PI3-kinase, and protein kinases, such as Src and Abelson (Niessen et al. 2011). These have the capacity to control many aspects of cytoskeletal activity at junctions. Rho-family GTPases—Rho, Rac and Cdc42—play major roles in the recruitment and coordination of the plethora of adaptor proteins at cadherin contacts. Rho-family GTPases localize to cadherin junctions, along with the guanine nucleotide exchange factors and GTPase-activating proteins that respectively activate and inhibit them (Braga and Yap 2005). The Rho-family GTPases influence multiple parameters of actin regulation, for instance, Rho regulates myosin II-dependent contractility (Vicente-Manzanares et al. 2009), Rac recruits cortactin (Weed et al. 1998) and activates WAVE (Yamazaki et al. 2007), while Cdc42 activates N-WASP (Miki et al. 1998). The functional outcome of Rho-family GTPase activation at cadherin junctions depends on cellular context, and different GTPases can act cooperatively or antagonistically. For instance, Rac1 and Cdc42 can both be activated by E-cadherin ligation and in turn activate Arp2/3 recruitment and F-actin accumulation, with Rac1 being the dominant player (Kraemer et al. 2007). Conversely, Rac stimulates the translocation of p190RhoGAP to cadherin junctions, where it interacts with p120-catenin and inhibits Rho, thereby suppressing contractility and maintaining junctional stability (Wildenberg et al. 2006). Also, the effects of Rho-family GTPases on actin remodeling are dependent on the level of activation, and intercellular adhesion is impaired both by dominant-negative and dominant–active forms of Rac and Cdc42 (Chu et al. 2004).

Phosphotyrosine signaling is another key regulatory pathway at cadherin-based contacts, which are enriched in tyrosine-phosphorylated proteins, as well as tyrosine kinases and phosphatases. An example of this is c-Src signaling which can support E-cadherin-based adhesion (McLachlan et al. 2007). c-Src phosphorylates and modulates the functions of a range of junctional actin regulatory proteins, including cortactin (Ren et al. 2009b) and vinculin (Ito et al. 1983). Like Rho-family GTPases, the effect of c-Src on junctions depends on the level of activity, with low and very high levels of c-Src reducing E-cadherin adhesion (McLachlan et al. 2007). Thus, the presence of multiple adaptor and regulatory proteins provides targets for signals that modulate adhesion and actin organization in a graded fashion (Pokutta and Weis 2002).

Furthermore, several key junctional signals, including Rac, Cdc42, PI3-kinase and Src protein kinase, can be acutely stimulated by cadherin ligation (Kovacs et al. 2002b; Kraemer et al. 2007; McLachlan et al. 2007). Thus cadherin adhesion is likely to be a major upstream cue that stimulates these signaling pathways when cells assemble or remodel contacts with one another. Such cadherin-activated cell signaling, coupled to intermolecular associations, may then provide a way in which cytoskeletal processes that involve an ensemble of proteins, such as actin assembly, can be efficiently directed to cadherin adhesion sites.

It should be emphasized, however, that individual signals can have multiple targets at cadherin adhesions. For example, potential Rac targets identified at junctions include nucleation promoting factors (WAVE2), scaffolding and adaptor proteins (cortactin), and other signals (PI3-kinase) (Bosse et al. 2007; Weed et al. 1998; Yamazaki et al. 2007). One reason for this may be to support a cytoskeletal process by coordinating several different molecules that need to work together in an ensemble. As well, different signals may be activated at cadherin junctions at different times, or they may overlap. As an example of the former, distinct zones of

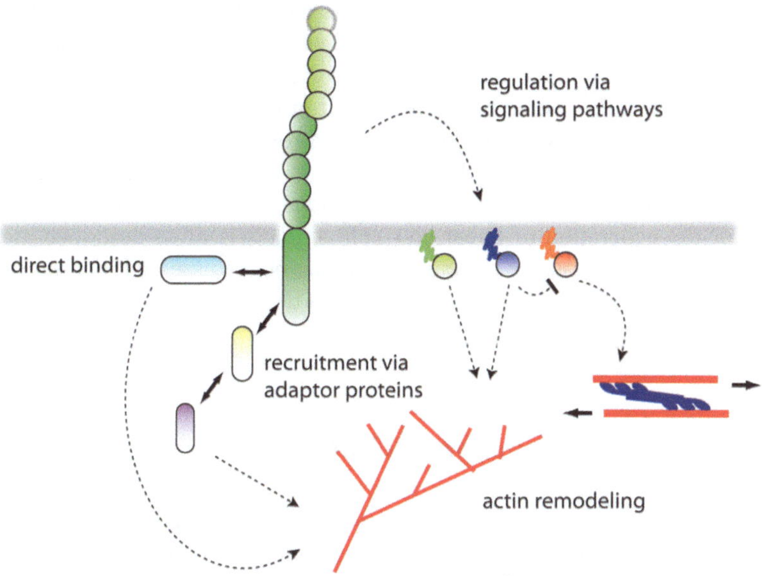

Fig. 6.3 Spatial coordination of actin regulation at cadherin adhesions. Cadherin ligation can recruit actin regulators through direct binding to cadherin, indirect binding via cadherin-binding adaptor proteins, or signaling pathways that mediate localization and activation of proteins involved in actin remodeling and actomyosin contractility

Rac and Rho signaling have been observed as isolated cells first form contacts with one another. Rac is thought to support lamellipodial activity, which may promote the spreading of cells upon one another, whereas Rho drives contractility (Yamada and Nelson 2007). In contrast, both myosin IIA and myosin IIB concentrate at the epithelial zonula adherens, but whereas myosin IIA is recruited in response to Rho signaling, myosin IIB depends on Src and Rap1 signaling (McLachlan and Yap 2011; Smutny et al. 2010). Ultimately, cytoskeletal regulation at cadherin junctions is likely to be a form of systems biology, where the functional outcome is determined by the architecture of the network of signals and effectors that are active in specific biological contexts (Fig. 6.3).

6.4 Functional Implications of Cadherin-Actin Coordination

How then do these diverse physical and functional interactions between cadherins and the actin cytoskeleton contribute to cadherin biology? Here it is useful to emphasize that classical cadherins have many biological outcomes. Many are morphogenetic and involve the regulation of cell–cell interactions to mediate tissue organization. Increasingly, however, cadherins are implicated in cell growth control, population dynamics and differentiation. How these may intersect with cadherin-actin coordination remains to be determined. Further, it should be noted that the term "adhesion" is commonly used to signify different things depending on the experimental context. Formally, it refers to the ability to resist detachment force, something that can be more readily measured in cell culture or with recombinant proteins than can be assessed in whole organisms or tissues. The term is also used to denote morphological changes where a change in resistance to detachment is often inferred, but not directly measured. For example, close apposition of cells to one another or the generation of linear cell–cell contacts is often described as "tight" adhesion, although resistance to detachment may not be assessed. For our purposes we will focus on cellular-level processes with a discussion of broader implications as appropriate.

6.4.1 Surface Organization of Cadherins

Classical cadherins do not distribute diffusely at cell–cell contacts. Instead, cadherins often organize laterally into a range of structures of varying size. These range from lateral clusters that typically exist on the sub-micron scale (Angres et al. 1996; Yap et al. 1997) to large structures that include "spot" adherens junctions in early *Drosophila* embryonic epithelia (Tepass and Hartenstein 1994) and the zonula adherens (Farquhar and Palade 1963) that can ring the apical junctions of many polarized epithelia (Kovacs et al. 2011; Meng et al. 2008; Smutny et al. 2010). These different patterns of lateral organization are likely to be interrelated. The apical zonula adherens often appears to comprise a ring of smaller clusters (Kovacs et al. 2011) and lateral clusters may move into the apical junctions through basal-to-apical cortical flow (Kametani and Takeichi 2007). Similarly, in *Drosophila* embryos adherens junctions are formed from the coalescence of cadherin clusters, and the apical positioning of these clusters is perturbed by actin depolymerization (McGill et al. 2009).

Multiple mechanisms contribute to these diverse forms of lateral organization, including *cis*-interactions between cadherin ectodomains (Harrison et al. 2011) and association with catenins (Yap et al. 1998). These are likely to constitute the precursors for larger-scale lateral organization that may then be built with contribution from the actomyosin cytoskeleton. Thus, lateral clustering in cells is modulated by several cytoskeletal proteins, that include Ena/VASP proteins (Scott et al. 2006) and myosin IIA (Shewan et al. 2005; Smutny et al. 2010). The integrity of the apical zonula adherens in simple mammalian epithelia requires contributions of myosin II isoforms (Smutny et al. 2010) and N-WASP, the latter acting through a post-nucleation pathway to stabilize nascent junctional actin filaments (Kovacs et al. 2011). In early embryonic *Drosophila* epithelia small patches of F-actin stabilize spot adherens junctions, whereas interactions with a dynamic F-actin network restrict their lateral diffusion, a process that involves α-catenin (Cavey et al. 2008). Thus the actin cytoskeleton is likely to regulate different aspects of cadherin organization through multiple effectors.

Lateral organization through these different cytoskeletal contributions may have several impacts on cadherin biology. Firstly, the binding strength of individual cadherin ectodomains is relatively weak, but lateral clustering enhances cooperative interactions that strengthen adhesion (Angres et al. 1996; Yap et al. 1997; Yap et al. 1998; Harrison et al. 2011). Potentiation of clustering may then account for the adhesive strengthening associated with myosin II (Shewan et al. 2005; Smutny et al. 2010). Furthermore, strengthening of E-cadherin-mediated intercellular adhesion requires an intact actin cytoskeleton (Angres et al. 1996), which demonstrates the importance of actin organization in E-cadherin binding interactions (Chu et al. 2004).

Secondly, the organization of clusters into larger junctional structures may further amplify adhesion. In particular, the zonula adherens has been implicated in morphogenetic events where changes in cell shape are coordinated with adhesive interactions to shape tissue organization. This is exemplified by the process of apical constriction during gastrulation, where contractile force exerted on apical adhesive junctions leads to the folding of epithelial sheets (Sawyer et al. 2010). This process requires functional adhesion (Dawes-Hoang et al. 2005), actomyosin contractility (Martin et al. 2009; Sawyer et al. 2011) and coupling of the adhesion system to the actomyosin cytoskeleton (Sawyer et al. 2009).

6.4.2 Mechanotransduction at Cell–Cell Junctions

As exemplified by the case of apical constriction, cell–cell junctions are sites where mechanical force is exerted by cells upon other cells. Beyond requiring mechanisms that physically couple adhesion systems to the cytoskeleton, it is increasingly apparent that the integration of adhesion and cytoskeleton does not simply generate apparatuses that passively resist force. Instead, they are sites of mechanotransduction, which incorporate active mechanisms that sense force being exerted upon the cells and instigate proportional responses to those forces (Schwartz 2010; Schwartz and DeSimone 2008).

Formally, mechanotransduction must involve receptors that allow cells to sense forces being exerted upon them, mechanisms that are activated to transduce the physical stimulus into molecular signals, and cytoskeletal response to those signals. Cell–cell adhesion molecules play central upstream roles in sensing force. Thus, E-cadherin has recently been identified as a force transduction mechanism in epithelial cells (Ladoux et al. 2010; le Duc et al. 2010; Liu et al. 2010; Maruthamuthu et al. 2011), whereas in vascular endothelial cells, VE-cadherin serves as an adaptor to link PE-CAM, an Ig superfamily adhesion molecule, to the VEGF receptor, which serves to elicit downstream cell signaling (Tzima et al. 2005). It is likely that many cell–cell adhesion molecules participate, directly or indirectly, in force-sensing.

The signaling responses to force are likely to include conventional intracellular signal transduction pathways (Liu et al. 2010) as well as force-induced changes in protein conformation. Of note, in the latter case stretching forces may induce conformational changes in adaptor proteins to expose cryptic binding sites, and thereby alter protein–protein interactions (del Rio et al. 2009). This latter model is especially interesting as it carries the capacity for local forces on cytoskeletal proteins to directly alter functionally relevant protein–protein interactions. Amongst cadherin-associated cytoskeletal proteins, putative tension transducers include α-catenin and vinculin. Vinculin is known to exist in an auto-inhibited folded conformation and must be "opened" for full biological activity (Carisey and Ballestrem 2011; Ziegler et al. 2006). Of note, vinculin has been implicated in E-cadherin mediated mechanotransduction, although its precise role remains to be elucidated (le Duc et al. 2010). Cryptic epitopes have also been identified for α-catenin, which are revealed in response to actomyosin contractility (Yonemura et al. 2010). Further, such conformational changes correlate with the recruitment of vinculin (Yonemura et al. 2010), which is known to bind α-catenin (Watabe-Uchida et al. 1998; Weiss et al. 1998).

Finally, besides modulating the force-dependent recruitment of actin regulators to cadherin junctions, changes in actomyosin contractility likely play a major role in the coordination of the cytoskeletal response to mechanical stimulation. Cadherin junctions concentrate non-muscle myosins in both invertebrate and vertebrate tissues (Bertet et al. 2004; Fernandez-Gonzalez et al. 2009; Miyake et al. 2006; Sawyer et al. 2009; Smutny et al. 2010). Contractile force generated by non-muscle myosin II increases adhesion stiffness, and produces a countering force in response to the initiating external mechanical signal (Smutny and Yap 2010). In addition, myosin II can cross-link and anchor F-actin bundles to each other, which stabilizes the actin cytoskeleton and allows it to resist mechanical disruption (Mangold et al. 2011; Vicente-Manzanares et al. 2009). While the mechanistic details remain elusive, it is likely that myosin activity is calibrated in response to the magnitude of force applied to the membrane and sensed by cadherin, which may involve regulation of Rho signaling and myosin phosphorylation.

6.4.3 Regulation of Cadherin Trafficking and Surface Stability

The surface expression of classical cadherins ultimately reflects a complex itinerary of membrane traffic to and from the cell surface, that is extensively reviewed elsewhere (Baum and Georgiou 2011; Bryant and Stow 2004; Niessen et al. 2011). The actin cytoskeleton can affect many steps in membrane traffic (Kaksonen et al. 2006) but its capacity to influence cadherin traffic is incompletely understood. Current evidence suggests roles that can potentially promote or inhibit cadherin function. Thus, actin integrity was necessary to inhibit clathrin-dependent E-cadherin endocytosis (Izumi et al. 2004), a process that also requires the actin-binding protein IQGAP and the small GTPases, Rac and Cdc42, signals that can be activated by cadherin homophilic ligation itself (Kovacs et al. 2002a; Noren et al. 2001). This suggests that actin regulation, perhaps in response to cadherin signaling itself, can stabilize cadherin at the cell surface by inhibiting its internalization. Conversely, both Cdc42 and N-WASP have been implicated in promoting cadherin internaliza-

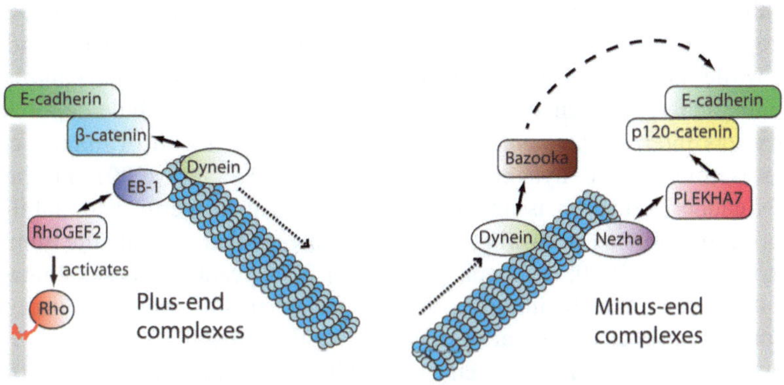

Fig. 6.4 Cadherin junctions and microtubules are physically linked and regulated by protein complexes at both plus- and minus ends

tion and turnover in Drosophila tissues (Georgiou et al. 2008; Leibfried et al. 2008). It is probable that local regulation of the actin cytoskeleton will exert diverse impacts on cadherin trafficking that depend closely on biological context, as it does for the many pathways that mediate membrane internalization in general (Howes et al. 2010; Kaksonen et al. 2006; Mayor and Pagano 2007).

6.5 Cadherins and Microtubules—an Emerging Model

While this chapter has primarily focused on the links between cadherin-based adhesions and the actin cytoskeleton, there is a recent and fast-growing body of research demonstrating the critical role microtubules play in cadherin junction regulation. Like actin microfilaments, microtubules are self-assembling polymers with defined polarities. α- and β-tubulin subunits are incorporated at plus ends and disassembled at minus ends, and each end binds distinct sets of regulatory proteins known as plus and minus end-binding proteins, respectively (Schroer 2001). In addition, motor proteins such as kinesin (plus end-directed) and dynein (minus end-directed) transport vesicles and protein complexes along microtubule tracks, or, when anchored, exert forces upon the microtubules themselves (Akhmanova et al. 2009).

Both plus and minus ends have been shown to associate with cadherin junctions and regulate their formation and maturation (Meng et al. 2008; Stehbens et al. 2006). A number of junctional protein complexes have been implicated in this process (Fig. 6.4). β-catenin can bind dynein, thereby potentially capturing and tethering microtubules at junctions, and inhibition of dynein disrupts cadherin adhesion assembly (Ligon and Holzbaur 2007; Ligon et al. 2001). Similarly, p120-catenin can bind PLEKHA7, which in turn binds Nezha, a minus-end binding protein, and inhibition of these proteins leads to loss of E-cadherin at the zonula adherens (Meng et al. 2008). Besides these direct associations between adhesion components and microtubule-binding proteins, the interaction of dynein with the *Drosophila* polarity regulator Bazooka is important for junctional orientation (Harris and Peifer 2005), while inhibition of plus end dynamics reduces E-cadherin accumulation at the zonula adherens (Stehbens et al. 2006). Similarly, microtubule depolymerization disrupts intercellular contacts, and conversely formation of mature contacts modulates microtubule dynamics (Waterman-Storer et al. 2000). Furthermore, E- or N-cadherin ligation is necessary and sufficient to recruit and stabilize microtubules, possibly by activating signaling pathways that regulate the minus-ends (Chausovsky et al. 2000; Stehbens et al. 2006). Thus, cadherin-based contacts and microtubules are linked by multiple regulatory mechanisms, which may act concurrently or within specific cellular or developmental contexts.

How does the cross-talk between cadherin junctions and microtubules affect cellular organization? Firstly, the physical binding between junctional components and microtubules may constitute a form of mechanical coupling that directly transmits forces between the zonula adherens and the microtubule cytoskeleton (Ligon et al. 2001; Meng et al. 2008). Secondly, microtubules deliver junctional components to the ZA, both by acting as tracks for vesicular transport, and also by trapping and concentrating freely diffusing proteins in the vicinity of junctions (Akhmanova et al. 2009). Thirdly, cadherin contacts regulate microtubule stability and act as cues for microtubule orientation (Chausovsky et al. 2000; den Elzen et al. 2009), which can produce global effects on cellular growth and organization.

6.6 Summary

In summary, the structural and functional linkages between cadherins and cytoskeletal components are complex and dynamic. These are subject to precise spatial and temporal regulation, which is important for coordinating the recruitment of specific ensembles of proteins required under different cellular contexts. Importantly, cadherin signaling shapes the actin cytoskeleton via the localized recruitment of actin regulators, while the actin cytoskeleton determines cadherin organization and mobility—and something similar may occur for the microtubule cytoskeleton. Thus, cadherins and the cytoskeleton are intricately coordinated systems which act in concert to regulate adhesion and junctional organization.

Acknowledgments Both authors were financially supported by the National Health and Medical Council of Australia.

References

Abe K, Takeichi M (2008) EPLIN mediates linkage of the cadherin-catenin complex to F-actin and stabilizes the circumferential actin belt. Proc Natl Acad Sci U S A 105:13–19

Achard V, Martiel J-L, Michelot A, Guerin C, Reymann A-C, Blanchoin L, Boujemaa-Paterski R (2010) A "Primer"-based mechanism underlies branched actin filament network formation and motility. Curr Biol 20:423–428

Akhmanova A, Stehbens SJ, Yap AS (2009) Touch, grasp, deliver and control: functional cross-talk between microtubules and cell adhesions. Traffic 10:268–274

Amann KJ, Pollard TD (2001) Direct real-time observation of actin filament branching mediated by Arp2/3 complex using total internal reflection fluorescence microscopy. Proc Natl Acad Sci U S A 98:15009–15013

Angres B, Barth A, Nelson WJ (1996) Mechanism for transition from initial to stable cell-cell adhesion: Kinetic analysis of E-cadherin-mediated adhesion using a quantitative adhesion assay. J Cell Biol 134:549–557

Ayalon O, Sabanai H, Lampugnani MG, Dejana E, Geiger B (1994) Spatial and temporal relationships between cadherins and PECAM-1 in cell-cell junctions of human endothelial-cells. J Cell Biol 126:247–258

Bakolitsa C, Cohen DM, Bankston LA, Bobkov AA, Cadwell GW, Jennings L, Critchley DR, Craig SW, Liddington RC (2004). Structural basis for vinculin activation at sites of cell adhesion. Nature 430:583–586

Baum B, Georgiou M (2011) Dynamics of adherens junctions in epithelial establishment, maintenance, and remodeling. J Cell Biol 192:907–917

Baum B, Perrimon N (2001) Spatial control of the actin cytoskeleton in Drosophila epithelial cells. Nat Cell Biol 3:883–890

Bear JE (2008) Follow the monomer. Cell 133:765–767

Bear JE, Svitkina TM, Krause M, Schafer DA, Loureiro JJ, Strasser GA, Maly IV, Chaga OY, Cooper JA, Borisy GG, Gertler FB (2002). Antagonism between Ena/VASP proteins and actin filament capping regulates fibroblast motility. Cell 109:509–521

Bernadskaya YY, Patel FB, Hsu H-T, Soto MC (2011) Arp2/3 promotes junction formation and maintenance in the Caenorhabditis elegans intestine by regulating membrane association of apical proteins. Mol Biol Cell 22:2886–2899

Bertet C, Sulak L, Lecuit T (2004) Myosin-dependent junction remodelling controls planar cell intercalation and axis elongation. Nature 429:667–671

Bosse T, Ehinger J, Czuchra A, Benesch S, Steffen A, Wu X, Schloen K, Niemann HH, Scita G, Stradal TEB, Brakebusch C, Rottner K (2007) Cdc42 and phosphoinositide 3-kinase drive Rac-mediated actin polymerization downstream of c-met in distinct and common pathways. Mol Cell Biol 27:6615–6628

Braga VMM, Yap AS (2005) The challenges of abundance: epithelial junctions and small GTPase signalling. Curr Opin Cell Biol 17:466–474

Braga VMM, Machesky LM, Hall A, Hotchin NA (1997) The small GTPases rho and rac are required for the establishment of cadherin-dependent cell-cell contacts. J Cell Biol 137:1421–1431.

Bryant DM, Stow JL (2004) The ins and outs of E-cadherin trafficking. Trends Cell Biol 14:427–434

Campellone KG, Welch MD (2010) A nucleator arms race: cellular control of actin assembly. Nat Rev Mol Cell Biol 11:237–251

Carisey A, Ballestrem C (2011) Vinculin, an adapter protein in control of cell adhesion signalling. Eur J Cell Biol 90:157–163

Carramusa L, Ballestrem C, Zilberman Y, Bershadsky AD (2007) Mammalian diaphanous-related formin Dia1 controls the organization of E-cadherin-mediated cell-cell junctions. J Cell Sci 120:3870–3882

Cavey M, Rauzi M, Lenne PF, Lecuit T (2008) A two-tiered mechanism for stabilization and immobilization of E-cadherin. Nature 453:751-U2

Chausovsky A, Bershadsky AD, Borisy GG (2000) Cadherin-mediated regulation of microtubule dynamics. Nat Cell Biol 2:797–804

Chesarone MA, DuPage AG, Goode BL (2010) Unleashing formins to remodel the actin and microtubule cytoskeletons. Nat Rev Mol Cell Biol 11:62–74

Chu YS, Thomas WA, Eder O, Pincet F, Perez E, Thiery JP, Dufour S (2004) Force measurements in E-cadherin-mediated cell doublets reveal rapid adhesion strengthened by actin cytoskeleton remodeling through Rac and Cdc42. J Cell Biol 167:1183–1194

Costa M, Raich W, Agbunag C, Leung B, Hardin J, Priess JR (1998) A putative catenin-cadherin system mediates morphogenesis of the Caenorhabditis elegans embryo. J Cell Biol 141:297–308

Dawes-Hoang RE, Parmar KM, Christiansen AE, Phelps CB, Brand AH, Wieschaus EF (2005) Folded gastrulation, cell shape change and the control of myosin localization. Dev 132:4165–4178

del Rio A, Perez-Jimenez R, Liu R, Roca-Cusachs P, Fernandez JM, Sheetz MP (2009) Stretching Single Talin Rod Molecules Activates Vinculin Binding. Sci 323:638–641

den Elzen N, Buttery CV, Maddugoda MP, Ren G, Yap AS (2009) Cadherin adhesion receptors orient the mitotic spindle during symmetric cell division in mammalian epithelia. Mol Biol Cell 20:3740–3750

Drees F, Pokutta S, Yamada S, Nelson WJ, Weis WI (2005) Alpha-catenin is a molecular switch that binds E-cadherin-beta-catenin and regulates actin-filament assembly. Cell 123:903–915

Drenckhahn D, Pollard TD (1986) Elongation of actin-filaments is a diffusion-limited reaction at the barbed end and is accelerated by inert macromolecules. J Biol Chem 261:2754–2758

El Sayegh TY, Arora PD, Laschinger CA, Laschinger CA, Lee W, Morrison C, Overall CM, Kapus A, McCulloch CAG (2004) Cortactin associates with N-cadherin adhesions and mediates intercellular adhesion strengthening in fibroblasts. J of Cell Sci 117:5117–5131

El Sayegh TY, Arora PD, Fan LZ, Laschinger CA, Greer PA, McCulloch CA, Kapus A (2005) Phosphorylation of N-cadherin-associated cortactin by Fer kinase regulates N-cadherin adhesion strength. Mol Biol Cell 16:5514–5527

Farquhar MG, Palade GE (1963) Junctional complexes in various epithelia. J Cell Biol 17:375–412

Fernandez-Gonzalez R, Simoes SdM, Roeper J-C, Eaton S, Zallen JA (2009) Myosin II dynamics are regulated by tension in intercalating cells. Deval Cell 17:736–743

Gates J, Nowotarski SH, Yin H, Mahaffey JP, Bridges T, Herrera C, Homem CCF, Janody F, Montell DJ, Peifer M (2009) Enabled and capping protein play important roles in shaping cell behavior during Drosophila oogenesis. Deval Biol 333:90–107

Geisbrecht ER, Montell DJ (2002) Myosin VI is required for E-cadherin-mediated border cell migration. Nat Cell Biol 4:616–620

Georgiou M, Marinari E, Burden J, Baum B (2008) Cdc42, Par6, and aPKC Regulate Arp2/3-mediated endocytosis to control local adherens junction stability. Curr Biol 18:1631–1638

Gloushankova NA, Alieva NA, Krendel MF, Bonder EM, Feder HH, Vasiliev JM, Gelfand IM (1997) Cell-cell contact changes the dynamics of lamellar activity in nontransformed epitheliocytes but net in their ras-transformed descendants. Proc Natl Acad Sci U S A 94:879–883

Goley ED, Welch MD (2006) The ARP2/3 complex: an actin nucleator comes of age. Nat Rev Mol Cell Biol 7:713–726

Hansen SD, Mullins RD (2010) VASP is a processive actin polymerase that requires monomeric actin for barbed end association. J Cell Biol 191:571–584

Hara T, Ishida H, Raziuddin R, Dorkhom S, Kamijo K, Miki T (2004) Novel kelch-like protein, KLEIP, is involved in actin assembly at cell-cell contact sites of Madin-Darby canine kidney cells. Mol Biol Cell 15:1172–1184

Harris TJC, Peifer M (2005) The positioning and segregation of apical cues during epithelial polarity establishment in Drosophila. J Cell Biol 170:813–823

Harrison OJ, Jin XS, Hong SJ, Bahna F, Ahlsen G, Brasch J, Wu YH, Vendome J, Felsovalyi K, Hampton CM, Troyanovsky RB, Ben-Shau A, Frank J, Troyanovsky SM, Shapiro L, Honig B (2011) The extracellular architecture of adherens junctions revealed by crystal structures of type I cadherins. Structure 19:244–256

Helwani FM, Kovacs EM, Paterson AD, Verma S, Ali RG, Fanning AS, Weed SA, Yap AS (2004) Cortactin is necessary for E-cadherin-mediated contact formation and actin reorganization. J Cell Biol 164:899–910

Higgs HN, Pollard TD (1999) Regulation of actin polymerization by Arp2/3 complex and WASp/Scar proteins. J Biol Chem 274:32531–32534

Howes MT, Mayor S, Parton RG (2010) Molecules, mechanisms, and cellular roles of clathrin-independent endocytosis. Curr Opin Cell Biol 22:519–527

Insall RH, Machesky LM (2009) Actin dynamics at the leading edge: From simple machinery to complex networks. Dev Cell 17:310–322
Ito S, Werth DK, Richert ND, Pastan I (1983) Vinculin phosphorylation by the Src kinase—Interaction of vinculin with phospholipid-vesicles. J Biol Chem 258:4626–4631
Ivanov AI, Hunt D, Utech M, Nusrat A, Parkos CA (2005) Differential roles for actin polymerization and a myosin II motor in assembly of the epithelial apical junctional complex. Mol Biol Cell 16:2636–2650
Izumi G, Sakisaka T, Baba T, Tanaka S, Morimoto K, Takai Y (2004) Endocytosis of E-cadherin regulated by Rac and Cdc42 small G proteins through IQGAP1 and actin filaments. J Cell Biol 166:237–248
Kaksonen M, Toret CP, Drubin DG. (2006) Harnessing actin dynamics for clathrin-mediated endocytosis. Nat Rev Mol Cell Biol 7:404–414
Kametani Y, Takeichi M (2007) Basal-to-apical cadherin flow at cell junctions. Nat Cell Biol 9:92-U118
Knudsen KA, Soler AP, Johnson KR, Wheelock MJ (1995) Interaction of alpha-actinin with the cadherin/catenin cell-cell adhesion complex via alpha-catenin. J Cell Biol 130:67–77
Kobielak A, Pasolli HA, Fuchs E (2004) Mammalian formin-1 participates in adherens junctions and polymerization of linear actin cables. Nat Cell Biol 6:21-U2
Kovacs EM, Ali RG, McCormack AJ, Yap AS (2002a) E-cadherin homophilic ligation directly signals through Rac and phosphatidylinositol 3-kinase to regulate adhesive contacts. J Biol Chem 277:6708–6718
Kovacs EM, Goodwin M, Ali RG, Paterson AD, Yap AS (2002b) Cadherin-directed actin assembly: E-cadherin physically associates with the Arp2/3 complex to direct actin assembly in nascent adhesive contacts. Curr Biol 12:379–382
Kovacs EM, Verma S, Ali RG, Ratheesh A, Hamilton NA, Akhmanova A, Yap AS (2011) N-WASP regulates the epithelial junctional actin cytoskeleton through a non-canonical post-nucleation pathway. Nat Cell Biol 13:934-U400
Kovar DR (2006) Molecular details of formin-mediated actin assembly. Curr Opin Cell Biol 18:11–17
Kovar DR, Pollard TD (2004) Progressing actin: sFormin as a processive elongation machine. Nat Cell Biol 6:1158–1159
Kraemer A, Goodwin M, Verma S, Yap AS, Ali RG (2007) Rac is a dominant regulator of cadherin-directed actin assembly that is activated by adhesive ligation independently of Tiam1. Am J Physiol Cell Physiol 292:C1061-C1069
Krause M, Dent EW, Bear JE, Loureiro JJ, Gertler FB (2003) ENA/VASP proteins: Regulators of the actin cytoskeleton and cell migration. Annu Rev Cell Dev Biol 19:541–564
Kwiatkowski AV, Maiden SL, Pokutta S, Choi H-J, Benjamin JM, Lynch AM, Nelson WJ, Weis WI, Hardin J (2010) In vitro and in vivo reconstitution of the cadherin-catenin-actin complex from Caenorhabditis elegans. Proc Natl Acad Sci U S A 107:14591–14596
Ladoux B, Anon E, Lambert M, Rabodzey A, Hersen P, Buguin A, Silberzan P, Mege R-M (2010) Strength Dependence of Cadherin-Mediated Adhesions. Biophys J 98:534–542
Lambert M, Choquet D, Mege RM (2002) Dynamics of ligand-induced, Rac1-dependent anchoring of cadherins to the actin cytoskeleton. J Cell Biol 157:469–479
le Duc Q, Shi Q, Blonk I, Sonnenberg A, Wang N, Leckband D, de Rooij J (2010) Vinculin potentiates E-cadherin mechanosensing and is recruited to actin-anchored sites within adherens junctions in a myosin II-dependent manner. J Cell Biol 189:1107–1115
Leibfried A, Fricke R, Morgan M, Bogdan S, Bellaiche Y (2008) Drosophila Cip4 and WASp define a branch of the Cdc42-Par6-aPKC pathway. Curr Biol 18:1639–1648
Li F, Higgs HN (2003) The mouse formin mDia1 is a potent actin nucleation factor regulated by autoinhibition. Curr Biol 13:1335–1340
Ligon LA, Holzbaur ELF (2007) Microtubules tethered at epithelial cell junctions by dynein facilitate efficient junction assembly. Traffic 8:808–819
Ligon LA, Karki S, Tokito M, Holzbaur ELF (2001) Dynein binds to beta-catenin and may tether microtubules at adherens junctions. Nat Cell Biol 3:913–917

Liu ZJ, Tan JL, Cohen DM, Yang MT, Sniadecki NJ, Ruiz SA, Nelson CM, Chen CS (2010) Mechanical tugging force regulates the size of cell-cell junctions. Proc Natl Acad Sci U S A 107:9944–9949

Machesky LM, Mullins RD, Higgs HN, Kaiser DA, Blanchoin L, May RC, Hall ME, Pollard TD (1999) Scar, a WASp-related protein, activates nucleation of actin filaments by the Arp2/3 complex. Proc Natl Acad Sci U S A 96:3739–3744

Maddugoda MP, Crampton MS, Shewan AM, Yap AS (2007) Myosin VI and vinculin cooperate during the morphogenesis of cadherin cell-cell contacts in mammalian epithelial cells. J Cell Biol 178:529–540

Mangold S, Wu SK, Norwood SJ, Collins BM, Hamilton NA, Thorn P, Yap AS (2011) Hepatocyte growth factor acutely perturbs actin filament anchorage at the epithelial zonula adherens. Curr Biol 21:503–507

Martin AC, Kaschube M, Wieschaus EF (2009) Pulsed contractions of an actin-myosin network drive apical constriction. Nature 457:495-U11

Martin AC, Gelbart M, Fernandez-Gonzalez R, Kaschube M, Wieschaus EF (2010) Integration of contractile forces during tissue invagination. J Cell Biol 188:735–749

Maruthamuthu V, Sabass B, Schwarz US, Gardel ML (2011) Cell-ECM traction force modulates endogenous tension at cell-cell contacts. Proc Natl Acad Sci U S A 108:4708–4713

Matsuzaki F, Mege RM, Jaffe SH, Friedlander DR, Gallin WJ, Goldberg JI, Cunningham BA, Edelman GM (1990) cDNAs of cell-adhesion molecules of different specificity induce changes in cell-shape and border formation in cultured S180 cells. J Cell Biol 110:1239–1252

Mayor S, Pagano RE (2007) Pathways of clathrin-independent endocytosis. Nat Rev Mol Cell Biol 8:603–612

McGill MA, McKinley RFA, Harris TJC (2009) Independent cadherin-catenin and Bazooka clusters interact to assemble adherens junctions. J Cell Biol 185:787–796

McLachlan RW, Yap AS (2011) Protein tyrosine phosphatase activity is necessary for E-Cadherin-activated src signaling. Cytoskeleton 68:32–43

McLachlan RW, Kraemer A, Helwani FM, Kovacs EM, Yap AS (2007) E-cadherin adhesion activates c-Src signaling at cell-cell contacts. Mol Biol Cell 18:3214–3223

Mellor H (2010) The role of formins in filopodia formation. Biochim Biophys Acta Mol Cell Res 1803:191–200

Meng W, Mushika Y, Ichii T, Takeichi M (2008) Anchorage of microtubule minus ends to adherens junctions regulates epithelial cell-cell contacts. Cell 135:948–959

Michelot A, Drubin DG (2011) Building distinct actin filament networks review in a common cytoplasm. Curr Biol 21:R560-R569

Miki H, Sasaki T, Takai Y, Takenawa T (1998) Induction of filopodium formation by a WASP-related actin-depolymerizing protein N-WASP. Nature 391:93–96

Miyaguchi K (2000) Ultrastructure of the zonula adherens revealed by rapid-freeze deep-etching. J Struct Biol 132:169–178

Miyake Y, Inoue N, Nishimura K, Kinoshita N, Hosoya H, Yonemura S (2006) Actomyosin tension is required for correct recruitment of adherens junction components and zonula occludens formation. Exp Cell Res 312:1637–1650

Moore PB, Huxley HE, Derosier DJ (1970) Three-dimensional reconstruction of F-actin, thin filaments and decorated thin filaments. J Mol Biol 50:279–295

Mullins RD, Heuser JA, Pollard TD (1998) The interaction of Arp2/3 complex with actin: Nucleation, high affinity pointed end capping, and formation of branching networks of filaments. Proc Natl Acad Sci U S A 95:6181–6186

Nagafuchi A, Takeichi M (1988) Cell binding function of E-cadherin is regulated by the cytoplasmic domain. Embo J 7:3679–3684

Nagafuchi A, Takeichi M (1989) Transmembrane control of cadherin-mediated cell-adhesion—A 94 KDa protein functionally associated with a specific region of the cytoplasmic domain of E-cadherin. Cell Regul 1:37–44

Nagafuchi A, Ishihara S, Tsukita S (1994) The roles of catenins in the cadherin-mediated cell-adhesion—functional-analysis of E-cadherin-alpha catenin fusion molecules. J Cell Biol 127:235–245

Nezami AG, Poy F, Eck MJ (2006) Structure of the autoinhibitory switch in formin mDia1. Struct 14:257–263

Niessen CM, Leckband D, Yap AS (2011) Tissue organization by Cadherin adhesion molecules: Dynamic molecular and cellular mechanisms of morphogenetic regulation. Physiol Rev 91:691–731

Noren NK, Niessen CM, Gumbiner BM, Burridge K (2001) Cadherin engagement regulates Rho family GTPases. J Biol Chem 276:33305–33308

Ono S (2007) Mechanism of depolymerization and severing of actin filaments and its significance in cytoskeletal dynamics. In: Jeon KW (ed) International Review of Cytology—a Survey of Cell Biology, vol 258, pp 1–82

Otani T, Ichii T, Aono S, Takeichi M (2006) Cdc42 GEF Tuba regulates the junctional configuration of simple epithelial cells. J Cell Biol 175:135–146

Otomo T, Tomchick DR, Otomo C, Panchal SC, Machius M, Rosen MK (2005) Structural basis of actin filament nucleation and processive capping by a formin homology 2 domain. Nature 433:488–494

Ozawa M, Hoschutzky H, Herrenknecht K, Kemler R (1990) A possible new adhesive site in the cell-adhesion molecule uvomorulin. Mec Dev 33:49–56

Padrick SB, Rosen MK (2010) Physical mechanisms of signal integration by WASP family proteins. In: Kornberg RD (ed) Annual review of biochemistry, vol 79, pp 707–735

Parton RG, Simons K (2007) The multiple faces of caveolae. Nat Rev Mol Cell Biol 8:185–194

Peng X, Cuff LE, Lawton CD, DeMali KA (2010) Vinculin regulates cell-surface E-cadherin expression by binding to beta-catenin. J Cell Sci 123:567–577

Pokutta S, Weis WI (2002) The cytoplasmic face of cell contact sites. Curr Opin Struct Biol 12:255–262

Pollard TD, Blanchoin L, Mullins RD (2000) Molecular mechanisms controlling actin filament dynamics in nonmuscle cells. Annu Rev Biophys Biomol Struct 29:545–576

Pollitt AY, Insall RH (2009) WASP and SCAR/WAVE proteins: The drivers of actin assembly. J Cell Sci 122:2575–2578

Ren G, Crampton MS, Yap AS (2009a) Cortactin: Coordinating adhesion and the actin cytoskeleton at cellular protrusions. Cell Motil Cytoskeleton 66:865–873

Ren G, Helwani FM, Verma S, McLachlan RW, Weed SA, Yap AS (2009b) Cortactin is a functional target of E-cadherin-activated src family kinases in MCF7 epithelial monolayers. J Biol Chem 284:18913–18922

Rimm DL, Koslov ER, Kebriaei P, Cianci CD, Morrow JS (1995) Alpha(1)(E)-catenin is an actin-binding and actin-bundling protein mediating the attachment of F-actin to the membrane adhesion complex. Proc Natl Acad Sci U S A 92:8813–8817

Rose R, Weyand M, Lammers M, Ishizaki T, Ahmadian MR, Wittinghofer A (2005) Structural and mechanistic insights into the interaction between Rho and mammalian Dia. Nature 435:513–518

Rottner K, Haenisch J, Campellone KG (2010) WASH, WHAMM and JMY: Regulation of Arp2/3 complex and beyond. Trends Cell Biol 20:650–661

Rouiller I, Xu XP, Amann KJ, Egile C, Nickell S, Nicastro D, Li R, Pollard TD, Volkmann N, Hanein D (2008) The structural basis of actin filament branching by the Arp2/3 complex. J Cell Biol 180:887–895

Sahai E, Marshall CJ (2002) ROCK and Dia have opposing effects on adherens junctions downstream of Rho. Nat Cell Biol 4:408–415

Sawyer JK, Harris NJ, Slep KC, Gaul U, Peifer M (2009) The Drosophila afadin homologue Canoe regulates linkage of the actin cytoskeleton to adherens junctions during apical constriction. J Cell Biol 186:57–73

Sawyer JM, Harrell JR, Shemer G, Sullivan-Brown J, Roh-Johnson M, Goldstein B (2010) Apical constriction: A cell shape change that can drive morphogenesis. Dev Biology 341:5–19

Sawyer JK, Choi W, Jung K-C, He L, Harris NJ, Peifer M (2011) A contractile actomyosin network linked to adherens junctions by Canoe/afadin helps drive convergent extension. Mol Biol Cell 22:2491–2508

Schroer TA (2001) Microtubules don and doff their caps: dynamic attachments at plus and minus ends. Curr Opin Cell Biol 13:92–96
Schwartz MA (2010) Integrins and extracellular matrix in mechanotransduction. Cold Spring Harb Perspect Biol 2:a005066
Schwartz MA, DeSimone DW (2008) Cell adhesion receptors in mechanotransduction. Curr Opin Cell Biol 20:551–556
Scott JA, Shewan AM, den Elzen NR, Loureiro JJ, Gertler FB, Yap AS (2006) Ena/VASP proteins can regulate distinct modes of actin organization at cadherin-adhesive contacts. Mol Biol Cell 17:1085–1095
Shewan AM, Maddugoda M, Kraemer A, Stehbens SJ, Verma S, Kovacs EM, Yap AS (2005) Myosin 2 is a key Rho kinase target necessary for the local concentration of E-cadherin at cell-cell contacts. Mol Biol Cell 16:4531–4542
Small JV, Rottner K, Hahne P, Anderson KI (1999) Visualising the actin cytoskeleton. Microsc Res Tech 47:3–17
Smutny M, Yap AS (2010) Neighborly relations: Cadherins and mechanotransduction. J Cell Biol 189:1075–1077
Smutny M, Cox HL, Leerberg JM, Kovacs EM, Conti MA, Ferguson C, Hamilton NA, Parton RG, Adelstein RS, Yap AS (2010) Myosin II isoforms identify distinct functional modules that support integrity of the epithelial zonula adherens. Nat Cell Biol 12:696–702
Smutny M, Wu SK, Gomez GA, MangoldS, Yap AS, Hamilton NA (2011) Multicomponent analysis of junctional movements regulated by Myosin II isoforms at the epithelial zonula adherens. Plos One 6:e22458
Stehbens SJ, Paterson AD, Crampton MS, Shewan AM, Ferguson C, Akhmanova A, Parton RG, Yap AS (2006) Dynamic microtubules regulate the local concentration of E-cadherin at cell-cell contacts. J Cell Sci 119:1801–1811
Svitkina TM, Bulanova EA, Chaga OY, Vignjevic DM, Kojima S, Vasiliev JM, Borisy GG (2003) Mechanism of filopodia initiation by reorganization of a dendritic network. J Cell Biol 160:409–421
Taguchi K, Ishiuchi T, Takeichi M (2011) Mechanosensitive EPLIN-dependent remodeling of adherens junctions regulates epithelial reshaping. J Cell Biol 194:643–656
Tepass U, Hartenstein V (1994) The development of cellular junctions in the Drosophila embryo. Dev Biology 161:563–96
Tzima E, Irani-Tehrani M, Kiosses WB, Dejana E, Schultz DA, Engelhardt B, Cao GY, DeLisser H, Schwartz MA (2005) A mechanosensory complex that mediates the endothelial cell response to fluid shear stress. Nature 437:426–431
Urban E, Jacob S, Nemethova M, Resch GP, Small JV (2010) Electron tomography reveals unbranched networks of actin filaments in lamellipodia. Nat Cell Biol 12:429-U36
Uruno T, Liu JL, Zhang PJ, Fan YX, Egile C, Li P, Mueller SC, Zhan X (2001) Activation of Arp2/3 complex-mediated actin polymerization by cortactin. Nat Cell Biol 3:259–266
Vasioukhin V, Bauer C, Yin M, Fuchs E (2000) Directed actin polymerization is the driving force for epithelial cell-cell adhesion. Cell 100:209–219
Verma S, Shewan AM, Scott JA, Helwani FM, den Elzen NR, Miki H, Takenawa T, Yap AS (2004) Arp2/3 activity is necessary for efficient formation of E-cadherin adhesive contacts. J Biol Chem 279:34062–34070
Vicente-Manzanares M, Ma XF, Adelstein RS, Horwitz AR (2009) Non-muscle myosin II takes centre stage in cell adhesion and migration. Nat Rev Mol Cell Biol 10:778–790
Wallar BJ, Alberts AS (2003) The formins: active scaffolds that remodel the cytoskeleton. Trends Cell Biol 13:435–446
Watabe M, Nagafuchi A, Tsukita S, Takeichi M (1994) Induction of polarized cell-cell association and retardation of growth by activation of the E-cadherin catenin adhesion system in a dispersed carcinoma line. J Cell Biol 127:247–256
Watabe-Uchida M, Uchida N, Imamura Y, Nagafuchi A, Fujimoto K, Uemura T, Vermeulen S, van Roy F, Adamson ED, Takeichi M (1998) Alpha-catenin-vinculin interaction functions to organize the apical junctional complex in epithelial cells. J Cell Biol 142:847–857

Waterman-Storer CM, Salmon WC, Salmon ED (2000) Feedback interactions between cell-cell adherens junctions and cytoskeletal dynamics in Newt lung epithelial cells. Mol Biol Cell 11:2471–2483

Weaver AM, Karginov AV, Kinley AW, Weed SA, Li Y, Parsons JT, Cooper JA (2001) Cortactin promotes and stabilizes Arp2/3-induced actin filament network formation. Curr Biol 11:370–374

Weaver AM, Heuser JE, Karginov AV, Lee WL, Parsons JT, Cooper JA (2002) Interaction of cortactin and N-WASp with Arp2/3 complex. Curr Biol 12:1270–1278

Weber A, Pennise CR, Pring M (1994) Dnase-1 increases the rate-constant of depolymerization at the pointed (-) end of actin filament. Biochem 33:4780–4786

Weed SA, Du YR, Parsons JT (1998) Translocation of cortactin to the cell periphery is mediated by the small GTPase Rac1. J Cell SciSci 111:2433–2443

Weis WI, Nelson WJ (2006) Re-solving the cadherin-catenin-actin conundrum. J Biol Chem 281:35593–35597

Weiss EE, Kroemker M, Rudiger AH, Jockusch BM, Rudiger M (1998) Vinculin is part of the cadherin-catenin junctional complex: Complex formation between alpha-catenin and vinculin. J Cell Biol 141:755–764

Wildenberg GA, Dohn MR, Carnahan RH, Davis MA, Lobdell NA, Settleman J, Reynolds AB (2006) p120-catenin and p190RhoGAP regulate cell-cell adhesion by coordinating antagonism between Rac and Rho. Cell 127:1027–1039

Xu YW, Moseley JB, Sagot I, Poy F, Pellman D, Goode BL, Eck MJ (2004) Crystal structures of a formin homology-2 domain reveal a tethered dimer architecture. Cell 116:711–723

Yamada S, Nelson WJ (2007) Localized zones of Rho and Rac activities drive initiation and expansion of epithelial cell-cell adhesion. J Cell Biol 178:517–527

Yamada S, Pokutta S, Drees F, Weis WI, Nelson WJ (2005) Deconstructing the cadherin-catenin-actin complex. Cell 123:889–901

Yamazaki D, Oikawa T, Takenawa T (2007) Rac-WAVE-mediated actin reorganization is required for organization and maintenance of cell-cell adhesion. J Cell Sci 120:86–100

Yap AS, Brieher WM, Pruschy M, Gumbiner BM (1997) Lateral clustering of the adhesive ectodomain: A fundamental determinant of cadherin function. Curr Biol 7:308–315

Yap AS, Niessen CM, Gumbiner BM (1998) The juxtamembrane region of the cadherin cytoplasmic tail supports lateral clustering, adhesive strengthening, and interaction with p129(ctn). J Cell Biol 141:779–789

Yonemura S, Itoh M, Nagafuchi A, Tsukita S (1995) Cell-to-cell adherens junction formation and actin filament organization—similarities and differences between non-polarized fibroblasts and polarized epithelial-cells. J Cell Sci 108:127–142

Yonemura S, Wada Y, Watanabe T, Nagafuchi A, Shibata M (2010) Alpha-catenin as a tension transducer that induces adherens junction development. Nat Cell Biol 12:533–542

Ziegler WH, Liddington RC, Critchley DR (2006) The structure and regulation of vinculin. Trends Cell Biol 16:453–460

Chapter 7
Immunoglobulin Superfamily Receptors and Adherens Junctions

Yohei Shimono, Yoshiyuki Rikitake, Kenji Mandai, Masahiro Mori and Yoshimi Takai

Abstract The immunogroblin (Ig) superfamily proteins characterized by the presence of Ig-like domains are involved in various cellular functions. The properties of the Ig-like domains to form rod-like structures and to bind specifically to other proteins make them ideal for cell surface receptors and cell adhesion molecules (CAMs). Ig-CAMs, nectins in mammals and Echinoid in *Drosophila*, are crucial components of cadherin-based adherens junctions in the epithelium. Nectins form cell–cell adhesion by their *trans*-interactions and recruit cadherins to the nectin-initiated cell–cell adhesion site to establish adherens junctions. Thereafter junction adhesion molecules, occludin, and claudins, are recruited to the apical side of adherens junctions to establish tight junctions. The recruitment of these molecules by nectins is mediated both by the direct and indirect interactions of afadin with many proteins, such as catenins, and zonula occludens proteins, and by the nectin-induced reorganization of the actin cytoskeleton. Nectins contribute to the formation of both homotypic and heterotypic types of cell–cell junctions, such as synapses in the brain, contacts between pigment and non-pigment cell layers of the ciliary epithelium in the eye, Sertoli cell-spermatid junctions in the testis, and sensory cells and supporting cells in the sensory organs. In addition, *cis*- and *trans*-interactions of nectins with various cell surface proteins, such as integrins, growth factor receptors, and nectin-like molecules (Necls) play important roles in the regulation of many cellular functions, such as cell polarization, movement, proliferation, differentiation, survival, and cell sorting. Furthermore, the Ig-CAMs are implicated in many human diseases including viral infections, ectodermal dysplasia, cancers, and Alzheimer's disease.

Y. Takai (✉) · Y. Shimono · Y. Rikitake · K. Mandai
Division of Molecular and Cellular Biology, Department of Biochemistry and Molecular Biology, Kobe University Graduate School of Medicine, Kobe 650-0017, Japan
e-mail: ytakai@med.kobe-u.ac.jp

Y. Rikitake
Division of Signal Transduction, Department of Biochemistry and Molecular Biology, Kobe University Graduate School of Medicine, Kobe 650-0017, Japan

M. Mori
Division of Neurophysiology, Department of Physiology and Cell Biology, Kobe University Graduate School of Medicine, Kobe 650-0017, Japan

T. Harris (ed.), *Adherens Junctions: From Molecular Mechanisms to Tissue Development and Disease,* Subcellular Biochemistry 60,
DOI 10.1007/978-94-007-4186-7_7, © Springer Science+Business Media Dordrecht 2012

7.1 Introduction

The immunogroblin (Ig) superfamily proteins, represented by the presence of Ig homology (Ig-like) domains, are the largest and most diverse superfamily found in vertebrates and invertebrates ranging from insects to human. Analysis of the human genome has revealed that the Ig-like domains encoded by 765 genes are most frequently observed protein domains in human genome (Lander et al. 2001). The Ig-like domains possess a characteristic Ig-fold that is typically formed by seven to nine anti-parallel β-sheets and are at least subdivided into three topological domain subtypes: C1 (constant 1), C2 (constant 2), and V (variable) (Bork et al. 1994; Williams and Barclay 1988). Their abilities to form rod-like structures when deployed in series and their properties to bind specifically to other proteins make them ideal protein domains as components of cell surface receptors and cell adhesion molecules (CAMs). Crystal structures of adhesive extracellular fragments from several homophilic Ig superfamily CAMs (e.g., Myelin Protein Zero (MP0), axonin-1/TAG-1, neural cell adhesion molecule (NCAM), and junction adhesion molecule (JAM)) have led to the idea that zipper and array-based assemblies could establish platforms for the organization of signaling complexes, anchorage points for cytoskeletal components, and cell–cell adhesion (Aricescu and Jones 2007).

Adherens junctions are a form of cell–cell adhesion structure observed in a variety of cell types including epithelial cells and fibroblasts, which is characterized by a pair of plasma membranes apposed with a distance of ~20–25 nm (Farquhar and Palade 1965; Leckband and Sivasankar 2000). The members of the cadherin family (cadherins) have been considered to be the core components of adherens junctions (Oda and Takeichi 2011; Yap et al. 1997), but it has become apparent that CAMs of the Ig superfamily, such as the members of the nectin family (nectins) in mammals, and Echinoid in *Drosophila*, are also important components of adherens junctions (Lecuit 2005; Takai and Nakanishi 2003; Wei et al. 2005). Other Ig superfamily proteins, such as Nephrin, and MP0, are also involved in the formation of specialized types of adherens junctions (Lehtonen et al. 2004; Menichella et al. 2001), but we do not describe them in this review. Nectin-1 and nectin-2 were originally identified to be the receptors for α-herpesvirus that facilitate attachment and entry of the virus into host cells (Geraghty et al. 1998; Warner et al. 1998). Then, they were characterized to serve as CAMs, followed by the characterization of nectin-3 and nectin-4 as members of the nectin family (Reymond et al. 2001; Satoh-Horikawa et al. 2000; Takahashi et al. 1999). Echinoid was identified as a putative cell adhesion molecule and a novel negative regulator of the EGF receptor signaling for the development of the *Drosophila* eye (Bai et al. 2001). Then, Echinoid was characterized as a component of adherens junctions, cooperating with DE-cadherin (Wei et al. 2005).

Nectins are involved in the formation of three types of cell–cell adhesion: symmetric homotypic, asymmetric homotypic, and heterotypic cell adhesion (Takai et al. 2003, 2008a, b; Takai and Nakanishi 2003). In addition to their involvement in the formation of adherens junctions, nectins regulate the formation of tight junctions, a type of cell–cell junctions that functions as a virtually impermeable barrier to fluid, by recruiting tight junction proteins, such as junctional adhesion molecules (JAMs), claudins, and, occludin, to the apical side of adherens junctions (Takai et al. 2008a).

Nectin-like molecules (Necls) are proteins with a structural similarity to nectins (Takai et al. 2003). Both nectins and Necls are associated with other membrane proteins, such as growth factor receptors, and integrins, and play roles not only in cell–cell adhesion but also in cell polarization, movement, proliferation, differentiation, survival and cell sorting; abnormalities of nectins and Necls are thus associated with many diseases, such as cancer (Kawano et al. 2009; Masuda et al. 2005; Morimoto et al. 2008), and Alzheimer's disease (Harold et al. 2009; Takei et al. 2009). Furthermore, congenital mutations of *NECTIN-1* and *NECTIN-4* are associated with disorders, such as human ectodermal dysplasia syndromes in which more than two ectodermal organs are affected (Brancati et al. 2010; Bustos et al. 1991; Sozen et al. 2001; Suzuki et al. 1998, 2000; Zlotogora et al. 1987), and congenital ocular defects and/or abnormalities (Lachke et al. 2011), suggesting that abnormality of nectins during development widely affects cell integrity and tissue formation. In this chapter, we describe a variety of cellular functions of nectins, afadin, and Necls, focusing on their relation to adherens junctions.

7.2 Nectins, Afadin, and Necls

Nectins and Necls comprise families with four and five members, respectively, and all of them are Ca^{2+}-independent Ig-CAMs (Takai et al. 2008a). They share three extracellular Ig-like domains comprising an amino-terminal variable region-like domain and two constant region-like domains, a transmembrane region, and a cytoplasmic region (Fig. 7.1a). Each member of Necls has other nomenclatures, such as CADM3, SynCAM3, or TSLL1, for Necl-1; CADM1, SynCAM, IGSF4, or TSLC1, for Necl-2; CADM2, or SynCAM2, for Necl-3; CADM4, SynCAM4, or TSLL2, for Necl-4; PVR, CD155, or TAGE4, for Necl-5. All the members of the nectin family, namely, nectin-1, nectin-2, nectin-3, and nectin-4, have two or three splice variants. Necl-1, -3, and -5 have two or three splice variants, and Necl-2 has 6 splice variants, but Necl-4 does not have a splice variant.

Nectins and Necls are classified by their binding properties to the actin filament (F-actin)-binding protein afadin: nectins bind afadin, whereas Necls do not (Takai et al. 2008a). The PDZ domain of afadin binds to the conserved motif (Glu/Ala-X-Tyr-Val) located at the cytoplasmic tails of nectin-1, nectin-2, and nectin-3. The PDZ domain of afadin binds to nectin-4, although the cytoplasmic tail of nectin-4 does not possess this motif.

The members of the nectin and Necl families *trans*-interact both homophilically and heterophilically with each other in various combinations (Fig. 7.1b) (Takai et al. 2008a). As discussed below, nectins and Necls also interact with other Ig superfamily proteins. In *Drosophila*, Echinoid is associated with Canoe, a *Drosophila* orthologue of afadin (Wei et al. 2005), and like nectins and Necls, Echinoid

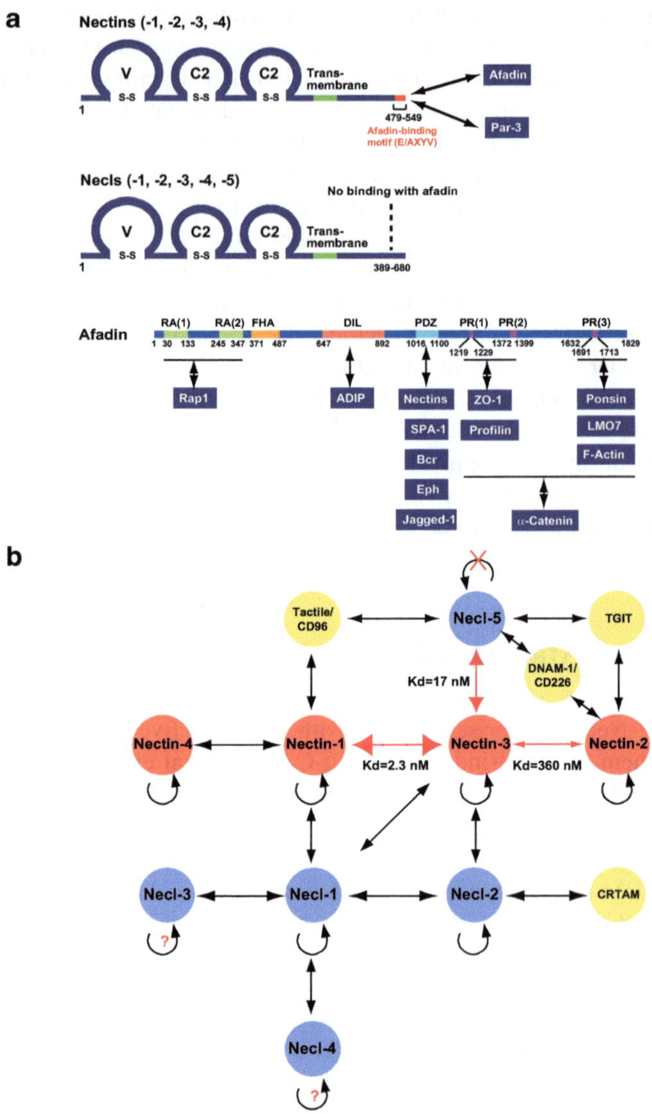

Fig. 7.1 Molecular structures and modes of *trans*-interactions of nectins, Necls, and their binding proteins. **a** Nectins and Necls share three Ig-like extracellular domains comprising an N-terminal variable region-like (V) domain and two constant region-like (C2) domains, a transmembrane region and a cytoplasmic domain. The conserved motif at the cytoplasmic tail of nectins binds the PDZ domain of afadin and the PDZ domain of Par-3. Necls do not bind afadin at their carboxyl terminal region. Afadin has multiple domains that bind other proteins and its carboxyl terminal region interacts with F-actin. In this way, afadin connects nectins to the actin cytoskeleton. RA, Ras-association domain; FHA, forkhead-associated domain; DIL, dilute domain; PDZ, PSD-95/Discs large/ZO-1 domain; PR, proline-rich domain; S-S, disulfide bond. **b** Schematic presentation of *trans*-interactions of nectins, Necls, and other Ig-like molecules. Homophilic (*looped arrows*) and heterophilic (*double-headed arrows*) *trans*-interactions are presented. The dissociation constant (Kd) values for a heterophilic *trans*-interaction are presented with the *red double-headed arrows*

trans-interacts homophilically with each other or heterophilically *trans*-interacts with Neuroglian, an L1-type CAM (Islam et al. 2003).

7.2.1 Trans-*Interactions of Nectins and Necls*

Cadherins first form homo-*cis*-dimers and then homo-*trans*-dimers in a calcium-dependent manner, causing cell–cell adhesion (Oda and Takeichi 2011). Similarly, nectins first form homo-*cis*-dimers, but different from cadherins, nectins then form both homo- and hetero-*trans*-dimers in a calcium-independent manner (Takai et al. 2008b). The hetero-*trans*-dimers (heterophilic *trans*-interactions) are formed between nectin-1 and nectin-3, between nectin-2 and nectin-3, and between nectin-1 and nectin-4 (Fig. 7.1b). Heterophilic *trans*-interactions of nectins are stronger than their homophilic *trans*-interactions (Satoh-Horikawa et al. 2000). The dissociation constant (K_d) between nectins are 2.3 nM for the interaction between nectin-1 and nectin-3, and 360 nM for that between nectin-2 and nectin-3. The K_d for the heterophilic interaction between nectin-3 and Necl-5 is 17 nM (Ikeda et al. 2003). Because of the lower K_d values of nectin interactions than that of E-cadherin interaction (~80 μM) (Koch et al. 1997), the interactions of nectins are more favorable for the formation of an initial contact between opposing cells. Ultra-structural studies have revealed that the first Ig-like domains of nectin-1 and nectin-3 are involved in the formation of the *trans*-dimers; that both their first and second Ig-like domains are involved in the formation of the *cis*-dimers; and that the four essential amino acid residues in the first Ig-like domain of nectin-1 are involved in the formation of the homo-*cis*-dimers (Narita et al. 2011).

It is uncertain whether Necls first form homo-*cis*-dimers, but they *trans*-interact in a calcium-independent manner, causing cell–cell adhesion, similar to nectins. The extracellular regions of Necls basically form both homo- and hetero-*trans*-dimers, whereas Necl-5 only forms hetero-*trans*-dimers with other nectins or Necls (Fig. 7.1b) (Takai et al. 2008a).

Nectins and Necls interact not only with the members of the nectin and Necl families, but also with other members of the Ig superfamily proteins (Fig. 7.1b), such as DNAM-1/CD226 (Bottino et al. 2003; Pende et al. 2005, 2006), T cell-activated increased late expression (Tactile)/CD96 (Fuchs et al. 2004), Class I-restricted T cell associated molecule (CRTAM) (Garay et al. 2010), and T cell immunoglobulin and ITIM domain (TIGIT) (Stanietsky et al. 2009). These interactions occur between cells of the immune system and are important for regulated immune responses. Thus, homophilic and heterophilic *trans*-interactions of nectins and Necls with each other, as well as with other Ig superfamily proteins, regulate the diverse cellular functions.

7.2.2 *Interaction of Nectins with Afadin and Par-3*

Afadin was originally identified as an F-actin-binding protein that is localized at adherens junctions, having a structure similar to the *AF-6* gene product that is an *ALL1* fusion partner involved in acute myeloid leukemia (Mandai et al. 1997; Prasad et al. 1993). Afadin contains multiple domains, which include two Ras-associated (RA) domains, a forkhead-associated (FHA) domain, a dilute (DIL) domain, a PDZ domain, and three proline-rich (PR) domains (Fig. 7.1a) (Takai et al. 2008a). Several splicing variants of afadin are known to date, including l-afadin and s-afadin (Mandai et al. 1997; Saito et al. 1998). l-Afadin is the largest splicing variant and ubiquitously expressed, whereas expression of s-afadin is relatively specific to the brain. s-Afadin lacks the third proline-rich domain and the F-actin-binding domain. In this chapter, l-afadin is simply referred to as afadin.

Afadin binds the small GTPase Rap1 through the RA domain (Boettner et al. 2000; Hoshino et al. 2005); the F-actin-binding protein afadin DIL domain-interacting protein (ADIP) through the DIL domain (Asada et al. 2003); the Rap1 GTPase-activating protein SPA-1 (Su et al. 2003), the protein kinase Bcr (Radziwill et al. 2003), Eph receptor tyrosine kinase (Buchert et al. 1999), and the Notch receptor ligand Jagged-1 (Hock et al. 1998) through the PDZ domain; the tumor suppressor LIM domain only 7 (LMO7) (Ooshio et al. 2004), the tight junction protein zonula occludens-1 (ZO-1) (Yamamoto et al. 1997), the actin-binding protein profilin (Boettner et al. 2000), the vinculin-binding protein ponsin (Mandai et al. 1999), α-catenin (Tachibana et al. 2000), and F-actin (Mandai et al. 1997) through the proline-rich region at its carboxyl terminal.

The cytoplasmic tail of nectins, a binding region of afadin, also serves as that of the cell polarity protein Par-3 (Fig. 7.1) (Ooshio et al. 2007; Takekuni et al. 2003). Par-3 is involved in the co-localization of afadin with nectins at the nectin-based cell–cell adhesion sites, and in the activation of Rac at adherens junctions (Ooshio et al. 2007). In cooperation with nectin and afadin, Par-3 regulates the formation of tight junctions and cell polarity in epithelial cells.

7.2.3 Echinoid and Canoe

7.2.3.1 Echinoid

The *Drosophila* CAM Echinoid is a member of the Ig superfamily with seven Ig-like domains, two fibronectin type III domains, a transmembrane region, and an intracellular region and is localized at adherens junctions. Although Echinoid is not considered to be an orthologue of nectins, this molecule is a component of adherens junctions that binds to Canoe, the *Drosophila* orthologue of afadin (Wei et al. 2005). Like nectins, Echinoid *trans*-interacts homophilically with each other or heterophilically with Neuroglian, an L1-type CAM (Islam et al. 2003). Echinoid also binds Bazooka, the *Drosophila* homolog of mammalian Par-3, through its carboxyl—terminal region and modulates the EGF and Notch signaling (Ahmed et al. 2003; Bai et al. 2001; Wei et al. 2005). Adherens junctions are absent between the cells either of which lacks Echinoid. Thus, Echinoid regulates cell–cell adhesion and cell sorting in *Drosophila* (Wei et al. 2005).

In addition, Echinoid regulates the generation of a contractile actomyosin cable required for epithelial morphogenesis and this action depends on the intracellular region of Echinoid (Laplante and Nilson 2006; Lin et al. 2007). During the process of the dorsal closure in *Drosophila* embryogenesis, an actin cable forms at the interface between the cells of the dorsal epidermis with Echinoid and the amnioserosa which lacks Echinoid. Furthermore, the planar polarized distribution of Echinoid in the dorsal-most epidermal cells in *Drosophila* regulates the distribution of the actin cytoskeleton at the leading edge, the localization of Bazooka, and contractile properties during the dorsal closure (Laplante and Nilson 2011).

7.2.3.2 Canoe

Canoe is the *Drosophila* orthologue of afadin and shares a similar domain organization (Gomperts 1996; Ponting 1995; Ponting and Benjamin 1996; Sheng 1996; Woods and Bryant 1993). In *Drosophila*, Canoe is genetically associated with Ras, JNK, Notch, and Wnt signaling pathways during the eye, bristle, and/or wing development (Carmena et al. 2006; Matsuo et al. 1999; Miyamoto et al. 1995; Takahashi et al. 1998). Canoe also binds Rap1 and Polychaetoid (Pyd), a *Drosophila* homologue of ZO-1 (Boettner et al. 2003; Takahashi et al. 1998). Canoe is involved in the formation of cell–cell junctions (Miyamoto et al. 1995; Takahashi et al. 1998), the development of the eye, bristle, and wing, morphogenic change during dorsal closure (Takahashi et al. 1998), and asymmetric division and cell fate choice in the nervous system and mesoderm (Speicher et al. 2008).

7.3 Formation of Cell–Cell Junctions by Nectins and Afadin

Nectins and afadin cooperatively recruit many proteins to the nectin-based cell–cell adhesion sites, leading to the formation of adherens junctions and tight junctions, and the establishment of cell polarity. As discussed below, these proteins include CAMs, such as cadherins, JAMs, claudins, occludin, and integrins; growth factor receptors; intracellular adaptor proteins, such as α-catenin, $p120^{ctn}$, PLEKHA7, ponsin, vinculin, ADIP, LMO7, α-actinin, and ZO-1; signaling molecules, such as c-Src, Rap, Rac, and Cdc42; and cell polarity protein complexes, such as the Par-3-Par-6-aPKC complex, the Crb3-Patj-Pals complex, and the Lgl2-Scribble-Par-1 complex.

7.3.1 Formation of Adherens Junctions

Core structural components of adherens junctions consist of the nectin–afadin and cadherin–catenin complexes. Nectins and afadin are characterized by their strict localization at adherens junctions in both epithelial cells and fibroblasts, whereas the localization of cadherins, α-catenin, β-catenin, and p120^{ctn} are not strictly limited to adherens junctions, and widely distributed along the lateral plasma membrane in epithelial cells (Mandai et al. 1997; Takahashi et al. 1999). Evidence thus far available indicates that after nectins initiate cell–cell adhesion, they recruit cadherins to the nectin-based cell–cell adhesion sites to form adherens junctions (Honda et al. 2003; Takai and Nakanishi 2003).

7.3.1.1 Initiation of Cell–Cell Adhesion

Studies with many cultured cell lines have revealed that nectins initiate the formation of adherens junctions between two neighboring cells before cadherins start to form cell–cell adhesion (Takai et al. 2008b). The complex of E-cadherin and α- and β-catenins is recruited to the nectin-based cell–cell adhesion sites through afadin without the *trans*-interaction of E-cadherin (Tachibana et al. 2000). Suppression of the formation of the nectin-1-based cell–cell adhesion inhibits the formation of the E-cadherin-based cell–cell adhesion (Honda et al. 2003). Moreover, a deficiency of afadin inhibits the formation of the cadherin-based adherens junctions and tight junctions (Ikeda et al. 1999). Thus, initiation of cell–cell contacts by the nectin–afadin system recruits the components of adherens junctions and tight junctions.

7.3.1.2 Cooperative Roles of Nectins and Afadin

The *trans*-interactions between Necl-5 and nectin-3, between nectins, and between nectins and afadin are mutually involved in the initiation and formation of adherens junctions and tight junctions. The Necl-5-nectin-3 *trans*-interaction first enhances the nectin–afadin interaction, which then enhances the nectin–nectin *trans*-interaction (Kurita et al. 2011). Interaction of Necl-5 with nectin-3 is described in more detail in the Chap. 7.4.3. Besides nectins, afadin directly binds α-catenin and other proteins involved in cell–cell adhesion and polarization. The knockout mice lacking each of the nectin family members do not always show obvious phenotypes because of the redundant and overlapped expression of the nectin family members in a cell. On the other hand, afadin knockout mice are embryonic lethal and this is accompanied by various developmental defects during and after gastrulation, including impaired migration of mesoderm and disorganization of the ectoderm with improperly organized adherens junctions and tight junctions (Ikeda et al. 1999; Zhadanov et al. 1999). In afadin knockdown cells, both cell–cell adhesion and the association of E-cadherin with F-actin and p120^{ctn} are impaired (Lorger and Moelling 2006; Sato et al. 2006), and the formation of both adherens junctions and tight junctions in these cells is restored by expression of full-length afadin (Ooshio et al. 2010). Therefore, afadin is essential for the proper structural organization of adherens junctions and tight junctions in polarized epithelial cells. Canoe, *Drosophila* orthologue of afadin, interacts with DE-cadherin (Sawyer et al. 2009), and in *canoe* mutants, apical constriction of mesodermal cells is impaired as a result of the actin cytoskeleton disconnecting from adherens junctions (Sawyer et al. 2009).

7.3.1.3 Activation of Intracellular Signaling and Reorganization of the Actin Cytoskeleton

The *trans*-interactions of nectins at the initial cell–cell contact sites first induce the activation of the tyrosine kinase c-Src (Fig. 7.2a) (Takai et al. 2008a, b). c-Src then induces the activation of Rap1 through Crk, an adaptor protein, and C3G, a guanine nucleotide-exchange factor (GEF) for Rap1. In addition, c-Src phosphorylates FRG, a GEF for Cdc42, and Vav2, a GEF for Rac. Cdc42 and Rac would bind their respective downstream effectors, such as IQGAP1, IRSp53/WAVE, NWASP, and WASP, and all of them are actin-binding proteins (Fig. 7.2a). Rap1, which is activated by *trans*-interactions of nectins at the initial cell–cell contact sites, subsequently binds to afadin (Hoshino et al. 2005). Then, afadin binds $p120^{ctn}$ that is associated with non-*trans*-interacting E-cadherin. The Rap1-dependent binding of afadin to $p120^{ctn}$ inhibits the endocytosis of E-cadherin, which then enhances both the accumulation of E-cadherin at the nectin-based cell–cell adhesion sites and the cell–cell adhesion activity of E-cadherin, leading to the establishment of adherens junctions (Hoshino et al. 2005; Sato et al. 2006). Nectins have an ability to associate with the actin cytoskeleton through afadin and its interaction with actin-binding proteins, such as α-catenin, EPLIN, vinculin, α-actinin, and ZO-1 (Fig. 7.2b) (Takai et al. 2008b). Moreover afadin interacts with α-catenin which binds to the reorganized actin cytoskeleton and recruits the cadherin-β-catenin complex (Hoshino et al. 2005; Sato et al. 2006). As a result of these nectin-induced intracellular signals and their cooperation with the cadherin–catenin system, enhanced recruitment of afadin to the cell–cell contact sites further facilitates the efficient formation of adherens junctions (Kurita et al. 2011). Thus, the signaling induced by nectins and cadherins facilitates the dynamic reorganization of the actin cytoskeleton to strengthen the clustering of the CAMs and their cell–cell adhesion activity.

PLEKHA7 was originally identified as a member of pleckstrin homology domain-containing, family A (Gerhard et al. 2004; Strausberg et al. 2002), and is strictly localized at adherens junctions like the nectin–afadin system (Fig. 7.2b) (Meng et al. 2008). The interaction of PLEKHA7 with both $p120^{ctn}$ and Nezha, which bind E-cadherin and the minus end of microtubles, respectively, connect the microtuble network to adherens junctions (Meng et al. 2008). We found that afadin binds and recruits PLEKHA7 to the nectin-based cell–cell adhesion sites (unpublished observation). The resulting ternary complex of afadin, PLEKHA7, and $p120^{ctn}$, is essential to form a circumferential and continuous belt-like structure of adherens junctions in Madin-Darby canine kidney cells (unpublished observation). Accordingly, the direct interaction of afadin with both α-catenin and PLEKHA7, recruitment of E-cadherin, and reorganization of the actin cytoskeleton cooperatively contribute to the establishment of adherens junctions.

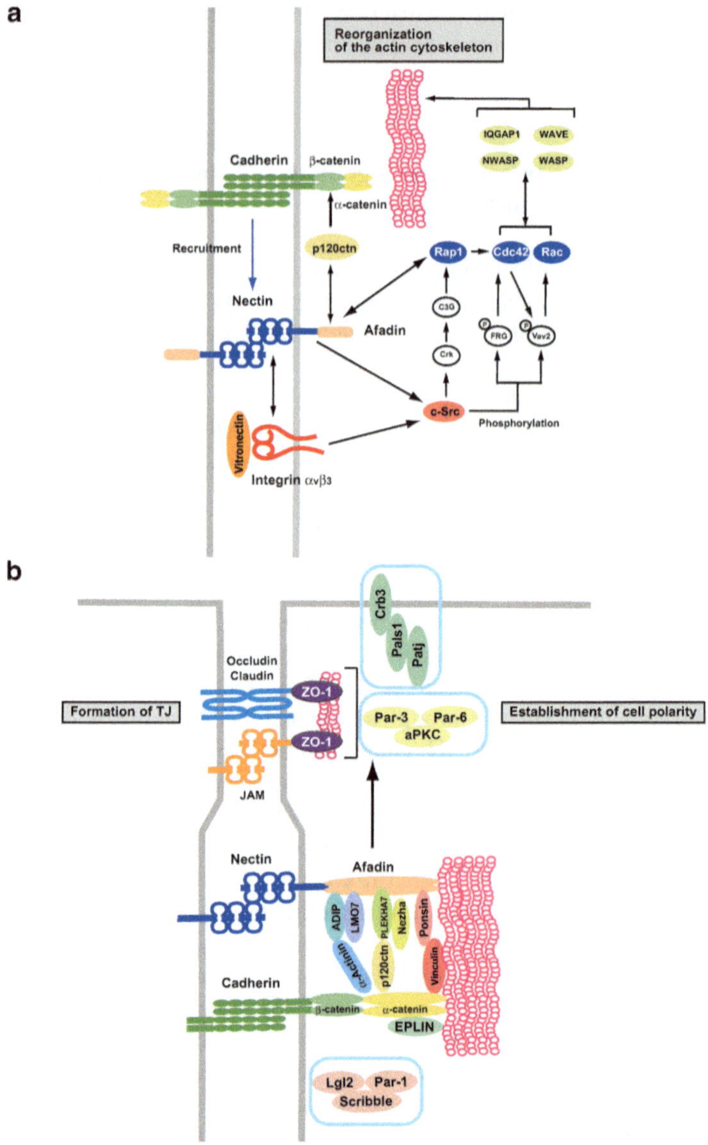

Fig. 7.2 Nectin-induced intracellular signaling for the formation of cell–cell junctions and the establishment of cell polarity. **a** Intracellular signaling for the formation of adherens junctions. The *trans*-interactions of nectins first induce the activation of c-Src and subsequently Rap1, Cdc42, and Rac at the initial cell–cell adhesion sites. The activation of c-Src is dependent on the activation of integrin $\alpha_V\beta_3$. The cadherin–catenin complex is then recruited to the nectin-based cell–cell adhesion sites through the direct and indirect interactions of afadin with the cadherin-binding proteins and the small G protein-induced reorganization of the actin cytoskeleton, eventually leading to the establishment of adherens junctions. **b** The formation of tight junctions and the recruitment of the cell polarity proteins. After or during the formation of adherens junctions, nectins recruit JAMs, and then recruit other proteins, including CAMs, such as cadherins, claudins, and occludin; intracellular adaptor proteins, such as ZO-1, ponsin, vinculin, ADIP, LMO7, α-actinin, EPLIN, and PLEKHA7; and cell polarity protein complexes, such as a Par-3-Par-6-aPKC complex, a Crb3-Patj-Pals complex, and a Lgl2-Scribble-Par-1 complex, to establish tight junctions and cell polarity

7.3.2 *Formation of Tight Junctions and Apicobasal Polarity*

In epithelial cells, apicobasal polarity is established by the formation of tight junctions at the apical side of adherens junctions. The nectin-induced intracellular signaling and the reorganization of the actin cytoskeleton in cooperation with afadin are important for the formation of both adherens junctions and tight junctions, as well as apicobasal polarity (Fig. 7.2b) (Takai et al. 2008a). After or during the formation of adherens junctions, nectins act through afadin to recruit ZO-1. The transient interaction of ZO-1 with afadin is necessary for the localization of JAMs and claudins at the apical side of adherens junctions (Kuramitsu et al. 2008). Nectins and afadin also function with Par-3 for the formation of both adherens junctions and tight junctions (Ooshio et al. 2007).

While many CAMs and cell polarity proteins required for apicobasal polarization have been identified, the cooperative roles of these proteins in this polarization are not completely understood. Intriguingly, exogenous expression of various combinations of CAMs and cell polarity proteins in fibroblasts that have adherens junctions but lack tight junctions, results in the establishment of both epithelial-like adherens junctions and tight junctions (unpublished observation). In addition to the CAMs that establish adherens junctions and tight junctions, simultaneous expression of cell polarity proteins, at least Par-3, aPKC, Par-6, Crb3, Pals1, and Patj, are necessary to properly localize tight junctions at the apical side of adherens junctions (unpublished observation). Of these cell polarity proteins, Crb3 is localized at the apical membrane and this localization appears to be essential for the localization of tight junctions at the apical side of adherens junctions. For better understanding of the molecular mechanism underlying the formation of apicobasal polarity, the most important issue to be addressed next is to clarify how Crb3 is transported to the apical membrane.

7.3.3 *Various Types of Cell–Cell Junctions*

According to their size and shape, F-actin-anchoring adherens junctions are classified into zonula adherens, fascia adherens, and puncta adherentia junctions: zonula adherens are observed in epithelial and other types of cells; fasciae adherentes connect mature mammalian cardiomyocytes; puncta adherentia junctions are characterized by a cluster of small plaque-bearing junctions from which relatively sparse F-actin projects into the cytoplasm. As discussed in the next paragraph below, nectins and afadin facilitate the formation of specialized adherens junctions localized at neural synapses in the brain, contacts between pigment and non-pigment cell layers of the ciliary epithelium in the eye, and Sertoli cell-spermatid junctions in the testis.

7.3.3.1 Puncta Adherentia Junctions in Neurons

The synapse, a special junction formed between axons and dendrites of neurons, contains two types of junctions: synaptic junctions and puncta adherentia junctions.

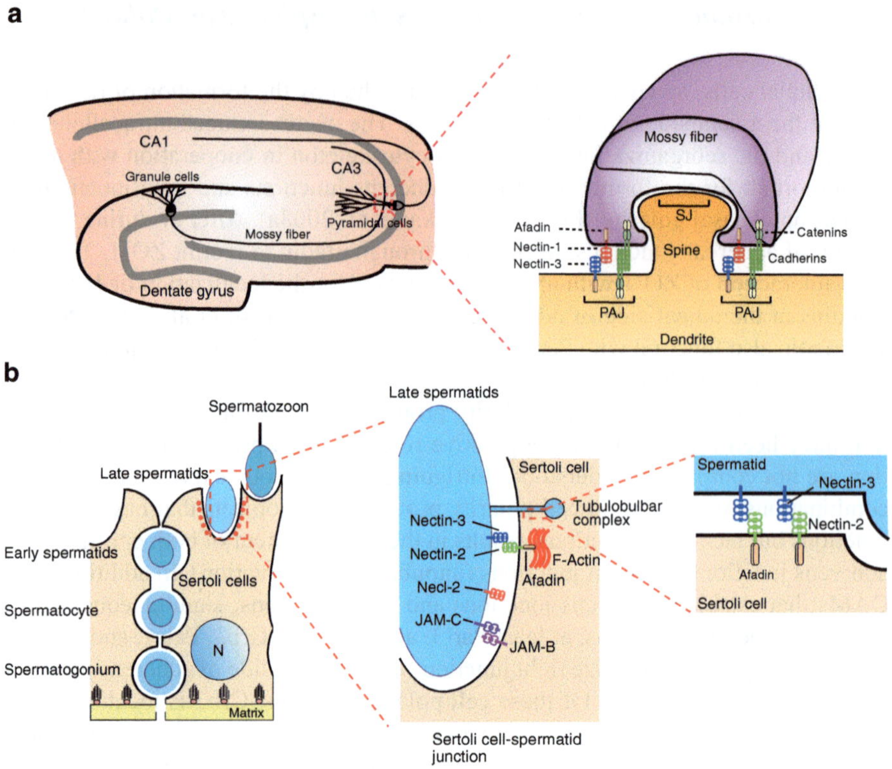

Fig. 7.3 Various types of nectin-based cell–cell junctions. **a** Puncta adherentia junctions in neurons. Synapse between a mossy fiber terminal of a granule cell and a dendrite of a pyramidal cell in the CA3 region of the hippocampus contains two types of junctions: synaptic junctions and puncta adherentia junctions. Nectin-1 and nectin-3 are asymmetrically localized at the mossy fiber terminal (presynaptic side) and at the dendrite of pyramidal cell (postsynaptic side), respectively, and form puncta adherentia junctions in cooperation with cadherins. PAJ, puncta adherentia junction; SJ, synaptic junction. **b** Sertoli cell-spermatid junctions in the testis. The unique cell–cell junctions between Sertoli cells as well as between Sertoli and germ cells are observed in the testis. The Sertoli cell-spermatid junctions are one of these junctions formed at the latter half of spermatogenesis. Nectin-2 and nectin-3 are asymmetrically localized in Sertoli cells and spermatids, respectively, and play critical roles in germ cell differentiation. JAM-B localized in Sertoli cells, and JAM-C and Necl-2 located in spermatids also involved in germ cell differentiation

Synaptic junctions function as neurotransmission sites, whereas puncta adherentia junctions are regarded as mechanical adhesion sites between axon terminals and their targets where adhesion molecules, such as cadherins, catenins, nectins, and NCAM are concentrated (Fig. 7.3a) (Fannon and Colman 1996; Yamada et al. 2003; Yamagata et al. 1995). These molecules are important for differentiation of neurons and synaptic plasticity. Puncta adherentia junctions are particularly well-developed in the CA3 region of the hippocampus. The synapses between the mossy fiber terminals and the dendrites of pyramidal cells in the CA3 region are postnatally formed and gradually remodeled to synaptic junctions and puncta adherentia junctions (Amaral and Dent 1981). N-cadherin and other classic cadherins, such as cadherin-8 and cadherin-11, exist at puncta adherentia junctions to mediate cell–cell adhesion at synapses. In addition, at puncta adherentia junctions formed between the mossy fiber terminals and the dendrites of the pyramidal cells in the CA3 region, nectin-1 and nectin-3 are asymmetrically localized at the presynaptic and the postsynaptic sides, respectively, whereas afadin is symmetrically localized at both sides (Mizoguchi et al. 2002; Takai et al. 2008a). By analogy with adherens junctions in epithelial cells and fibroblasts, the formation of the nectin-based cell–cell adhesion may be followed by the recruitment of the N-cadherin-catenin complex to form synapses, which are segregated into synaptic junctions and puncta adherentia junctions. Consistently, the number of puncta adherentia junctions in the CA3 region of the hippocampus of the adult brain is reduced and a mossy fiber trajectory is abnormal in nectin-1 knockout and nectin-3 knockout mice (Honda et al. 2006). It is well known that an axon specifically binds dendrites during the formation of synapses and dendrites hardly form stable adhesions with each other. Although the underlying molecular mechanism has been unclear, the *trans*-interaction of nectin-1, preferentially localized in an axon, with nectin-3, preferentially localized in dendrites, are critical for the ordered association of an axon with dendrites (Togashi et al. 2006).

7.3.3.2 Contacts between Pigment and Non-pigment Cell Layers of the Ciliary Epithelium in the Eye

The ciliary epithelia are located around the lens of the eye and consist of two layers, the pigment and non-pigment epithelia, making the ciliary process that produces aqueous humor (Raviola and Raviola 1978). In the normal eyes, the apices of the pigment and non-pigment epithelia are apposed and contact each other. The apex–apex contact between the pigment and non-pigment cell layers of the ciliary epithelium in the eye is impaired in both nectin-1 knockout and nectin-3 knockout mice that show a virtually identical ocular phenotype, microphthalmia (Inagaki et al. 2005). However, nectin-1 knockout and nectin-3 knockout mice show no impairment of the apico-lateral junctions between the pigment epithelial cells where nectin-1, nectin-2, and nectin-3 are localized, or of the apicolateral junctions between the non-pigment epithelial cells where nectin-2 and nectin-3, but not nectin-1, are localized. Thus, the heterophilic *trans*-interaction between nectin-1 and nectin-3 plays an important role in establishing the apex–apex contact in the ciliary epithelium in the eye.

7.3.3.3 Sertoli Cell-Spermatid Junctions in the Testis

In the testis, unique cell–cell junctions between Sertoli cells as well as between Sertoli and germ cells, and desmosome-like junctions between Sertoli and germ cells provide mechanical adhesion of germ cells onto Sertoli cells and play a critical role in germ cell morphogenesis and differentiation (Fig. 7.3b). Spermatogenic cells are embraced and cultivated by Sertoli cells during spermatogenesis. In the latter half of spermatogenesis, spermatids form prominent cell–cell junctions with Sertoli cells called Sertoli cell-spermatid junctions. In contrast to the Sertoli–Sertoli junctions that contain both adherens junctions and tight junctions, the Sertoli cell-spermatid junctions contain neither of these junctions (Takai et al. 2008b). The junctions depend mainly on the *trans*-interaction between nectin-2 in Sertoli cells and nectin-3 in spermatids (Cheng and Mruk 2002; Ozaki-Kuroda et al. 2002). In fact, nectin-2 knockout and nectin-3 knockout mice show abnormalities in spermatogenesis, resulting in male infertility (Inagaki et al. 2006; Mueller et al. 2003; Ozaki-Kuroda et al. 2002). The importance of the heterophilic interaction of nectins is further demonstrated by using transplantation assays in which the nectin-2 homozygous knockout and nectin-2 heterozygous knockout spermatogonia are introduced to the nectin-2 heterozygous knockout and nectin-2 homozygous knockout testis, respectively (Ozaki-Kuroda et al. 2002). Spermatogenesis is observed only when the nectin-2 homozygous knockout spermatogonia is introduced to the nectin-2 heterozygous knockout testis, confirming the importance of the *trans*-interaction between nectin-3 in spermatids and nectin-2 in Sertoli cells for normal spermatogenesis. In addition, other members of the Ig superfamily proteins, such as JAMs, and Necl-2, have been shown to be expressed in the testis. At the Sertoli cell-spermatid junctions, JAM-B and JAM-C are localized (Gliki et al. 2004). JAM-A and JAM-B are located at the Sertoli cell–Sertoli cell junctions and function as blood-testis barrier (Gliki et al. 2004; Mruk and Cheng 2004). Necl-2 is exclusively expressed in spermatogenic cells, and not in Sertoli cells, and Necl-2 knockout mice are defective in spermatogenesis (Wakayama and Iseki 2009).

7.3.3.4 Desmosomes

Desmosomes are spot-like structures for cell–cell adhesion that anchor cytoplasmic intermediate filaments, such as keratin, desmin, and vimentin (Brooke et al. 2012). They are observed in various types of cells and are most abundant in epithelial cells. In addition to adherens junctions and tight junctions, nectins are involved in the formation and maintenance of desmosomes (Barron et al. 2008; Yoshida et al. 2010). The desmosoms between stratum intermedium and ameloblasts in the developing teeth, which express nectin-3 and nectin-1, respectively, are significantly reduced in both nectin-1 knockout and nectin-1 and nectin-3 compound knockout mice.

7.4 Interactions of Nectins and Necls with Other Membrane Proteins

In addition to the *trans*-interactions with the members of the nectin and Necl families, nectins and Necls *trans*-interact with other members of the Ig superfamily proteins and *cis*-interact with other membrane proteins, such as integrins, and growth factor receptors. These interactions are critically involved in a variety of cellular functions, including cell adhesion, polarization, movement, proliferation, differentiation, survival, and cell sorting.

7.4.1 Interaction of Nectins with Integrin $\alpha_v\beta_3$ and Cross-talk Between Cell–Cell Junctions and Cell-Matrix Junctions

Integrins are transmembrane proteins composed of two non-covalently associated glycoprotein subunits called α and β. They form cell-matrix junctions, such as focal adhesions, focal complexes, and hemidesmosomes (Alberts 2002). Integrins positively or negatively regulate the formation and stability of adherens junctions. During embryonic development, integrins promote epithelial cell remodeling by reducing the interaction of cell–cell adhesion molecules at adherens junctions (Monier-Gavelle and Duband 1997). On the other hand, integrins induce the functional polarization of the cells and reinforce the cadherin-based adherens junctions (Schreider et al. 2002). However, a mechanism underlying a cross-talk between cell–cell adhesion and cell-matrix adhesion had remained unclear.

Integrin $\alpha_v\beta_3$ is expressed in various types of cells, including epithelial cells, osteoblasts, and vascular endothelial cells, and is overexpressed in tumors, and serves as a receptor for vitronectin. The activation of integrin $\alpha_V\beta_3$ is necessary during the initial stage of the nectin-induced formation of adherens junctions (Takai et al. 2008b). Nectin-1 and nectin-3, but not nectin-2, physically interact with both the active and inactive forms of integrin $\alpha_v\beta_3$ at cell–cell adhesion sites (Sakamoto et al. 2006). Talin, a protein that connects integrins to the actin cytoskeleton, binds the cytoplasmic tail of the integrin β_3 subunit, changes the intracellular conformation of integrin $\alpha_v\beta_3$, and increases its affinity for vitronectin (Tadokoro et al. 2003). The active form of integrin $\alpha_v\beta_3$ is essential for the activation of c-Src by nectins, which in turn is critical for the formation of adherens junctions (Fukuhara et al. 2004). After adherens junctions are established, integrin $\alpha_v\beta_3$ becomes inactive, but still continues to be co-localized with nectins. The inactivation of integrin $\alpha_v\beta_3$ is beneficial for the maintenance of adherens junctions because the sustained activation of integrin $\alpha_v\beta_3$ renders cells highly motile, which tends to disrupt cell–cell junctions. The interaction of nectin-1 and nectin-3 with integrin $\alpha_v\beta_3$ seems to be specific because other integrins so far tested do not interact with these nectins. Phosphatidylinositol-phosphate kinase type Iγ90 (PIPKIγ90) is involved in the activation of integrins. The *trans*-interactions of nectins enhance the activity of protein-tyrosine-phosphatase μ (PTPμ), which dephosphorylates PIPKIγ90 and thus suppresses the interaction between talin and integrin $\alpha_v\beta_3$ (Sakamoto et al. 2008). Nectins potentially interact with PTPμ through their extracellular regions and enhance the phosphatase activity of PTPμ decreasing the phosphorylation of PIPKIγ90. In this way, nectins function in the inactivation of integrin $\alpha_v\beta_3$ at adherens junctions. Studies of the interaction of nectins with integrins give us better understanding of the molecular mechanism underlying cross-talk between cell–cell adhesion and cell-matrix adhesion.

7.4.2 Interaction of Necl-5 with Growth Factor Receptor and Integrin $\alpha_v\beta_3$ and Cell Movement and Proliferation

Prior to the formation of cell–cell contacts and junctions where nectins and afadin are primarily involved, cells move in response to chemoattractants, such as platelet-derived growth factor (PDGF), and vascular endothelial growth factor (VEGF). During cell movement, cells form protrusions such as lamellipodia, and filopodia at the leading edge and peripheral ruffles over the lamellipodia to move to the direction of higher concentrations of the chemoattractants (Ronnstrand and Heldin 2001). Nectins are not observed at the leading edge. Instead, Necl-5 is preferentially accumulated there. Necl-5 in cooperation with PDGF receptor and integrin $\alpha_V\beta_3$ plays a pivotal role in the dynamics of the leading edge (Fig. 7.4a) (Takai et al. 2008b). Growth factor receptors and integrins synergistically interact and regulate various intracellular signaling pathways (Comoglio et al. 2003). In addition to the activation of PDGF receptor and integrin $\alpha_V\beta_3$, Necl-5 is essential for the formation of leading edge structures by enhancing the signals mediated by PDGF receptors and integrins. In moving cells, Necl-5 is localized at the leading edge and preferentially regulates the interaction between PDGF receptor and integrin $\alpha_V\beta_3$ by forming a ternary complex or any combination of binary complexes (Amano et al. 2008; Minami et al. 2007). These complexes transduce signals to activate Rap1 and Rac for cell movement, and Ras for cell proliferation.

In addition, afadin has been identified as a key player in persistent directional cell movement by facilitating clustering of the Necl-5-PDGF receptor-integrin $\alpha_V\beta_3$ complex presumably in a positive feedback amplification manner (Fig. 7.4a) (Miyata et al. 2009a). The formation and disassembly of leading edge structures continuously occur at the leading edges of moving cells and is tightly regulated by the actions of the small G proteins, such as Rap1, Rac, and Rho (Hall 1998). In order to keep cells moving, each member of the Rho family small G proteins are cyclically activated and inactivated. Afadin and its binding protein ADIP regulate the cyclical activation and inactivation of Rap1, Rac, and Rho at the leading edges (Fukumoto et al. 2011; Miyata et al. 2009b). Collectively, afadin plays an important role in the regulation of directionality of cell movement and cyclical activation and inactivation of the small G proteins.

7.4.3 Contact Inhibition of Cell Movement and Proliferation

A cell ceases to migrate after contact with another cell. This phenomenon, so called contact inhibition of cell movement, was originally described in fibroblasts (Abercrombie and Heaysman 1953, 1954). On the other hand, when proliferating cells become confluent, they cease to proliferate. This phenomenon is referred to as contact inhibition of cell proliferation (Fisher and Yeh 1967). Many mechanisms for contact inhibition of cell movement and proliferation have been proposed, but these are not fully understood. We found a novel mechanism of this contact inhibition involving the interaction of Necl-5 and nectin-3.

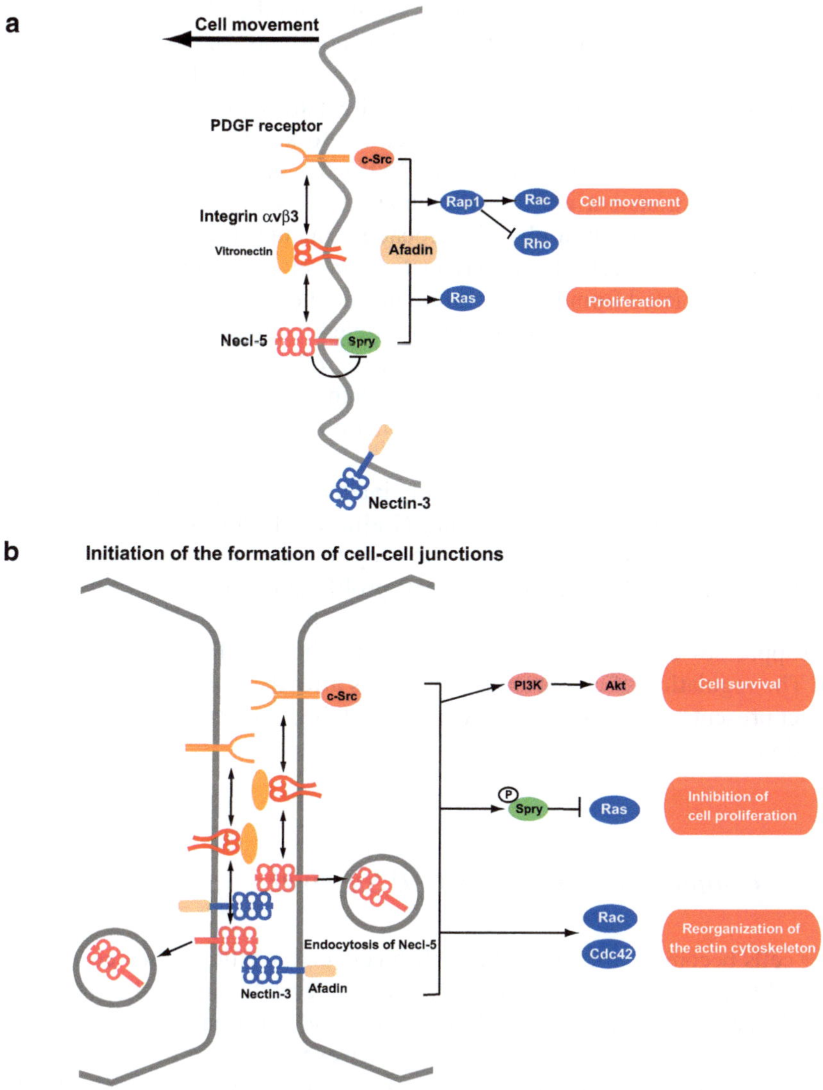

Fig. 7.4 Regulation of cell movement and proliferation. **a** Interaction of Necl-5 with growth factor receptor and integrin $\alpha_v\beta_3$ at the leading edge of a directionally moving cell. Necl-5 preferentially regulates the interaction between PDGF receptor and integrin $\alpha_v\beta_3$ and the clustering of the Necl-5-PDGF receptor-integrin $\alpha_v\beta_3$ complex plays a pivotal role in signal transduction at the leading edge. Afadin is a key player in persistent directional cell movement and enhances the clustering of the Necl-5-PDGF receptor-integrin $\alpha_V\beta_3$ complex. These complexes transduce signals to activate Rap1 and Rac for cell movement, and Ras for cell proliferation. Nectins are not observed at the leading edge. Necl-5 interacts with Spry2 and suppresses its function. **b** Contact inhibition of cell proliferation and cell survival. When moving cells come into contact with each other, Necl-5 first heterophilically interacts in *trans* with nectin-3 on the cell surface of the apposing cell and induces the activation of Cdc42 and Rac, both of which reorganize the actin cytoskeleton. Then, Necl-5 is internalized from the cell surface by endocytosis. Spry, released from Necl-5, is phosphorylated by c-Src and suppresses cell proliferation by inhibiting the activation of Ras. The down-regulation of Necl-5 leads to reductions in cell movement and proliferation. Furthermore, nectin-3 and afadin play a crucial role in the PDGF-induced cell survival by preventing apoptosis through the activation of the phosphatidylinositol 3-kinase (PI3K)-Akt signaling

When moving cells come into contact with each other, Necl-5 on the cell surface heterophilically interacts *in trans* with nectin-3 on the opposing cell surface to initiate the formation of cell–cell junctions (Fig. 7.4b) (Takai et al. 2008b). The *trans*-interaction between nectin-3 and Necl-5 induces the activation of Cdc42 and Rac, both of which reorganize the actin cytoskeleton and increase cell–cell adhesion (Fujito et al. 2005; Sato et al. 2005). The *trans*-interaction of nectin-3 with Necl-5 is transient and down-regulation of Necl-5 from the cell surface occurs by endocytosis in a clathrin-dependent manner (Fujito et al. 2005). The down-regulation of Necl-5 leads to reductions in cell movement and proliferation. Nectin-3, dissociated from Necl-5, is retained on the cell surface and subsequently *trans*-interacts with nectin-1, which most efficiently interacts with nectin-3 (Ikeda et al. 2003). As discussed above in the Chap. 7.3.1, these *trans*-interactions of nectins induce the recruitment of cadherins to the nectin-based cell–cell adhesion sites and eventually establish adherens junctions. Cell movement terminates in this way.

Necl-5 also physically and functionally interacts with Sprouty (Spry) and regulates the PDGF-induced Ras signaling (Kajita et al. 2007). When cells do not contact other cells, Necl-5 interacts with Spry to prevent its function and inhibits the PDGF-induced Ras signaling for cell proliferation (Fig. 7.4). Spry, released by the down-regulation of Necl-5 on the cell surface, is phosphorylated by c-Src and suppresses cell proliferation by inhibiting the PDGF-induced activation of Ras. The regulation of cell proliferation status after the establishment of cell–cell contact presents one of the mechanisms underlying contact inhibition of cell proliferation.

7.4.4 Contact Inhibition and Cell Survival

After cells become confluent and establish cell–cell junctions, they cease to move and proliferate but continue to survive. Survival, growth, and proliferation of animal cells are dependent on extracellular signals and nutrition. Limitation of growth factors leads to decline of nutrient transporter expression on the cell surface and perturbation of mitochondrial physiology. These processes are considered to be related to induction of cell death by apoptosis and/or autophagy (Lum et al. 2005). Nectin-based cell–cell adhesion plays a crucial role in the PDGF-induced cell survival by preventing apoptosis through the activation of the phosphatidylinositol 3-kinase (PI3K)-Akt signaling (Fig. 7.4b) (Takai et al. 2008a). In NIH3T3 cells, nectin-3 and PDGF receptor are co-localized at cell–cell adhesion sites. The PDGF-induced phosphorylation of Akt is attenuated by knockdown of nectin-3 or afadin. Likewise, afadin regulates the VEGF-induced phosphorylation of Akt and survival of endothelial cells (Tawa et al. 2010). Collectively, the interaction of the nectin–afadin system with growth factor receptors links cell–cell adhesion to the cell survival signaling.

7.5 The Ig-like Receptors and Cell Sorting

Organs and tissues in mammals are composed of different types of cells which homotypically and heterotypically adhere to each other. It is known that cells in certain organs and tissues, such as the oviduct, and the auditory epithelia of the cochlea in the inner ear, are arranged in a well-organized manner. Mechanisms that regulate these well-organized cell arrangements had been completely unknown; however, it has become evident that the Ig superfamily proteins play important roles in these arrangements.

7.5.1 Lattice Patterning in Drosophila *Retinal Cells*

The formation of the multicellular hexagonal lattice in the *Drosophila* retina is the result of a cell sorting process (Tepass and Harris 2007). At least two cell–cell adhesion systems, Hibris-Roughest and cadherin systems are involved in the formation of adherens junctions between developing retinal cells (Cordero et al. 2007; Hayashi and Carthew 2004). The cell rearrangements that occur during the formation of the lattice require precisely orchestrated changes in adhesive interactions between retinal cells (Fig. 7.5a). At the beginning, the primary pigment cells become distinct and come in contact to form a concentric ring around the cone cells. The cone cells express high levels of Delta, a ligand for the Notch receptor, and the Notch-Delta signaling appears to enlarge and differentiate the uncommitted cells into the primary pigment cells. As the primary pigment cells enlarge and form the ommatidia, all remaining cells are pushed and the interommatidial precursor cells are constrained into a lattice array. The secondary pigment and tertiary pigment cells arise from the interommatidial precursor cells. The interommatidial precursor cells are initially arranged in multiple rows between the forming ommatidia and then sort themselves to form a single row. Remodeling of adhesive contacts between neighboring cells is the first event in the pattern formation of the interommatidial precursor cells (Carthew 2007). A failure of the primary pigment cells to pattern the interommatidial precursor cell lattice results in a rough eye phenotype.

The genes associated with this patterning are the Ig-like superfamily transmembrane proteins, Roughest and Hibris. Roughest protein is localized within the interommatidial precursor cells specifically at the interfaces between the primary pigment cells and the interommatidial precursor cells. Hibris is expressed in the primary pigment cells and it is likely to be localized at adherens junctions of the primary pigment cells where it directly interacts with Roughest protein localized at the interommatidial precursor cells. An increase or a decrease in the expression levels of the Roughest protein prevents the normal sorting and lattice formation of the interommatidial precursor cells (Ramos et al. 1993; Reiter et al. 1996; Tanenbaum et al. 2000; Wolff and Ready 1991). Reducing the expression levels of Hibris also causes sorting defects (Bao and Cagan 2005). Both *in vivo* and biochemical

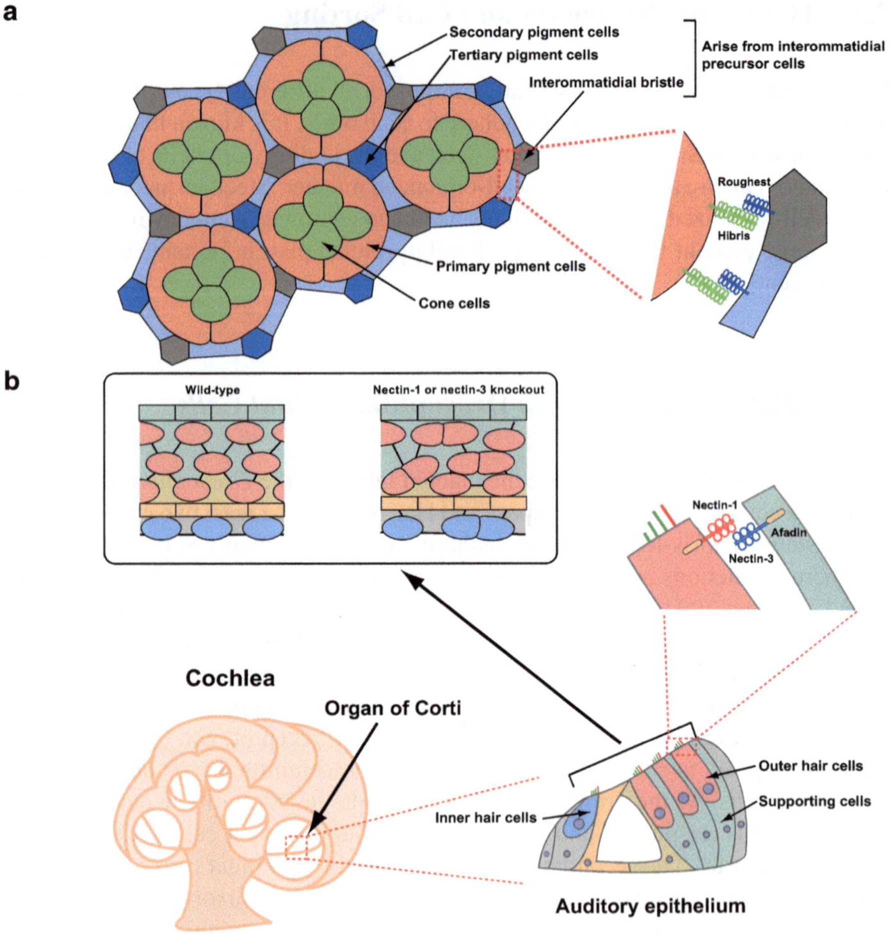

Fig. 7.5 Roles of the Ig-CAMs in cell sorting. **a** Formation of the lattice pattern in *Drosophila* retinal cells. The multicellular hexagonal lattice is formed as a result of a cell sorting process. This process is mediated by the Ig-like superfamily transmembrane proteins, Roughest and Hibris, expressed in the interommatidial precursor cells and the primary pigment cells, respectively. The secondary pigment, tertiary pigment cells, and interommatidal bristle arise from the interommatidial precursor cells. **b** Formation of a checkerboard-like pattern in the auditory epithelial cells. The auditory epithelium is located at the Organ of Corti in the cochlea. The checkerboard-like pattern observed in the auditory epithelia is formed by hair cells, consisting of inner hair cells and outer hair cells, and supporting cells as a result of a cell sorting process. This process is mediated by nectin-1 and nectin-3, expressed in hair cells and supporting cells, respectively (left in the box). In the auditory epithelium of the nectin-1 or nectin-3 knockout mice, hair cells are aberrantly attached and this checkerboard-like pattern is impaired (right in the box)

experiments indicate that Hibris is a Roughest-binding partner and that Roughest preferentially interacts with Hibris rather than interacts with itself (Bao and Cagan 2005). It is considered that the highly adhesive heterogeneous interaction between Roughest and Hibris nucleates a core layer and excludes weakly adhesive cells. Thus, the cells with less Roughest are outcompeted to become the secondary pigment and tertiary pigment cells (Carthew 2007). Roughest and Hibris are *Drosophila* homologues of Neph1 and nephrin, respectively, that mediate specific Ca^{2+}-independent adhesive interactions between different cell types (Dworak et al. 2001). In the mammalian kidney, Neph1 and nephrin function to maintain selective permeability of the glomerulus (Kestila et al. 1998; Liu et al. 2003). In *C. elegans*, an interaction of Synaptogenesis Abnormal (Syg)-1 and Syg-2, Roughest and Hibris homologues, respectively, is essential for synaptogenesis (Shen and Bargmann 2003; Shen et al. 2004).

7.5.2 Checkerboard-like Pattering of the Auditory Epithelial Cells

Cell sorting occurs during the developmental process in which different kinds of cells are generated and arranged in complex and elaborate patterns to form tissues and organs. The checkerboard-like pattern is observed in certain tissues such as the oviduct, and the auditory epithelium. In the auditory epithelia of the cochlea, the hair and supporting cells, expressing nectin-1 and nectin-3, respectively, exhibit a checkerboard-like assembly (Fig. 7.5b). Although the lateral inhibition mediated by the Notch-Delta signaling contributes to the formation of a checkerboard-like pattern, genetic inactivation of the Notch signaling does not impair the checkerboard-like pattern itself (Lanford et al. 1999). This checkerboard-like pattern of the hair and supporting cells is impaired and hair cells are aberrantly attached to each other in the nectin-1 or nectin-3 knockout mice (Togashi et al. 2011). In contrast to cadherins that mainly connect cells expressing the same cadherins through their homophilic interactions, nectins have ability to interact both homophilically and heterophilically. Thus, the ability of nectins to facilitate interaction between heterogeneous cells mediates the formation of checkerboard-like cell arrangement. In the aberrantly attached hair cells, the orientation and shapes of the sensory hair bundles on the surface of the hair cells are perturbed, suggesting that planar cell polarity is disturbed in these cells (unpublished observation). The apical junctional complexes composed of adherens junctions and tight junctions are associated with the establishment of apicobasal polarity. The abnormally broad distribution of the apical junctional complexes is observed at the boundaries between the hair cells in nectin-3 knockout mice (unpublished observation). In addition, localization of the planar cell polarity core molecules is perturbed in these cells. These observations suggest that nectin-mediated establishment of the apicobasal polarity regulates the planar cell polarity of hair cells. N-Cadherin and E-cadherin are expressed in a mutually exclusive pattern in the inner hair cells and in the outer hair cells, respectively (Simonneau et al. 2003). However, the involvement of cadherin systems in the formation of checkerboard-like cell arrangement in the auditory epithelium is not elucidated.

7.6 Nectins and Necls in Human Diseases

Considering the involvement of nectins, afadin, and Necls in a variety of fundamental cellular processes, it is reasonable that the Ig superfamily proteins and their interacting proteins are related to and/or have causative roles in various human disorders. In this chapter, we focus on the roles of nectins, afadin, and Necls in human diseases. Other examples of the Ig-CAMs involved in human diseases, such as the coxsackievirus-adenovirus receptor (CAR), and NCAM-1, are described elsewhere (Bergelson et al. 1997; Thoulouze et al. 1998; Tomko et al. 1997). It is quite reasonable that various human disorders are caused by or are related to nectins, afadin, and Necls that regulate various cellular function including not only cell–cell adhesion but also cell polarization, movement, proliferation, differentiation, survival and cell sorting.

7.6.1 Entry Receptors for Viruses

Many viruses depend on the adhesive properties of the Ig-like superfamily proteins to mediate virus attachment and subsequent entry into the host cells. Among the members of the nectin family, nectin-1 and nectin-2 were originally isolated as entry receptors for viruses (Aoki et al. 1997; Eberle et al. 1995; Lopez et al. 1995; Morrison and Racaniello 1992). They were initially considered to be receptors identical to poliovirus receptors and named poliovirus receptor-related proteins (PRR), but later it was shown that they were not related to the poliovirus infection. Instead, it was proved that nectin-1 and nectin-2, also termed as PVRL1 and PVRL 2 (provirus receptor-related 1 and 2), serve as receptors for envelope glycoprotein D of herpes simplex viruses (HSV)-1 and HSV-2. In this way, nectin-1 and nectin-2, expressed in human cells of epithelial and neural origins, mediate infection of HSV-1 and HSV-2 (Geraghty et al. 1998). The interaction of nectin-1 with afadin increases the efficiency of cell–cell spread, but not entry, of HSV-1, and does not affect the binding of glycoprotein D, a viral component mediating entry of HSV-1 into host cells, to nectin-1 (Sakisaka et al. 2001). In addition, NCAM-1 is a receptor for rabies virus, JAM-A for reovirus, signaling lymphocyte-activation molecule (SLAM) for measles virus (Dermody et al. 2009) and nectin-4 also for measles virus (Muhlebach et al. 2011; Noyce et al. 2011).

7.6.2 Ectodermal Dysplasia

Ectodermal dysplasia syndromes are congenital disorders characterized by abnormalities in two or more ectodermal organs such as teeth, hair, epidermis, and several exocrine glands. Cleft lip/palate-ectodermal dysplasia syndrome (CLPED) is a severe human ectodermal dysplasia syndrome and CLPED patients are clinically characterized by unusual faces, cleft lip/palate, dental anomalies with reduced numbers of teeth (hypodontia), syndactyly of the fingers and toes, functional abnormalities of sweat and salivary glands, kinky and sparse hair (hypotrichosis), and thickening of palm skin (palmoplantar hyperkeratosis), and in some cases, mental retardation (Bustos et al. 1991; Suzuki et al. 2000; Zlotogora 1994; Zlotogora et al. 1987). Mutations in *NECTIN-1* have been identified as the cause of this disease (Sozen et al. 2001; Suzuki et al. 1998). The mutations identified in CLPED lead to truncated protein that lacks the transmembrane region and carboxyl-terminus (Sozen et al. 2001; Suzuki et al. 2000). However, the cleft lip/palate is not clearly observed in the nectin-1 knockout mice (Yoshida et al. 2010, 2012). The truncated form of the nectin-1 expressed in the CLPED patients may interfere with the functions of other nectin-1-interacting proteins, such as nectin-3, nectin-4, and/or Necl-1, in a dominant-negative manner. Mutations in *NECTIN-4* have been identified as the cause of ectodermal dysplasia-syndactyly syndrome (EDSS), another form of ectodermal dysplasia, in which hair and tooth abnormalities, loss of hair (alopecia) and cutaneous syndactyly are observed. Consistently, nectin-4 is highly expressed in hair follicles and the separating digits, and the mutated nectin-4 loses its capability to bind nectin-1 (Brancati et al. 2010).

It has been shown that nectin-3 protein expression is significantly reduced in patients with a balanced translocation between chromosome 1 and 3, manifesting in severe bilateral congenital cataracts, central nervous system abnormalities, and mild developmental delay (Lachke et al. 2011). Although different from the ectodermal dysplasia syndromes, this is consistent with the fact that nectin-3 knockout mice exhibit lens and other ocular defects (Inagaki et al. 2005; Lachke et al. 2011).

7.6.3 Cancers

Afadin is structurally similar to the *AF-6* gene product which is originally identified as a fusion partner of *ALL-1*, human homologue of trithorax, in human acute leukemia (Prasad et al. 1993). Knockdown of afadin expression significantly increases migration and invasion of breast cancer cells (Fournier et al. 2011). Loss of afadin expression in breast cancer is associated with the adverse prognosis and an increased risk of metastatic relapse (Letessier et al. 2007). Necl-2 expression is down-regulated in various cancer cells and Necl-2 serves as a tumor suppressor gene (Masuda et al. 2005). The *cis*-interaction of Necl-2 with the extracellular region of ErbB3 reduces the ligand-induced ErbB2-catalyzed tyrosine phosphorylation of ErbB3 and suppresses cancer cell movement and survival (Kawano et al. 2009; Masuda et al. 2005). In addition, Necl-2 interacts with integrin $\alpha_6\beta_4$ and suppresses disruption of hemidesmosome-like structures (Mizutani et al. 2011). Expression of both Necl-1 and Necl-4 suppresses the tumorigenicity of colon cancer cells (Raveh et al. 2009). Necl-4 is a novel tumor suppressor in renal clear cell carcinoma (Nagata et al. 2012). Absence or marked reduction of Necl-4 expression is frequently observed in renal clear cell carcinoma cell lines and surgically resected renal clear cell carcinoma. On the other hand, Necl-5 stimulates Ras signaling to enhance cell proliferation (Kakunaga et al. 2004) and Rac signaling to enhance cell migration (Minami et al. 2010). Necl-5 is associated with unfavorable prognosis in human lung cancer patients (Nakai et al. 2010) and the interaction of Necl-5 with DNAM-1 facilitates the binding between Necl-5-expressing cancer cells and platelets and enhances lung metastasis of the cancer cells (Morimoto et al. 2008).

7.6.4 Alzheimer's Disease

A single nucleotide polymorphism (SNP) in the 3'UTR region of *NECTIN-2* (*PVRL2*) is one of the thirteen genome-wide significant SNPs that map within or close to the *APOE* (Apolipoprotein E) locus on chromosome 19, whose polymorphic expression is widely associated with Alzheimer's disease (Harold et al. 2009). The association of SNPs within the *NECTIN-2* and late-onset Alzheimer's disease is also observed in Japanese patients (Takei et al. 2009). These results suggest that together with known association of *APOE* with Alzheimer's disease, *NECTIN-2* may have implications for predisposition to this disease.

7.7 Future Perspectives

We have described in this chapter that the Ig superfamily proteins, nectins and Necls, mediate a variety of cellular processes, such as initiation of cell–cell adhesion, activation of intracellular signaling, reorganization of the actin cytoskeleton, and regulation of cell movement proliferation and differentiation. In addition, nectins and Necls are proved to be widely involved in other cellular processes, such as the formation of apicobasal polarity, tight junctions, and heterotypic cell junctions, contact inhibition of cell movement and proliferation, and cell sorting. Even though we present that nectins are involved in the formation of heterotypic and asymmetric homotypic cell–cell adhesions in some tissues and organs, such as the testis, the brain, the eyes, and the auditory epithelium, far more complex cell–cell adhesions are observed in other tissues and organs. In addition, heterotypic cell–cell adhesions are involved in the regulation of stem cell behavior, such as symmetric and asymmetric cell division, and subsequent differentiation (Gonzalez-Reyes 2003; Speicher et al. 2008). Moreover, accumulating evidence suggests that tumor-associated stroma plays a role in metastasis of cancer cells (Gupta and Massague 2006), and one may consider cancer-stroma interaction as a form of heterotypic cell–cell adhesion observed under a pathological condition. Thus, it is necessary to characterize the roles of the Ig superfamily CAMs and their interacting proteins, such as nectins, Necls, and afadin, in heterotypic and asymmetric homotypic cell–cell adhesion to promote our understanding of normal development and regeneration of tissues and organs, as well as their involvement in human diseases, such as neurological/psychiatric disorders, disorders in sensory and reproductive organs, and cancers.

Acknowledgments We thank present and previous faculty members, colleagues and collaborators, for their great contributions and excellent achievements.

References

Abercrombie M, Heaysman JE (1953) Observations on the social behaviour of cells in tissue culture. I. Speed of movement of chick heart fibroblasts in relation to their mutual contacts. Exp Cell Res 5:111–131

Abercrombie M, Heaysman JE (1954) Observations on the social behaviour of cells in tissue culture. II. Monolayering of fibroblasts. Exp Cell Res 6:293–306

Ahmed A, Chandra S, Magarinos M, Vaessin H (2003) Echinoid mutants exhibit neurogenic phenotypes and show synergistic interactions with the Notch signaling pathway. Development 130:6295–6304

Alberts B (2002) Molecular biology of the cell, 4th edn. Garland Science, New York, p. 1464

Amano H, Ikeda W, Kawano S, Kajita M, Tamaru Y, Inoue N, Minami Y, Yamada A, Takai Y (2008) Interaction and localization of Necl-5 and PDGF receptor beta at the leading edges of moving NIH3T3 cells: Implications for directional cell movement. Genes Cells 13:269–284

Amaral DG, Dent JA (1981) Development of the mossy fibers of the dentate gyrus: I. A light and electron microscopic study of the mossy fibers and their expansions. J Comp Neurol 195:51–86

Aoki J, Koike S, Asou H, Ise I, Suwa H, Tanaka T, Miyasaka M, Nomoto A (1997) Mouse homolog of poliovirus receptor-related gene 2 product, mPRR2, mediates homophilic cell aggregation. Exp Cell Res 235:374–384

Aricescu AR, Jones EY (2007) Immunoglobulin superfamily cell adhesion molecules: zippers and signals. Curr Opin Cell Biol 19:543–550

Asada M, Irie K, Morimoto K, Yamada A, Ikeda W, Takeuchi M, Takai Y (2003) ADIP, a novel Afadin- and alpha-actinin-binding protein localized at cell-cell adherens junctions. J Biol Chem 278:4103–4111

Bai J, Chiu W, Wang J, Tzeng T, Perrimon N, Hsu J (2001) The cell adhesion molecule Echinoid defines a new pathway that antagonizes the Drosophila EGF receptor signaling pathway. Development 128:591–601

Bao S, Cagan R (2005) Preferential adhesion mediated by Hibris and Roughest regulates morphogenesis and patterning in the Drosophila eye. Dev Cell 8:925–935

Barron MJ, Brookes SJ, Draper CE, Garrod D, Kirkham J, Shore RC, Dixon MJ (2008) The cell adhesion molecule nectin-1 is critical for normal enamel formation in mice. Hum Mol Genet 17:3509–3520

Bergelson JM, Cunningham JA, Droguett G, Kurt-Jones EA, Krithivas A, Hong JS, Horwitz MS, Crowell RL, Finberg RW (1997) Isolation of a common receptor for Coxsackie B viruses and adenoviruses 2 and 5. Science 275:1320–1323

Boettner B, Govek EE, Cross J, Van Aelst L (2000) The junctional multidomain protein AF-6 is a binding partner of the Rap1A GTPase and associates with the actin cytoskeletal regulator profilin. Proc Natl Acad Sci U S A 97:9064–9069

Boettner B, Harjes P, Ishimaru S, Heke M, Fan HQ, Qin Y, Van Aelst L, Gaul U (2003) The AF-6 homolog canoe acts as a Rap1 effector during dorsal closure of the Drosophila embryo. Genetics 165:159–169

Bork P, Holm L, Sander C (1994) The immunoglobulin fold. Structural classification, sequence patterns and common core. J Mol Biol 242:309–320

Bottino C, Castriconi R, Pende D, Rivera P, Nanni M, Carnemolla B, Cantoni C, Grassi J, Marcenaro S, Reymond N, Vitale M, Moretta L, Lopez M, Moretta A (2003) Identification of PVR (CD155) and nectin-2 (CD112) as cell surface ligands for the human DNAM-1 (CD226) activating molecule. J Exp Med 198:557–567

Brancati F, Fortugno P, Bottillo I, Lopez M, Josselin E, Boudghene-Stambouli O, Agolini E, Bernardini L, Bellacchio E, Iannicelli M, Rossi A, Dib-Lachachi A, Stuppia L, Palka G, Mundlos S, Stricker S, Kornak U, Zambruno G, Dallapiccola B (2010) Mutations in PVRL4, encoding cell adhesion molecule nectin-4, cause ectodermal dysplasia-syndactyly syndrome. Am J Hum Genet 87:265–273

Brooke MA, Nitoiu D, Kelsell DP (2012) Cell-cell connectivity: desmosomes and disease. J Pathol 226:158–171

Buchert M, Schneider S, Meskenaite V, Adams MT, Canaani E, Baechi T, Moelling K, Hovens CM (1999) The junction-associated protein AF-6 interacts and clusters with specific Eph receptor tyrosine kinases at specialized sites of cell-cell contact in the brain. J Cell Biol 144:361–371

Bustos T, Simosa V, Pinto-Cisternas J, Abramovits W, Jolay L, Rodriguez L, Fernandez L, Ramela M (1991) Autosomal recessive ectodermal dysplasia: I. An undescribed dysplasia/malformation syndrome. Am J Med Genet 41:398–404

Carmena A, Speicher S, Baylies M (2006) The PDZ protein Canoe/AF-6 links Ras-MAPK, Notch and Wingless/Wnt signaling pathways by directly interacting with Ras, Notch and Dishevelled. PLoS One 1:e66

Carthew RW (2007) Pattern formation in the Drosophila eye. Curr Opin Genet Dev 17:309–313

Cheng CY, Mruk DD (2002) Cell junction dynamics in the testis: Sertoli-germ cell interactions and male contraceptive development. Physiol Rev 82:825–874

Comoglio PM, Boccaccio C, Trusolino L (2003) Interactions between growth factor receptors and adhesion molecules: breaking the rules. Curr Opin Cell Biol 15:565–571

Cordero JB, Larson DE, Craig CR, Hays R, Cagan R (2007) Dynamic decapentaplegic signaling regulates patterning and adhesion in the Drosophila pupal retina. Development 134:1861–1871

Dermody TS, Kirchner E, Guglielmi KM, Stehle T (2009) Immunoglobulin superfamily virus receptors and the evolution of adaptive immunity. PLoS Pathog 5:e1000481

Dworak HA, Charles MA, Pellerano LB, Sink H (2001) Characterization of Drosophila hibris, a gene related to human nephrin. Development 128:4265–4276

Eberle F, Dubreuil P, Mattei MG, Devilard E, Lopez M (1995) The human PRR2 gene, related to the human poliovirus receptor gene (PVR), is the true homolog of the murine MPH gene. Gene 159:267–272

Fannon AM, Colman DR (1996) A model for central synaptic junctional complex formation based on the differential adhesive specificities of the cadherins. Neuron 17:423–434

Farquhar MG, Palade GE (1965) Cell junctions in amphibian skin. J Cell Biol 26:263–291

Fisher HW, Yeh J (1967) Contact inhibition in colony formation. Science 155:581–582

Fournier G, Cabaud O, Josselin E, Chaix A, Adelaide J, Isnardon D, Restouin A, Castellano R, Dubreuil P, Chaffanet M, Birnbaum D, Lopez M (2011) Loss of AF6/afadin, a marker of poor outcome in breast cancer, induces cell migration, invasiveness and tumor growth. Oncogene 30:3862–3874

Fuchs A, Cella M, Giurisato E, Shaw AS, Colonna M (2004) Cutting edge: CD96 (tactile) promotes NK cell-target cell adhesion by interacting with the poliovirus receptor (CD155). J Immunol 172:3994–3998

Fujito T, Ikeda W, Kakunaga S, Minami Y, Kajita M, Sakamoto Y, Monden M, Takai Y (2005) Inhibition of cell movement and proliferation by cell-cell contact-induced interaction of necl-5 with nectin-3. J Cell Biol 171:165–173

Fukuhara T, Shimizu K, Kawakatsu T, Fukuyama T, Minami Y, Honda T, Hoshino T, Yamada T, Ogita H, Okada M, Takai Y (2004) Activation of Cdc42 by trans interactions of the cell adhesion molecules nectins through c-Src and Cdc42-GEF FRG. J Cell Biol 166:393–405

Fukumoto Y, Kurita S, Takai Y, Ogita H (2011) Role of the scaffold protein ADIP in platelet-derived growth factor-induced cell movement by activating Rac through Vav2. J Biol Chem 286:43537–43548

Garay E, Patino-Lopez E, Islas S, Alarcon L, Canche-Pool E, Valle-Rios R, Medina-Contreras O, Granados G, Chavez-Munguia B, Juaristi E, Ortiz-Navarrete V, Gonzalez-Mariscal L (2010) CRTAM: a molecule involved in epithelial cell adhesion. J Cell Biochem 111:111–122

Geraghty RJ, Krummenacher C, Cohen GH, Eisenberg RJ, Spear PG (1998) Entry of alphaherpesviruses mediated by poliovirus receptor-related protein 1 and poliovirus receptor. Science 280:1618–1620

Gerhard DS, Wagner L, Feingold EA, Shenmen CM, Grouse LH, Schuler G, Klein SL, Old S, Rasooly R, Good P, Guyer M, Peck AM, Derge JG, Lipman D, Collins FS, Jang W, Sherry S, Feolo M, Misquitta L, Lee E, Rotmistrovsky K, Greenhut SF, Schaefer CF, Buetow K, Bonner TI, Haussler D, Kent J, Kiekhaus M, Furey T, Brent M, Prange C, Schreiber K, Shapiro N, Bhat NK, Hopkins RF, Hsie F, Driscoll T, Soares MB, Casavant TL, Scheetz TE, Brown-stein MJ, Usdin TB, Toshiyuki S, Carninci P, Piao Y, Dudekula DB, Ko MS, Kawakami K, Suzuki Y, Sugano S, Gruber CE, Smith MR, Simmons B, Moore T, Waterman R, Johnson SL, Ruan Y, Wei CL, Mathavan S, Gunaratne PH, Wu J, Garcia AM, Hulyk SW, Fuh E, Yuan Y, Sneed A, Kowis C, Hodgson A, Muzny DM, McPherson J, Gibbs RA, Fahey J, Helton E, Ketteman M, Madan A, Rodrigues S, Sanchez A, Whiting M, Madari A, Young AC, Wetherby KD, Granite SJ, Kwong PN, Brinkley CP, Pearson RL, Bouffard GG, Blakesly RW, Green ED, Dickson MC, Rodriguez AC, Grimwood J, Schmutz J, Myers RM, Butterfield YS, Griffith M, Griffith OL, Krzywinski MI, Liao N, Morin R, Palmquist D (2004) The status, quality, and expansion of the NIH full-length cDNA project: the mammalian gene collection (MGC). Genome Res 14:2121–2127

Gliki G, Ebnet K, Aurrand-Lions M, Imhof BA, Adams RH (2004) Spermatid differentiation requires the assembly of a cell polarity complex downstream of junctional adhesion molecule-C. Nature 431:320–324

Gomperts SN (1996) Clustering membrane proteins: it's all coming together with the PSD-95/SAP90 protein family. Cell 84:659–662

Gonzalez-Reyes A (2003) Stem cells, niches and cadherins: a view from Drosophila. J Cell Sci 116:949–954

Gupta GP, Massague J (2006) Cancer metastasis: building a framework. Cell 127:679–695

Hall A (1998) Rho GTPases and the actin cytoskeleton. Science. 279:509–514

Harold D, Abraham R, Hollingworth P, Sims R, Gerrish A, Hamshere ML, Pahwa JS, Moskvina V, Dowzell K, Williams A, Jones N, Thomas C, Stretton A, Morgan AR, Lovestone S, Powell J, Proitsi P, Lupton MK, Brayne C, Rubinsztein DC, Gill M, Lawlor B, Lynch A, Morgan K, Brown KS, Passmore PA, Craig D, McGuinness B, Todd S, Holmes C, Mann D, Smith AD, Love S, Kehoe PG, Hardy J, Mead S, Fox N, Rossor M, Collinge J, Maier W, Jessen F, Schurmann B, Van Den Bussche H, Heuser I, Kornhuber J, Wiltfang J, Dichgans M, Frolich L, Hampel H, Hull M, Rujescu D, Goate AM, Kauwe JS, Cruchaga C, Nowotny P, Morris JC, Mayo K, Sleegers K, Bettens K, Engelborghs S, De Deyn PP, Van Broeckhoven C, Livingston G, Bass NJ, Gurling H, McQuillin A, Gwilliam R, Deloukas P, Al-Chalabi A, Shaw CE, Tsolaki M, Singleton AB, Guerreiro R, Muhleisen TW, Nothen MM, Moebus S, Jockel KH, Klopp N, Wichmann HE, Carrasquillo MM, Pankratz VS, Younkin SG, Holmans PA, O'Donovan M, Owen MJ, Williams J (2009) Genome-wide association study identifies variants at CLU and PICALM associated with Alzheimer's disease. Nat Genet 41:1088–1093

Hayashi T, Carthew RW (2004) Surface mechanics mediate pattern formation in the developing retina. Nature 431:647–652

Hock B, Bohme B, Karn T, Yamamoto T, Kaibuchi K, Holtrich U, Holland S, Pawson T, Rubsamen-Waigmann H, Strebhardt K (1998) PDZ-domain-mediated interaction of the Eph-related receptor tyrosine kinase EphB3 and the ras-binding protein AF6 depends on the kinase activity of the receptor. Proc Natl Acad Sci U S A 95:9779–9784

Honda T, Shimizu K, Kawakatsu T, Yasumi M, Shingai T, Fukuhara A, Ozaki-Kuroda K, Irie K, Nakanishi H, Takai Y (2003) Antagonistic and agonistic effects of an extracellular fragment of nectin on formation of E-cadherin-based cell-cell adhesion. Genes Cells 8:51–63

Honda T, Sakisaka T, Yamada T, Kumazawa N, Hoshino T, Kajita M, Kayahara T, Ishizaki H, Tanaka-Okamoto M, Mizoguchi A, Manabe T, Miyoshi J, Takai Y. (2006) Involvement of nectins in the formation of puncta adherentia junctions and the mossy fiber trajectory in the mouse hippocampus. Mol Cell Neurosci 31:315–325

Hoshino T, Sakisaka T, Baba T, Yamada T, Kimura T, Takai Y (2005) Regulation of E-cadherin endocytosis by nectin through afadin, Rap1, and p120ctn. J Biol Chem 280:24095–24103

Ikeda W, Nakanishi H, Miyoshi J, Mandai K, Ishizaki H, Tanaka M, Togawa A, Takahashi K, Nishioka H, Yoshida H, Mizoguchi A, Nishikawa S, Takai Y (1999) Afadin: a key molecule essential for structural organization of cell-cell junctions of polarized epithelia during embryogenesis. J Cell Biol 146:1117–1132

Ikeda W, Kakunaga S, Itoh S, Shingai T, Takekuni K, Satoh K, Inoue Y, Hamaguchi A, Morimoto K, Takeuchi M, Imai T, Takai Y (2003) Tage4/nectin-like molecule-5 heterophilically trans-interacts with cell adhesion molecule nectin-3 and enhances cell migration. J Biol Chem 278:28167–28172

Inagaki M, Irie K, Ishizaki H, Tanaka-Okamoto M, Morimoto K, Inoue E, Ohtsuka T, Miyoshi J, Takai Y (2005) Roles of cell-adhesion molecules nectin 1 and nectin 3 in ciliary body development. Development 132:1525–1537

Inagaki M, Irie K, Ishizaki H, Tanaka-Okamoto M, Miyoshi J, Takai Y (2006) Role of cell adhesion molecule nectin-3 in spermatid development. Genes Cells 11:1125–1132

Islam R, Wei SY, Chiu WH, Hortsch M, Hsu JC (2003) Neuroglian activates Echinoid to antagonize the Drosophila EGF receptor signaling pathway. Development 130:2051–2059

Kajita M, Ikeda W, Tamaru Y, Takai Y (2007) Regulation of platelet-derived growth factor-induced Ras signaling by poliovirus receptor Necl-5 and negative growth regulator Sprouty2. Genes Cells 12:345–357

Kakunaga S, Ikeda W, Shingai T, Fujito T, Yamada A, Minami Y, Imai T, Takai Y (2004) Enhancement of serum- and platelet-derived growth factor-induced cell proliferation by Necl-5/Tage4/poliovirus receptor/CD155 through the Ras-Raf-MEK-ERK signaling. J Biol Chem 279:36419–36425

Kawano S, Ikeda W, Kishimoto M, Ogita H, Takai Y (2009) Silencing of ErbB3/ErbB2 signaling by immunoglobulin-like Necl-2. J Biol Chem 284:23793–23805

Kestila M, Lenkkeri U, Mannikko M, Lamerdin J, McCready P, Putaala H, Ruotsalainen V, Morita T, Nissinen M, Herva R, Kashtan CE, Peltonen L, Holmberg C, Olsen A, Tryggvason K (1998) Positionally cloned gene for a novel glomerular protein–nephrin–is mutated in congenital nephrotic syndrome. Mol Cell 1:575–582.

Koch AW, Pokutta S, Lustig A, Engel J (1997) Calcium binding and homoassociation of E-cadherin domains. Biochemistry 36:7697–7705

Kuramitsu K, Ikeda W, Inoue N, Tamaru Y, Takai Y (2008) Novel role of nectin: implication in the co-localization of JAM-A and claudin-1 at the same cell-cell adhesion membrane domain. Genes Cells 13:797–805

Kurita S, Ogita H, Takai Y (2011) Cooperative role of the nectin-nectin and nectin-afadin interactions in the formation of the nectin-based cell-cell adhesion. J Biol Chem 286:36297–36303

Lachke SA, Higgins AW, Inagaki M, Saadi I, Xi Q, Long M, Quade BJ, Talkowski ME, Gusella JF, Fujimoto A, Robinson ML, Yang Y, Duong QT, Shapira I, Motro B, Miyoshi J, Takai Y, Morton CC, Maas RL (2011) The cell adhesion gene PVRL3 is associated with congenital ocular defects. Hum Genet 131:235–250

Lander ES, Linton LM, Birren B, Nusbaum C, Zody MC, Baldwin J, Devon K, Dewar K, Doyle M, FitzHugh W, Funke R, Gage D, Harris K, Heaford A, Howland J, Kann L, Lehoczky J, LeVine R, McEwan P, McKernan K, Meldrim J, Mesirov JP, Miranda C, Morris W, Naylor J, Raymond C, Rosetti M, Santos R, Sheridan A, Sougnez C, Stange-Thomann N, Stojanovic N, Subramanian A, Wyman D, Rogers J, Sulston J, Ainscough R, Beck S, Bentley D, Burton J, Clee C, Carter N, Coulson A, Deadman R, Deloukas P, Dunham A, Dunham I, Durbin R, French L, Grafham D, Gregory S, Hubbard T, Humphray S, Hunt A, Jones M, Lloyd C, McMurray A, Matthews L, Mercer S, Milne S, Mullikin JC, Mungall A, Plumb R, Ross M, Shownkeen R, Sims S, Waterston RH, Wilson RK, Hillier LW, McPherson JD, Marra MA, Mardis ER, Fulton LA, Chinwalla AT, Pepin KH, Gish WR, Chissoe SL, Wendl MC, Delehaunty KD, Miner TL, Delehaunty A, Kramer JB, Cook LL, Fulton RS, Johnson DL, Minx PJ, Clifton SW, Hawkins T, Branscomb E, Predki P, Richardson P, Wenning S, Slezak T, Doggett N, Cheng JF, Olsen A,

Lucas S, Elkin C, Uberbacher E, Frazier M (2001) Initial sequencing and analysis of the human genome. Nature 409:860–921

Lanford PJ, Lan Y, Jiang R, Lindsell C, Weinmaster G, Gridley T, Kelley MW (1999) Notch signalling pathway mediates hair cell development in mammalian cochlea. Nat Genet 21:289–292

Laplante C, Nilson LA (2006) Differential expression of the adhesion molecule Echinoid drives epithelial morphogenesis in Drosophila. Development 133:3255–3264

Laplante C, Nilson LA (2011) Asymmetric distribution of Echinoid defines the epidermal leading edge during Drosophila dorsal closure. J Cell Biol 192:335–348

Leckband D, Sivasankar S (2000) Mechanism of homophilic cadherin adhesion. Curr Opin Cell Biol 12:587–592

Lecuit T (2005) Cell adhesion: sorting out cell mixing with echinoid? Curr Biol 15:R505–507

Lehtonen S, Lehtonen E, Kudlicka K, Holthofer H, Farquhar MG (2004) Nephrin forms a complex with adherens junction proteins and CASK in podocytes and in Madin-Darby canine kidney cells expressing nephrin. Am J Pathol 165:923–936

Letessier A, Garrido-Urbani S, Ginestier C, Fournier G, Esterni B, Monville F, Adelaide J, Geneix J, Xerri L, Dubreuil P, Viens P, Charafe-Jauffret E, Jacquemier J, Birnbaum D, Lopez M, Chaffanet M (2007) Correlated break at PARK2/FRA6E and loss of AF-6/Afadin protein expression are associated with poor outcome in breast cancer. Oncogene 26:298–307

Lin HP, Chen HM, Wei SY, Chen LY, Chang LH, Sun YJ, Huang SY, Hsu JC (2007) Cell adhesion molecule Echinoid associates with unconventional myosin VI/Jaguar motor to regulate cell morphology during dorsal closure in Drosophila. Dev Biol 311:423–433

Liu G, Kaw B, Kurfis J, Rahmanuddin S, Kanwar YS, Chugh SS (2003) Neph1 and nephrin interaction in the slit diaphragm is an important determinant of glomerular permeability. J Clin Invest 112:209–221

Lopez M, Eberle F, Mattei MG, Gabert J, Birg F, Bardin F, Maroc C, Dubreuil P (1995) Complementary DNA characterization and chromosomal localization of a human gene related to the poliovirus receptor-encoding gene. Gene 155:261–265

Lorger M, Moelling K (2006) Regulation of epithelial wound closure and intercellular adhesion by interaction of AF6 with actin cytoskeleton. J Cell Sci 119:3385–3398

Lum JJ, Bauer DE, Kong M, Harris MH, Li C, Lindsten T, Thompson CB (2005) Growth factor regulation of autophagy and cell survival in the absence of apoptosis. Cell 120:237–248

Mandai K, Nakanishi H, Satoh A, Obaishi H, Wada M, Nishioka H, Itoh M, Mizoguchi A, Aoki T, Fujimoto T, Matsuda Y, Tsukita S, Takai Y (1997) Afadin: a novel actin filament-binding protein with one PDZ domain localized at cadherin-based cell-to-cell adherens junction. J Cell Biol 139:517–528

Mandai K, Nakanishi H, Satoh A, Takahashi K, Satoh K, Nishioka H, Mizoguchi A, Takai Y (1999) Ponsin/SH3P12: an l-afadin- and vinculin-binding protein localized at cell-cell and cell-matrix adherens junctions. J Cell Biol 144:1001–1017

Masuda M, Kikuchi S, Maruyama T, Sakurai-Yageta M, Williams YN, Ghosh HP, Murakami Y (2005) Tumor suppressor in lung cancer (TSLC)1 suppresses epithelial cell scattering and tubulogenesis. J Biol Chem 280:42164–42171

Matsuo T, Takahashi K, Suzuki E, Yamamoto D (1999) The Canoe protein is necessary in adherens junctions for development of ommatidial architecture in the Drosophila compound eye. Cell Tissue Res 298:397–404

Meng W, Mushika Y, Ichii T, Takeichi M. (2008) Anchorage of microtubule minus ends to adherens junctions regulates epithelial cell-cell contacts. Cell 135:948–959

Menichella DM, Arroyo EJ, Awatramani R, Xu T, Baron P, Vallat JM, Balsamo J, Lilien J, Scarlato G, Kamholz J, Scherer SS, Shy ME (2001) Protein zero is necessary for E-cadherin-mediated adherens junction formation in Schwann cells. Mol Cell Neurosci 18:606–618

Minami Y, Ikeda W, Kajita M, Fujito T, Amano H, Tamaru Y, Kuramitsu K, Sakamoto Y, Monden M, Takai Y (2007) Necl-5/poliovirus receptor interacts in cis with integrin alphaVbeta3 and regulates its clustering and focal complex formation. J Biol Chem 282:18481–18496

Minami A, Mizutani K, Waseda M, Kajita M, Miyata M, Ikeda W, Takai Y (2010) Necl-5/PVR enhances PDGF-induced attraction of growing microtubules to the plasma membrane of the leading edge of moving NIH3T3 cells. Genes Cells 15:1123–1135

Miyamoto H, Nihonmatsu I, Kondo S, Ueda R, Togashi S, Hirata K, Ikegami Y, Yamamoto D (1995) Canoe encodes a novel protein containing a GLGF/DHR motif and functions with Notch and scabrous in common developmental pathways in Drosophila. Genes Dev 9:612–625

Miyata M, Ogita H, Komura H, Nakata S, Okamoto R, Ozaki M, Majima T, Matsuzawa N, Kawano S, Minami A, Waseda M, Fujita N, Mizutani K, Rikitake Y, Takai Y (2009a) Localization of nectin-free afadin at the leading edge and its involvement in directional cell movement induced by platelet-derived growth factor. J Cell Sci 122:4319–4329

Miyata M, Rikitake Y, Takahashi M, Nagamatsu Y, Yamauchi Y, Ogita H, Hirata K, Takai Y (2009b) Regulation by afadin of cyclical activation and inactivation of Rap1, Rac1, and RhoA small G proteins at leading edges of moving NIH3T3 cells. J Biol Chem 284:24595–24609

Mizoguchi A, Nakanishi H, Kimura K, Matsubara K, Ozaki-Kuroda K, Katata T, Honda T, Kiyohara Y, Heo K, Higashi M, Tsutsumi T, Sonoda S, Ide C, Takai Y (2002) Nectin: an adhesion molecule involved in formation of synapses. J Cell Biol 156:555–565

Mizutani K, Kawano S, Minami A, Waseda M, Ikeda W, Takai Y. (2011) Interaction of nectin-like molecule 2 with integrin α6β4 and inhibition of the disassembly of integrin α6β4 from hemidesmosomes. J Biol Chem 286(42):36667–36676

Monier-Gavelle F, Duband JL (1997) Cross talk between adhesion molecules: control of N-cadherin activity by intracellular signals elicited by β1 and β3 integrins in migrating neural crest cells. J Cell Biol 137:1663–1681

Morimoto K, Satoh-Yamaguchi K, Hamaguchi A, Inoue Y, Takeuchi M, Okada M, Ikeda W, Takai Y, Imai T (2008) Interaction of cancer cells with platelets mediated by Necl-5/poliovirus receptor enhances cancer cell metastasis to the lungs. Oncogene 27:264–273

Morrison ME, Racaniello VR (1992) Molecular cloning and expression of a murine homolog of the human poliovirus receptor gene. J Virol 66:2807–2813

Mruk DD, Cheng CY (2004) Sertoli-Sertoli and Sertoli-germ cell interactions and their significance in germ cell movement in the seminiferous epithelium during spermatogenesis. Endocr Rev 25:747–806

Mueller S, Rosenquist TA, Takai Y, Bronson RA, Wimmer E (2003) Loss of nectin-2 at Sertoli-spermatid junctions leads to male infertility and correlates with severe spermatozoan head and midpiece malformation, impaired binding to the zona pellucida, and oocyte penetration. Biol Reprod 69:1330–1340

Muhlebach MD, Mateo M, Sinn PL, Prufer S, Uhlig KM, Leonard VH, Navaratnarajah CK, Frenzke M, Wong XX, Sawatsky B, Ramachandran S, McCray PB, Cichutek K, von Messling V, Lopez M, Cattaneo R (2011) Adherens junction protein nectin-4 is the epithelial receptor for measles virus. Nature

Nagata M, Sakurai-Yageta M, Yamada D, Goto A, Ito A, Fukuhara H, Kume H, Morikawa T, Fukayama M, Homma Y, Murakami Y (2012) Aberrations of a cell adhesion molecule CADM4 in renal clear cell carcinoma. Int J Cancer 130:1329–1337

Nakai R, Maniwa Y, Tanaka Y, Nishio W, Yoshimura M, Okita Y, Ohbayashi C, Satoh N, Ogita H, Takai Y, Hayashi Y (2010) Overexpression of Necl-5 correlates with unfavorable prognosis in patients with lung adenocarcinoma. Cancer Sci 101:1326–1330

Narita H, Yamamoto Y, Suzuki M, Miyazaki N, Yoshida A, Kawai K, Iwasaki K, Nakagawa A, Takai Y, Sakisaka T (2011) Crystal Structure of the cis-Dimer of Nectin-1: implications for the architecture of cell-cell junctions. J Biol Chem 286:12659–12669

Noyce RS, Bondre DG, Ha MN, Lin LT, Sisson G, Tsao MS, Richardson CD (2011) Tumor cell marker PVRL4 (nectin 4) is an epithelial cell receptor for measles virus. PLoS Pathog 7:e1002240

Oda H, Takeichi M (2011) Evolution: structural and functional diversity of cadherin at the adherens junction. J Cell Biol 193:1137–1146

Ooshio T, Irie K, Morimoto K, Fukuhara A, Imai T, Takai Y (2004) Involvement of LMO7 in the association of two cell-cell adhesion molecules, nectin and E-cadherin, through afadin and alpha-actinin in epithelial cells. J Biol Chem 279:31365–31373

Ooshio T, Fujita N, Yamada A, Sato T, Kitagawa Y, Okamoto R, Nakata S, Miki A, Irie K, Takai Y (2007) Cooperative roles of Par-3 and afadin in the formation of adherens and tight junctions. J Cell Sci 120:2352–2365

Ooshio T, Kobayashi R, Ikeda W, Miyata M, Fukumoto Y, Matsuzawa N, Ogita H, Takai Y (2010) Involvement of the interaction of afadin with ZO-1 in the formation of tight junctions in Madin-Darby canine kidney cells. J Biol Chem 285:5003–5012

Ozaki-Kuroda K, Nakanishi H, Ohta H, Tanaka H, Kurihara H, Mueller S, Irie K, Ikeda W, Sakai T, Wimmer E, Nishimune Y, Takai Y (2002) Nectin couples cell-cell adhesion and the actin scaffold at heterotypic testicular junctions. Curr Biol 12:1145–1150

Pende D, Spaggiari GM, Marcenaro S, Martini S, Rivera P, Capobianco A, Falco M, Lanino E, Pierri I, Zambello R, Bacigalupo A, Mingari MC, Moretta A, Moretta L (2005) Analysis of the receptor-ligand interactions in the natural killer-mediated lysis of freshly isolated myeloid or lymphoblastic leukemias: evidence for the involvement of the Poliovirus receptor (CD155) and Nectin-2 (CD112). Blood 105:2066–2073

Pende D, Castriconi R, Romagnani P, Spaggiari GM, Marcenaro S, Dondero A, Lazzeri E, Lasagni L, Martini S, Rivera P, Capobianco A, Moretta L, Moretta A, Bottino C (2006) Expression of the DNAM-1 ligands, Nectin-2 (CD112) and poliovirus receptor (CD155), on dendritic cells: relevance for natural killer-dendritic cell interaction. Blood 107:2030–2036

Ponting CP (1995) AF-6/cno: neither a kinesin nor a myosin, but a bit of both. Trends Biochem Sci 20:265–266

Ponting CP, Benjamin DR (1996) A novel family of Ras-binding domains. Trends Biochem Sci 21:422–425

Prasad R, Gu Y, Alder H, Nakamura T, Canaani O, Saito H, Huebner K, Gale RP, Nowell PC, Kuriyama K (1993) Cloning of the ALL-1 fusion partner, the AF-6 gene, involved in acute myeloid leukemias with the t(6;11) chromosome translocation. Cancer Res 53:5624–5628

Radziwill G, Erdmann RA, Margelisch U, Moelling K (2003) The Bcr kinase downregulates Ras signaling by phosphorylating AF-6 and binding to its PDZ domain. Mol Cell Biol 23:4663–4672

Ramos RG, Igloi GL, Lichte B, Baumann U, Maier D, Schneider T, Brandstatter JH, Frohlich A, Fischbach KF (1993) The irregular chiasm C-roughest locus of Drosophila, which affects axonal projections and programmed cell death, encodes a novel immunoglobulin-like protein. Genes Dev 7:2533–2547

Raveh S, Gavert N, Spiegel I, Ben-Ze'ev A (2009) The cell adhesion nectin-like molecules (Necl) 1 and 4 suppress the growth and tumorigenic ability of colon cancer cells. J Cell Biochem 108:326–336

Raviola G, Raviola E (1978) Intercellular junctions in the ciliary epithelium. Invest Ophthalmol Vis Sci 17:958–981

Reiter C, Schimansky T, Nie Z, Fischbach KF (1996) Reorganization of membrane contacts prior to apoptosis in the Drosophila retina: the role of the IrreC-rst protein. Development 122:1931–1940

Reymond N, Fabre S, Lecocq E, Adelaide J, Dubreuil P, Lopez M (2001) Nectin4/PRR4, a new afadin-associated member of the nectin family that trans-interacts with nectin1/PRR1 through V domain interaction. J Biol Chem 276:43205–43215

Ronnstrand L, Heldin CH (2001) Mechanisms of platelet-derived growth factor-induced chemotaxis. Int J Cancer 91:757–762

Saito S, Matsushima M, Shirahama S, Minaguchi T, Kanamori Y, Minami M, Nakamura Y (1998) Complete genomic structure DNA polymorphisms, and alternative splicing of the human AF-6 gene. DNA Res 5:115–120

Sakamoto Y, Ogita H, Hirota T, Kawakatsu T, Fukuyama T, Yasumi M, Kanzaki N, Ozaki M, Takai Y (2006) Interaction of integrin $\alpha v\beta 3$ with nectin. Implication in cross-talk between cell-matrix and cell-cell junctions. J Biol Chem 281:19631–19644

Sakamoto Y, Ogita H, Komura H, Takai Y (2008) Involvement of nectin in inactivation of integrin αvβ3 after the establishment of cell-cell adhesion. J Biol Chem 283:496–505

Sakisaka T, Taniguchi T, Nakanishi H, Takahashi K, Miyahara M, Ikeda W, Yokoyama S, Peng YF, Yamanishi K, Takai Y (2001) Requirement of interaction of nectin-1alpha/HveC with afadin for efficient cell-cell spread of herpes simplex virus type 1. J Virol 75:4734–4743

Sato T, Irie K, Okamoto R, Ooshio T, Fujita N, Takai Y (2005) Common signaling pathway is used by the trans-interaction of Necl-5/Tage4/PVR/CD155 and nectin, and of nectin and nectin during the formation of cell-cell adhesion. Cancer Sci 96:578–589

Sato T, Fujita N, Yamada A, Ooshio T, Okamoto R, Irie K, Takai Y (2006) Regulation of the assembly and adhesion activity of E-cadherin by nectin and afadin for the formation of adherens junctions in Madin-Darby canine kidney cells. J Biol Chem 281:5288–5299

Satoh-Horikawa K, Nakanishi H, Takahashi K, Miyahara M, Nishimura M, Tachibana K, Mizoguchi A, Takai Y (2000) Nectin-3, a new member of immunoglobulin-like cell adhesion molecules that shows homophilic and heterophilic cell-cell adhesion activities. J Biol Chem 275:10291–10299

Sawyer JK, Harris NJ, Slep KC, Gaul U, Peifer M (2009) The Drosophila afadin homologue Canoe regulates linkage of the actin cytoskeleton to adherens junctions during apical constriction. J Cell Biol 186:57–73

Schreider C, Peignon G, Thenet S, Chambaz J, Pincon-Raymond M (2002) Integrin-mediated functional polarization of Caco-2 cells through E-cadherin–actin complexes. J Cell Sci 115:543–552

Shen K, Bargmann CI (2003) The immunoglobulin superfamily protein SYG-1 determines the location of specific synapses in C. elegans. Cell 112:619–630

Shen K, Fetter RD, Bargmann CI (2004) Synaptic specificity is generated by the synaptic guidepost protein SYG-2 and its receptor, SYG-1. Cell 116:869–881

Sheng M (1996) PDZs and receptor/channel clustering: rounding up the latest suspects. Neuron 17:575–578

Simonneau L, Gallego M, Pujol R (2003) Comparative expression patterns of T-, N-, E-cadherins, β-catenin, and polysialic acid neural cell adhesion molecule in rat cochlea during development: implications for the nature of Kolliker's organ. J Comp Neurol 459:113–126

Sozen MA, Suzuki K, Tolarova MM, Bustos T, Fernandez Iglesias JE, Spritz RA (2001) Mutation of PVRL1 is associated with sporadic, non-syndromic cleft lip/palate in northern Venezuela. Nat Genet 29:141–142

Speicher S, Fischer A, Knoblich J, Carmena A (2008) The PDZ protein Canoe regulates the asymmetric division of Drosophila neuroblasts and muscle progenitors. Curr Biol 18:831–837

Stanietsky N, Simic H, Arapovic J, Toporik A, Levy O, Novik A, Levine Z, Beiman M, Dassa L, Achdout H, Stern-Ginossar N, Tsukerman P, Jonjic S, Mandelboim O (2009) The interaction of TIGIT with PVR and PVRL2 inhibits human NK cell cytotoxicity. Proc Natl Acad Sci U S A 106:17858–17863

Strausberg RL, Feingold EA, Grouse LH, Derge JG, Klausner RD, Collins FS, Wagner L, Shenmen CM, Schuler GD, Altschul SF, Zeeberg B, Buetow KH, Schaefer CF, Bhat NK, Hopkins RF, Jordan H, Moore T, Max SI, Wang J, Hsieh F, Diatchenko L, Marusina K, Farmer AA, Rubin GM, Hong L, Stapleton M, Soares MB, Bonaldo MF, Casavant TL, Scheetz TE, Brownstein MJ, Usdin TB, Toshiyuki S, Carninci P, Prange C, Raha SS, Loquellano NA, Peters GJ, Abramson RD, Mullahy SJ, Bosak SA, McEwan PJ, McKernan KJ, Malek JA, Gunaratne PH, Richards S, Worley KC, Hale S, Garcia AM, Gay LJ, Hulyk SW, Villalon DK, Muzny DM, Sodergren EJ, Lu X, Gibbs RA, Fahey J, Helton E, Ketteman M, Madan A, Rodrigues S, Sanchez A, Whiting M, Young AC, Shevchenko Y, Bouffard GG, Blakesley RW, Touchman JW, Green ED, Dickson MC, Rodriguez AC, Grimwood J, Schmutz J, Myers RM, Butterfield YS, Krzywinski MI, Skalska U, Smailus DE, Schnerch A, Schein JE, Jones SJ, Marra MA (2002) Generation and initial analysis of more than 15,000 full-length human and mouse cDNA sequences. Proc Natl Acad Sci U S A 99:16899–16903

Su L, Hattori M, Moriyama M, Murata N, Harazaki M, Kaibuchi K, Minato N (2003) AF-6 controls integrin-mediated cell adhesion by regulating Rap1 activation through the specific recruitment of Rap1GTP and SPA-1. J Biol Chem 278:15232–15238

Suzuki K, Bustos T, Spritz RA (1998) Linkage disequilibrium mapping of the gene for Margarita Island ectodermal dysplasia (ED4) to 11q23. Am J Hum Genet 63:1102–1107

Suzuki K, Hu D, Bustos T, Zlotogora J, Richieri-Costa A, Helms JA, Spritz RA (2000) Mutations of PVRL1, encoding a cell-cell adhesion molecule/herpesvirus receptor, in cleft lip/palate-ectodermal dysplasia. Nat Genet 25:427–430

Tachibana K, Nakanishi H, Mandai K, Ozaki K, Ikeda W, Yamamoto Y, Nagafuchi A, Tsukita S, Takai Y (2000) Two cell adhesion molecules, nectin and cadherin, interact through their cytoplasmic domain-associated proteins. J Cell Biol 150:1161–1176

Tadokoro S, Shattil SJ, Eto K, Tai V, Liddington RC, de Pereda JM, Ginsberg MH, Calderwood DA (2003) Talin binding to integrin beta tails: a final common step in integrin activation. Science 302:103–106

Takahashi K, Matsuo T, Katsube T, Ueda R, Yamamoto D (1998) Direct binding between two PDZ domain proteins Canoe and ZO-1 and their roles in regulation of the jun N-terminal kinase pathway in Drosophila morphogenesis. Mech Dev 78:97–111

Takahashi K, Nakanishi H, Miyahara M, Mandai K, Satoh K, Satoh A, Nishioka H, Aoki J, Nomoto A, Mizoguchi A, Takai Y (1999) Nectin/PRR: an immunoglobulin-like cell adhesion molecule recruited to cadherin-based adherens junctions through interaction with Afadin, a PDZ domain-containing protein. J Cell Biol 145:539–549

Takai Y, Nakanishi H (2003) Nectin and afadin: novel organizers of intercellular junctions. J Cell Sci 116:17–27

Takai Y, Irie K, Shimizu K, Sakisaka T, Ikeda W (2003) Nectins and nectin-like molecules: roles in cell adhesion, migration, and polarization. Cancer Sci 94:655–667

Takai Y, Ikeda W, Ogita H, Rikitake Y (2008a) The immunoglobulin-like cell adhesion molecule nectin and its associated protein afadin. Annu Rev Cell Dev Biol 24:309–342

Takai Y, Miyoshi J, Ikeda W, Ogita H (2008b) Nectins and nectin-like molecules: roles in contact inhibition of cell movement and proliferation. Nat Rev Mol Cell Biol 9:603–615

Takei N, Miyashita A, Tsukie T, Arai H, Asada T, Imagawa M, Shoji M, Higuchi S, Urakami K, Kimura H, Kakita A, Takahashi H, Tsuji S, Kanazawa I, Ihara Y, Odani S, Kuwano R (2009) Genetic association study on in and around the APOE in late-onset Alzheimer disease in Japanese. Genomics 93:441–448

Takekuni K, Ikeda W, Fujito T, Morimoto K, Takeuchi M, Monden M, Takai Y (2003) Direct binding of cell polarity protein PAR-3 to cell-cell adhesion molecule nectin at neuroepithelial cells of developing mouse. J Biol Chem 278:5497–5500

Tanenbaum SB, Gorski SM, Rusconi JC, Cagan RL (2000) A screen for dominant modifiers of the irreC-rst cell death phenotype in the developing Drosophila retina. Genetics 156:205–217

Tawa H, Rikitake Y, Takahashi M, Amano H, Miyata M, Satomi-Kobayashi S, Kinugasa M, Nagamatsu Y, Majima T, Ogita H, Miyoshi J, Hirata K, Takai Y (2010) Role of afadin in vascular endothelial growth factor- and sphingosine 1-phosphate-induced angiogenesis. Circ Res 106:1731–1742

Tepass U, Harris KP (2007) Adherens junctions in Drosophila retinal morphogenesis. Trends Cell Biol 17:26–35

Thoulouze MI, Lafage M, Schachner M, Hartmann U, Cremer H, Lafon M (1998) The neural cell adhesion molecule is a receptor for rabies virus. J Virol 72:7181–7190

Togashi H, Miyoshi J, Honda T, Sakisaka T, Takai Y, Takeichi M (2006) Interneurite affinity is regulated by heterophilic nectin interactions in concert with the cadherin machinery. J Cell Biol 174:141–151

Togashi H, Kominami K, Waseda M, Komura H, Miyoshi J, Takeichi M, Takai Y (2011) Nectins establish a checkerboard-like cellular pattern in the auditory epithelium. Science 333(6046):1144–1147

Tomko RP, Xu R, Philipson L (1997) HCAR and MCAR: the human and mouse cellular receptors for subgroup C adenoviruses and group B coxsackieviruses. Proc Natl Acad Sci U S A 94:3352–3356

Wakayama T, Iseki S (2009) Role of the spermatogenic-Sertoli cell interaction through cell adhesion molecule-1 (CADM1) in spermatogenesis. Anat Sci Int 84:112–121

Warner MS, Geraghty RJ, Martinez WM, Montgomery RI, Whitbeck JC, Xu R, Eisenberg RJ, Cohen GH, Spear PG (1998). A cell surface protein with herpesvirus entry activity (HveB) confers susceptibility to infection by mutants of herpes simplex virus type 1, herpes simplex virus type 2, and pseudorabies virus. Virology 246:179–189

Wei SY, Escudero LM, Yu F, Chang LH, Chen LY, Ho YH, Lin CM, Chou CS, Chia W, Modolell J, Hsu JC (2005) Echinoid is a component of adherens junctions that cooperates with DE-Cadherin to mediate cell adhesion. Dev Cell 8:493–504

Williams AF, Barclay AN (1988) The immunoglobulin superfamily–domains for cell surface recognition. Annu Rev Immunol 6:381–405

Wolff T, Ready DF (1991) Cell death in normal and rough eye mutants of Drosophila. Development 113:825–839

Woods DF, Bryant PJ (1993) ZO-1, DlgA and PSD-95/SAP90: homologous proteins in tight, septate and synaptic cell junctions. Mech Dev 44:85–89

Yamada A, Irie K, Deguchi-Tawarada M, Ohtsuka T, Takai Y (2003) Nectin-dependent localization of synaptic scaffolding molecule (S-SCAM) at the puncta adherentia junctions formed between the mossy fibre terminals and the dendrites of pyramidal cells in the CA3 area of the mouse hippocampus. Genes Cells 8:985–994

Yamagata M, Herman JP, Sanes JR (1995) Lamina-specific expression of adhesion molecules in developing chick optic tectum. J Neurosci 15:4556–4571

Yamamoto T, Harada N, Kano K, Taya S, Canaani E, Matsuura Y, Mizoguchi A, Ide C, Kaibuchi K (1997) The Ras target AF-6 interacts with ZO-1 and serves as a peripheral component of tight junctions in epithelial cells. J Cell Biol 139:785–795

Yap AS, Brieher WM, Gumbiner BM (1997) Molecular and functional analysis of cadherin-based adherens junctions. Annu Rev Cell Dev Biol 13:119–146

Yoshida M, Shimono Y, Togashi H, Matsuzaki K, Miyoshi J, Mizoguchi A, Komori T, Takai Y (2012) Periderm cells covering palatal shelves have tight junctions and their desquamation reduces the polarity of palatal shelf epithelial cells in palatogenesis. Genes Cells (in press)

Yoshida T, Miyoshi J, Takai Y, Thesleff I (2010) Cooperation of nectin-1 and nectin-3 is required for normal ameloblast function and crown shape development in mouse teeth. Dev Dyn 239:2558–2569

Zhadanov AB, Provance DW, Jr Speer CA, Coffin JD, Goss D, Blixt JA, Reichert CM, Mercer JA(1999) Absence of the tight junctional protein AF-6 disrupts epithelial cell-cell junctions and cell polarity during mouse development. Curr Biol 9:880–888

Zlotogora J (1994) Syndactyly, ectodermal dysplasia, and cleft lip/palate. J Med Genet 31:957–959

Zlotogora J, Zilberman Y, Tenenbaum A, Wexler MR (1987) Cleft lip and palate, pili torti, malformed ears, partial syndactyly of fingers and toes, and mental retardation: a new syndrome? J Med Genet 24:291–293

Chapter 8
Signaling from the Adherens Junction

Abbye E. McEwen, David E. Escobar and Cara J. Gottardi

Abstract The cadherin/catenin complex organizes to form a structural Velcro that joins the cytoskeletal networks of adjacent cells. Functional loss of this complex arrests the development of normal tissue organization, and years of research have gone into teasing out how the physical structure of adhesions conveys information to the cell interior. Evidence that most cadherin-binding partners also localize to the nucleus to regulate transcription supports the view that cadherins serve as simple stoichiometric inhibitors of nuclear signals. However, it is also clear that cadherin-based adhesion initiates a variety of molecular events that can ultimately impact nuclear signaling. This chapter discusses these two modes of cadherin signaling in the context of tissue growth and differentiation.

8.1 Introduction

To those new to the field of cell–cell adhesion, one only needs to watch a movie of a developing embryo or migrating monolayer of cells in culture to recognize the remarkably fluid yet coordinated nature of cell–cell adhesions. Indeed, observing such cell behaviors brings to mind two clear questions: How is cell–cell adhesion regulated and how is the state of cell contact communicated to the cell's interior? A central role for the cadherin/catenin adhesive complex in these cell behaviors was initially inferred from early studies showing that embryonic tissues fail to undergo normal morphogenesis in the presence of antibodies to the extracellular domain of E-cadherin (Gallin et al. 1986; Hirai et al. 1989). This result implied that cells fail to send morphogenetic signals when cadherin function is perturbed. In this chapter, we

Abbye E. McEwen and David E. Escobar contributed equally

C. J. Gottardi (✉)
Northwestern University Feinberg School of Medicine,
McGaw Pavilion Suite M-323 240 E Huron, Chicago, IL 60611, USA
e-mail: c-gottardi@northwestern.edu

A. E. McEwen · D. E. Escobar
The Integrated Graduate Program in the Life Sciences, Department of Medicine,
Northwestern University Feinberg School of Medicine, Chicago, IL 60611, USA

T. Harris (ed.), *Adherens Junctions: From Molecular Mechanisms to Tissue Development and Disease,* Subcellular Biochemistry 60,
DOI 10.1007/978-94-007-4186-7_8, © Springer Science+Business Media Dordrecht 2012

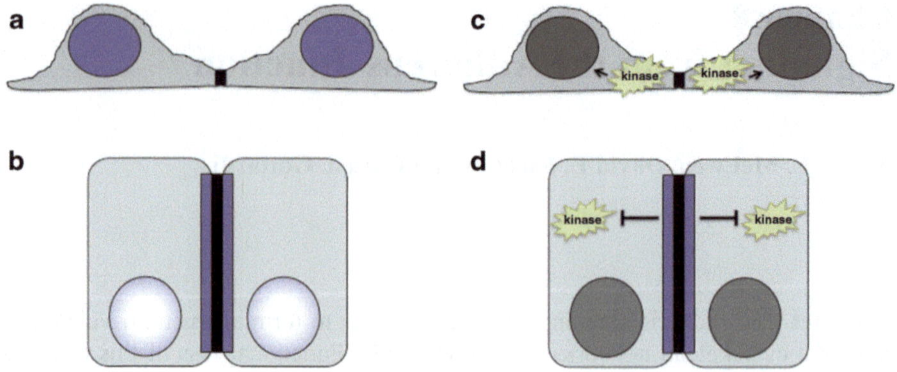

Fig. 8.1 General models of cadherin signaling to the nucleus. Cadherins interact with dual-localization proteins (e.g., β-catenin, Plakoglobin and p120 ctn) that functionally link cadherins to the cortical cytoskeleton and also control the activation of DNA-binding factors in the nucleus. The model presented in **a** and **b** reflects evidence that cells with greater cadherin abundance (*black bar*, **b**) can sequester, and thereby inhibit, the transcriptional co-activator function of these dual-localization proteins (shown as *purple color*) better than cells with fewer cadherins (*black bar*, **a**). The model presented in **c** and **d** reflects evidence that E-cadherin in densely packed epithelial monolayers can inhibit signaling from diverse growth factor receptor kinases (**d**) better than cells with less mature contacts (**c**)

focus on the nature of these signals, particularly those that impact gene expression. Other chapters in this volume address how cadherins signal more locally to alter the cortical actin cytoskeleton, which ultimately impacts the adhesive and mechanical properties of the cell (see Chaps. 6, 7 and 10).

Two models of cadherin signaling are presented, generally referred to as "transcriptional co-activator sequestration" versus "kinase inhibition" models (Fig. 8.1). For reasons that are largely historical in nature, the former mode is better appreciated since most cytoplasmic "peripheral" components of the cadherin complex (i.e., catenins) also localize to the nucleus to directly impact gene expression. Evidence that cadherins interact with transcriptional co-activators has long suggested a simple way to coordinate adhesion with changes in transcription, however there are problems with this model that merit deeper discussion. It is also clear that cadherin-based adhesion can strongly impact various growth factor receptor kinase signaling cascades, although clear molecular models for explaining these findings have yet to emerge. By discussing the differences between these two modes of cadherin signaling, we hope to build a conceptual framework for thinking about adhesion signaling.

8.2 β-Catenin is a Dual-Function Adhesion/Transcriptional Co-Activator Protein

The idea that cadherins might signal to the nucleus was first inspired by the discovery that β-catenin, originally identified as a stoichiometric co-precipitating partner of cadherins, was found to be highly homologous to Armadillo, a fly protein required for proper segmentation or "patterning" of the ventral epidermis (McCrea and Gumbiner 1991). At that time, Armadillo was one of a small number of components known to transduce a signal initiated by a secreted factor known as wingless (Wg), where a mutation in the fly β-catenin gene phenocopied the loss of a Wg signal (Peifer and Wieschaus 1990). Wg (and its vertebrate homologues, Wnts 1–19) are now widely appreciated as being used throughout embryonic development and adult tissue homeostasis to activate a repertoire of cell- and context-dependent genes that direct distinct cellular fates (Cadigan and Peifer 2009). The discovery that a cadherin-associated molecule also served an essential role in Wg/Wnt signal transduction led to one early hypothesis that plasma membrane-to-nuclear signaling occurred via β-catenin at the adherens junction. However, studies in both *Drosophila* and *Xenopus* systems later indicated that it was a cadherin-independent pool of β-catenin that was essential for transducing Wnt signals. For example, in the absence of a Wnt signal, most of the β-catenin is found associated with cadherins at cell contacts. In cells receiving a Wnt signal, however, a cytoplasmic/nuclear pool of β-catenin was also observed by immunfluorescence and biochemical fractionation methods (Funayama et al. 1995; Peifer et al. 1994; Schneider et al. 1996). Consistent with its nuclear localization, β-catenin was ultimately found to interact with LEF/TCF-type DNA-binding factors (Behrens et al. 1996; Molenaar et al. 1996), where β-catenin serves an essential co-activator function (Hecht et al. 1999) by recruiting components required for chromatin remodeling and RNA polymerase activation (reviewed in (Willert and Jones 2006)).

While formation of this binary transcription complex is the ultimate downstream step of Wnt signaling, it became clear that a large number of pathway components appear dedicated to generating a cadherin-free, nuclear signaling pool of β-catenin. Indeed, a convergence of genetic epistasis, biochemical and human cancer studies led to a rapid ordering of receptor complex and midstream players in this pathway (reviewed in (van Amerongen and Nusse 2009); Fig. 8.2). We now appreciate that a secreted Wg/Wnt acts through cell surface receptors of the Frizzled (Fz) and Low-density lipoprotein (LDL) Receptor Related Protein (LRP) families. Fz receptors are seven-pass transmembrane proteins that topologically (and to some degree, functionally) resemble G-protein coupled receptors (Wang et al. 2006). The ultimate consequence of Frizzled/LRP5/6 co-receptor activation is the inhibition of a multi-protein kinase/scaffold complex that controls the phosphorylation-dependent destruction of β-catenin not otherwise bound with high affinity to cadherins (MacDonald et al. 2009).

8.3 Cadherins as Stoichiometric Inhibitors of β-Catenin Signaling

The existence of two compositionally distinct pools of β-catenin in the cell, one associated with the plasma membrane as an integral part of the cadherin core complex, and the other a cytoplasmic/nuclear pool that serves to transduce a membrane-to-nuclear signal, raises intriguing questions as to whether or not adhesion and

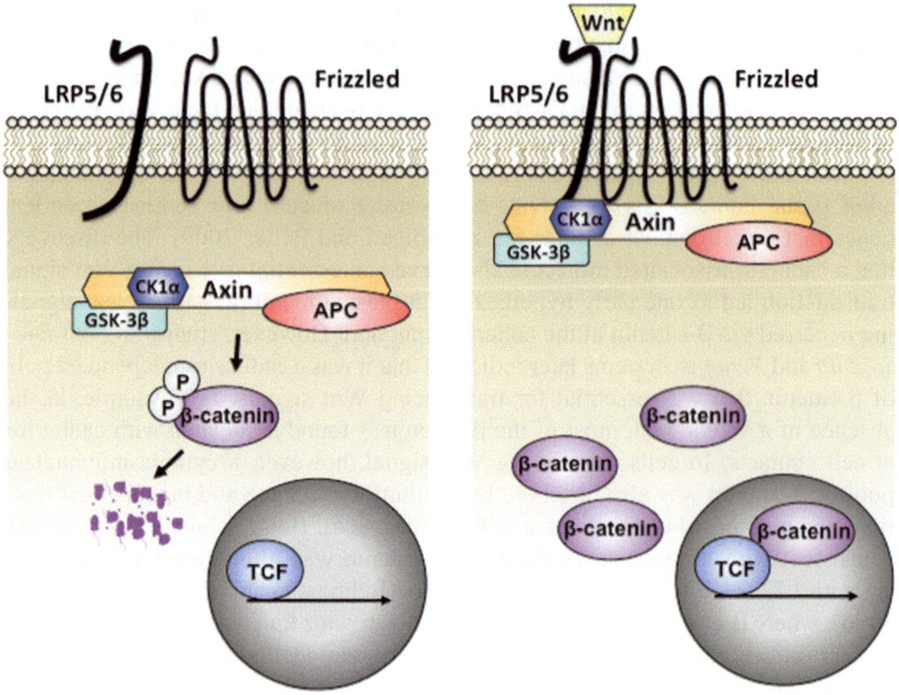

Fig. 8.2 Wnt signaling pathway. In the absence of Wnt (*left*), cytosolic β-catenin is continually phosphorylated by casein kinase 1α (CK1) and glycogen synthase kinase 3β (GSK3β) within an Axin1 scaffold complex. This phosphorylation allows β-catenin to be recognized by a specific E3 ligase (βTrCP, not shown), which catalyzes the ubiquitylation and rapid degradation of β-catenin. The adenomatous polyposis coli (APC) tumor suppressor participates in the phospho-destruction of β-catenin by antagonizing β-catenin de-phosphorylation by phosphatases. During Wnt activation (*right*), GSK3β activity is inhibited directly by Lrp5/6, which allows β-catenin to accumulate, enter the nucleus, interact with LEF/TCF family members and promote transcription

Wnt signaling are coordinated through use of this common component, β-catenin. Indeed, experimental manipulations have revealed that cadherin expression and β-catenin signaling are interrelated. For example, forced expression of the cadherin can antagonize β-catenin signaling activity in a number of systems (Fagotto et al. 1996; Heasman et al. 1994; Orsulic et al. 1999; Sanson et al. 1996). Conversely, reductions in cadherin protein levels can enhance β-catenin signaling in certain contexts (Ciruna and Rossant 2001; Cox et al. 1996). Since cadherin can bind β-catenin directly (Jou et al. 1995), it is generally appreciated that cadherins inhibit β-catenin by sequestering the cytosolic signaling pool to membranes, preventing its access to the nuclear compartment. Indeed, biochemical and crystallographic evidence that β-catenin binds cadherins or TCFs through an overlapping binding interface (Graham et al. 2000; Huber and Weis 2001) rationalizes how cadherins can function as stoichiometric inhibitors of β-catenin/TCF signaling (Gottardi and Gumbiner 2001). It is important to recognize, however, that the ability of a cadherin to impact β-catenin signaling requires a baseline of active Wnt signaling. For example, introduction of E-cadherin in L929 cadherin-negative fibroblasts that are not receiving a Wnt signal shows that E-cadherin-mediated adhesion has little effect on gene expression (Kuphal and Behrens 2006), despite the well appreciated phenomenon that cadherins robustly upregulate and associate with β-catenin in this system (Ozawa et al. 1990). Moreover, epithelial cancers that have lost E-cadherin expression by various means fail to show a concomitant upregulation in β-catenin signaling (Caca et al. 1999; Herzig et al. 2007; van de Wetering et al. 2001). In some cell culture models, targeted loss of E-cadherin is associated with loss or down-regulation of β-catenin (Hendriksen et al. 2008), presumably because there are no other β-catenin-binding cadherins in these systems (e.g., N-cadherin or P-cadherin), and loss of this major high affinity β-catenin-binding partner leads to β-catenin elimination by the phospho-destruction complex. Indeed, isothermal calorimetry affinity measurements can rationalize this observation, as β-catenin binds the cadherin with anywhere from 28- to 190-fold higher affinity than to components of the destruction complex, APC and Axin (Choi et al. 2006). Thus taken altogether, the ability of a cadherin to limit β-catenin signaling is contextual and occurs only when cells are actively engaged in Wnt signaling.

The aforementioned studies combined with evidence that the affinity of β-catenin for the phospho-form of cadherin likely present in cells is ~570-fold over the estimated β-catenin/TCF binding affinity (Choi et al. 2006) suggest that the cadherin might serve as an effective "sink" for β-catenin, so that β-catenin levels would have to rise beyond a threshold of cadherin expression in order to signal. However outside of the gain- and loss-of-function perturbation experiments discussed above, evidence that cadherin levels are modulated *in vivo* to set thresholds for Wnt signals is formally lacking. Quantitative microarray studies of Wnt-activated cells expressing different levels of cadherin, for example, might be informative for testing this principle. Alternatively, one might predict some cell types to be more sensitive to Wnt signals than others due to differences in cadherin abundance. Studies from our group, however, indicate this is not the case for primary lung fibroblasts and alveolar epithelial cells, which show similar levels of cadherin-bound β-catenin despite differences in expression of cadherin subtypes (Flozak and Gottardi, unpublished observation). Given that immune cell differentiation is known to be controlled by Wnt/β-catenin signaling (e.g., TCF (T-cell factor) was originally shown to be important in T cell development (Verbeek et al. 1995)), and that immune cells lack robust cadherin expression, one wonders whether immune cells might be most sensitive to Wnt signals.

Mathematical modeling studies indicate that changes in the *rate* of cadherin protein synthesis, rather than its turnover, are expected to have the most direct consequence on Wnt signaling (van Leeuwen et al. 2007). While there are a few signals that have been shown to increase E-cadherin transcription in both cell culture (e.g., Wnt7a, (Ohira et al. 2003); WT-1, (Hosono et al. 2000)) and developmental models (Montell et al. 1992; Niewiadomska et al. 1999; Shimamura and Takeichi 1992), it is unclear whether these increases in cadherin synthesis are used to dampen endogenous Wnt/β-catenin signals, in addition to providing enhanced cell–cell adhesiveness required for certain morphogenetic events. Interestingly, TCF-binding sites have been identified in the E-cadherin promoter (Huber et al. 1996), raising the possibility of a negative feedback mechanism where β-catenin nuclear signaling could activate E-cadherin transcription, which would in turn inhibit β-catenin signaling through sequestration. While activation of β-catenin signaling has been associated with the upregulation of E-cadherin in mouse intestine (Wong et al. 1998) and a Drosophila cell line (Yanagawa et al. 1997), the universality of this feedback mechanism is unclear. For example, it is also appreciated that the presence of TCF sites in promoters are not always associated with transcriptional upregulation (Blauwkamp et al. 2008). Indeed, one study showed that a Lef/TCF site in the E-cadherin promoter could interact with other factors to inhibit E-cadherin transcription during hair follicle development (Jamora et al. 2003).

If there are few instances where E-cadherin levels are elevated above its baseline for differentiative and morphogenetic purposes, there is clear evidence that E-cadherin is subject to potent negative regulation by transcriptional repressors that drive epithelial-mesenchymal transitions (EMT) during development and disease, such as Snail/Slug family zinc-finger transcription factors (Nieto 2002), the basic helix-loop-helix (bHLH) transcription factor, Twist (Yang et al. 2004) and ZEB1&2 (Korpal et al. 2008). Since these transcriptional regulators respond to a range of growth factor signaling pathways, including transforming growth factor beta (TGFβ) -1 and −2, bone morphogenetic proteins (BMPs), and fibroblast growth factors (FGFs) (reviewed in (Christofori 2006; Thiery et al. 2009)), it is easy to see how a number of signaling pathways could sensitize cells to Wnt signals by repressing a major negative regulator of β-catenin, E-cadherin. However, it is important to bear in mind that a phenomenon known as "cadherin-switching" typically accompanies EMTs, where the epithelial-specific E-cadherin (Epithelial-cadherin) is downregulated and replaced by the mesenchymal-specific N-cadherin (Neural-cadherin) (Wheelock et al. 2008). While N-cadherin and E-cadherin contribute to distinct adhesive activities (e.g., N-cadherin promotes while E-cadherin antagonizes cell motility and invasion (Chen et al. 1997; Fedor-Chaiken et al. 2003b; Kim et al. 2000; Nieman et al. 1999)), their abilities to bind β-catenin and antagonize Wnt signals appear identical (Gottardi et al. 2001; Sadot et al. 1998). Thus if E-cadherin downregulation during EMT is indeed a way to sensitize cells to Wnt signals, the upregulation of N-cadherin would have to be delayed for a sufficient temporal window so that Wnt signals are not buffered by another β-catenin-binding cadherin.

8.4 Evidence for β-Catenin "Release" from the Junction and Nuclear Signaling?

While the aforementioned studies present compelling evidence that changes in *cadherin biosynthesis* impact β-catenin nuclear signaling function, there has remained much interest in whether the cadherin-associated pool of β-catenin is ever "released" for the purposes of nuclear signaling (e.g., (Kam and Quaranta 2009)). Indeed it is quite appealing to imagine that changes in cadherin-engagement, junctional organization or some other aspect of adhesion might be communicated from the extracellular to cytoplasmic domain of the cadherin, resulting in molecular changes that lead to the release of β-catenin into the cytosol. There are many reasons this idea is attractive. For starters, it is worth noting that the precise contribution of β-catenin to adhesion has always been less apparent than its role as a transcriptional co-activator of Wnt signals. This may be in part because β-catenin is an essential component for Wnt signal transduction, while β-catenin adhesive function can be compensated by the highly homologous desmosomal component, Plakoglobin (Bierkamp et al. 1996; Haegel et al. 1995; Huber et al. 1997; Huelsken et al. 2000; Nieset et al. 1997). In addition, β-catenin's role in adhesion has always been over-shadowed by cadherin and α-catenin, which provide essential homophilic recognition and actin-binding functions to the cadherin/catenin adhesive complex. The appeal of this release model is also driven by evidence that phosphorylations that impact the β-catenin/cadherin binding interface can substantially impact the affinity of these two proteins *in vitro* (reviewed in (Daugherty and Gottardi 2007)), raising the possibility that kinases and phosphatases could modulate β-catenin release. Lastly, evidence that the cadherin bound pool of β-catenin is generally much more abundant than the cytosolic fraction stabilized by Wnts further contributes to the notion that cadherins harbor a pool of β-catenin used for signaling. Given estimates that the N-terminally unphosphorylated signaling forms of β-catenin may be small, even relative to the stabilized pool (Hendriksen et al. 2008; Maher et al. 2010), it is possible to rationalize that a small level of β-catenin release from the cadherin (which might be difficult to detect using standard and typically inefficient co-immunoprecipitation analysis) could be freed into the signaling pool. However it is important to bear in mind that one mechanistic point seems inescapable for the "release model" to be true, and that is that β-catenin must be diverted from the pathway that constitutively destroys the cadherin-free pool. In other words, mechanisms that promote β-catenin release from cadherin would either need to be coupled with a Wnt, or Wnt-like signal that inhibits the GSK3-dependent destruction of β-catenin. Alternatively, β-catenin would need to be released from the membrane in a form that would be protected from degradation, such as associated with the cadherin cytoplasmic domain (Simcha et al. 2001). This latter model will be discussed further in the context of cadherin ectodomain shedding and cytoplasmic domain processing below.

It should be noted that evidence that β-catenin signaling is sensitive to the protein synthesis inhibitor, cycloheximide, has long been interpreted to imply that only a *newly synthesized* form of β-catenin (i.e., rather than a previously synthesized pool coming from, for example, the cadherin complex) contributes to signaling (Willert et al. 2002). However, data from a cell-free Xenopus extract system that contains nuclei capable of responding to Wnt signals came to an opposite conclusion, finding that β-catenin signaling activity could in fact be recruited from a pre-existing (cycloheximide-insensitive) pool (Nelson and Gumbiner 1999). This finding has been recently supported in a cell culture-based study (Howard et al. 2011). Since interpretation of cycloheximide-based experiments can be problematic (Hanna et al. 2003; Liu et al. 2008), newer methods and insights may be required. Thus, to the extent that there remains debate over the source of the β-catenin signaling pool, the debate centers on whether β-catenin comes directly from the ribosome or via other multi-protein complexes (e.g., β-catenin/cadherin, β-catenin/APC, or β-catenin/Axin).

8.5 Cadherin-Based Adhesion can Limit β-Catenin Signaling Catalytically

If there is indeed a preexisting pool of β-catenin poised to signal, data from our lab and others suggest that the β-catenin phospho-destruction complex may be where to look (Faux et al. 2010; Harris and Nelson 2010; Hendriksen et al. 2008; Maher et al. 2010). For example, our lab has found that N-terminal phospho-forms of β-catenin (required for inhibition and degradation of β-catenin, (Liu et al. 2002)) can accumulate and co-localize with Axin and APC at cell–cell contacts, in a complex that is largely excluded from the cadherin/catenin complex (Maher et al. 2009). The implications of detecting these β-catenin phospho-forms at cell junctions are manifold. First, the ability to readily detect N-terminally phosphorylated β-catenin, when prevailing models suggest these phospho-forms are short-lived species, indicates that these forms are not as tightly coupled with degradation as previously expected, and raises the possibility that competing, N-terminal de-phosphorylation events could effectively "release" β-catenin for signaling. Second, evidence that N-terminal phospho-forms co-localize with Axin/APC at cadherin contacts, but are not obviously associated with the cadherin complex, raises the possibility that cadherin-based membrane dynamics might impact β-catenin signaling *indirectly* through modulating the activity of the β-catenin phospho-destruction complex. Indeed hints for such a model were already supported by studies in flies, where a single point mutation in APC that impacts its localization to adherens junctions was sufficient to compromise APC's ability to promote β-catenin degradation (Jarrett et al. 2001; McCartney et al. 1999; Yu et al. 1999). Using a cell line that allowed us to more robustly capture phospho-forms of β-catenin that were considered transient intermediates, we learned that cadherins can promote the N-terminal "inhibitory" phosphorylation of β-catenin. In normal cells, we also found that cadherin-based adhesion itself, rather than changes in cadherin abundance, can limit the accumulation of β-catenin induced by Wnts through enhancing the rate of β-catenin destruction (Maher et al. 2009) (Fig. 8.3).

How the phospho-destruction complex is localized to cell contacts, and how cadherins promote β-catenin N-terminal phosphorylation within this complex are not understood, but some recent studies are shedding molecular light on this area (Hay et al. 2009; Tanneberger et al. 2011). Indeed, it is worth noting that a wide variety of molecular components have been shown to affect β-catenin signaling. For example, a *Drosophila* genetic screen for enhancers of β-catenin signaling revealed that loss of proteins necessary for the establishment and maintenance of epithelial

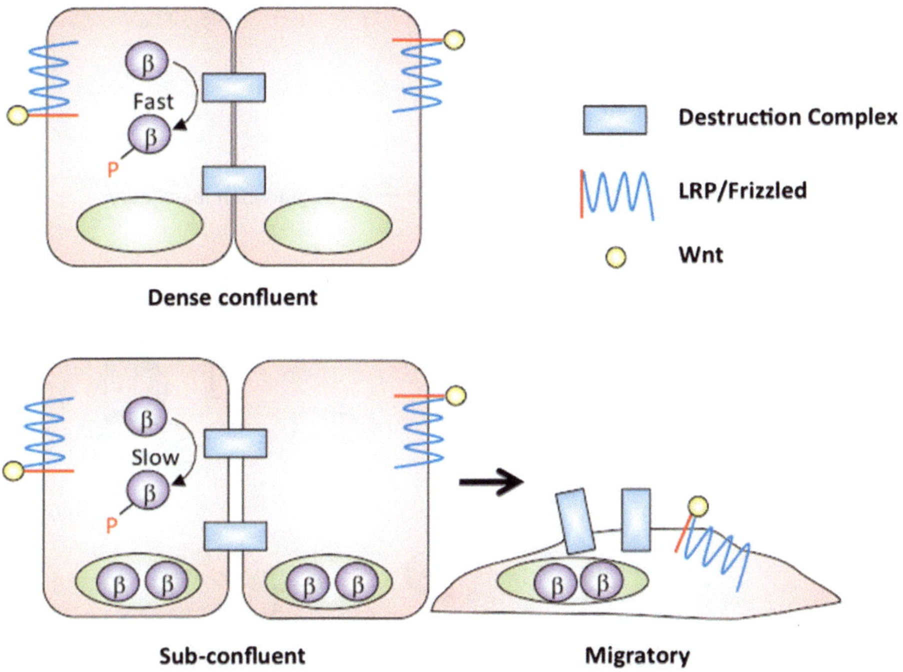

Fig. 8.3 Density-dependent turnover of cytosolic β-catenin. In densely confluent cells, cadherins promote a faster turnover of β-catenin than in less adhesive (sub-confluent) cells. This may explain why cells migrating adjacent to a wound appear sensitized to Wnt signals. (Figure adapted from Maher et al. 2009)

polarity, such as the Fat cadherin, Stardust and Dlg, could enhance β-catenin signaling (Greaves et al. 1999). More recently, components required for primary cilium structure/function and machineries that control planar polarity and convergence/extension movements have been shown to antagonize β-catenin signaling at the level of Disheveled (Corbit et al. 2008; Schwarz-Romond et al. 2002). We reason that a simple framework for explaining these effects is to recognize that the β-catenin phospho-destruction complex is "tunable," and subject to a number of signaling inputs that ultimately control the rate at which β-catenin is consumed by the destruction complex. Cadherin-based adhesion may be the upstream "master cue" that polarity components and non-canonical Wnt signaling inputs depend upon for inhibition of β-catenin signaling via the destruction complex.

Overall, it appears that cadherins can inhibit β-catenin signaling in two ways: one as a stoichiometric binding partner that sequesters β-catenin from the nucleus, the other through a catalytic mechanism that impacts the rate at which β-catenin is consumed by the phospho-destruction complex (Figs. 8.3 and 8.4). Why might the cell need two modes for inhibiting β-catenin signaling by cadherins? Perhaps each mode is responsible for different degrees of inhibition. Whereas changes in cadherin biosynthesis that accompany EMT allow for robust β-catenin signals that

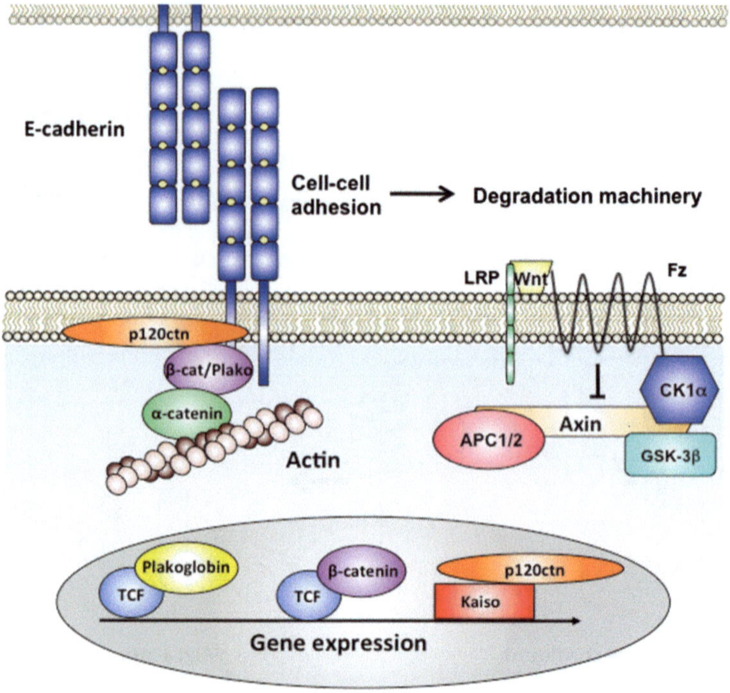

Fig. 8.4 Armadillo family proteins in cadherin-based adhesion and nuclear signaling. The cadherin cytoplasmic domain binds directly to three distinct armadillo-repeat proteins, β-catenin, Plakoglobin and p120ctn. These proteins play an obligate role in cadherin-based adhesion (*Left*). To varying degrees, cytoplasmic and nuclear pools of these catenins are generated by Wnt signals, which favors catenin activation of transcription. Cadherins appear to antagonize nuclear catenin functions via both stoichiometric sequestration (*left*) and catalytic destruction models (*right*)

alter cell fate, changes in cadherin-based adhesion associated with epithelial sheet wound closure may allow for a more modest regulation that impacts cell behaviors like motility and proliferation.

8.6 Cadherin Tail Clipping and Nuclear Signaling: The Notch Paradigm

Notch is a transmembrane protein that engages another transmembrane "ligand" on an adjacent cell (Delta) and Notch/Delta pairing is required for activating genes that impact neurogenesis (Louvi and Artavanis-Tsakonas 2006). In contrast to other ways that nuclear signals are conveyed from the plasma membrane, such as growth factor receptor signaling, which typically involves a cascade of kinase activation events and numerous intermediates, or the examples described above for cadherin signaling via β-catenin, the Notch cytoplasmic domain *directly* activates gene targets in the nucleus (Fortini 2002; Schroeter et al. 1998; Struhl and Adachi 1998). Specifically, the Notch cytodomain gains access to the nuclear compartment through a regulated intra-membrane cleavage event that liberates the cytodomain from the plasma membrane. Sequences within Notch favor its nuclear targeting and localization to Notch-regulated promoters required to inhibit neurogenesis (Bray 2006). Consistent with this Notch signaling paradigm, there is clear evidence that E-cadherin is subject to both matrix metalloprotease (MMP)-mediated ectodomain and gamma secretase intra-membrane cleavage events (Lochter et al. 1997; Marambaud et al. 2002; Maretzky et al. 2005). One consequence of cadherin cytodomain cleavage has been demonstrated for N-cadherin, where the cytodomain inhibits numerous transcriptional targets by binding CREB-binding protein (CBP) and targeting it for proteosomal destruction (Marambaud et al. 2002). A second consequence of E-cadherin cytodomain cleavage appears to be an increase in β-catenin signaling (Maretzky et al. 2005), although the extent to which the signaling pool of β-catenin is liberated from a cleaved, cadherin cytodomain, or generated by a parallel Wnt signal that inhibits the destruction of newly synthesized β-catenin is not clear. Given evidence that the cadherin cytodomain is a potent inhibitor of β-catenin turnover by the phospho-destruction complex (Simcha et al. 2001) (through binding residues in β-catenin that overlap with those that engage phospho-destruction components, APC and Axin, (Ha et al. 2004; Xing et al. 2003, 2004)), it is easy to see how cadherin cytodomain clipping could potentially liberate a substantial pool of β-catenin/cadherin complexes into the cytosol. However, it is unclear how efficiently β-catenin can be displaced from the cadherin cytodomain given affinity measurements for the two proteins in the picomolar range (Choi et al. 2006), along with evidence that cadherin cytoplasmic domain-stabilized β-catenin shows no obvious signaling in reporter assays (Carien Niessen, personal communication). Thus if cadherin cytodomain clipping emerges as a way to generate a β-catenin nuclear signal, future mechanistic studies will be required to distinguish between β-catenin being released from the cadherin tail versus being stabilized by a Wnt or Wnt-like signal. It is also worth noting that a clipped cadherin cytodomain may have consequences for one of the other dual-localization catenins, p120-catenin (p120ctn) (Ferber et al. 2008).

8.7 Armadillo-Repeat Catenin Proteins in Adhesion and Transcription

While β-catenin is the best-known example of a dual-function adhesion-nuclear signaling protein, it is important to recognize that other catenins appear to follow the same paradigm. For example, Plakoglobin (also known as γ-catenin), which is highly homologous to β-catenin and typically associated with desmosomal cadherins, can interact with E-cadherin under conditions where β-catenin is limiting (Huelsken et al. 2000). Like β-catenin, Plakoglobin can also interact with TCF family DNA binding proteins and impact gene expression (Kolligs et al. 1999; Simcha et al.

1998; Zhurinsky et al. 2000), although Plakoglobin appears to bind a distinct region on TCF that may differentially impact target gene expression (Miravet et al. 2002; Solanas et al. 2004). p120ctn, in addition to binding and stabilizing E-cadherin at the cell surface (see Chap. 9), can also independently interact with Kaiso (Daniel and Reynolds 1999), a DNA binding factor of the POZ family. Kaiso functions as a transcriptional repressor, and p120ctn appears to either promote Kaiso release or prevent its recruitment to DNA binding sites (Kelly et al. 2004; Kim et al. 2004; Ruzov et al. 2004). Interestingly, the *proximity* of p120ctn and β-catenin binding regions within the cadherin cytoplasmic domain is shared by some Wnt-regulated promoters, which can be co-regulated by proximal TCF- and Kaiso-binding elements (Park et al. 2006). These data imply that changes in the rate of cadherin synthesis or adhesion could doubly impact the expression of gene targets co-regulated by β-catenin (or Plakoglobin)/TCF and p120ctn/Kaiso. Like β-catenin, both p120ctn and Plakoglobin also contain a similar N-terminal GSK3-sequence that controls the level of cadherin-free Plakoglobin/p120ctn in cells and allows their modest stabilization by Wnts (Hong et al. 2010). Thus Wnt signals may stabilize a family of catenin proteins that can impact gene expression, and cadherin-based adhesion may limit their signaling through both stoichiometric sequestration and catalytic phospho-destruction models (Fig. 8.4).

Is there a way to rationalize the observation that three distinct catenin proteins, β-catenin, p120ctn and Plakoglobin, play dual roles in transcription and cadherin-based adhesion? One theme in the organization of signal transduction pathways is that different pathways tend to rely on distinct protein-protein binding mechanisms in the service of transducing membrane to nuclear signals. For example, receptor tyrosine kinase signaling uses src-homology 2 (SH2) and phospho-tyrosine binding interactions, while the Hippo/Warts pathway uses WW and PPxY domain interactions at multiple levels of the pathway to transduce signals (Salah and Aqeilan 2011). In this light, it is important to recognize that that β-catenin, Plakoglobin and p120ctn are all armadillo-repeat proteins. The armadillo repeat is a 42 amino acid motif that forms a triplet of alpha helices. When multiple repeats are brought together, as in the catenins, these triple helices stack to form a superhelix of helices that forms a groove into which nearly all arm-repeat protein ligands fit. Interestingly, the core nuclear import machinery proteins, Importins α and β, contain armadillo repeats, and the structurally related HEAT repeats, respectively (Andrade et al. 2001). These repeating-units form versatile protein-protein binding interfaces that allow importins to drive the recognition and nuclear accumulation of a seemingly diverse set of ligands (Coates 2003). Given that β-catenin is imported into the nucleus independently of a classic nuclear localization signal or the known importins, and interacts directly with the nuclear pore (Fagotto et al. 1998; Suh and Gumbiner 2003), it seems likely that p120ctn and Plakoglobin (and other junction-localized armadillo-repeat proteins localized to desmosomal cadherins, such as plakophilins) interact similarly with the nuclear pore complex to mediate their own nuclear/cytoplasmic shuttling (Henderson 2000; Karnovsky and Klymkowsky 1995; Krieghoff et al. 2006; Mertens et al. 1996). In light of this apparent structural conservation (Andrade et al. 2001), it has been reasoned that these armadillo repeat junction/nuclear proteins and the nuclear importins evolved from a common ancestor. Given evidence that catenin-binding cadherins are only found in metazoans ((Hulpiau and van Roy 2011); Chap. 2), it appears that cell–cell adhesive cadherins evolved to co-opt these nuclear signaling proteins for both structural and signaling purposes.

8.8 E-cadherin Mutations in Human Tumors and Implications for Critical Functions

An understanding of the relationship between cadherin protein expression, cell–cell adhesion and β-catenin signaling has been aided by studies that have sought to tease out how these factors contribute to E-cadherin tumor suppressor function. Numerous studies have reported loss of, or mutation within, the E-cadherin/catenin complex across a wide variety of epithelial cancers (reviewed by (Berx and van Roy 2009)). In most cases, E-cadherin loss correlates with the invasive component of a given tumor, suggesting that E-cadherin loss of function might promote tumor progression, local invasion and metastasis (Birchmeier et al. 1996; Vleminckx et al. 1991). Indeed in a well-defined mouse model of pancreatic islet cell cancer (Perl et al. 1998), E-cadherin is lost as tumors progress from adenoma to carcinoma, and forced expression of E-cadherin holds tumors at the adenoma stage compared with control mice. Conversely, forced down regulation of endogenous E-cadherin expression increases the number of metastases and tumors detected at the carcinoma stage, indicating that the down regulation of E-cadherin constitutes a key rate-limiting step in tumor progression. Germline mutations in E-cadherin have been found to be associated with a familial form of gastric cancer (Hereditary diffuse gastric carcinoma, HDGC), underscoring its importance as a true tumor suppressor gene (Berx et al. 1998). With regards to molecular mechanisms, it was long reasoned that the tumor/invasion suppressor activities of E-cadherin would be mediated through maintaining cell–cell adhesion. However, evidence that β-catenin is an oncogene, and that constitutive signaling is associated with numerous cancers (Giles et al. 2003; Howe and Brown 2004; Polakis 2000; Takigawa and Brown 2008) raises the possibility that an equally important role for E-cadherin in tumor suppression might be through antagonizing the nuclear signaling activity of β-catenin. Interestingly, restoring cadherin-negative epithelial cancer lines with forms of the cadherin that can rescue cell–cell adhesion independently of binding β-catenin, versus a form of the cadherin that can associate with β-catenin but not mediate adhesion reveals that an ability to bind β-catenin is most critical to E-cadherin's function as a growth or invasion suppressor (Gottardi et al. 2001; Wong and Gumbiner 2003). Remarkably, rescuing adhesive function with a well-studied E-cadherin-α-catenin fusion construct is not sufficient to mediate growth or invasion suppression. Interestingly, the requirement for β-catenin binding is not dependent on whether cells are receiving a Wnt signal (Wong and Gumbiner 2003), indicating that a form of E-cadherin that binds β-catenin is critical to its tumor suppressor function *regardless* of whether tumor cells rely on Wnt/β-catenin signaling.

While such domain analyses indicate that E-cadherin binding to β-catenin is most critical to its tumor suppressive and invasive activities *in vitro* (Wong and Gumbiner 2003), mutations in E-cadherin associated with breast and gastric cancers *in vivo* do not reveal the β-catenin binding domain as a mutational hotspot, as has been found for other factors that bind and inhibit β-catenin signaling like APC and Axin (Berx et al. 1998). Instead, the E-cadherin mutations that are widely distributed in breast cancer are truncations that occur in the extracellular domain, while gastric cancer reveals a mutational hotspot in the third cadherin repeat in the extracellular domain. Remarkably, mutations that delete the β-catenin binding domain are rare. This broad distribution of E-cadherin mutations along the entire coding sequence strongly suggests that cadherin signaling cannot be all about the inhibition of β-catenin signaling. As will be discussed below, cadherins are required for epithelial polarity and impact a number of growth-factor receptor signaling pathways, functions that depend on the entire full-length protein.

8.9 Transmitting Diverse Signals from Cadherin-Based Contacts

If we expand the view of cadherin signaling beyond the core complex and the nuclear functions of armadillo repeat proteins, β-catenin, Plakoglobin and p120 ctn, what emerges is a view of cadherin signaling that ultimately encompasses what it means to be a multicellular tissue. As the major cell adhesion system in epithelia, the E-cadherin/catenin complex is essential for establishing the close cell contacts that so many other junction and juxtacrine signaling molecules depend upon, from tight and gap junctions to membrane anchored signaling pairs like Notch/Delta or Ephrins and Eph-receptors (Fagotto and Gumbiner 1996; Ferreira et al. 2011; Zantek et al. 1999). Indeed as a master regulator of epithelial cell polarization (Nejsum and Nelson 2007, 2009), which entails the formation of distinct apical and basolateral membrane domains, one can readily see how most cell contact-dependent functions ultimately depend on E-cadherin, which formally places the cadherin "upstream" of, and responsible for, the transmission of numerous and diverse signals (Fig. 8.5). Because of this, the following sections aim to focus on the more proximal or direct targets of E-cadherin signaling.

The two phenomena most often examined in the context of cadherin signaling are contact inhibition of cell movement and contact inhibition of proliferation (reviewed in (Takai et al. 2008)). The former is readily observed when single epithelial cells join and become immobilized within a pre-existing colony of cells. This phenomenon is thought to depend on membrane-proximal cadherin-signals, which coordinate Rho-family GTPases and their regulators and effectors to change the

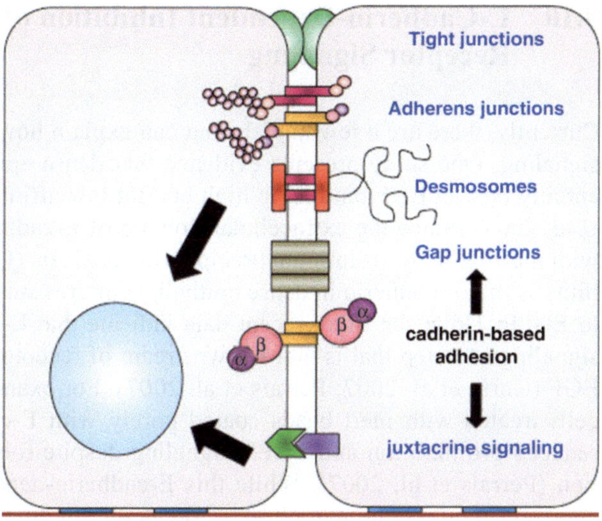

Fig. 8.5 E-cadherin is a master initiator of cell–cell contact, junction formation and epithelial polarity. E-cadherin-based adhesion is required for the establishment of diverse cell–cell junctions (e.g., tight junction, zonula adherens junction, desmosomes, gap junctions), as well as signals that require membrane-anchored ligand/receptor interactions (e.g., Notch/Delta, Ephrin/EphR). From this more global viewpoint, "E-cadherin signaling" encompasses signals coming from all of these complexes

dynamic organization of actin between adjacent cells (Mayor and Carmona-Fontaine 2010), and is discussed in detail in Chaps. 6 and 10. In contrast, contact inhibition of proliferation involves membrane-proximal events that ultimately lead to changes in gene expression and nuclear events required for mitosis. While these two phenomena almost certainly share similar molecular underpinnings, it is important to recognize that the time courses for these two phenomena are quite different. For example, contact inhibition of cell movement occurs in minutes, while inhibition of proliferation takes days and depends on cell density more than formal cell "contact" (Takai et al. 2008), indicating that these processes are molecularly distinct. Because cell proliferation depends on growth factors, and the time course of contact-dependent inhibition of proliferation strongly correlates with the inhibition of growth factor receptor tyrosine kinase activity (Takahashi and Suzuki 1996), there has been a longstanding interest in the relationship between E-cadherin and growth factor receptor tyrosine kinases, in particular, Epidermal growth factor receptor (EGFR). Indeed, while early studies showed that EGFR can colocalized with E-cadherin at apically-localized adherens junctions (Chen et al. 2002) and co-associate in immunoprecipitation assays (e.g., (Hoschuetzky et al. 1994)), a more intimate relationship between these two proteins was supported by evidence that EGFR-activation could promote the phosphorylation of tyrosine residues in cadherin-associated catenins (Daniel and Reynolds 1997; Hoschuetzky et al. 1994), suggesting that the cadherin/catenin complex may be a proximal target of EGFR signaling. Consistent with this idea, EGFR and E-cadherin genetically interact during eye development in flies, where loss of EGFR function can phenocopy E-cadherin overexpression, while a constitutively active form of EGFR worsens a weak mutant allele of E-cadherin (Dumstrei et al. 2002). Altogether, these data indicate that E-cadherin is both a downstream target *and* an upstream inhibitor of EGFR signaling.

8.10 E-Cadherin-Dependent Inhibition of Growth Factor Receptor Signaling

Currently, there are a few models that can explain how E-cadherin inhibits EGFR signaling. One study presents evidence that dense epithelial cell cultures preferentially restrict EGF binding to high but not low affinity sites on the EGFR (Qian et al. 2004). Since the extracellular domain of E-cadherin is sufficient to interact with the EGFR by co-immunoprecipitation analysis (Qian et al. 2004), one possibility is that E-cadherin in dense epithelial cultures sterically hinders EGF binding to EGFR. However, more recent data indicate that E-cadherin can inhibit EGFR signaling at a step that is more downstream of receptor binding and activation by EGF (Curto et al. 2007; Perrais et al. 2007). For example, E-cadherin expressing cells treated with inert beads coated purely with E-cadherin-ectodomains show reduced proliferation and EGFR signaling despite robust EGFR phospho-activation (Perrais et al. 2007). While this E-cadherin-dependent inhibition of EGFR signaling requires the cytoplasmic, β-catenin-binding domain of E-cadherin and associated catenins (Perrais et al. 2007), molecular details of this inhibition remain unclear.

Some mechanistic hints may be provided by studies of the neurofibromatosis 2 (Nf2) tumor suppressor protein, also known as Merlin. Merlin is a member of the ezrin radixin and moesin (ERM) family of membrane/cytoskeleton linking proteins (reviewed in (Fehon et al. 2010)), and loss of Merlin results in a loss of density-dependent inhibition of cell proliferation in numerous cell types (Curto et al. 2007; Lallemand et al. 2003). Interestingly, Merlin can block the endocytosis of ligand-bound EGFR specifically in dense cell cultures, where EGFR internalization is known to be required for a full signaling response (Sorkin and von Zastrow 2009). Merlin can also be found to co-immunoprecipitate with both E-cadherin/catenin and EGFR complexes in dense but not sparse cultures, through making a direct interaction with α-catenin and an indirect interaction with EGFR through NHERF-1 (Curto et al. 2007; Gladden et al. 2010). Thus the E-cadherin/catenin complex can work with Merlin to shut-down EGFR signaling by preventing is internalization into an endocytic compartment from which it signals. Curiously, while E-cadherin can inhibit different classes of receptor tyrosine kinases (RTKs) (e.g., IGF-1R, c-Met receptor, ErB2-4; (Qian et al. 2004; Vermeer et al. 2003), Merlin is selective for the EGFR (Curto et al. 2007), raising the possibility that the cadherin/catenin complex uses molecules functionally analogous to Merlin to limit signaling from distinct RTKs.

Alternatively, E-cadherin's general role in the establishment of a junctional barrier (Fig. 8.5) might limit access of apically localized growth factors, such as Heregulin α, from their basolaterally-localized ErB2-4 receptors (Vermeer et al. 2003). Not all receptor complexes are regulated by density or cadherin expression, such as heterotrimeric G protein-coupled, lysophosphatidic acid (LPA) and muscarinic receptors (Qian et al. 2004), indicating that there is some specificity with regards to density-dependent downregulation of receptor signaling. It is also important to

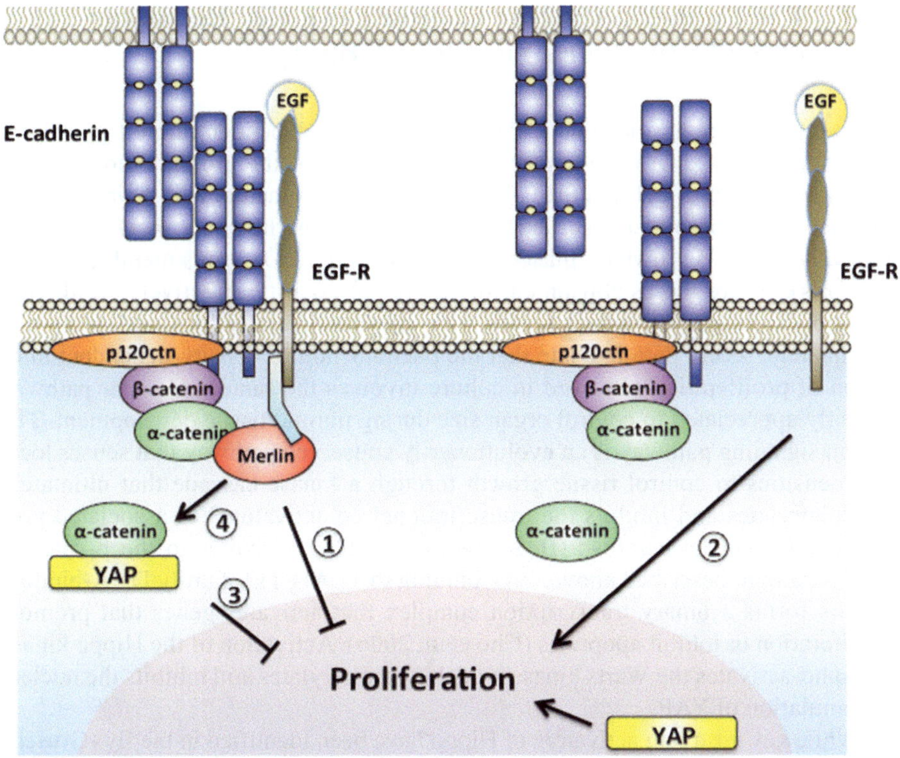

Fig. 8.6 E-cadherin and density-dependent inhibition of proliferation. E-cadherin in densely packed epithelial monolayers can inhibit access of EGF to the EGFR as well as downstream signaling from the EGFR (**1**) compared with less mature contacts (**2**) E-cadherin engagement can also limit the nuclear accumulation of YAP through a poorly defined mechanism that requires α-catenin (**3** and **4**)

recognize that initial stages of cell contact formation are actually accompanied by an increase in EGFR activity (Fedor-Chaiken et al. 2003a; Pece and Gutkind 2000) and activation of other kinases/signaling events that promote proliferation and survival (Cadigan and Liu 2006; Goodwin et al. 2003; Nelson and Chen 2003), while only later stage "dense" contacts are associated with down-regulation of these same signals (reviewed in (Brunton et al. 2004)). Lastly, cadherin-subtypes can impact RTKs differentially. For example, while N-cadherin sustains FGF-receptor signals by preventing their endocytosis-mediated downregulation (Suyama et al. 2002), VE-cadherin attenuates some but not all effectors of VEGFR2 signaling (Carmeliet et al. 1999; Grazia Lampugnani et al. 2003; Rahimi and Kazlauskas 1999). Taken altogether, it is clear that the relationship between cadherins, cell contact and signaling from diverse growth factor receptors is complex and depends on the type and maturity of the contact (Fig. 8.6).

8.11 The Cadherin/Catenin Complex as a Key Regulator of the Hippo/Warts Signaling Pathway

Historically, there has been concern that the density-dependent inhibition of proliferation observed *in vitro* (and examined in most of the aforementioned studies) is an artifact of cell culture. *In vivo*, most growth factors and their receptors interact at the basolateral surfaces of epithelial cells, however *in vitro* studies typically use cells grown on plastic dishes (as opposed to porous membrane filter supports), where the establishment of junctional polarity effectively "seals-off" the growth factor-rich apical media from the basolaterally-localized receptors. Fortunately, recent data indicate that the phenomenon of density-dependent inhibition of proliferation observed in culture involves the same molecular pathway recently appreciated to control organ size during normal tissue development. The Hippo signaling pathway is an evolutionarily conserved pathway that senses local cell densities to control tissue growth through a kinase cascade that ultimately phosphorylates and inhibits the transcriptional co-activator, Yes-associated protein (YAP) and its paralogs (Beausoleil et al. 2004). Similar to the β-catenin/TCF paradigm described above, YAP binding to TEAD/TEF family DNA-binding factors forms a binary transcription complex that activates genes that promote proliferation or inhibit apoptosis (Cho et al. 2006). Activation of the Hippo kinase phospho-activates the Warts kinase, which phosphorylates and inhibits the nuclear accumulation of YAP.

While key upstream activators of Hippo have been identified in the fly (Grusche et al. 2010), three recent studies indicate that the cadherin/catenin complex is a key regulator of the Hippo pathway in mammals. For example, forced expression of E-cadherin in cancer cells that previously silenced the E-cadherin gene restores the density-dependent exclusion of YAP from the nucleus (Kim et al. 2011). More important, the ability of E-cadherin ectodomain-coated beads to inhibit epithelial cell proliferation is lost upon knock-down of Hippo pathway signaling components or overexpression by YAP (Kim et al. 2011). How cadherin engagement limits the nuclear accumulation of YAP is not well understood, but two independent studies suggest that α-catenin functionally and physically participates in a complex that restrains the nuclear accumulation of YAP (Schlegelmilch et al. 2011; Silvis et al. 2011). Curiously, since YAP is generally not observed to accumulate at dense cell–cell contacts, and the vast majority of cytosolic α-catenin behaves as a monomer by gel filtration chromatography (Drees et al. 2005), suggesting that most of this pool of α-catenin is largely not associated with another protein, it may be more likely that α-catenin participates in the transmission of a signal that ultimately impacts the activity of central kinases in the Hippo/Warts signaling cassette, rather than serving as a stoichiometric inhibitor of YAP. Regardless of this issue, it is clear that the cadherin/catenin complex can convey growth inhibitory signals from dense contacts, through seemingly close physical interactions with both EGFR and Hippo pathway components (Fig. 8.6).

8.12 Summary and Perspectives

When we consider the various signals generated from cadherin-based adhesion receptors, the challenge has not been in appreciating the link between adhesion and signaling, but rather in understanding how an adhesive structure that lacks core enzymatic activity conveys information to the cell's interior. Evidence that most cadherin-binding partners belong to the armadillo family of proteins, many of which also associate with DNA binding factors in the nucleus, reveals a seemingly simple way to coordinate changes in gene expression with changes in the abundance of adhesive structures. Through this mode, cadherins can be viewed as simple stoichiometric inhibitors of catenin nuclear signals. However, it is also clear that cadherin-based adhesion in dense cell arrangements can impact a number of distinct molecular pathways required for tissue growth and proliferation. Mechanistically, the contribution of cadherin-based adhesion to these signals is less clear but appears to depend less on cadherin/catenin protein levels (which don't substantively change with cell density) than an organization that broadly impacts the activity of kinases involved in proliferation (e.g., growth factor receptor tyrosine kinases and Hippo/Warts signaling). Through this mode, cadherins can be viewed as inhibitors of kinase signaling cascades. The question is, "What is the arrangement or organization of a cadherin/catenin complex that broadly shuts down the activities of diverse kinases?" While a common mechanism may be unlikely, we speculate that the organization of cadherin-based adhesions into higher order junctional arrangements may be an important feature of this mode cadherin signaling (Niessen and Gottardi 2008). We look forward to future studies that aim to better define the organization of the cadherin/catenin complex into junctions, and how this organization conveys nuclear signals.

References

Andrade MA, Petosa C, O'Donoghue SI, Muller CW, Bork P (2001) Comparison of ARM and HEAT protein repeats. J Mol Biol 309:1–18

Beausoleil SA, Jedrychowski M, Schwartz D, Elias JE, Villen J, Li J, Cohn MA, Cantley LC, Gygi SP (2004) Large-scale characterization of HeLa cell nuclear phosphoproteins. Proc Natl Acad Sci USA 101:12130–12135

Behrens J, von Kries JP, Kuhl M, Bruhn L, Wedlich D, Grosschedl R, Birchmeier W (1996) Functional interaction of beta-catenin with the transcription factor LEF-1. Nature 382:638–642

Berx G, van Roy F (2009) Involvement of members of the cadherin superfamily in cancer. Cold Spring Harb Perspect Biol 1:a003129

Berx G, Becker KF, Hofler H, van Roy F (1998) Mutations of the human E-cadherin (CDH1) gene. Hum Mutat 12:226–237

Bierkamp C, McLaughlin KJ, Schwarz H, Huber O, Kemler R (1996) Embryonic heart and skin defects in mice lacking plakoglobin. Dev Biol 180:780–785

Birchmeier W, Behrens J, Weidner KM, Hulsken J, Birchmeier C (1996) Epithelial differentiation and the control of metastasis in carcinomas. Curr Top Microbiol Immunol 213 (Pt 2):117–135

Blauwkamp TA, Chang MV, Cadigan KM (2008) Novel TCF-binding sites specify transcriptional repression by Wnt signalling. EMBO J 27:1436–1446
Bray SJ (2006) Notch signalling: a simple pathway becomes complex. Nature reviews. Mol Cell Biol 7:678–689
Brunton VG, MacPherson IR, Frame MC (2004) Cell adhesion receptors, tyrosine kinases and actin modulators: a complex three-way circuitry. Biochimica et biophysica acta 1692:121–144
Caca K, Kolligs FT, Ji X, Hayes M, Qian J, Yahanda A, Rimm DL, Costa J, Fearon ER (1999) Beta- and gamma-catenin mutations, but not E-cadherin inactivation, underlie T-cell factor/lymphoid enhancer factor transcriptional deregulation in gastric and pancreatic cancer. Cell Growth Differ 10:369–376
Cadigan KM, Liu YI (2006) Wnt signaling: complexity at the surface. J Cell Sci 119:395–402
Cadigan KM, Peifer M (2009) Wnt signaling from development to disease: insights from model systems. Cold Spring Harb Perspect Biol 1:a002881
Carmeliet P, Lampugnani MG, Moons L, Breviario F, Compernolle V, Bono F, Balconi G, Spagnuolo R, Oosthuyse B, Dewerchin M, Zanetti A, Angellilo A, Mattot V, Nuyens D, Lutgens E, Clotman F, de Ruiter MC, Gittenberger-de Groot A, Poelmann R, Lupu F, Herbert JM, Collen D, Dejana E (1999) Targeted deficiency or cytosolic truncation of the VE-cadherin gene in mice impairs VEGF-mediated endothelial survival and angiogenesis. Cell 98:147–157
Chen H, Paradies NE, Fedor-Chaiken M, Brackenbury R (1997) E-cadherin mediates adhesion and suppresses cell motility via distinct mechanisms. J Cell Sci 110 (Pt 3):345–356
Chen MC, Solomon TE, Kui R, Soll AH (2002) Apical EGF receptors regulate epithelial barrier to gastric acid: endogenous TGF-alpha is an essential facilitator. Am J Physiol Gastrointest Liver Physiol 283:G1098–1106
Cho E, Feng Y, Rauskolb C, Maitra S, Fehon R, Irvine KD (2006) Delineation of a Fat tumor suppressor pathway. Nat Genet 38:1142–1150
Choi HJ, Huber AH, Weis WI (2006) Thermodynamics of beta-catenin-ligand interactions: the roles of the N- and C-terminal tails in modulating binding affinity. J Biol Chem 281:1027–1038
Christofori G (2006) New signals from the invasive front. Nature 441:444–450
Ciruna B, Rossant J (2001) FGF signaling regulates mesoderm cell fate specification and morphogenetic movement at the primitive streak. Dev Cell 1:37–49
Coates JC (2003) Armadillo repeat proteins: beyond the animal kingdom. Trends Cell Biol 13:463–471
Corbit KC, Shyer AE, Dowdle WE, Gaulden J, Singla V, Chen MH, Chuang PT, Reiter JF (2008) Kif3a constrains beta-catenin-dependent Wnt signalling through dual ciliary and non-ciliary mechanisms. Nat Cell Biol 10:70–76
Cox RT, Kirkpatrick C, Peifer M (1996) Armadillo is required for adherens junction assembly, cell polarity, and morphogenesis during Drosophila embryogenesis. J Cell Biol 134:133–148
Curto M, Cole BK, Lallemand D, Liu CH, McClatchey AI (2007) Contact-dependent inhibition of EGFR signaling by Nf2/Merlin. J Cell Biol 177:893–903
Daniel JM, Reynolds AB (1997) Tyrosine phosphorylation and cadherin/catenin function. BioEssays 19:883–891
Daniel JM, Reynolds AB (1999) The catenin p120(ctn) interacts with Kaiso, a novel BTB/POZ domain zinc finger transcription factor. Mol Cell Biol 19:3614–3623
Daugherty RL, Gottardi CJ (2007) Phospho-regulation of Beta-catenin adhesion and signaling functions. Physiology (Bethesda) 22:303–309
Drees F, Pokutta S, Yamada S, Nelson WJ, Weis WI (2005) Alpha-catenin is a molecular switch that binds E-cadherin-beta-catenin and regulates actin-filament assembly. Cell 123:903–915
Dumstrei K, Wang F, Shy D, Tepass U, Hartenstein V (2002) Interaction between EGFR signaling and DE-cadherin during nervous system morphogenesis. Development 129:3983–3994
Fagotto F, Gumbiner BM (1996) Cell contact-dependent signaling. Dev Biol 180:445–454
Fagotto F, Funayama N, Gluck U, Gumbiner BM (1996) Binding to cadherins antagonizes the signaling activity of beta-catenin during axis formation in Xenopus. J Cell Biol 132:1105–1114
Fagotto F, Gluck U, Gumbiner BM (1998) Nuclear localization signal-independent and importin/karyopherin-independent nuclear import of beta-catenin. Curr Biol 8:181–190

Faux MC, Coates JL, Kershaw NJ, Layton MJ, Burgess AW (2010) Independent interactions of phosphorylated beta-catenin with E-cadherin at cell–cell contacts and APC at cell protrusions. PloS one 5:e14127

Fedor-Chaiken M, Hein PW, Stewart JC, Brackenbury R, Kinch MS (2003a) E-cadherin binding modulates EGF receptor activation. Cell Commun Adhes 10:105–118

Fedor-Chaiken M, Meigs TE, Kaplan DD, Brackenbury R (2003b) Two regions of cadherin cytoplasmic domains are involved in suppressing motility of a mammary carcinoma cell line. J Biol Chem 278:52371–52378

Fehon RG, McClatchey AI, Bretscher A (2010) Organizing the cell cortex: the role of ERM proteins. Nat Rev Mol Cell Biol 11:276–287

Ferber EC, Kajita M, Wadlow A, Tobiansky L, Niessen C, Ariga H, Daniel J, Fujita Y (2008) A role for the cleaved cytoplasmic domain of E-cadherin in the nucleus. J Biol Chem 283:12691–12700

Ferreira AC, Suriano G, Mendes N, Gomes B, Wen X, Carneiro F, Seruca R, Machado JC (2011) E-cadherin impairment increases cell survival through Notch-dependent upregulation of Bcl-2. Hum Mol Genet

Fortini ME (2002) Gamma-secretase-mediated proteolysis in cell-surface-receptor signalling. Nat Rev Mol Cell Biol 3:673–684

Funayama N, Fagotto F, McCrea P, Gumbiner B (1995) Embryonic axis induction by the armadillo repeat domain of beta-catenin: evidence for intracellular signaling. J Cell Biol 128:959–968

Gallin WJ, Chuong CM, Finkel LH, Edelman GM (1986) Antibodies to liver cell adhesion molecule perturb inductive interactions and alter feather pattern and structure. Proc Natl Acad Sci USA 83:8235–8239

Giles RH, van Es JH, Clevers H (2003) Caught up in a Wnt storm: Wnt signaling in cancer. Biochim Biophys Acta 1653:1–24

Gladden AB, Hebert AM, Schneeberger EE, McClatchey AI (2010) The NF2 tumor suppressor, Merlin, regulates epidermal development through the establishment of a junctional polarity complex. Dev Cell 19:727–739

Goodwin M, Kovacs EM, Thoreson MA, Reynolds AB, Yap AS (2003) Minimal mutation of the cytoplasmic tail inhibits the ability of E-cadherin to activate Rac but not phosphatidylinositol 3-kinase: direct evidence of a role for cadherin-activated Rac signaling in adhesion and contact formation. J Biol Chem 278:20533–20539

Gottardi CJ, Gumbiner BM (2001) Adhesion signaling: how beta-catenin interacts with its partners. Curr Biol 11:R792–794

Gottardi CJ, Wong E, Gumbiner BM (2001) E-cadherin suppresses cellular transformation by inhibiting beta-catenin signaling in an adhesion-independent manner. J Cell Biol 153:1049–1060

Graham TA, Weaver C, Mao F, Kimelman D, Xu W (2000) Crystal structure of a beta-catenin/Tcf complex. Cell 103:885–896

Grazia Lampugnani M, Zanetti A, Corada M, Takahashi T, Balconi G, Breviario F, Orsenigo F, Cattelino A, Kemler R, Daniel TO, Dejana E (2003) Contact inhibition of VEGF-induced proliferation requires vascular endothelial cadherin, beta-catenin, and the phosphatase DEP-1/CD148. J Cell Biol 161:793–804

Greaves S, Sanson B, White P, Vincent JP (1999) A screen for identifying genes interacting with armadillo, the Drosophila homolog of beta-catenin. Genetics 153:1753–1766

Grusche FA, Richardson HE, Harvey KF (2010) Upstream regulation of the hippo size control pathway. Curr Biol 20:R574–582

Ha NC, Tonozuka T, Stamos JL, Choi HJ, Weis WI (2004) Mechanism of phosphorylation-dependent binding of APC to beta-catenin and its role in beta-catenin degradation. Mol Cell 15:511–521

Haegel H, Larue L, Ohsugi M, Fedorov L, Herrenknecht K, Kemler R (1995) Lack of beta-catenin affects mouse development at gastrulation. Development 121:3529–3537

Hanna J, Leggett DS, Finley D (2003) Ubiquitin depletion as a key mediator of toxicity by translational inhibitors. Mol Cell Biol 23:9251–9261

Harris ES, Nelson WJ (2010) Adenomatous polyposis coli regulates endothelial cell migration independent of roles in beta-catenin signaling and cell–cell adhesion. Mol Biol Cell 21:2611–2623

Hay E, Laplantine E, Geoffroy V, Frain M, Kohler T, Muller R, Marie PJ (2009) N-cadherin interacts with axin and LRP5 to negatively regulate Wnt/beta-catenin signaling, osteoblast function, and bone formation. Mol Cell Biol 29:953–964

Heasman J, Crawford A, Goldstone K, Garner-Hamrick P, Gumbiner B, McCrea P, Kintner C, Noro C, Wylie C (1994) Overexpression of cadherins and underexpression of beta-catenin inhibit dorsal mesoderm induction in early Xenopus embryos. Cell 79:791–803

Hecht A, Litterst CM, Huber O, Kemler R (1999) Functional characterization of multiple transactivating elements in beta-catenin, some of which interact with the TATA-binding protein in vitro. J Biol Chem 274:18017–18025

Henderson BR (2000) Nuclear-cytoplasmic shuttling of APC regulates beta-catenin subcellular localization and turnover. Nat Cell Biol 2:653–660

Hendriksen J, Jansen M, Brown CM, Van Der Velde H, van Ham M, Galjart N, Offerhaus GJ, Fagotto F, Fornerod M (2008) Plasma membrane recruitment of dephosphorylated beta-catenin upon activation of the Wnt pathway. J Cell Sci 121:1793–1802

Herzig M, Savarese F, Novatchkova M, Semb H, Christofori G (2007) Tumor progression induced by the loss of E-cadherin independent of beta-catenin/Tcf-mediated Wnt signaling. Oncogene 26:2290–2298

Hirai Y, Nose A, Kobayashi S, Takeichi M (1989) Expression and role of E- and P-cadherin adhesion molecules in embryonic histogenesis. II. Skin morphogenesis. Development 105:271–277

Hong JY, Park JI, Cho K, Gu D, Ji H, Artandi SE, McCrea PD (2010) Shared molecular mechanisms regulate multiple catenin proteins: canonical Wnt signals and components modulate p120-catenin isoform-1 and additional p120 subfamily members. J Cell Sci 123:4351–4365

Hoschuetzky H, Aberle H, Kemler R (1994) Beta-catenin mediates the interaction of the cadherin-catenin complex with epidermal growth factor receptor. J Cell Biol 127:1375–1380

Hosono S, Gross I, English MA, Hajra KM, Fearon ER, Licht JD (2000) E-cadherin is a WT1 target gene. J Biol Chem 275:10943–10953

Howard S, Deroo T, Fujita Y, Itasaki N (2011) A positive role of cadherin in Wnt/beta-catenin signalling during epithelial-mesenchymal transition. PloS one 6:e23899

Howe LR, Brown AM (2004) Wnt signaling and breast cancer. Cancer Biol Ther 3:36–41

Huber AH, Weis WI (2001) The structure of the beta-catenin/E-cadherin complex and the molecular basis of diverse ligand recognition by beta-catenin. Cell 105:391–402

Huber O, Korn R, McLaughlin J, Ohsugi M, Herrmann BG, Kemler R (1996) Nuclear localization of beta-catenin by interaction with transcription factor LEF-1. Mech Dev 59:3–10

Huber O, Krohn M, Kemler R (1997) A specific domain in alpha-catenin mediates binding to beta-catenin or plakoglobin. J Cell Sci 110:1759–1765

Huelsken J, Vogel R, Brinkmann V, Erdmann B, Birchmeier C, Birchmeier W (2000) Requirement for beta-catenin in anterior-posterior axis formation in mice. J Cell Biol 148:567–578

Hulpiau P, van Roy F (2011) New insights into the evolution of metazoan cadherins. Mol Biol Evol 28:647–657

Jamora C, DasGupta R, Kocieniewski P, Fuchs E (2003) Links between signal transduction, transcription and adhesion in epithelial bud development. Nature 422:317–322

Jarrett CR, Blancato J, Cao T, Bressette DS, Cepeda M, Young PE, King CR, Byers SW (2001) Human APC2 localization and allelic imbalance. Cancer Res 61:7978–7984

Jou TS, Stewart DB, Stappert J, Nelson WJ, Marrs JA (1995) Genetic and biochemical dissection of protein linkages in the cadherin-catenin complex. Proc Natl Acad Sci USA 92:5067–5071

Kam Y, Quaranta V (2009) Cadherin-bound beta-catenin feeds into the Wnt pathway upon adherens junctions dissociation: evidence for an intersection between beta-catenin pools. PloS one 4:e4580

Karnovsky A, Klymkowsky MW (1995) Anterior axis duplication in Xenopus induced by the over-expression of the cadherin-binding protein plakoglobin. Proc Natl Acad Sci USA 92:4522–4526

Kelly KF, Spring CM, Otchere AA, Daniel JM (2004) NLS-dependent nuclear localization of p120ctn is necessary to relieve Kaiso-mediated transcriptional repression. J Cell Sci 117:2675–2686

Kim JB, Islam S, Kim YJ, Prudoff RS, Sass KM, Wheelock MJ, Johnson KR (2000) N-Cadherin extracellular repeat 4 mediates epithelial to mesenchymal transition and increased motility. J Cell Biol 151:1193–1206

Kim NG, Koh E, Chen X, Gumbiner BM (2011) E-cadherin mediates contact inhibition of proliferation through Hippo signaling-pathway components. Proc Natl Acad Sci USA 108:11930–11935

Kim SW, Park JI, Spring CM, Sater AK, Ji H, Otchere AA, Daniel JM, McCrea PD (2004) Non-canonical Wnt signals are modulated by the Kaiso transcriptional repressor and p120-catenin. Nat Cell Biol 6:1212–1220

Kolligs FT, Hu G, Dang CV, Fearon ER (1999) Neoplastic transformation of RK3E by mutant beta-catenin requires deregulation of Tcf/Lef transcription but not activation of c-myc expression. Mol Cell Biol 19:5696–5706

Korpal M, Lee ES, Hu G, Kang Y (2008) The miR-200 family inhibits epithelial-mesenchymal transition and cancer cell migration by direct targeting of E-cadherin transcriptional repressors ZEB1 and ZEB2. J Biol Chem 283:14910–14914

Krieghoff E, Behrens J, Mayr B (2006) Nucleo-cytoplasmic distribution of beta-catenin is regulated by retention. J Cell Sci 119:1453–1463

Kuphal F, Behrens J (2006) E-cadherin modulates Wnt-dependent transcription in colorectal cancer cells but does not alter Wnt-independent gene expression in fibroblasts. Exp Cell Res 312:457–467

Lallemand D, Curto M, Saotome I, Giovannini M, McClatchey AI (2003) NF2 deficiency promotes tumorigenesis and metastasis by destabilizing adherens junctions. Genes Dev 17:1090–1100

Liu C, Li Y, Semenov M, Han C, Baeg GH, Tan Y, Zhang Z, Lin X, He X (2002) Control of beta-catenin phosphorylation/degradation by a dual-kinase mechanism. Cell 108:837–847

Liu YC, Li F, Handler J, Huang CR, Xiang Y, Neretti N, Sedivy JM, Zeller KI, Dang CV (2008) Global regulation of nucleotide biosynthetic genes by c-Myc. PloS one 3:e2722

Lochter A, Galosy S, Muschler J, Freedman N, Werb Z, Bissell MJ (1997) Matrix metalloproteinase stromelysin-1 triggers a cascade of molecular alterations that leads to stable epithelial-to-mesenchymal conversion and a premalignant phenotype in mammary epithelial cells. J Cell Biol 139:1861–1872

Louvi A, Artavanis-Tsakonas S (2006) Notch signalling in vertebrate neural development. Nat Rev Neurosci 7:93–102

MacDonald BT, Tamai K, He X (2009) Wnt/beta-catenin signaling: components, mechanisms, and diseases. Dev Cell 17:9–26

Maher MT, Flozak AS, Stocker A, Chenn A, Gottardi CJ (2009) Activity of the β-catenin phosphodestruction at cell–cell contacts is enhanced by cadherin-based adhesion. J Cell Biol 186:219–228

Maher MT, Mo R, Flozak AS, Peled ON, Gottardi CJ (2010) Beta-catenin phosphorylated at serine 45 is spatially uncoupled from beta-catenin phosphorylated in the GSK3 domain: implications for signaling. PloS one 5:e10184

Marambaud P, Shioi J, Serban G, Georgakopoulos A, Sarner S, Nagy V, Baki L, Wen P, Efthimiopoulos S, Shao Z, Wisniewski T, Robakis NK (2002) A presenilin-1/gamma-secretase cleavage releases the E-cadherin intracellular domain and regulates disassembly of adherens junctions. EMBO J 21:1948–1956

Maretzky T, Reiss K, Ludwig A, Buchholz J, Scholz F, Proksch E, de Strooper B, Hartmann D, Saftig P (2005) ADAM10 mediates E-cadherin shedding and regulates epithelial cell–cell adhesion, migration, and beta-catenin translocation. Proc Natl Acad Sci USA 102:9182–9187

Mayor R, Carmona-Fontaine C (2010) Keeping in touch with contact inhibition of locomotion. Trends Cell Biol 20:319–328

McCartney BM, Dierick HA, Kirkpatrick C, Moline MM, Baas A, Peifer M, Bejsovec A (1999) Drosophila APC2 is a cytoskeletally-associated protein that regulates wingless signaling in the embryonic epidermis. J Cell Biol 146:1303–1318

McCrea PD, Gumbiner BM (1991) Purification of a 92-kDa cytoplasmic protein tightly associated with the cell–cell adhesion molecule E-cadherin (uvomorulin). Characterization and extractability of the protein complex from the cell cytostructure. J Biol Chem 266:4514–4520

Mertens C, Kuhn C, Franke WW (1996) Plakophilins 2a and 2b: constitutive proteins of dual location in the karyoplasm and the desmosomal plaque. J Cell Biol 135:1009–1025

Miravet S, Piedra J, Miro F, Itarte E, Garcia de Herreros A, Dunach M (2002) The transcriptional factor Tcf-4 contains different binding sites for beta-catenin and plakoglobin. J Biol Chem 277:1884–1891

Molenaar M, van de Wetering M, Oosterwegel M, Peterson-Maduro J, Godsave S, Korinek V, Roose J, Destree O, Clevers H (1996) XTcf-3 transcription factor mediates beta-catenin-induced axis formation in Xenopus embryos. Cell 86:391–399

Montell DJ, Rorth P, Spradling AC (1992) slow border cells, a locus required for a developmentally regulated cell migration during oogenesis, encodes Drosophila C/EBP. Cell 71:51–62

Nejsum LN, Nelson WJ (2007) A molecular mechanism directly linking E-cadherin adhesion to initiation of epithelial cell surface polarity. J Cell Biol 178:323–335

Nejsum LN, Nelson WJ (2009) Epithelial cell surface polarity: the early steps. Front Biosci 14:1088–1098

Nelson CM, Chen CS (2003) VE-cadherin simultaneously stimulates and inhibits cell proliferation by altering cytoskeletal structure and tension. J Cell Sci 116:3571–3581

Nelson RW, Gumbiner BM (1999) A cell-free assay system for beta-catenin signaling that recapitulates direct inductive events in the early xenopus laevis embryo. J Cell Biol 147:367–374

Nieman MT, Prudoff RS, Johnson KR, Wheelock MJ (1999) N-cadherin promotes motility in human breast cancer cells regardless of their E-cadherin expression. J Cell Biol 147:631–644

Nieset JE, Redfield AR, Jin F, Knudsen KA, Johnson KR, Wheelock MJ (1997) Characterization of the interactions of alpha-catenin with alpha-actinin and beta-catenin/plakoglobin. J Cell Sci 110 (Pt 8):1013–1022

Niessen CM, Gottardi CJ (2008) Molecular components of the adherens junction. Biochim Biophys Acta 1778:562–571

Nieto MA (2002) The snail superfamily of zinc-finger transcription factors. Nat Rev Mol Cell Biol 3:155–166

Niewiadomska P, Godt D, Tepass U (1999) DE-Cadherin is required for intercellular motility during Drosophila oogenesis. J Cell Biol 144:533–547

Ohira T, Gemmill RM, Ferguson K, Kusy S, Roche J, Brambilla E, Zeng C, Baron A, Bemis L, Erickson P, Wilder E, Rustgi A, Kitajewski J, Gabrielson E, Bremnes R, Franklin W, Drabkin HA (2003) WNT7a induces E-cadherin in lung cancer cells. Proc Natl Acad Sci U S A 100:10429–10434

Orsulic S, Huber O, Aberle H, Arnold S, Kemler R (1999) E-cadherin binding prevents beta-catenin nuclear localization and beta-catenin/LEF-1-mediated transactivation. J Cell Sci 112:1237–1245

Ozawa M, Ringwald M, Kemler R (1990) Uvomorulin-catenin complex formation is regulated by a specific domain in the cytoplasmic region of the cell adhesion molecule. Proc Natl Acad Sci USA 87:4246–4250

Park JI, Ji H, Jun S, Gu D, Hikasa H, Li L, Sokol SY, McCrea PD (2006) Frodo links Dishevelled to the p120-catenin/Kaiso pathway: distinct catenin subfamilies promote Wnt signals. Dev Cell 11:683–695

Pece S, Gutkind JS (2000) Signaling from E-cadherins to the MAPK pathway by the recruitment and activation of epidermal growth factor receptors upon cell–cell contact formation. J Biol Chem 275:41227–41233

Peifer M, Wieschaus E (1990) The segment polarity gene armadillo encodes a functionally modular protein that is the Drosophila homolog of human plakoglobin. Cell 63:1167–1176

Peifer M, Sweeton D, Casey M, Wieschaus E (1994) wingless signal and Zeste-white 3 kinase trigger opposing changes in the intracellular distribution of Armadillo. Development 120:369–380
Perl AK, Wilgenbus P, Dahl U, Semb H, Christofori G (1998) A causal role for E-cadherin in the transition from adenoma to carcinoma. Nature 392:190–193
Perrais M, Chen X, Perez-Moreno M, Gumbiner BM (2007) E-cadherin homophilic ligation inhibits cell growth and epidermal growth factor receptor signaling independently of other cell interactions. Mol Biol Cell 18:2013–2025
Polakis P (2000) Wnt signaling and cancer. Genes Dev 14:1837–1851
Qian X, Karpova T, Sheppard AM, McNally J, Lowy DR (2004) E-cadherin-mediated adhesion inhibits ligand-dependent activation of diverse receptor tyrosine kinases. Embo J 23:1739–1748
Rahimi N, Kazlauskas A (1999) A role for cadherin-5 in regulation of vascular endothelial growth factor receptor 2 activity in endothelial cells. Mol Biol Cell 10:3401–3407
Ruzov A, Dunican DS, Prokhortchouk A, Pennings S, Stancheva I, Prokhortchouk E, Meehan RR (2004) Kaiso is a genome-wide repressor of transcription that is essential for amphibian development. Development 131:6185–6194
Sadot E, Simcha I, Shtutman M, Ben-Ze'ev A, Geiger B (1998) Inhibition of beta-catenin-mediated transactivation by cadherin derivatives. Proc Natl Acad Sci USA 95:15339–15344
Salah Z, Aqeilan RI (2011) WW domain interactions regulate the Hippo tumor suppressor pathway. Cell Death Dis 2:e172
Sanson B, White P, Vincent JP (1996) Uncoupling cadherin-based adhesion from wingless signalling in Drosophila. Nature 383:627–630
Schlegelmilch K, Mohseni M, Kirak O, Pruszak J, Rodriguez JR, Zhou D, Kreger BT, Vasioukhin V, Avruch J, Brummelkamp TR, Camargo FD (2011) Yap1 acts downstream of alpha-catenin to control epidermal proliferation. Cell 144:782–795
Schneider S, Steinbeisser H, Warga RM, Hausen P (1996) Beta-catenin translocation into nuclei demarcates the dorsalizing centers in frog and fish embryos. Mech Dev 57:191–198
Schroeter EH, Kisslinger JA, Kopan R (1998) Notch-1 signalling requires ligand-induced proteolytic release of intracellular domain. Nature 393:382–386
Schwarz-Romond T, Asbrand C, Bakkers J, Kuhl M, Schaeffer HJ, Huelsken J, Behrens J, Hammerschmidt M, Birchmeier W (2002) The ankyrin repeat protein Diversin recruits Casein kinase Iepsilon to the beta-catenin degradation complex and acts in both canonical Wnt and Wnt/JNK signaling. Genes Dev 16:2073–2084
Shimamura K, Takeichi M (1992) Local and transient expression of E-cadherin involved in mouse embryonic brain morphogenesis. Development 116:1011–1019
Silvis ML, Mosher TJ, Smetana BS, Chinchilli VM, Flemming DJ, Walker EA, Black KP (2011) High prevalence of pelvic and hip magnetic resonance imaging findings in asymptomatic collegiate and professional hockey players. Am J Sports Med 39:715–721
Simcha I, Shtutman M, Salomon D, Zhurinsky J, Sadot E, Geiger B, Ben-Ze'ev A (1998) Differential nuclear translocation and transactivation potential of beta-catenin and plakoglobin. J Cell Biol 141:1433–1448
Simcha I, Kirkpatrick C, Sadot E, Shtutman M, Polevoy G, Geiger B, Peifer M, Ben-Ze'ev A (2001) Cadherin sequences that inhibit beta-catenin signaling: a study in yeast and mammalian cells. Mol Biol Cell 12:1177–1188
Solanas G, Miravet S, Casagolda D, Castano J, Raurell I, Corrionero A, de Herreros AG, Dunach M (2004) beta-Catenin and plakoglobin N- and C-tails determine ligand specificity. J Biol Chem 279:49849–49856
Sorkin A, von Zastrow M (2009) Endocytosis and signalling: intertwining molecular networks. Nature reviews. Molecular cell biology 10:609–622
Struhl G, Adachi A (1998) Nuclear access and action of notch in vivo. Cell 93:649–660
Suh EK, Gumbiner BM (2003) Translocation of beta-catenin into the nucleus independent of interactions with FG-rich nucleoporins. Exp Cell Res 290:447–456
Suyama K, Shapiro I, Guttman M, Hazan RB (2002) A signaling pathway leading to metastasis is controlled by N-cadherin and the FGF receptor. Cancer cell 2:301–314

Takahashi K, Suzuki K (1996) Density-dependent inhibition of growth involves prevention of EGF receptor activation by E-cadherin-mediated cell–cell adhesion. Exp Cell Res 226:214–222
Takai Y, Miyoshi J, Ikeda W, Ogita H (2008) Nectins and nectin-like molecules: roles in contact inhibition of cell movement and proliferation. Nature reviews Mol Cell Biol 9:603–615
Takigawa Y, Brown AM (2008) Wnt signaling in liver cancer. Curr Drug Targets 9:1013–1024
Tanneberger K, Pfister AS, Kriz V, Bryja V, Schambony A, Behrens J (2011) Structural and functional characterization of the Wnt inhibitor APC membrane recruitment 1 (Amer1). J Biol Chem 286:19204–19214
Thiery JP, Acloque H, Huang RY, Nieto MA (2009) Epithelial-mesenchymal transitions in development and disease. Cell 139:871–890
van Amerongen R, Nusse R (2009) Towards an integrated view of Wnt signaling in development. Development 136:3205–3214
van de Wetering M, Barker N, Harkes IC, Van Der Heyden M, Dijk NJ, Hollestelle A, Klijn JG, Clevers H, Schutte M (2001) Mutant E-cadherin breast cancer cells do not display constitutive Wnt signaling. *Cancer Res*. 61:278–284
van Leeuwen IM, Byrne HM, Jensen OE, King JR (2007) Elucidating the interactions between the adhesive and transcriptional functions of beta-catenin in normal and cancerous cells. J Theor Biol 247:77–102
Verbeek S, Izon D, Hofhuis F, Robanus-Maandag E, te Riele H, van de Wetering M, Oosterwegel M, Wilson A, MacDonald HR, Clevers H (1995) An HMG-box-containing T-cell factor required for thymocyte differentiation. Nature 374:70–74
Vermeer PD, Einwalter LA, Moninger TO, Rokhlina T, Kern JA, Zabner J, Welsh MJ (2003) Segregation of receptor and ligand regulates activation of epithelial growth factor receptor. Nature 422:322–326
Vleminckx K, Vakaet L Jr, Mareel M, Fiers W, van Roy F (1991) Genetic manipulation of E-cadherin expression by epithelial tumor cells reveals an invasion suppressor role. Cell 66:107–119
Wang HY, Liu T, Malbon CC (2006) Structure-function analysis of Frizzleds. Cell Signal 18:934–941
Wheelock MJ, Shintani Y, Maeda M, Fukumoto Y, Johnson KR (2008) Cadherin switching. J Cell Sci 121:727–735
Willert J, Epping M, Pollack JR (2002) A transcriptional response to Wnt protein in human embryonic carcinoma cells. BMC Dev Biol 2:8
Willert K, Jones KA (2006) Wnt signaling: is the party in the nucleus? Genes Dev 20:1394–1404
Wong AS, Gumbiner BM (2003) Adhesion-independent mechanism for suppression of tumor cell invasion by E-cadherin. J Cell Biol 161:1191–1203
Wong MH, Rubinfeld B, Gordon JI (1998) Effects of forced expression of an NH2-terminal truncated beta-Catenin on mouse intestinal epithelial homeostasis. J Cell Biol 141:765–777
Xing Y, Clements WK, Kimelman D, Xu W (2003) Crystal structure of a beta-catenin/axin complex suggests a mechanism for the beta-catenin destruction complex. Genes Dev 17:2753–2764
Xing Y, Clements WK, Trong I Le, Hinds TR, Stenkamp R, Kimelman D, Xu W (2004) Crystal structure of a beta-catenin/APC complex reveals a critical role for APC phosphorylation in APC function. Mol Cell 15:523–533
Yanagawa S, Lee JS, Haruna T, Oda H, Uemura T, Takeichi M, Ishimoto A (1997) Accumulation of Armadillo induced by Wingless, Dishevelled, and dominant-negative Zeste-White 3 leads to elevated DE-cadherin in Drosophila clone 8 wing disc cells. J Biol Chem 272:25243–25251
Yang J, Mani SA, Donaher JL, Ramaswamy S, Itzykson RA, Come C, Savagner P, Gitelman I, Richardson A, Weinberg RA (2004) Twist, a master regulator of morphogenesis, plays an essential role in tumor metastasis. Cell 117:927–939
Yu X, Waltzer L, Bienz M (1999) A new Drosophila APC homologue associated with adhesive zones of epithelial cells. Nat Cell Biol 1:144–151
Zantek ND, Azimi M, Fedor-Chaiken M, Wang B, Brackenbury R, Kinch MS (1999) E-cadherin regulates the function of the EphA2 receptor tyrosine kinase. Cell Growth Differ 10:629–638
Zhurinsky J, Shtutman M, Ben-Ze'ev A (2000) Differential mechanisms of LEF/TCF family-dependent transcriptional activation by beta-catenin and plakoglobin. Mol Cell Biol 20:4238–4252

Chapter 9
Adherens Junction Turnover: Regulating Adhesion Through Cadherin Endocytosis, Degradation, and Recycling

Andrew P. Kowalczyk and Benjamin A. Nanes

Abstract Adherens junctions are important mediators of intercellular adhesion, but they are not static structures. They are regularly formed, broken, and rearranged in a variety of situations, requiring changes in the amount of cadherins, the main adhesion molecule in adherens junctions, present at the cell surface. Thus, endocytosis, degradation, and recycling of cadherins are crucial for dynamic regulation of adherens junctions and control of intercellular adhesion. In this chapter, we review the involvement of cadherin endocytosis in development and disease. We discuss the various endocytic pathways available to cadherins, the adaptors involved, and the sorting of internalized cadherin for recycling or lysosomal degradation. In addition, we review the regulatory pathways controlling cadherin endocytosis and degradation, including regulation of cadherin endocytosis by catenins, cadherin ubiquitination, and growth factor receptor signaling pathways. Lastly, we discuss the proteolytic cleavage of cadherins at the plasma membrane.

9.1 Introduction

Cell contacts are not static structures. They are regularly formed, broken, and rearranged both during normal physiological processes and in disease states. In order to allow for dynamic changes in cell contact strength, adherens junctions must themselves be plastic. A key mechanism for modulating adhesion strength is the adjustment of the amount of cadherin, the main adhesion molecule in adherens junctions, present at the plasma membrane (unless otherwise noted, we use 'cadherin' to mean classical cadherins, the cadherin subfamily which forms adherens junctions). Cadherin levels are determined by the balance between endocytosis and degradation,

A. P. Kowalczyk (✉)
Department of Cell Biology, Emory University School of Medicine, Atlanta, GA 30332, USA

Department of Dermatology, Emory University School of Medicine, Atlanta, GA 30332, USA
e-mail: akowalc@emory.edu

B. A. Nanes
Department of Cell Biology, Emory University School of Medicine, Atlanta, GA 30332, USA

Graduate Program in Biochemistry, Cell and Developmental Biology,
Emory University School of Medicine, Atlanta, GA 30332, USA

T. Harris (ed.), *Adherens Junctions: From Molecular Mechanisms to Tissue Development and Disease,* Subcellular Biochemistry 60,
DOI 10.1007/978-94-007-4186-7_9, © Springer Science+Business Media Dordrecht 2012

which remove cadherin from the plasma membrane, and synthesis and recycling, which increase the amount of cadherin available. Transcriptional regulation of cadherins also plays an important role in development and disease (Peinado et al. 2004). However, because the metabolic half-life of cadherins is long, approximately five to ten hours in cultured cells (McCrea and Gumbiner 1991; Shore and Nelson 1991), transcriptional regulation cannot account for more rapid changes in adhesion strength. As we discuss in this chapter, endocytosis, degradation, and recycling of cadherins are crucial for dynamic regulation of adherens junctions and control of intercellular adhesion.

Cadherins are named for their calcium-dependent adhesion. Depletion of extracellular calcium disrupts adherens junctions (Kartenbeck et al. 1982), and it was this process that provided the first evidence that cadherin turnover might play a role in the dynamic control of cell adhesion. Classic electron microscopy and immunofluorescence studies demonstrated that, subsequent to calcium depletion, cadherins are removed from cell junctions by endocytosis (Kartenbeck et al. 1991; Mattey and Garrod 1986). Cadherin endocytosis plays a role in physiological processes as well. For example, cells undergoing mitosis often appear to adopt a rounded morphology, suggesting that they have become detached from their neighbors. Cadherin endocytosis was found to accompany mitosis-related cell rounding, decreasing the junctional pool of cadherin to allow for decreased adhesion, even as the total amount of cadherin expression remained constant (Bauer et al. 1998). More recent work suggests that cadherin endocytosis is a particularly important mechanism for the disassembly of cadherin-based adhesive contacts (Troyanovsky et al. 2006). The significance of cadherin internalization to the dynamic regulation of cell-cell adhesion is now well established. Cadherin endocytosis has been observed in a large variety of developmental and disease processes, and in recent years, tremendous progress has been made toward understanding the molecular mechanisms involved in cadherin internalization and degradation.

In this chapter, we review the evidence for the involvement of cadherin endocytosis during development and its mis-regulation in disease. We also discuss the rapidly accumulating body of work detailing the trafficking pathways involved in cadherin endocytosis. Both clathrin-dependent and clathrin-independent pathways have been implicated, and several endocytic adaptors which interact with cadherins have been identified. In addition, we consider the process of sorting internalized cadherin for recycling or degradation and how the regulation of cadherin recycling may be used to control adherens junction turnover. Regulation of cadherin endocytosis by catenins is also important, and we review the effects of catenins on cadherin internalization. p120-catenin in particular has gained prominence as a "set-point" for cadherin levels, but α- and β-catenins may have important roles as well. We also review the evidence for cadherin ubiquitination as a signal for adherens junction turnover and the ubiquitin ligases which have been found to target cadherins and affect cadherin trafficking. In order to further consider the regulation of cadherin internalization, we discuss the many growth factor signaling pathways that affect cadherin trafficking. Interestingly, in some cases the connection is bidirectional, with growth factor signaling altering cadherin trafficking and cadherins modulating growth factor receptor signaling. Finally, we briefly discuss another important mechanism for adherens junction turnover, the proteolytic degradation of cadherins at the plasma membrane.

9.2 Cadherin Endocytosis in Development and Disease

Perhaps the best examples of the importance of cadherin endocytosis and the dynamic regulation of adherens junctions come from tissue patterning and development. Initially, cadherins were observed to control tissue patterning by facilitating cell sorting based on the type of cadherin expressed (Nose et al. 1988). However, Steinberg and Takeichi also demonstrated that varying the expression level of a single cadherin could also be used as a mechanism for cell sorting (Steinberg and Takeichi 1994). Thus, the prominent role of cadherin endocytosis in development should come as no surprise. For example, during epithelial-mesenchymal transitions, cells decrease the expression level of cadherins through a process involving cadherin internalization (Miller and McClay 1997). Cadherin internalization has also been reported during gastrulation in a variety of organisms (Oda et al. 1998; Ogata et al. 2007), where it may be controlled by Wnt signaling (Ulrich et al. 2005). Other developmental processes where cadherin internalization is important include nervous system development, where both the Rab5-dependent endocytosis and Rab11-mediated recycling of N-cadherin are required for neuronal patterning (Kawauchi et al. 2010). Two lines of investigation also demonstrate the importance of cadherin endocytosis for developmental processes involving planar cell polarity. First, convergent extension in *Xenopus* embryos typically involves the coordinated down-regulation of C-cadherin in response to mesoderm-inducing signals (Brieher and Gumbiner 1994; Zhong et al. 1999). Inhibiting dynamin in *Xenopus* embryos blocks C-cadherin endocytosis, disrupting convergent extension (Jarrett et al. 2002). Second, in *Drosophila*, planar-polarized endocytosis of DE-cadherin mediates cell intercalation necessary for germ band extension, and blocking cadherin endocytosis prevents this critical developmental process (Levayer et al. 2011). Thus, cadherin internalization plays a key role in a variety of developmental processes.

Of course, processes which play important roles in development often contribute to disease when they are activated inappropriately. Cadherin internalization is no exception, and loss of cell adhesion is a key requirement for cancer metastasis. Loss of adhesion in many types of cancer is often attributed to decreased E-cadherin expression (Hirohashi 1998). While this is most often due to decreased synthesis, there is some evidence that increased cadherin endocytosis may also play a role. One recent study found that a non-junctional, presumably internalized, E-cadherin expression pattern was associated with poor survival in nasopharyngeal cancer (Xie et al. 2010). Another found Src-dependent E-cadherin internalization with shear stress in an oropharyngeal cancer cell line (Lawler et al. 2009). Increased E-cadherin internalization has also been found in a mouse model of UV-irradiation-induced squamous cell carcinoma (Brouxhon et al. 2007). As discussed below, there is also considerable evidence for the involvement of cancer-associated signaling molecules, such as receptor tyrosine kinases and v-Src, in cadherin internalization.

Cadherin endocytosis may play a role in other disease processes as well. For example, internalization of E-cadherin by pancreatic acinar cells was found to be increased in an experimental model of acute pancreatitis (Lerch et al. 1997). Acute pancreatitis is classically associated with significant pancreatic edema, and increased cadherin endocytosis leading to loss of epithelial integrity is an attractive pathophysiological mechanism. Another disease process in which cadherin endocytosis has been implicated is the autoimmune blistering disease pemphigus vulgaris. Auto-antibodies from pemphigus patients cause increased internalization of the desmosomal cadherin desmoglein 3, which may contribute to loss of epithelial integrity and blister formation (Calkins et al. 2006; Delva et al. 2008). Intriguingly, cadherin endocytosis may also be involved in infectious processes. The bacterium *Listeria monocytogenes* appears to hijack a constitutive cadherin endocytic pathway in order to gain entry to cells, a key contributor to the pathogen's virulence (Veiga and Cossart 2005). The potential involvement of cadherin endocytosis in such a variety of diseases makes it a tempting target for new therapies, though it remains to be seen whether aberrant cadherin internalization in disease can be inhibited without affecting cadherin endocytosis necessary for normal biological processes. Turning these discoveries into a new generation of anti-cancer drugs will certainly require a better understanding of the molecular mechanisms and regulation of adherens junction turnover.

9.3 Cadherin Trafficking Pathways

Understanding the pathways cadherins use to move in and out of adherens junctions has been a major research focus over the past decade (Chiasson and Kowalczyk 2008). This work has significantly increased our understanding of how cadherins are internalized and how they are selected for degradation or for recycling back to the plasma membrane. Trafficking pathways essentially control the rate of cadherin turnover; the higher the rate of cadherin endocytosis and the higher the proportion of endocytosed cadherin selected for degradation rather than recycling, the lower the amount of cadherin that will be available to form adherens junctions. We review the clathrin-dependent endocytosis of cadherins and the adaptor proteins involved, as well as several clathrin-independent endocytic pathways and pathways involved in the recycling of internalized cadherin (Fig. 9.1).

9.3.1 Clathrin-Mediated Endocytosis

Cadherin internalization occurs through several distinct endocytic pathways. Of them, most work has focused on clathrin-mediated endocytosis, which is also the

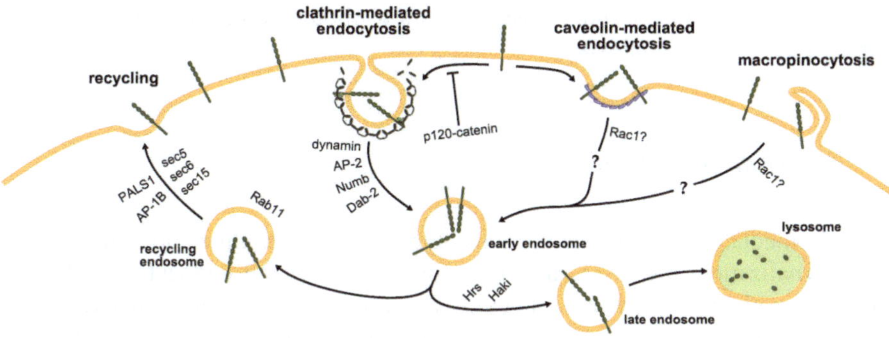

Fig. 9.1 Cadherin trafficking pathways. Cadherin internalization can occur through either clathrin-mediated, caveolin-mediated, or macropinocytosis-like pathways. Internalized cadherin is then sorted either for lysosomal degradation or recycling back to the plasma membrane

endocytic pathway understood in the greatest detail (Bonifacino and Traub 2003). Proteins are targeted for clathrin-mediated endocytosis by the binding of adaptor protein complexes. Once bound, adaptor proteins recruit other components of the endocytic machinery and cluster into clathrin-coated pits. Clathrin-coated pits containing proteins targeted for endocytosis then undergo dynamin-mediated scission from the plasma membrane, budding off to form endocytic vesicles. Internalized proteins can be sorted for recycling back to the plasma membrane or sorted to the lysosome for degradation.

Cadherin was first recognized to undergo clathrin-mediated endocytosis by Le and colleagues, who observed constitutive clathrin-mediated endocytosis and recycling of E-cadherin in MDCK cells (Le et al. 1999). We also found that endocytosis of VE-cadherin in endothelial cells occurs through a clathrin-mediated pathway ultimately resulting in degradation of the cadherin by the lysosome (Xiao et al. 2003b). Furthermore, clathrin-mediated endocytosis appears to be responsible for two types of growth factor-induced cadherin internalization, FGF-mediated internalization of E-cadherin (Bryant et al. 2005) and VEGF-mediated internalization of VE-cadherin (Gavard and Gutkind 2006). Interestingly, clathrin-mediated endocytosis of E-cadherin may be related to the cadherin's adhesive state. Izumi and colleagues isolated adherens junction-containing membrane from rat liver and, using a reconstitution system, observed budding of E-cadherin into clathrin-coated vesicles with electron microscopy and biochemical fractionation. Adding antibody against the extracellular domain of E-cadherin, which blocks *trans* interactions, to the reconstitution system increased the amount of cadherin which entered clathrin-coated vesicles, while adding E-cadherin extracellular domain fragments decreased recruitment of cadherin to clathrin-coated vesicles. They also found that *trans* interaction-mediated inhibition of cadherin endocytosis involved activation of the small G-proteins Rac and Cdc42, as well as the actin-binding protein IQGAP1 (Izumi et al. 2004). In addition, exposing an intestinal epithelial cell line to low-calcium conditions, which disrupts cadherin *trans* interactions, results in the clathrin mediated endocytosis of E-cadherin, along with other adherens junction and tight junction components, into a unique syntaxin-4-positive compartment (Ivanov et al. 2004).

Thus, clathrin-mediated endocytosis appears to modulate cadherin function in a variety of biological contexts.

9.3.2 Endocytic Adaptors

Clathrin-mediated endocytosis depends on adaptors to recognize proteins targeted for internalization and to recruit other components of the endocytic machinery. Identifying clathrin-mediated endocytosis as a pathway for cadherin internalization raises the question of what endocytic adaptors might recognize cadherins. One likely candidate is the adaptor protein complex AP-2, which commonly recognizes cargo proteins with a tyrosine- or dileucine-based motif (Traub 2003). E-cadherin contains a putative dileucine-based AP-2 binding motif in its cytoplasmic tail, and mutating those residues disrupts the normal basolateral localization of E-cadherin (Miranda et al. 2001) and prevents E-cadherin clathrin-mediated endocytosis (Miyashita and Ozawa 2007b). This motif is also present in many other classical cadherins, including N- and P-cadherins. It is not, however, present in VE-cadherin or in *Drosophila* DE-cadherin. Nonetheless, the VE-cadherin cytoplasmic tail is sufficient to mediate clathrin-dependent endocytosis when attached to an unrelated transmembrane protein, strongly suggesting that cadherins may contain other endocytic adaptor binding sequences as well (Xiao et al. 2005). In recent years, more direct evidence for the involvement of AP-2 in the clathrin-mediated endocytosis of cadherins has begun to accumulate. We found that internalization of VE-cadherin is clathrin-, dynamin-, and AP-2-dependent and that AP-2 both co-localizes with VE-cadherin and co-immunoprecipitates with the VE-cadherin cytoplasmic tail (Chiasson et al. 2009). An AP-2 subunit was also found to co-immunoprecipitate with the E-cadherin cytoplasmic tail (Sato et al. 2011). Interestingly, Levayer and colleagues also found that AP-2- and clathrin-mediated endocytosis of DE-cadherin is crucial for the establishment of planar cell polarity in germ band extension. Polarized distributions of Dia and Myosin-II induce planar DE-cadherin clustering in junctions perpendicular to the developing long axis of the germ band. DE-cadherin clustering recruits AP-2 and clathrin to these junctions, leading to the preferential endocytosis of DE-cadherin from perpendicular junctions and the relative accumulation of DE-cadherin in junctions parallel to the germ band axis (Levayer et al. 2011).

However, the question of what endocytic adaptors are important for cadherin endocytosis remains incompletely resolved. It is not yet clear that AP-2 interacts directly with cadherins. It is also possible that other endocytic adaptors may be involved depending on the biological context. Mice null for Dab-2, another adaptor protein associated with clathrin-mediated endocytosis, support this possibility. They exhibit loss of apical-basal polarized distribution of E-cadherin, as well as the LDL receptor-related protein megalin, in the developing endoderm (Yang et al. 2007). Several reports also suggest a role for the endocytic adaptor Numb in cadherin internalization. In radial glial cells, Numb co-immunoprecipitates with cadherins, and Numb depletion disrupts adherens junctions (Rasin et al. 2007). Numb also binds to E-cadherin in epithelial cell lines and mediates endocytosis of cadherins specifically from the apical surface, contributing to the lateral localization of cadherins in adherens junctions (Lau and McGlade 2011; Wang et al. 2009). This polarization is due to localized phosphorylation and inactivation of Numb at lateral membranes by the PAR polarity complex member aPKC (Sato et al. 2011). Consequently, the role of adaptor proteins in cadherin endocytosis remains an exciting area for future discovery.

9.3.3 Clathrin-Independent Endocytic Pathways

Cadherin turnover has also been associated with clathrin-independent endocytic pathways, though considerably less work has been done in this area compared to clathrin-mediated cadherin endocytosis. Studies have suggested that cadherin endocytosis may occur through both caveolin-mediated and macropinocytosis-like pathways. Akhtar and colleagues found that a dominant-active form of the small GTPase Rac1 could disrupt cell-cell adhesion in keratinocytes. This was associated with the endocytosis of E-cadherin through a pathway that appeared to be distinct from the uptake of transferrin, which is clathrin-mediated, and through structures that co-localized with caveolin (Akhtar and Hotchin 2001). Further evidence for caveolin-mediated cadherin endocytosis was provided by Lu and colleagues, who demonstrated that EGF signaling could disrupt cell-cell adhesion by triggering the caveolin-mediated internalization of E-cadherin, a mechanism which may be relevant to epithelial-to-mesenchymal transition in cancers (Lu et al. 2003). In contrast, Bryant and colleagues characterized the EGF-induced internalization of E-cadherin in a breast carcinoma cell line, in which E-cadherin was internalized along with the cadherin-binding proteins p120 and β-catenin, as Rac1-modulated macropinocytosis, rather than caveolin-mediated (Bryant et al. 2007). It is not clear if the EGF-related mechanisms described by Lu and Bryant are in fact different and, if they are, how they can be reconciled. However, Paterson and colleagues have observed E-cadherin endocytosis that is both clathrin- and caveolin-independent, but dynamin-dependent. This pathway, which they identify as similar to macropinocytosis, appears to affect cadherin that is not engaged in *trans* interactions in an adherens junction (Paterson et al. 2003). Lastly, the desmosomal cadherin desmoglein 3 undergoes lipid-raft-mediated endocytosis, though it is unclear if this pathway is available to classical cadherins as well (Delva et al. 2008). Though some of the specific details of the clathrin-independent pathways remain unclear, it appears that both clathrin-dependent and clathrin-independent endocytic pathways play a role in cadherin turnover.

9.3.4 Recycling Pathways

Not all molecules that enter an endocytic pathway face immediate degradation in the lysosome. Some are sorted and recycled back to the plasma membrane. Recycling pathways are particularly important for cadherins, and the choice between degradation and recycling can help fine-tune the amount of cadherin present at adherens junctions and the strength of cell-cell adhesion. The first suggestion of the importance of a recycling pathway to cadherin trafficking came from the discovery that E-cadherin does not travel directly from the Golgi complex to the cell surface, but transits first through Rab11-positive recycling endosomes (Lock and Stow 2005). Interestingly, while expressing dominant-negative Rab11 blocked delivery of wild type E-cadherin to the plasma membrane, an E-cadherin mutant lacking the dileucine motif important for clathrin-mediated endocytosis traffics to the plasma membrane without impediment, though it is mislocalized to the apical surface (Lock and Stow 2005; Miranda et al. 2001). In contrast, *Drosophila* DE-cadherin traffics through Rab11-positive endosomes and inhibiting Rab-11 disrupts the integrity of the embryonic ectoderm, even though DE-cadherin lacks the dileucine motif (Roeth et al. 2009). In addition to acting as way stations for newly synthesized cadherin on its way to the plasma membrane, Rab11-positive recycling endosomes can also sort internalized cadherin for recycling back to the cell surface. In fact, Classen and colleagues found that Rab11 recycling of cadherin mediates the rearrangements in cell-cell contacts seen in the hexagonal packing of *Drosophila* wing disk cells (Classen et al. 2005). Desclozeaux and colleagues also found that cadherin recycling is necessary for maintaining adherens junctions and epithelial polarity and that disrupting the recycling endosome with dominant-negative Rab11 prevented MDCK cells from forming cysts when grown in three-dimensional culture (Desclozeaux et al. 2008).

Additional work has begun to illuminate the molecular mechanisms responsible for cadherin recycling. In particular, components of the exocyst complex appear to be critical. Sec5, sec6, and sec15 are all required for DE-cadherin trafficking from recycling endosomes to the plasma membrane (Langevin et al. 2005). Depletion of the scaffolding protein PALS1 also causes the mislocalization of the exocyst complex and disrupts recycling of E-cadherin (Wang et al. 2007). Recently, Guichard and colleagues identified Rab11- and exocyst complex-mediated recycling of cadherins as a target of the pathogen *Bacillus anthracis*, highlighting its pathophysiological importance. *B. anthracis*, the causative agent of anthrax, produces two different toxins, lethal factor and edema factor, which both inhibit the exocyst complex through independent mechanisms. This results in the loss of cadherin from adherens junctions, potentially contributing to the toxin-mediated epithelial and vascular disruption which occurs with *B. anthracis* infection (Guichard et al. 2010). In addition to the exocyst complex, another potential mediator of cadherin recycling is the adaptor protein complex AP-1B, which usually mediates recycling of basolaterally targeted proteins. Ling and colleagues found that AP-1B interacts with E-cadherin through phosphatidylinositol-4-phosphate 5-kinase type Iγ (PIPKIγ), which binds directly to the E-cadherin cytoplasmic tail near the β-catenin binding site (Ling et al. 2007). Interestingly, an E-cadherin mutation at the PIPKIγ binding site is associated with familial diffuse gastric cancer (Yabuta et al. 2002).

Our understanding of cadherin recycling remains incomplete. Though many of the important components of the cadherin recycling pathway have been identified, the list is likely to grow further. Furthermore, although we review below some evidence that ubiquitination may trigger the selection of cadherin for degradation rather than recycling (Palacios et al. 2005), the regulation of the cadherin recycling pathways remains, for now, only partially elucidated.

9.4 Regulation of Cadherin Endocytosis by Catenins

Given the importance of cadherin endocytosis for the proper maintenance and dynamic regulation of cell-cell adhesion, identifying the regulatory mechanisms controlling cadherin internalization and recycling has become a significant research focus. Much attention has been paid to the catenins, the cytoplasmic binding partners of cadherins, which stabilize adherens junctions and link them to the actin cytoskeleton (Delva and Kowalczyk 2009). These include α-catenin, β-catenin, and p120-catenin. β-catenin binds to the C-terminal catenin-binding domain of cadherins and, along with α-catenin, helps link the cadherin to the actin cytoskeleton. p120-catenin binds to the juxtamembrane domain, N-terminal to the β-catenin binding site, and stabilizes cadherin at the adherens junction. All three catenins contribute to the regulation of adherens junctions.

9.4.1 p120-Catenin

p120-Catenin (p120) plays a key role as an inhibitor of cadherin turnover and as a "set-point" for cadherin expression levels (Fig. 9.2). A member of the armadillo family of proteins, p120 binds to the juxtamembrane domain of cadherins (Reynolds 2007). Ireton and colleagues discovered that epithelial morphology in a colon carcinoma cell line lacking p120 could be restored with exogenous p120 expression. Furthermore, p120 rescue of epithelial morphology required p120 binding to E-cadherin. The mechanism of this activity involved increased E-cadherin protein levels and half-life without changes to E-cadherin mRNA levels (Ireton et al. 2002). Those results, which strongly suggested that p120 binding to cadherin is necessary to prevent rapid cadherin turnover, were confirmed by studies directly demonstrating that loss of p120 results in cadherin endocytosis (Davis et al. 2003; Xiao et al. 2003a). Importantly, p120 acts not only as an inhibitor of cadherin endocytosis, but as a "set-point" for cadherin expression (Fig. 9.2a). Expressing cadherin mutants which compete for p120 binding results in the endocytosis of endogenous cadherin, while cadherin mutants which cannot bind to p120 lack this activity (Xiao et al. 2003a, 2005). This raises the interesting possibility that p120 might serve as a master regulator of cadherin levels in cells. For example, increased expression of one cadherin might, through competition for p120 binding, cause increased turnover and down-regulation of other cadherins in the cell. Exactly this dynamic has been reported to occur in two studies of cells expressing multiple cadherin types. In A431

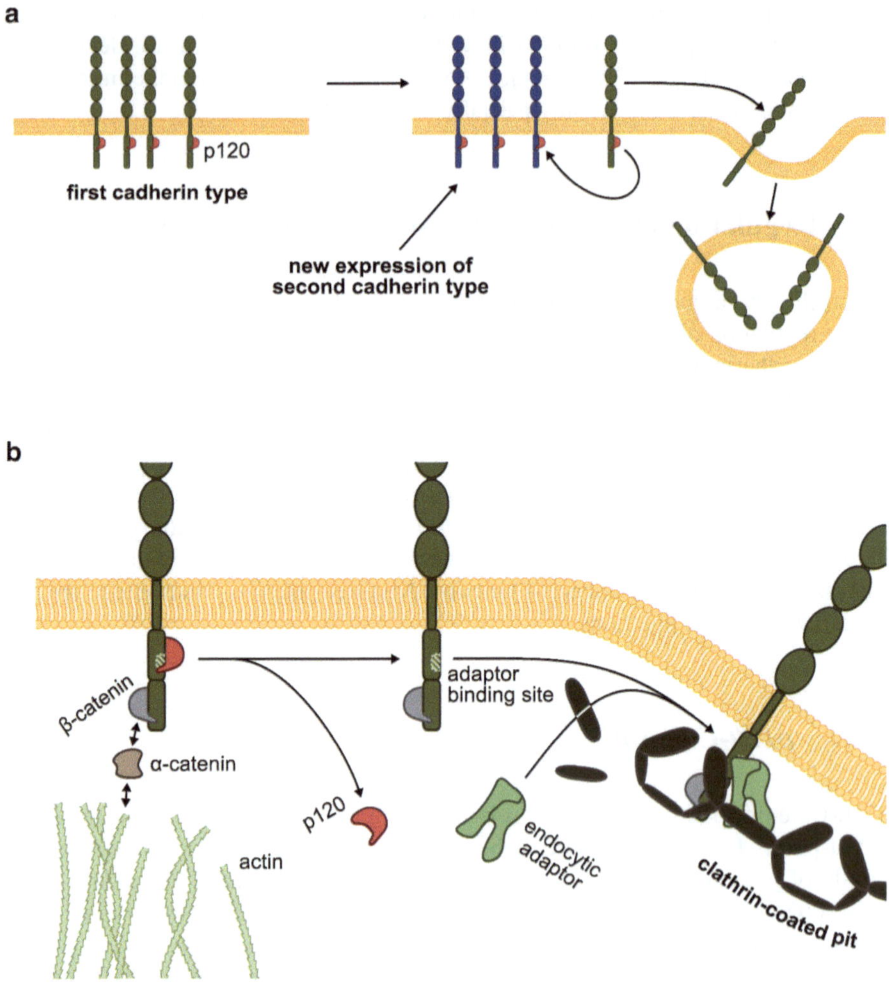

Fig. 9.2 p120-Catenin regulates cadherin endocytosis. **a** p120 acts as a "set-point" for cadherin levels. Increased expression of a second cadherin type competes for p120 binding, causing the internalization of the first cadherin type. This activity allows p120 to serve as a master regulator of cadherin expression in cells (Ferreri et al. 2008; Maeda et al. 2006; Xiao et al. 2003a, 2005). **b** p120 binds to cadherins and masks an endocytic adaptor binding site. When p120 dissociates from the cadherin, the adaptor binding site is exposed, allowing the endocytic adaptor to bind to the cadherin, triggering cadherin endocytosis (Chiasson et al. 2009; Ishiyama et al. 2010)

cells, exogenously expressing R-cadherin caused the endocytosis and down-regulation of endogenous E- and P-cadherins (Maeda et al. 2006). Similarly, in endothelial cells, which express both VE- and N-cadherins, but which rely primarily on VE-cadherin to form adherens junctions, altering expression levels of one cadherin inversely affects protein levels of the other cadherin (Ferreri et al. 2008).

Multiple mechanisms have been proposed to explain how p120 regulates cadherin turnover. Cadherin internalization mediated by p120 loss is clathrin-dependent, as discussed above in more detail (Xiao et al. 2005). Clathrin-dependent endocytosis requires an adaptor protein to bind to cargo and recruit other components of the endocytic machinery. p120 binding to the cadherin cytoplasmic domain could potentially mask the binding site of such an endocytic adaptor. Alternatively, p120 could regulate cadherin turnover by locally modifying actin dynamics through its well-described role as an inhibitor of the small GTPase RhoA (Anastasiadis 2007). For example, cells exogenously expressing high levels of p120 display increased actin branching and the formation of long dendritic spines (Anastasiadis et al. 2000; Noren et al. 2000; Reynolds et al. 1996). It has become increasingly clear however, that p120 binding to cadherins, not p120 inhibition of RhoA, is the mechanism of p120-mediated cadherin stabilization (Fig. 9.2b). First, our lab and others have shown that p120 binding to cadherin is an absolute requirement for p120-mediated cadherin stabilization (Ireton et al. 2002; Miyashita and Ozawa 2007b; Xiao et al. 2005). We also demonstrated that inhibition of RhoA signaling is insufficient to block cadherin endocytosis and that cadherin can also be stabilized by a p120 mutant unable to inhibit Rho (Chiasson et al. 2009). Neither of these observations support a role for RhoA in p120 regulation of cadherin endocytosis. Lastly, our observation that p120 prevents VE-cadherin from clustering into AP-2- and clathrin-enriched membrane domains directly supports the hypothesis that p120 masks an endocytic adaptor binding site on the cadherin cytoplasmic tail (Chiasson et al. 2009). This model received additional support from the recently published crystal structure of a portion of the E-cadherin cytoplasmic domain in complex with p120. The E-cadherin-p120 interface contains both static and dynamic binding regions, an interaction which could support binding competition or regulated exchange with an endocytic adaptor protein (Ishiyama et al. 2010).

Numerous studies of animal models have underscored the physiological importance of p120 to adherens junction regulation, at least in mammals. p120 binding to cadherin is apparently dispensable in *Drosophila* and *C. elegans* (Myster et al. 2003; Pacquelet et al. 2003; Pettitt et al. 2003). However, p120 binding is critical for adherens junction stability in mice. Numerous tissue-specific p120-null mouse models have been developed, and all of them display disrupted cadherin-mediated cell adhesion (summarized in Table 9.1). The reasons for the different requirements for p120 in mammals and invertebrates remains unknown. Though, as outlined above, cadherin trafficking pathways in *Drosophila* appear similar to those in mammalian systems, there may be significant differences in their regulation. Interestingly, the p120 sub-family of catenins is considerably larger in vertebrates than in invertebrates, with additional members including p0071, δ-catenin/NPRAP, ARVCF, and the plakophilins (Hatzfeld 2005). These observations suggest that vertebrate tissue patterning requires additional levels of control over cadherin trafficking, with both the expanded role of vertebrate p120 and the expanded size of the vertebrate p120 sub-family serving as points of regulation not present in simpler organisms.

Table 9.1 Tissue-specific p120-null mouse models display phenotypes characteristic of decreased cadherin levels and impaired intercellular adhesion

Tissue/Cell Type	Phenotype	Reference
Salivary gland	E-cadherin levels reduced; acinar development blocked	(Davis and Reynolds 2006)
Skin	Reduced levels of cadherins and other adherens junction proteins; chronic inflammation due to NFκB activation	(Perez-Moreno et al. 2006)
Hippocampal neurons	Decreased cadherin levels; fewer synapses	(Elia et al. 2006)
Endothelium	VE-cadherin and N-cadherin levels reduced; vascular patterning defects and hemorrhaging	(Oas et al. 2010)
Intestinal epithelium	Down-regulation of adheres junction proteins; compromised barrier function	(Smalley-Freed et al. 2010)
Oropharyngeal epithelium	Decreased E-cadherin expression; development of invasive squamous cell carcinoma	(Stairs et al. 2011)
Kidney	Decreased cadherin levels; impaired tubule morphogenesis; development of cystic kidney disease	(Marciano et al. 2011)

9.4.2 β-Catenin and α-Catenin

Another cytoplasmic binding partner of cadherins is β-catenin, which binds to the C-terminal portion of the cadherin cytoplasmic tail, termed the catenin-binding domain. β-catenin plays an important role in adherens junction structure, contributing to the link between cadherins and the actin cytoskeleton (Hartsock and Nelson 2008). β-Catenin binding to cadherins is clearly important for its ability to recruit α-catenin, which, through a mechanism that is not fully understood, links cadherins to actin (Yamada et al. 2005). In fact, this may be the primary role of β-catenin in adherens junctions, since mutant cadherin which cannot bind to β-catenin but is fused to α-catenin forms junctions that are apparently normal (Nagafuchi et al. 1994; Pacquelet and Rorth 2005). Further support for the hypothesis that β-catenin stabilizes adherens junctions through the recruitment of α-catenin comes from a knock-in mouse model recently created by Schulte and colleagues with a mutant VE-cadherin which does not bind to β-catenin but is fused to α-catenin replacing the wild-type VE-cadherin gene. The mutant mice are viable, though they are not born at Mendelian frequencies, and are resistant to inflammatory stimuli that trigger increased vascular permeability in wild-type mice, suggesting supra-physiological stabilization of their endothelial adherens junctions (Schulte et al. 2011).

Though β-catenin clearly has an important role in adherens junction regulation, its role in cadherin trafficking is far from clear. One report does suggest that β-catenin is required for proper cadherin localization and that disrupting β-catenin binding to cadherins results in cadherin accumulation in intracellular compartments (Chen et al. 1999). However, other studies have yielded conflicting results, though several studies have found at least circumstantial evidence for a β-catenin role in cadherin trafficking. First, Dupre-Crochet and colleagues found that casein kinase

1 (CK1) inhibition stabilizes adherens junctions, while CK1 over-expression disrupts adherens junctions. CK1 phosphorylates E-cadherin, primarily on a serine residue within the catenin binding domain. They also found that a phosphomimetic mutation at that site weakens β-catenin binding to E-cadherin and increases E-cadherin internalization (Dupre-Crochet et al. 2007). Second, Tai and colleagues report that in cultured hippocampal neurons, NMDA inhibits N-cadherin turnover and causes β-catenin to accumulate in dendritic spines. Both effects are related to β-catenin phosphorylation (Tai et al. 2007). Lastly, Sharma and colleagues report that β-catenin is internalized by macropinocytosis in cultured fibroblasts, and that internalized β-catenin co-localizes with N-cadherin. This process appears to be mediated by IQGAP1 binding to β-catenin (Sharma and Henderson 2007). These three accounts are somewhat contradictory. The first two suggest that β-catenin binding to cadherin inhibits its endocytosis, while the last one suggests that β-catenin binding has a role in mediating cadherin endocytosis. Complicating things further, Miyashita and Ozawa report that, while β-catenin binding to E-cadherin may affect E-cadherin localization, the mechanism is unrelated to cadherin turnover. They find that an E-cadherin mutant which cannot bind to β-catenin is mislocalized to an intracellular compartment. However, this mislocalization occurs even with the co-expression of dominant-negative dynamin, which blocks all dynamin-mediated endocytosis. Interestingly, mislocalization of the non-β-catenin-binding mutant cadherin is dependent on the dileucine motif important for clathrin-mediated internalization of E-cadherin; mutant cadherin which cannot bind β-catenin and lacks the dileucine motif traffics to the plasma membrane and does not accumulate intracellularly (Miyashita and Ozawa 2007a). Given the conflicting evidence, more work is needed to understand how β- and α-catenin-mediated cytoskeletal linkages might affect cadherin endocytosis, as well as any other effects that β-catenin binding to cadherins might have on cadherin trafficking.

9.5 Regulation of Cadherin Endocytosis and Degradation by Ubiquitination

Cadherin ubiquitination also plays an important role in regulating cadherin turnover. Proteins are selected for ubiquitination through interaction with E3 ubiquitin ligase proteins which recruit E2 ubiquitin conjugating enzymes charged with ubiquitin and catalyze the transfer of ubiquitin to the target molecule, usually on lysine residues. Ubiquitin molecules can be attached singly or linked together to form a poly-ubiquitin chain. While poly-ubiquitination is usually associated with targeting intracellular proteins for degradation by the 26S proteasome, mono-ubiquitination can also trigger the endocytosis and lysosomal degradation of membrane proteins (Clague and Urbe 2010). Because of its association with endocytosis and degradation, cadherin ubiquitination has been an attractive candidate process for regulating cadherin turnover. Additionally, as a posttranslational modification, cadherin ubiquitination could potentially be influenced by a variety of signaling pathways, ensuring ample control points for the modulation of cadherin endocytosis and degradation. Circumstantial support for a role for ubiquitination in cadherin turnover comes from studies showing that proteasome inhibitors such as MG-132 can block cadherin endocytosis, though the mechanism of this effect remains unclear (Xiao et al. 2003b). In fact, a significant body of work has now developed to establish the importance of ubiquitination in cadherin turnover.

The first ubiquitin ligase identified to target cadherin was Hakai, a c-Cbl-like protein with phosphotyrosine-binding, RING finger, and proline-rich domains characterized by Fujita and colleagues. Hakai associates with and ubiquitinates E-cadherin, causing its internalization. Interestingly, this function is dependent on Src-mediated phosphorylation of E-cadherin at two specific tyrosine residues in the juxtamembrane domain (Fujita et al. 2002). This both explains the previously reported ability of v-Src to transform cultured epithelial cells to a fibroblastic phenotype (Behrens et al. 1993) and provides a potential explanation for the ability of p120 to inhibit cadherin internalization, since p120 binding could mask or prevent the phosphorylation of the E-cadherin tyrosine residues required for Hakai binding. However, these tyrosine residues are not conserved in all classical cadherins. P-cadherin contains only one of the two tyrosine residues, and N- and VE-cadherins lack both of them. Hakai-mediated down-regulation of cadherins therefore may not play a role at all adherens junctions.

Further work by Palacios and colleagues has clarified the mechanism of Hakai-induced E-cadherin turnover. Hakai-mediated ubiquitination of E-cadherin may not directly trigger E-cadherin internalization, since an E-cadherin mutant that cannot interact with Hakai can still be internalized. However, Hakai-mediated ubiquitination of E-cadherin changes the destination of E-cadherin once it has been internalized, redirecting it from a recycling pathway to degradation in the lysosome (Palacios et al. 2005). This redirection requires Hrs, a ubiquitin-interacting protein with a role in shuttling mono-ubiquitinated cargo to the lysosome (Palacios et al. 2005; Toyoshima et al. 2007). Studies have also linked Hakai to developmental and disease processes. Hakai is essential for the maintenance of epithelial integrity in *Drosophila*, though its interaction with DE-cadherin is considerably different than the interaction of mammalian Hakai with E-cadherin. *Drosophila* Hakai can interact with DE-cadherin based on the extracellular and transmembrane portions of the cadherin without the intracellular portion (Kaido et al. 2009). Because Hakai is a cytoplasmic protein, it is not clear how this interaction can occur without the assistance of another protein. Hakai has also been linked to disease in some human colorectal carcinomas, where elevated Slit-Robo signaling induces an epithelial to mesenchymal transformation by recruiting Hakai to ubiquitinate E-cadherin, causing its down-regulation. Elevated Slit-Robo expression is also associated with increased risk of metastasis and decreased survival (Zhou et al. 2011). Though the function of Hakai may be limited to only a subset of adherens junctions, it clearly plays an important role.

Hakai is not the only ubiquitin ligase that has been connected to adherens junction turnover. The ubiquitin ligase MDM 2 also ubiquitinates and causes the degradation of E-cadherin, and in human breast carcinoma specimens, increased

MDM 2 expression was associated with decreased E-cadherin protein levels (Yang et al. 2006). A third ubiquitin ligase, the viral protein K5, has also been shown to target VE-cadherin (Mansouri et al. 2008). K5 is expressed by human herpesvirus-8 (HHV-8), which causes the angioproliferative neoplasm Kaposi sarcoma. K5 is thought to play a role in the virus's ability to evade the host immune response by ubiquitinating and causing the internalization of immune recognition components such as the class I major histocompatibility complex. The increased vascular permeability associated with Kaposi sarcoma may be due to a similar mechanism inducing the endocytosis and down-regulation of VE-cadherin (Qian et al. 2008). Because K5 is a member of the membrane-associated RING-CH (MARCH) family of ubiquitin ligases, which includes several human proteins expressed in a variety of tissues (Nathan and Lehner 2009), it is possible that HHV-8 may be appropriating a more generally important cellular mechanism for cadherin regulation involving endogenous MARCH proteins.

9.6 Growth Factor Signaling and Cadherin Endocytosis

Cell-cell junctions are fundamental links between a cell and its environment. It is not a surprise then, that adherens junctions are not regulated only by intracellular processes, but also by intercellular cues. A variety of growth factor signaling pathways have been tied to the dynamic regulation of cadherin endocytosis, including hepatocyte growth factor (HGF), epithelial growth factor (EGF), vascular endothelial growth factor (VEGF), fibroblast growth factor (FGF), and transforming growth factor β (TGFβ). Many of these pathways affect cadherin trafficking and catenin binding, which are discussed above in more detail.

The first growth factor receptor associated with cadherin endocytosis was the HGF receptor, c-Met (Fig. 9.3a). HGF is also called scatter factor for its ability to stimulate epithelial cell motility. Treatment of cultured cells with HGF or a small molecule HGF receptor agonist causes the co-endocytosis of the HGF receptor and associated E-cadherin (Kamei et al. 1999). This effect requires the activation of the small GTPase Arf6 (Palacios et al. 2001). Additionally, HGF signaling causes Numb, an endocytic adaptor which may play a role in establishing the lateral localization of cadherins by facilitating their specific endocytosis from the apical surface, to decouple from E-cadherin and associate with aPKC and Par6 instead, disrupting cell polarity (Wang et al. 2009). Thus, HGF appears to cause both the general down-regulation of cadherin and the disruption of adherens junction polarity. Both effects are consistent with the ability of HGF to induce a fibroblast-like phenotype. However, the cause of HGF-mediated cell scattering remains in dispute, since, in MDCK cells, HGF enhances integrin-mediated interactions with the extracellular matrix which pull the cells apart, but does not appear to disrupt E-cadherin mediated adhesion (de Rooij et al. 2005). More work will be needed to understand the functional importance and precise mechanism of HGF-mediated cadherin endocytosis.

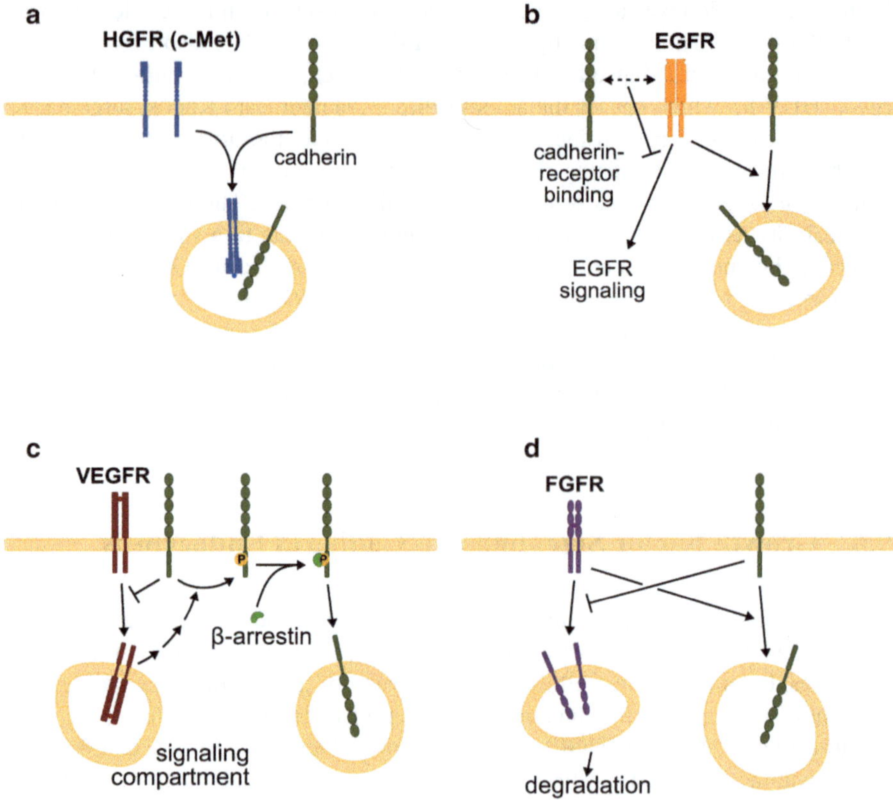

Fig. 9.3 Growth factor signaling pathways influence cadherin endocytosis. **a** HGF activation of c-Met causes co-endocytosis of the receptor with E-cadherin (Kamei et al. 1999; Palacios et al. 2001). **b** EGFR over-expression induces E-cadherin endocytosis and E-cadherin binding to the receptor inhibits EGFR signaling (Bremm et al. 2008; Bryant et al. 2007; Lu et al. 2003). **c** VEGFR activation triggers the phosphorylation of VE-cadherin through a Src, Vav2, Rac, and PAK signaling cascade. Phosphorylated VE-cadherin recruits β-arrestin and triggers clathrin-mediated endocytosis of the cadherin (Gavard and Gutkind 2006; Gavard et al. 2008). Conversely, VE-cadherin inhibits the internalization of VEGFR into signaling compartments upon ligand binding (Lampugnani et al. 2006). **d** FGFR activation induces E-cadherin endocytosis, and cadherins inhibit the endocytosis and degradation of FGFR, forming a negative feedback loop (Bryant et al. 2005, 2007; Suyama et al. 2002)

EGF signaling has also been tied to cadherin endocytosis (Fig. 9.3b). The effect of EGF receptor signaling is notable because it causes cadherin internalization through a clathrin-independent pathway. As discussed above, however, beyond clathrin independence, there is disagreement over which endocytic pathway is involved. Lu and colleagues reported that EGF receptor over-expression caused E-cadherin internalization through a caveolin-mediated pathway (Lu et al. 2003). In contrast, Bryant and colleagues reported that EGF induced E-cadherin internalization through macropinocytosis (Bryant et al. 2007). More work will need to be done to sort out these conflicting findings. Interestingly, the relationship between the EGF receptor and cadherins appears to be bidirectional. Certain mutations in the extracellular domain of E-cadherin are associated with decreased formation of E-cadherin-EGF receptor complexes, resulting in increased EGF receptor signaling in both cultured cells and human gastric carcinoma samples (Bremm et al. 2008). This finding suggests that while EGF signaling can cause E-cadherin endocytosis, E-cadherin can inhibit EGF signaling. Clearly, adherens junctions are not simply acted upon by signaling pathways, but are active participants in them as well.

A third growth factor associated with cadherin internalization is VEGF, an important growth factor in vasculogenesis and angiogenesis, which increases vascular permeability by disrupting endothelial cell-cell junctions (Fig. 9.3c). Gavard and Gutkind demonstrated that VEGF signaling causes the Src-mediated phosphorylation of VE-cadherin, resulting in the recruitment of β-arrestin and the subsequent clathrin-mediated endocytosis of VE-cadherin (Gavard and Gutkind 2006). The pathway is interrupted by angiopoietin-1, which strengthens vascular integrity and decreases permeability. Angiopoietin-1 inhibits Src activation by the VEGF receptor, counteracting VEGF-mediated cadherin internalization (Gavard et al. 2008). As with E-cadherin and the EGF receptor, the relationship between VE-cadherin and the VEGF receptor is bidirectional. In cell culture, confluent endothelial cells are resistant to the effects of VEGF, an effect which requires both VE-cadherin and β-catenin (Lampugnani et al. 2003). VE-cadherin association with the VEGF receptor prevents VEGF receptor internalization in response to VEGF binding. When internalized in response to VEGF binding, the VEGF receptor is not degraded. Rather, it enters an endosomal signaling compartment where it activates the MAP kinase pathway. Thus, by preventing VEGF receptor endocytosis, VE-cadherin can inhibit VEGF signaling (Lampugnani et al. 2006).

A similar two-way interaction also occurs between cadherins and the FGF receptor (Fig. 9.3d). FGF activation of the FGF receptor induces macropinocytosis of E-cadherin (Bryant et al. 2005, 2007). Conversely, increased expression of E- or N-cadherin inhibits internalization of the FGF receptor (Bryant et al. 2005; Suyama et al. 2002). In contrast to the VEGF receptor however, internalization of ligand-bound FGF receptor serves to shut off FGF signaling, primarily through subsequent degradation of the receptor. Thus, FGF signaling down-regulates cadherins and cadherins support FGF signaling, essentially forming a negative-feedback loop. Lastly, cadherin trafficking can be affected by TGFβ signaling. TGFβ and Raf-1 synergistically induce E-cadherin endocytosis and epithelial to mesenchymal transition in mammary epithelial cells (Janda et al. 2006). Interestingly, TGFβ- and Raf-1-induced cadherin internalization is associated with cadherin ubiquitination.

The large variety of growth factor signaling pathways affecting cadherin endocytosis clearly indicates the importance of the dynamic and coordinated regulation of cadherin internalization and intercellular adhesion. More work is needed, however, to understand how these disparate pathways are interrelated in different biological contexts. The potential for two-way communication between growth factor receptors and adherens junctions is particularly intriguing, and the full potential of these mechanisms has yet to be explored.

9.7 Cadherin Shedding

In this chapter, we have focused mainly on down-regulation of adherens junctions through the removal of cadherin from the cell surface. However, this is not the only mechanism available for reducing the amount of cadherin available to form adhesive contacts. In some situations, cadherins may be proteolytically cleaved while they remain at the plasma membrane. This process, often termed cadherin "shedding," can lead to the release of cadherin extracellular domains from the cell or fragments of the cadherin cytoplasmic tail into the cytoplasm, with potential effects beyond loss of adhesion.

Released fragments of cadherin extracellular domains were first identified as factors that inhibited cell adhesion in conditioned medium from a breast cancer cell line (Damsky et al. 1983; Wheelock et al. 1987). Inducing E-cadherin shedding in cell culture can also promote cell invasion into a collagen substrate (Noe et al. 2001). Consequently, there has been considerable excitement for the possible involvement of cadherin shedding in the loss of intercellular adhesion in cancer and the use of cadherin extracellular domain fragments as tumor biomarkers. However, results from observational studies have been mixed (reviewed in De Wever et al. 2007). While serum levels of E-cadherin extracellular domains are elevated approximately three-fold in patients with several types of cancer, there is no correlation with disease progression. It is also possible that increased cadherin shedding detected in these studies is related to general inflammatory processes rather than to the tumor specifically (Pittard et al. 1996). In addition to possible roles in cancer and inflammation, cadherin shedding appears to be involved in several developmental processes. N-cadherin is cleaved during chick retinal development, where, counter-intuitively, the truncated product promotes cell adhesion and neurite development (Paradies and Grunwald 1993). N-cadherin shedding has also been reported in neural crest delamination and in adult neurons (Marambaud et al. 2003; Shoval et al. 2007). Lastly, in response to Eph/ephrin signaling, E-cadherin shedding plays a role in cell sorting (Solanas et al. 2011). Given the variety of processes in which it has been implicated, cadherin shedding appears to have an important role in development. However, more work will need to be done to understand the role cadherin shedding in more detail and in additional developmental processes.

Many of the proteases responsible for cadherin shedding have been identified. Members of the "a disintegrin and metalloprotease" (ADAM) family, and ADAM10 in particular, appear to be an important generators of free E-cadherin and N-cadherin extracellular domain fragments (Maretzky et al. 2005; Reiss et al. 2005). Interestingly, EGFR-mediated down-regulation of desmosomal cadherins appears to occur, at least in part, through ADAM proteases, a result suggesting how cadherin shedding might be connected to signaling pathways (Klessner et al. 2009). A variety of other proteases have also been implicated in cadherin shedding, including matrix metalloproteinases and kallikreins (Klucky et al. 2007; McGuire et al. 2003; Noe et al. 2001). Still other proteases, including caspases and presenilin, can cleave cadherins intracellularly, releasing a soluble cadherin fragment into the cytoplasm (Marambaud et al. 2002). Interestingly, these intracellular fragments can traffic to the nucleus, potentially affecting a variety of transcription factors (Ferber et al. 2008). The relationship of intracellular cadherin proteolysis to extracellular cadherin shedding is not yet understood, but, in addition to modulating intercellular adhesion, these mechanisms have the potential to integrate adherens junctions with cell signaling networks.

9.8 Summary and Future Perspectives

Cadherin endocytosis and degradation play crucial roles in the dynamic control of intercellular adhesion. By adjusting the rate of cadherin internalization, cells are able to quickly modify the strength of their adherens junctions, rearranging their relationship with their environment. This process is absolutely critical during development, and, as we have seen, cadherin endocytosis and degradation have been linked to a growing number of developmental processes in a variety of species. A particularly exciting area of current research focuses on planar-cell-polarized endocytosis of cadherin as a mechanism for the establishment of planar polarization of an epithelial layer. The role of cadherin endocytosis during development may turn out to be more complicated—and more important—than simply allowing cells to switch between epithelial and mesenchymal phenotypes. The mis-regulation of cadherin endocytosis also appears to be increasingly important in disease processes, and, consequently, as a possible therapeutic target. However, our understanding remains incomplete, and devising a new generation of anti-cancer drugs targeting cadherin endocytosis will require further work.

In addition to contributing to our understanding of the role of cadherin internalization in development and disease, recent work has also advanced our understanding of the molecular mechanisms underlying cadherin endocytosis. In particular, we have learned a great deal about clathrin-mediated cadherin endocytosis and its contribution to adherens junction dynamics. However, more needs to be done in order to characterize the clathrin-independent endocytic pathways that cadherins can enter, as well as to better understand which pathways are active in different biological contexts. Furthermore, while several endocytic adaptors have been associated with adherens junction turnover, the nature of the interactions between these adaptors and cadherins remains largely unknown. In order to unwind the pathways regulating cadherin endocytosis, it will be necessary to more precisely identify the cadherin domains which drive their removal from the cell membrane. Do cadherin endocytic signals overlap with the p120 binding site, allowing p120 to compete with endocytic adaptors for cadherin binding, thus stabilizing cadherins at the cell membrane, as has been proposed? Furthermore, how does cadherin shedding relate to cadherin internalization? Answers to these questions must await further investigation.

In addition to better understanding the molecular mechanisms of cadherin endocytosis, another important focus of future research will be the signaling pathways that allow for its dynamic regulation. One possibility is raised by studies supporting the role of α- and β-catenins in cadherin regulation. Since α- and β-catenins link cadherins to the actin cytoskeleton, might this link play some role in cadherin trafficking? For now, the evidence is unclear. A second possibility is that cadherin ubiquitination may be used as signal to promote cadherin endocytosis. Several ubiquitin ligases have been found to mediate the ubiquitination and down-regulation of cadherins. However, based on what is known so far, the scope of each of the pathways identified remains limited to specific biological contexts. Further research will be needed to determine whether cadherin ubiquitination is a broadly applicable mechanism that regulates cell-cell adhesion. Finally, the many growth factor signaling pathways implicated in cadherin endocytosis suggest several opportunities to link intercellular contacts to intercellular signaling. The possibility that this relationship might be bidirectional, allowing growth factors to affect cadherin endocytosis and cadherins to affect growth factor signaling pathways, is particularly exciting. Still, it will take more work to integrate the disparate pathways that have been identified.

Though our understanding of cadherin internalization and degradation and the mechanisms that regulate them is far from complete, much has been learned in the decades since cadherin endocytosis was first observed in response to calcium depletion. Cadherin endocytosis is now recognized as an important factor in the dynamic control of intercellular adhesion. It remains an active area of research, with the promise to further our understanding of the ever-changing adhesive interactions between cells and the implications of adherens junction dynamics for development and disease.

Acknowledgements We would like to thank Victor Faundez as well as members of the Kowalczyk lab for insightful and engaging conversations during the preparation of this manuscript. We would also like to acknowledge funding from the National Institutes of Health (R01AR050501 and R01AR048266 to APK). BAN was supported by a fellowship from the American Heart Association (11PRE7590097).

References

Akhtar N, Hotchin NA (2001) RAC1 regulates adherens junctions through endocytosis of E-cadherin. Mol Biol Cell 12:847–862

Anastasiadis PZ (2007) p120-ctn: a nexus for contextual signaling via Rho GTPases. Biochim Biophys Acta 1773:34–46

Anastasiadis PZ, Moon SY, Thoreson MA, Mariner DJ, Crawford HC, Zheng Y, Reynolds AB (2000) Inhibition of RhoA by p120 catenin. Nat Cell Biol 2:637–644

Bauer A, Lickert H, Kemler R, Stappert J (1998) Modification of the E-cadherin-catenin complex in mitotic madin-darby canine kidney epithelial cells. J Biol Chem 273:28314–28321

Behrens J, Vakaet L, Friis R, Winterhager E, Van Roy F, Mareel MM, Birchmeier W (1993) Loss of epithelial differentiation and gain of invasiveness correlates with tyrosine phosphorylation of the E-cadherin/beta-catenin complex in cells transformed with a temperature-sensitive v-SRC gene. J Cell Biol 120:757–766

Bonifacino JS, Traub LM (2003) Signals for sorting of transmembrane proteins to endosomes and lysosomes. Annu Rev Biochem 72:395–447

Bremm A, Walch A, Fuchs M, Mages J, Duyster J, Keller G, Hermannstadter C, Becker KF, Rauser S, Langer R, von Weyhern CH, Hofler H, Luber B (2008) Enhanced activation of epidermal growth factor receptor caused by tumor-derived E-cadherin mutations. Cancer Res 68:707–714

Brieher WM, Gumbiner BM (1994) Regulation of C-cadherin function during activin induced morphogenesis of Xenopus animal caps. J Cell Biol 126:519–527

Brouxhon S, Kyrkanides S, O'Banion MK, Johnson R, Pearce DA, Centola GM, Miller JN, McGrath KH, Erdle B, Scott G, Schneider S, VanBuskirk J, Pentland AP (2007) Sequential downregulation of E-cadherin with squamous cell carcinoma progression: loss of E-cadherin via a prostaglandin E2-EP2 dependent posttranslational mechanism. Cancer Res 67:7654–7664

Bryant DM, Wylie FG, Stow JL (2005) Regulation of endocytosis, nuclear translocation, and signaling of fibroblast growth factor receptor 1 by E-cadherin. Mol Biol Cell 16:14–23

Bryant DM, Kerr MC, Hammond LA, Joseph SR, Mostov KE, Teasdale RD, Stow JL (2007) EGF induces macropinocytosis and SNX1-modulated recycling of E-cadherin. J Cell Sci 120:1818–1828

Calkins CC, Setzer SV, Jennings JM, Summers S, Tsunoda K, Amagai M, Kowalczyk AP (2006) Desmoglein endocytosis and desmosome disassembly are coordinated responses to pemphigus autoantibodies. J Biol Chem 281:7623–7634

Chen YT, Stewart DB, Nelson WJ (1999) Coupling assembly of the E-cadherin/beta-catenin complex to efficient endoplasmic reticulum exit and basal-lateral membrane targeting of E-cadherin in polarized MDCK cells. J Cell Biol 144:687–699

Chiasson CM, Kowalczyk AP (2008) Cadherin trafficking and junction dynamics. Cell Junctions. Wiley-VCH Verlag GmbH & Co. KGaA, Germany, pp 251–270

Chiasson CM, Wittich KB, Vincent PA, Faundez V, Kowalczyk AP (2009) p120-catenin inhibits VE-cadherin internalization through a Rho-independent mechanism. Mol Biol Cell 20:1970–1980

Clague MJ, Urbe S (2010) Ubiquitin: same molecule, different degradation pathways. Cell 143:682–685

Classen AK, Anderson KI, Marois E, Eaton S (2005) Hexagonal packing of Drosophila wing epithelial cells by the planar cell polarity pathway. Dev Cell 9:805–817

Damsky CH, Richa J, Solter D, Knudsen K, Buck CA (1983) Identification and purification of a cell surface glycoprotein mediating intercellular adhesion in embryonic and adult tissue. Cell 34:455–466

Davis MA, Reynolds AB (2006) Blocked acinar development, E-cadherin reduction, and intraepithelial neoplasia upon ablation of p120-catenin in the mouse salivary gland. Dev Cell 10:21–31

Davis MA, Ireton RC, Reynolds AB (2003) A core function for p120-catenin in cadherin turnover. J Cell Biol 163:525–534

de Rooij J, Kerstens A, Danuser G, Schwartz MA, Waterman-Storer CM (2005) Integrin-dependent actomyosin contraction regulates epithelial cell scattering. J Cell Biol 171:153–164

De Wever O, Derycke L, Hendrix A, De Meerleer G, Godeau F, Depypere H, Bracke M (2007) Soluble cadherins as cancer biomarkers. Clin Exp Metastasis 24:685–697

Delva E, Kowalczyk AP (2009) Regulation of cadherin trafficking. Traffic 10:259–267

Delva E, Jennings JM, Calkins CC, Kottke MD, Faundez V, Kowalczyk AP (2008) Pemphigus vulgaris IgG-induced desmoglein-3 endocytosis and desmosomal disassembly are mediated by a clathrin- and dynamin-independent mechanism. J Biol Chem 283:18303–18313

Desclozeaux M, Venturato J, Wylie FG, Kay JG, Joseph SR, Le HT, Stow JL (2008) Active Rab11 and functional recycling endosome are required for E-cadherin trafficking and lumen formation during epithelial morphogenesis. Am J Physiol Cell Physiol 295:C545–556

Dupre-Crochet S, Figueroa A, Hogan C, Ferber EC, Bialucha CU, Adams J, Richardson EC, Fujita Y (2007) Casein kinase 1 is a novel negative regulator of E-cadherin-based cell-cell contacts. Mol Cell Biol 27:3804–3816

Elia LP, Yamamoto M, Zang K, Reichardt LF (2006) p120 Catenin regulates dendritic spine and synapse development through Rho-family GTPases and cadherins. Neuron 51:43–56

Ferber EC, Kajita M, Wadlow A, Tobiansky L, Niessen C, Ariga H, Daniel J, Fujita Y (2008) A role for the cleaved cytoplasmic domain of E-cadherin in the nucleus. J Biol Chem 283:12691–12700

Ferreri DM, Minnear FL, Yin T, Kowalczyk AP, Vincent PA (2008) N-cadherin levels in endothelial cells are regulated by monolayer maturity and p120 availability. Cell Commun Adhes 15:333–349

Fujita Y, Krause G, Scheffner M, Zechner D, Leddy HE, Behrens J, Sommer T, Birchmeier W (2002) Hakai, a c-Cbl-like protein, ubiquitinates and induces endocytosis of the E-cadherin complex. Nat Cell Biol 4:222–231

Gavard J, Gutkind JS (2006) VEGF controls endothelial-cell permeability by promoting the [beta]-arrestin-dependent endocytosis of VE-cadherin. Nat Cell Biol 8:1223–1234

Gavard J, Patel V, Gutkind JS (2008) Angiopoietin-1 prevents VEGF-induced endothelial permeability by sequestering src through mDia. Dev Cell 14:25–36

Guichard A, McGillivray SM, Cruz-Moreno B, van Sorge NM, Nizet V, Bier E (2010) Anthrax toxins cooperatively inhibit endocytic recycling by the Rab11/Sec15 exocyst. Nature 467:854–858

Hartsock A, Nelson WJ (2008) Adherens and tight junctions: structure, function and connections to the actin cytoskeleton. Biochim Biophys Acta 1778:660–669

Hatzfeld M (2005) The p120 family of cell adhesion molecules. Eur J Cell Biol 84:205–214

Hirohashi S (1998) Inactivation of the E-cadherin-mediated cell adhesion system in human cancers. Am J Pathol 153:333–339

Ireton RC, Davis MA, van Hengel J, Mariner DJ, Barnes K, Thoreson MA, Anastasiadis PZ, Matrisian L, Bundy LM, Sealy L, Gilbert B, van Roy F, Reynolds AB (2002) A novel role for p120 catenin in E-cadherin function. J Cell Biol 159:465–476

Ishiyama N, Lee SH, Liu S, Li GY, Smith MJ, Reichardt LF, Ikura M (2010) Dynamic and static interactions between p120 catenin and E-cadherin regulate the stability of cell-cell adhesion. Cell 141:117–128

Ivanov AI, Nusrat A, Parkos CA (2004) Endocytosis of epithelial apical junctional proteins by a clathrin-mediated pathway into a unique storage compartment. Mol Biol Cell 15:176–188

Izumi G, Sakisaka T, Baba T, Tanaka S, Morimoto K, Takai Y (2004) Endocytosis of E-cadherin regulated by Rac and Cdc42 small G proteins through IQGAP1 and actin filaments. J Cell Biol 166:237–248

Janda E, Nevolo M, Lehmann K, Downward J, Beug H, Grieco M (2006) Raf plus TGFbeta-dependent EMT is initiated by endocytosis and lysosomal degradation of E-cadherin. Oncogene 25:7117–7130

Jarrett O, Stow JL, Yap AS, Key B (2002) Dynamin-dependent endocytosis is necessary for convergent-extension movements in Xenopus animal cap explants. Int J Dev Biol 46:467–473

Kaido M, Wada H, Shindo M, Hayashi S (2009) Essential requirement for RING finger E3 ubiquitin ligase Hakai in early embryonic development of drosophila. Genes Cells 14:1067–1077

Kamei T, Matozaki T, Sakisaka T, Kodama A, Yokoyama S, Peng YF, Nakano K, Takaishi K, Takai Y (1999) Coendocytosis of cadherin and c-Met coupled to disruption of cell-cell adhesion in MDCK cells–regulation by Rho, Rac and Rab small G proteins. Oncogene 18:6776–6784

Kartenbeck J, Schmelz M, Franke WW, Geiger B (1991) Endocytosis of junctional cadherins in bovine kidney epithelial (MDBK) cells cultured in low Ca2+ion medium. J Cell Biol 113:881–892

Kartenbeck J, Schmid E, Franke WW, Geiger B (1982) Different modes of internalization of proteins associated with adhaerens junctions and desmosomes: experimental separation of lateral contacts induces endocytosis of desmosomal plaque material. EMBO J 1:725–732

Kawauchi T, Sekine K, Shikanai M, Chihama K, Tomita K, Kubo K, Nakajima K, Nabeshima Y, Hoshino M (2010) Rab GTPases-dependent endocytic pathways regulate neuronal migration and maturation through N-cadherin trafficking. Neuron 67:588–602

Klessner JL, Desai BV, Amargo EV, Getsios S, Green KJ (2009) EGFR and ADAMs cooperate to regulate shedding and endocytic trafficking of the desmosomal cadherin desmoglein 2. Mol Biol Cell 20:328–337

Klucky B, Mueller R, Vogt I, Teurich S, Hartenstein B, Breuhahn K, Flechtenmacher C, Angel P, Hess J (2007) Kallikrein 6 induces E-cadherin shedding and promotes cell proliferation, migration, and invasion. Cancer Res 67:8198–8206

Lampugnani MG, Zanetti A, Corada M, Takahashi T, Balconi G, Breviario F, Orsenigo F, Cattelino A, Kemler R, Daniel TO, Dejana E (2003) Contact inhibition of VEGF-induced proliferation requires vascular endothelial cadherin, beta-catenin, and the phosphatase DEP-1/CD148. J Cell Biol 161:793–804

Lampugnani MG, Orsenigo F, Gagliani MC, Tacchetti C, Dejana E (2006) Vascular endothelial cadherin controls VEGFR-2 internalization and signaling from intracellular compartments. J Cell Biol 174:593–604

Langevin J, Morgan MJ, Sibarita JB, Aresta S, Murthy M, Schwarz T, Camonis J, Bellaiche Y (2005) Drosophila exocyst components Sec5, Sec6, and Sec15 regulate DE-Cadherin trafficking from recycling endosomes to the plasma membrane. Dev Cell 9:365–376

Lau KM, McGlade CJ (2011) Numb is a negative regulator of HGF dependent cell scattering and Rac1 activation. Exp Cell Res 317:539–551

Lawler K, O'Sullivan G, Long A, Kenny D (2009) Shear stress induces internalization of E-cadherin and invasiveness in metastatic oesophageal cancer cells by a Src-dependent pathway. Cancer Sci 100:1082–1087

Le TL, Yap AS, Stow JL (1999) Recycling of E-cadherin: a potential mechanism for regulating cadherin dynamics. J Cell Biol 146:219–232

Lerch MM, Lutz MP, Weidenbach H, Muller-Pillasch F, Gress TM, Leser J, Adler G (1997) Dissociation and reassembly of adherens junctions during experimental acute pancreatitis. Gastroenterology 113:1355–1366

Levayer R, Pelissier-Monier A, Lecuit T (2011) Spatial regulation of Dia and Myosin-II by RhoGEF2 controls initiation of E-cadherin endocytosis during epithelial morphogenesis. Nat Cell Biol 13:529–540

Ling K, Bairstow SF, Carbonara C, Turbin DA, Huntsman DG, Anderson RA (2007) Type I gamma phosphatidylinositol phosphate kinase modulates adherens junction and E-cadherin trafficking via a direct interaction with mu 1B adaptin. J Cell Biol 176:343–353

Lock JG, Stow JL (2005) Rab11 in recycling endosomes regulates the sorting and basolateral transport of E-cadherin. Mol Biol Cell 16:1744–1755

Lu Z, Ghosh S, Wang Z, Hunter T (2003) Downregulation of caveolin-1 function by EGF leads to the loss of E-cadherin, increased transcriptional activity of beta-catenin, and enhanced tumor cell invasion. Cancer Cell 4:499–515

Maeda M, Johnson E, Mandal SH, Lawson KR, Keim SA, Svoboda RA, Caplan S, Wahl 3rd JK, Wheelock MJ, Johnson KR (2006) Expression of inappropriate cadherins by epithelial tumor cells promotes endocytosis and degradation of E-cadherin via competition for p120(ctn). Oncogene 25:4595–4604

Mansouri M, Rose PP, Moses AV, Fruh K (2008) Remodeling of endothelial adherens junctions by Kaposi's sarcoma-associated herpesvirus. J Virol 82:9615–9628

Marambaud P, Shioi J, Serban G, Georgakopoulos A, Sarner S, Nagy V, Baki L, Wen P, Efthimiopoulos S, Shao Z, Wisniewski T, Robakis NK (2002) A presenilin-1/gamma-secretase cleavage releases the E-cadherin intracellular domain and regulates disassembly of adherens junctions. EMBO J 21:1948–1956

Marambaud P, Wen PH, Dutt A, Shioi J, Takashima A, Siman R, Robakis NK (2003) A CBP binding transcriptional repressor produced by the PS1/epsilon-cleavage of N-cadherin is inhibited by PS1 FAD mutations. Cell 114:635–645

Marciano DK, Brakeman PR, Lee CZ, Spivak N, Eastburn DJ, Bryant DM, Beaudoin 3rd GM, Hofmann I, Mostov KE, Reichardt LF (2011) p120 Catenin is required for normal renal tubulogenesis and glomerulogenesis. Development 138:2099–2109

Maretzky T, Reiss K, Ludwig A, Buchholz J, Scholz F, Proksch E, de Strooper B, Hartmann D, Saftig P (2005) ADAM10 mediates E-cadherin shedding and regulates epithelial cell-cell adhesion, migration, and beta-catenin translocation. Proc Natl Acad Sci U S A 102:9182–9187

Mattey DL, Garrod DR (1986) Splitting and internalization of the desmosomes of cultured kidney epithelial cells by reduction in calcium concentration. J Cell Sci 85:113–124

McCrea PD, Gumbiner BM (1991) Purification of a 92-kDa cytoplasmic protein tightly associated with the cell-cell adhesion molecule E-cadherin (uvomorulin). Characterization and extractability of the protein complex from the cell cytostructure. J Biol Chem 266:4514–4520

McGuire JK, Li Q, Parks WC (2003) Matrilysin (matrix metalloproteinase-7) mediates E-cadherin ectodomain shedding in injured lung epithelium. Am J Pathol 162:1831–1843

Miller JR, McClay DR (1997) Characterization of the role of cadherin in regulating cell adhesion during sea urchin development. Dev Biol 192:323–339

Miranda KC, Khromykh T, Christy P, Le TL, Gottardi CJ, Yap AS, Stow JL, Teasdale RD. (2001) A dileucine motif targets E-cadherin to the basolateral cell surface in madin-darby canine kidney and LLC-PK1 epithelial cells. J Biol Chem 276:22565–22572

Miyashita Y, Ozawa M (2007a) A dileucine motif in its cytoplasmic domain directs beta-catenin-uncoupled E-cadherin to the lysosome. J Cell Sci 120:4395–4406

Miyashita Y, Ozawa M (2007b) Increased internalization of p120-uncoupled E-cadherin and a requirement for a dileucine motif in the cytoplasmic domain for endocytosis of the protein. J Biol Chem 282:11540–11548

Myster SH, Cavallo R, Anderson CT, Fox DT, Peifer M (2003) Drosophila p120catenin plays a supporting role in cell adhesion but is not an essential adherens junction component. J Cell Biol 160:433–449

Nagafuchi A, Ishihara S, Tsukita S (1994) The roles of catenins in the cadherin-mediated cell adhesion: functional analysis of E-cadherin-alpha catenin fusion molecules. J Cell Biol 127:235–245

Nathan JA, Lehner PJ (2009) The trafficking and regulation of membrane receptors by the RING-CH ubiquitin E3 ligases. Exp Cell Res 315:1593–1600

Noe V, Fingleton B, Jacobs K, Crawford HC, Vermeulen S, Steelant W, Bruyneel E, Matrisian LM, Mareel M (2001) Release of an invasion promoter E-cadherin fragment by matrilysin and stromelysin-1. J Cell Sci 114:111–118

Noren NK, Liu BP, Burridge K, Kreft B (2000) p120 Catenin regulates the actin cytoskeleton via Rho family GTPases. J Cell Biol 150:567–580

Nose A, Nagafuchi A, Takeichi M (1988) Expressed recombinant cadherins mediate cell sorting in model systems. Cell 54:993–1001

Oas RG, Xiao K, Summers S, Wittich KB, Chiasson CM, Martin WD, Grossniklaus HE, Vincent PA, Reynolds AB, Kowalczyk AP (2010) p120-Catenin is required for mouse vascular development. Circ Res 106:941–951

Oda H, Tsukita S, Takeichi M (1998) Dynamic behavior of the cadherin-based cell-cell adhesion system during drosophila gastrulation. Dev Biol 203:435–450

Ogata S, Morokuma J, Hayata T, Kolle G, Niehrs C, Ueno N, Cho KW (2007) TGF-beta signaling-mediated morphogenesis: modulation of cell adhesion via cadherin endocytosis. Genes Dev 21:1817–1831

Pacquelet A, Rorth P (2005) Regulatory mechanisms required for DE-cadherin function in cell migration and other types of adhesion. J Cell Biol 170:803–812

Pacquelet A, Lin L, Rorth P (2003) Binding site for p120/delta-catenin is not required for drosophila E-cadherin function in vivo. J Cell Biol 160:313–319

Palacios F, Price L, Schweitzer J, Collard JG, D'Souza-Schorey C (2001) An essential role for ARF6-regulated membrane traffic in adherens junction turnover and epithelial cell migration. EMBO J 20:4973–4986

Palacios F, Tushir JS, Fujita Y, D'Souza-Schorey C (2005) Lysosomal targeting of E-cadherin: a unique mechanism for the down-regulation of cell-cell adhesion during epithelial to mesenchymal transitions. Mol Cell Biol 25:389–402

Paradies NE, Grunwald GB (1993) Purification and characterization of NCAD90, a soluble endogenous form of N-cadherin, which is generated by proteolysis during retinal development and retains adhesive and neurite-promoting function. J Neurosci Res 36:33–45

Paterson AD, Parton RG, Ferguson C, Stow JL, Yap AS (2003) Characterization of E-cadherin endocytosis in isolated MCF-7 and chinese hamster ovary cells: the initial fate of unbound E-cadherin. J Biol Chem 278:21050–21057

Peinado H, Portillo F, Cano A (2004) Transcriptional regulation of cadherins during development and carcinogenesis. Int J Dev Biol 48:365–375

Perez-Moreno M, Davis MA, Wong E, Pasolli HA, Reynolds AB, Fuchs E (2006) p120-Catenin mediates inflammatory responses in the skin. Cell 124:631–644

Pettitt J, Cox EA, Broadbent ID, Flett A, Hardin J (2003) The Caenorhabditis elegans p120 catenin homologue, JAC-1, modulates cadherin-catenin function during epidermal morphogenesis. J Cell Biol 162:15–22

Pittard AJ, Banks RE, Galley HF, Webster NR (1996) Soluble E-cadherin concentrations in patients with systemic inflammatory response syndrome and multiorgan dysfunction syndrome. Br J Anaesth 76:629–631

Qian L-W, Greene W, Ye F, Gao S-J (2008) Kaposi's sarcoma-associated herpesvirus disrupts adherens junctions and increases endothelial permeability by inducing degradation of VE-cadherin. J Virol 82:11902–11912

Rasin MR, Gazula VR, Breunig JJ, Kwan KY, Johnson MB, Liu-Chen S, Li HS, Jan LY, Jan YN, Rakic P, Sestan N (2007) Numb and numbl are required for maintenance of cadherin-based adhesion and polarity of neural progenitors. Nat Neurosci 10:819–827

Reiss K, Maretzky T, Ludwig A, Tousseyn T, de Strooper B, Hartmann D, Saftig P (2005) ADAM10 cleavage of N-cadherin and regulation of cell-cell adhesion and beta-catenin nuclear signalling. EMBO J 24:742–752

Reynolds AB (2007) p120-Catenin: past and present. Biochim Biophys Acta 1773:2–7

Reynolds AB, Daniel JM, Mo YY, Wu J, Zhang Z (1996) The novel catenin p120cas binds classical cadherins and induces an unusual morphological phenotype in NIH3T3 fibroblasts. Exp Cell Res 225:328–337

Roeth JF, Sawyer JK, Wilner DA, Peifer M (2009) Rab11 helps maintain apical crumbs and adherens junctions in the drosophila embryonic ectoderm. PLoS One 4(10):e7634. doi:10.1371/journal.pone.0007634

Sato K, Watanabe T, Wang S, Kakeno M, Matsuzawa K, Matsui T, Yokoi K, Murase K, Sugiyama I, Ozawa M, Kaibuchi K (2011) Numb controls E-cadherin endocytosis through p120 catenin with aPKC. Mol Biol Cell 22(17):3103–19

Schulte D, Kuppers V, Dartsch N, Broermann A, Li H, Zarbock A, Kamenyeva O, Kiefer F, Khandoga A, Massberg S, Vestweber D (2011) Stabilizing the VE-cadherin-catenin complex blocks leukocyte extravasation and vascular permeability. EMBO J 30(20):4157–4170

Sharma M, Henderson BR (2007) IQ-domain GTPase-activating protein 1 regulates beta-catenin at membrane ruffles and its role in macropinocytosis of N-cadherin and adenomatous polyposis coli. J Biol Chem 282:8545–8556

Shore EM, Nelson WJ (1991) Biosynthesis of the cell adhesion molecule uvomorulin (E-cadherin) in madin-darby canine kidney epithelial cells. J Biol Chem 266:19672–19680

Shoval I, Ludwig A, Kalcheim C (2007) Antagonistic roles of full-length N-cadherin and its soluble BMP cleavage product in neural crest delamination. Development 134:491–501

Smalley-Freed WG, Efimov A, Burnett PE, Short SP, Davis MA, Gumucio DL, Washington MK, Coffey RJ, Reynolds AB (2010) p120-catenin is essential for maintenance of barrier function and intestinal homeostasis in mice. J Clin Invest 120:1824–1835

Solanas G, Cortina C, Sevillano M, Batlle E (2011) Cleavage of E-cadherin by ADAM10 mediates epithelial cell sorting downstream of EphB signalling. Nat Cell Biol 13:1100–1107

Stairs DB, Bayne LJ, Rhoades B, Vega ME, Waldron TJ, Kalabis J, Klein-Szanto A, Lee JS, Katz JP, Diehl JA, Reynolds AB, Vonderheide RH, Rustgi AK (2011) Deletion of p120-catenin results in a tumor microenvironment with inflammation and cancer that establishes it as a tumor suppressor gene. Cancer Cell 19:470–483

Steinberg MS, Takeichi M (1994) Experimental specification of cell sorting, tissue spreading, and specific spatial patterning by quantitative differences in cadherin expression. Proc Natl Acad Sci U S A 91:206–209

Suyama K, Shapiro I, Guttman M, Hazan RB (2002) A signaling pathway leading to metastasis is controlled by N-cadherin and the FGF receptor. Cancer Cell 2:301–314

Tai CY, Mysore SP, Chiu C, Schuman EM (2007) Activity-regulated N-cadherin endocytosis. Neuron 54:771–785

Toyoshima M, Tanaka N, Aoki J, Tanaka Y, Murata K, Kyuuma M, Kobayashi H, Ishii N, Yaegashi N, Sugamura K (2007) Inhibition of tumor growth and metastasis by depletion of vesicular sorting protein Hrs: its regulatory role on E-cadherin and beta-catenin. Cancer Res 67:5162–5171

Traub LM (2003) Sorting it out: AP-2 and alternate clathrin adaptors in endocytic cargo selection. J Cell Biol 163:203–208

Troyanovsky RB, Sokolov EP, Troyanovsky SM (2006) Endocytosis of cadherin from intracellular junctions is the driving force for cadherin adhesive dimer disassembly. Mol Biol Cell 17:3484–3493

Ulrich F, Krieg M, Schotz EM, Link V, Castanon I, Schnabel V, Taubenberger A, Mueller D, Puech PH, Heisenberg CP (2005) Wnt11 functions in gastrulation by controlling cell cohesion through Rab5c and E-cadherin. Dev Cell 9:555–564

Veiga E, Cossart P (2005) Listeria hijacks the clathrin-dependent endocytic machinery to invade mammalian cells. Nat Cell Biol 7:894–900

Wang Q, Chen XW, Margolis B (2007) PALS1 regulates E-cadherin trafficking in mammalian epithelial cells. Mol Biol Cell 18:874–885

Wang Z, Sandiford S, Wu C, Li SS (2009) Numb regulates cell-cell adhesion and polarity in response to tyrosine kinase signalling. EMBO J 28:2360–2373

Wheelock MJ, Buck CA, Bechtol KB, Damsky CH (1987) Soluble 80-kd fragment of cell-CAM 120/80 disrupts cell-cell adhesion. J Cell Biochem 34:187–202

Xiao K, Allison DF, Buckley KM, Kottke MD, Vincent PA, Faundez V, Kowalczyk AP (2003a) Cellular levels of p120 catenin function as a set point for cadherin expression levels in microvascular endothelial cells. J Cell Biol 163:535–545

Xiao K, Allison DF, Kottke MD, Summers S, Sorescu GP, Faundez V, Kowalczyk AP (2003b) Mechanisms of VE-cadherin processing and degradation in microvascular endothelial cells. J Biol Chem 278:19199–19208

Xiao K, Garner J, Buckley KM, Vincent PA, Chiasson CM, Dejana E, Faundez V, Kowalczyk AP (2005) p120-Catenin regulates clathrin-dependent endocytosis of VE-cadherin. Mol Biol Cell 16:5141–5151

Xie LQ, Bian LJ, Li Z, Li Y, Li ZX, Li B (2010) Altered expression of E-cadherin by hepatocyte growth factor and effect on the prognosis of nasopharyngeal carcinoma. Ann Surg Oncol 17:1927–1936

Yabuta T, Shinmura K, Tani M, Yamaguchi S, Yoshimura K, Katai H, Nakajima T, Mochiki E, Tsujinaka T, Takami M, Hirose K, Yamaguchi A, Takenoshita S, Yokota J (2002) E-cadherin gene variants in gastric cancer families whose probands are diagnosed with diffuse gastric cancer. Int J Cancer 101:434–441

Yamada S, Pokutta S, Drees F, Weis WI, Nelson WJ (2005) Deconstructing the cadherin-catenin-actin complex. Cell 123:889–901

Yang DH, Cai KQ, Roland IH, Smith ER, Xu XX (2007) Disabled-2 is an epithelial surface positioning gene. J Biol Chem 282:13114–13122

Yang JY, Zong CS, Xia W, Wei Y, Ali-Seyed M, Li Z, Broglio K, Berry DA, Hung MC (2006) MDM 2 promotes cell motility and invasiveness by regulating E-cadherin degradation. Mol Cell Biol 26:7269–7282

Zhong Y, Brieher WM, Gumbiner BM (1999) Analysis of C-cadherin regulation during tissue morphogenesis with an activating antibody. J Cell Biol 144:351–359

Zhou WJ, Geng ZH, Chi S, Zhang W, Niu XF, Lan SJ, Ma L, Yang X, Wang LJ, Ding YQ, Geng JG (2011) Slit-Robo signaling induces malignant transformation through hakai-mediated E-cadherin degradation during colorectal epithelial cell carcinogenesis. Cell Res 21:609–626

Part IV
How AJs Affect Cell Behaviour and Multicellular Development

Chapter 10
Adherens Junctions During Cell Migration

Sandrine Etienne-Manneville

Abstract Migration is a key cellular process, involved during morphogenetic movements as well as in the adult where it participates in immune cell trafficking, wound healing or tumour invasion. As they migrate, cells interact with a microenvironment composed of extracellular matrix and neighbouring cells. Cell–cell adhesions ensure tissue integrity while they allow migration of single or grouped cells within this tissue. Cadherin and nectin-based adherens junctions are key players in intercellular interactions. They are used as adhesive complexes whose mechanical properties improve cell coordination during collective migration and promote cell motility on cadherin substrates. In addition, adherens junctions transduce signals that actively participate in the control of directed cell migration, by providing polarity cues and also participating in contact inhibition of motility.

10.1 Introduction

Cell migration is essential for embryogenesis and in the adult where it contributes to immune surveillance, tissue renewal, wound healing as well as to tumor dissemination. Depending on the context, cells migrate alone, in chains or in sheets, in a random or in a directed manner, in a 2D or 3D environment and in response to a variety of extracellular factors.

The common basic mechanisms of cell mobility are now well understood and rely on a repetition of a sequence of well characterized steps. To migrate efficiently, cells must first determine the position of their leading and trailing edges. This initial polarization leads to the formation of a leading protrusion involving cytoskeleton remodelling and in particular actin polymerization, together with the formation of new adhesions with the surrounding substrate. This step is followed by the contraction of the cell body and the detachment of the cell rear from the substrate which finally allows the net displacement of the cell. It is important to distinguish two fundamentally different types of migration, mesenchymal and amoeboid migration. During ameboid migration, cell adhesion to the substrate is minimal and extracellular matrix

S. Etienne-Manneville (✉)
Institut Pasteur, Cell Polarity and Migration Group and CNRS URA 2582,
25 Rue du Dr Roux, 75724, Paris Cedex 15, France
e-mail: sandrine.etienne-manneville@pasteur.fr

T. Harris (ed.), *Adherens Junctions: From Molecular Mechanisms to Tissue Development and Disease*, Subcellular Biochemistry 60,
DOI 10.1007/978-94-007-4186-7_10, © Springer Science+Business Media Dordrecht 2012

degradation is not required. In this case, the major driving force is actomyosin contractility which squeezes the cell body and helps it pass between the matrix fibers. In contrast, cells migrating in a mesenchymal manner interact with the surrounding substrate which is used as a physical support and remodel the matrix as they move through it. In these conditions, cells must establish and regulate contacts with their microenvironment to make their way through the surrounding tissue. New contacts with the substrate are formed at the front of the cell, where membrane protrusion is mainly caused by actin polymerization which leads to the extension of a lamellipodia.

During migration in a 2D or 3D environment, cells can interact with the extracellular matrix using integrin-mediated adhesions and/or with neighbouring cells using intercellular adhesion molecules (Friedl and Wolf 2009; Rorth 2009). Cell–cell interactions can have three distinct functions during cell migration (Friedl and Gilmour 2009; Moh and Shen 2009; Weijer 2009). First, they can support cell migration. For instance, during neuronal migration, neuronal cells use radial glial cells as a substrate. Second, cell–cell interactions are essential to maintain tissue architecture during collective migration of cell groups, chains or sheets. In this case, cells linked by intercellular interactions migrate together in a coordinated manner. Third, intercellular interactions can induce contact inhibition and either serve as a brake for cell migration or change the cell direction.

Regulation of cell migration by intercellular contacts involves a wide variety of adhesion molecules and adhesion complexes. Amongst these, the adherens junctions (AJs) are prominent players. AJs are observed in a variety of cell types in most animal species and play a major role in maintaining the physical association between cells. In addition, AJs mediate crucial intracellular signals that directly participate in the regulation of major cellular functions, such as proliferation, differentiation or migration (Etienne-Manneville 2011; Harris and Tepass 2010; Perez-Moreno et al. 2003). Classical cadherins, together with nectins, are major components of AJs. They generally mediate strong cell–cell adhesion through calcium-dependent homophilic interactions (Smutny and Yap 2010). Classical cadherins show distinct tissue distribution patterns and were originally named from the tissue in which they are predominantly expressed (Takeichi 1988). However, the expression pattern of each cadherin is generally less restricted than what was initially thought and many cell types can express several types of cadherins, allowing them to interact with various different cell types. The 20 Type I classical cadherins, including E-, N-, P- and R-cadherins, share a common domain organization which, in vertebrates, comprises an extracellular domain formed of five cadherin repeats (EC), a single transmembrane domain and a highly conserved cytoplasmic tail (Oda and Takeichi 2011). In contrast to cadherins, the four members of the nectin family are calcium-independent immunoglobulin-like molecules (Nakanishi and Takai 2004; Takai et al. 2003). Nectins participate in AJs through homophilic or heterophilic interactions (Takahashi et al. 1999).

Cadherins are not uniformly distributed over the cell surface but cluster in specific domains of the plasma membrane, which serve as signalling platforms. The

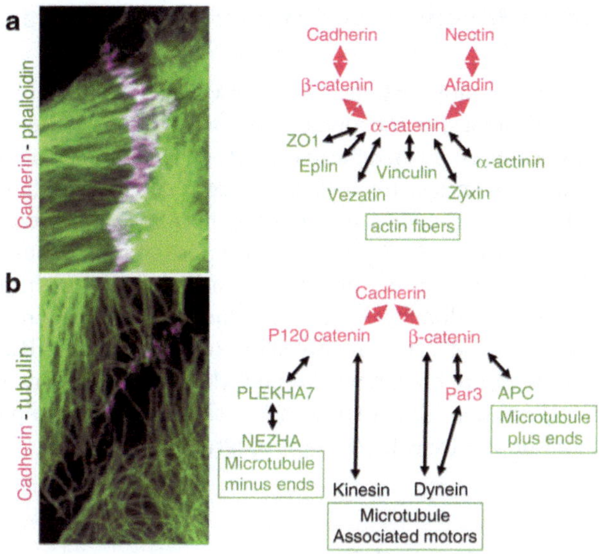

Fig. 10.1 Adherens junction connections with the cytoskeleton. **a** AJs are connected to the actin cytoskeleton. *Left panel* cadherin (*purple*) and F-actin (*green*) costaining shows actin fibres running parallel and perpendicular to the plasma membrane of contacting astrocytes. *Right panel* cadherin-catenin complexes interact with several actin-associated proteins (*green*). The black double headed arrows indicate molecular interactions. **b** AJs are connected to microtubules. *Left panel* Cadherin (*purple*) and microtubules (*green*) costaining shows microtubules closely associated with AJs in adjacent astrocytes. *Right panel* cadherin-catenin complexes interact with several microtubule-associated proteins (*green*) and microtubule-associated motors (*black*). The black double headed arrows indicate molecular interactions

juxtamembrane portion of the cadherin cytoplasmic domain binds p120-catenin, and the carboxy-terminal half associates with β-catenin and γ-catenin (plakoglobin), which bind to α-catenin. These catenins interact with a number of cytoplasmic proteins that can affect the dynamics and strength of cadherin-mediated adhesion and mediate a wide variety of intracellular signalling pathways (Meng and Takeichi 2009; Niessen and Gottardi 2008). The cytoplasmic domain of Nectins associates with the cytoplasmic protein AF6/afadin. Afadin can interact with α-catenin to provide a physical connection between the cadherin and the nectin complexes (Tachibana et al. 2000) (Fig. 10.1). Through their numerous interacting partners, cadherins directly connect with the cytoskeleton or signal through polarity proteins, RhoGTPases, tyrosine kinases and lipid modifications to control cytoskeleton dynamics, membrane trafficking and cell motility in regions that can be immediately adjacent or distant from AJs (Etienne-Manneville 2011; Harris and Tepass 2010).

This chapter will give an overview of the different functions of AJs during cell migration, whether AJs are used to promote, guide, coordinate or inhibit cell migration

10.2 Adherens Junctions as a Support for Cell Migration

Cadherin-based AJs not only serve as a glue between adjacent cells, they also functionally couple their adhesive functions to cytoskeleton regulation. This coupling allows cells to use cadherin as an adhesive substrate which promotes cell migration. During *Drosophila* oogenesis, for instance, border cells which have delaminated from the epithelium, use E-cadherin to migrate between nurse cells and reach the oocyte (Niewiadomska et al. 1999). In Zebrafish, loss of function experiments have revealed that N-cadherin controls neuronal positioning and pathfinding by regulating neuroectodermal cell adhesion and cell movement during neurulation (Lele et al. 2002). In mice, N-cadherin is involved in axonal outgrowth (Riehl et al. 1996), fasciculation and dendritic branching (Yu and Malenka 2003; Zhu and Luo 2004). Thus cells can migrate between other cells by using cadherin as a physical support for movement and/or as a transducer of pro-migratory signals.

10.2.1 AJs in Mechanical Coupling

In culture models, N-cadherin-coated substrates promote growth cone migration which, in this case, relies on the dynamic interplay between cadherin-mediated cell interactions and the underlying actin cytoskeleton (Bard et al. 2008; Giannone et al. 2009). On N-cadherin substrates, cadherin-based AJs support the mechanical stresses associated with cell migration in a similar manner to integrin-mediated focal adhesions when a cell migrates on an extracellular matrix. Cadherin-mediated interactions can generate both the protrusive and the contractile forces that are required for cell motility. The forces supported by AJs during cell migration are of the same order of magnitude as those transmitted at focal adhesions (Ganz et al. 2006; Ladoux et al. 2010).

AJs mechanically couple the cadherin substrate and the actomyosin contractile network. Actin bundles are commonly found in association with AJs and several physical links connect AJs to actin fibers that run parallel or perpendicular to the plasma membrane. When cells migrate on a cadherin substrate, this mechanical coupling is likely to involve multiple actin binding proteins that associate directly or indirectly with cadherin-associated catenins. Growth cone velocity correlates with the mechanical coupling between N-cadherin and the F-actin flow and depends on α-catenin (Bard et al. 2008), indicating that α-catenin which directly interacts with cadherin-bound β-catenin is crucial for AJs coupling to the actin cytoskeleton (Rimm et al. 1995).Although, biochemical evidence do not support a role of β-catenin-bound α-catenin in a direct interaction with actin filaments (Drees et al. 2005; Yamada et al. 2005), catenins may indirectly couple cadherin to actin fibers through their multiple actin-associated partners including EPLIN, Vezatin, vinculin, α-actinin, Formin, ZO1 (Abe and Takeichi 2008; Kobielak and Fuchs 2004; Kobielak et al. 2004; Sousa et al. 2004); see for review (Harris and Tepass 2010)).

In addition, AJ coupling to actin can also involve the nectin-afadin complex since afadin directly interacts with both actin and α-catenin (Pokutta et al. 2002; Sawyer et al. 2009) (Fig. 10.1A).

The formation of new cadherin-mediated contacts induces the assembly of an actin meshwork underneath the plasma membrane. The formation of actin fibers parallel to the membrane supports a flow of cadherin clusters towards forming adhesion sites (Kametani and Takeichi 2007). Contractile actin cables connect to the AJ complex and the increased myosin-mediated contractility at the contact zone between two cells strengthens cadherin adhesions (Adams and Nelson 1998; Chu et al. 2004). Like integrin-mediated focal adhesions, AJs can function as molecular clutches that allow actin-driven membrane protrusion (Giannone et al. 2009). Actin-driven membrane protrusion is followed by the activation of Rho, which recruits Myosin II (Liu et al. 2010; Yamada and Nelson 2007) and consolidates AJs. The formation of strong AJ-actin connections that act against the reaction force exerted by the plasma membrane on the polymerizing actin filaments, allows actin polymerization to promote membrane extension. The formation of new adhesive contacts as the membrane reaches free surfaces of a cadherin-covered substrate is further enhanced by directed vesicular transport of new E-cadherin molecules towards the forming junction (Kametani and Takeichi 2007). What remains unclear is how, when the cadherin substrate is actually a cell, cadherin clusters in the guiding cell resist the tension exerted by the migrating cell and what differentiates the two sides of the AJs to promote the migration of one of the two cells compared to that of the other.

10.2.2 AJs as a Signalling Platform Controlling Cytoskeleton Rearrangements

In addition to their direct connection to the cytoskeleton, AJs exhibit signalling functions that influence cell motility. Cadherins do not have any catalytic activity, but they can, upon cell–cell adhesion, promote the recruitment and/or the activation of several signalling effectors. AJ formation triggers the activation of the tyrosine kinase Src which, in turn, leads to the recruitment of PI3K (Phosphoinositide 3' kinase) to forming cell contacts (Perez et al. 2008; Singleton et al. 2005). PI3K may participate in the regulation of the actin cytoskeleton by locally modulating PIP2 levels and by phosphorylating the actin-regulatory protein cortactin (Ren et al. 2009). Cadherin signalling controls cytoskeleton rearrangement via the regulation of Rho GTPases. Cadherin engagement modulates Rac, Cdc42 and Rho activity to promote polymerization of the actin network, enlarge the region of cell–cell contact and maturate the AJs (Perez et al. 2008). Classical cadherins do not all have the same effect on the regulation of Rho GTPases. N-cadherin engagement, for instance, has been shown to activate Rho but not Rac (Charrasse et al. 2002) and VE-cadherin signals via RhoC to promote actomyosin contractility (Abraham et al. 2009). The unique morphology of growth cones migrating on various cadherin substrates may reflect signalling by a specific combination of Rho GTPases (Oblander and Brady-Kalnay 2010). However, the exact mechanisms by which classical cadherins differentially regulate RhoGTPases still need to be clarified. The precise and coordinated timing of Rac and Rho activity is crucial for the formation of stable AJs, and may result from the recruitment of p190RhoGAP by p120-catenin (Anastasiadis et al. 2000). p120-catenin is a regulator of N-cadherin dependent neural crest cell migration (Ciesiolka et al. 2004). Depletion of p120-catenin can be rescued by expression of a dominant negative Rac suggesting that Rho GTPase signalling is crucial for cadherin-mediated control of cell migration. Regulation of Rac can also occur via Tiam-1, an exchange factor that bind to the polarity protein Par-3 which is recruited to AJs (Hordijk et al. 1997; Iden and Collard 2008).

AJs functions as signalling platforms which also include other transmembrane receptors. Cadherin/catenin complexes interact with Receptor Tyrosine Kinases (RTK) , which may contribute their own signalling capacity to the regulation of cell migration. For instance, E-cadherin engagement leads to ligand-independent activation of the EGFR (Epidermal Growth Factor Receptor) which contributes to the activation of signalling pathways such as the PI3K and the MAPK (mitogen activated protein kinase) pathway (Kovacs et al. 2002; Pece and Gutkind 2000). Similarly, N-cadherin directly associates with FGFR1 (Fibroblast Growth Factor Receptor 1) via its two Ig domains. This leads to the activation of the Erk signalling cascade which promotes cell migration (Suyama et al. 2002). The tyrosine kinase activity of FGFR is required for neurite outgrowth induced by N-cadherin (Kolkova et al. 2000). FGFR activity has also been involved in cadherin-11 driven induction of neurite outgrowth (Boscher and Mege 2008). In endothelial cells, VE-cadherin, which contributes by both mechanical and signalling functions to the development of the vascular system (Carmeliet et al. 1999), positively regulates VEGFR-2 signalling which can, for instance, affect cytoskeletal organization via PI3K (Taddei et al. 2008). In addition, VE-cadherin binds to several components of the TGF-β receptor complex to induce TGF-β signalling (Rudini et al. 2008).

Variations in the combination of associated proteins and in the signalling properties of different cadherins is likely to be responsible for differential adhesion and migration, in particular in the central nervous system where the combined expression of cadherins plays a key role in the aggregation and the sorting of precursor cells (Redies 2000).

10.2.3 Cadherins as Directional Sensors

In addition to modulating the migration capacity of cells, cadherin-mediated interactions also provide polarity cues that influence cell polarity and the direction cell migration. The role of cadherins as directional sensors has been highlighted by studies in the central nervous system, where several classical cadherins, expressed in axonal tracts, play a key role in growth cone navigation and circuit formation

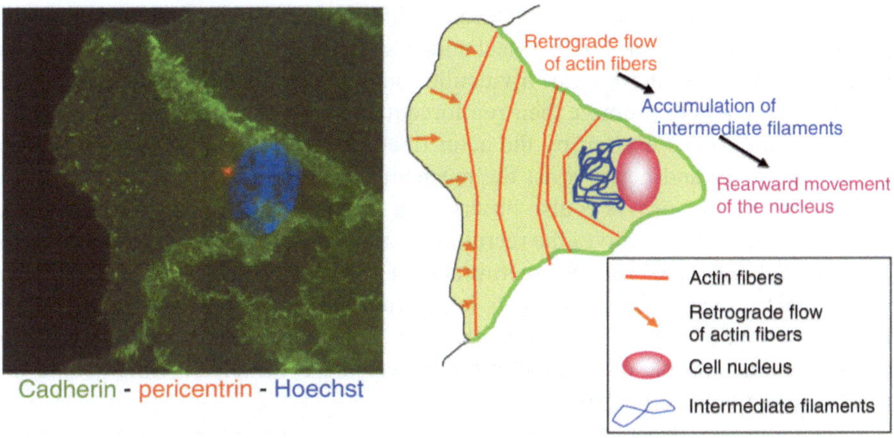

Fig. 10.2 Adherens junctions and polarized intracellular organization. *Left panel* cadherin (*green*), centrosome (*red*) and nucleus (*blue*) co-staining shows the orientation of the nucleus-centrosome axis towards the free cell edge in a polarized astrocyte. *Right panel* schematic showing how anisotropic cell–cell interactions lead to the polarized organization of the actin and intermediate filament network and nuclear positioning

(Honig et al. 1998; Iwai et al. 1997). During neuronal migration from the ventricular zone to the neocortical plate, directed migration of multipolar neurons is initiated by a Reelin-induced increase of N-cadherin at the plasma membrane, suggesting that N-cadherin mediated interactions are responsible for neuronal polarization (Jossin and Cooper 2011). Accordingly, the orientation of neuronal migration is strongly perturbed by expression of dominant negative N-cadherin mutants that prevent N-cadherin signalling from the plasma membrane. The exact mechanism responsible for the role of cadherin in directional sensing remains unclear. However two hypotheses can be drawn from recent observations.

First, the distribution of cadherin-mediated interactions at the cell periphery could be sufficient to promote cell orientation (Fig. 10.2). When glial cells are plated on asymmetric N-cadherin coated micropatterns, the centrosome, together with the Golgi apparatus, move in front of the nucleus towards the AJ-free cell edge (Dupin et al. 2011). Thus, the position of cadherins expressed on radially oriented neuronal processes or radial glia, may be sensed as environmental directional cues by migrating multipolar cells. It is also possible that N-cadherin interacts with other cell surface receptors that respond to directional signals from the cortical plates (Jossin and Cooper 2011).

Second, the molecular machinery that forms AJs has recently been shown to be a mechanosensor, which may guide cells on a cadherin substrate depending on its physical properties and in particular on its rigidity. When the rigidity of a cadherin-coated substrate increases, the forces exerted by the cell on the cadherin-mediated contacts also increase (Ladoux et al. 2010). Rigidity sensing may thus contribute to direct cell migration on cadherin substrates. Mechanosensing by AJs relies on a dynamic interplay between AJs and the actin cytoskeleton (Maruthamuthu et al. 2010) and in particular on α-catenin, which can undergo a stress-induced conformational change (Yonemura et al. 2010). Mechanical stretching of the protein leads to its unfolding and the unmasking of a vinculin binding site. Vinculin is thus recruited temporarily to AJs to promote their reinforcement and maturation (Yonemura et al. 2010). As the forces exerted by the actin fibers on the AJ complex increase, AJs become stronger and better resist the mechanical stress (Ladoux et al. 2010). An increase in the size of AJs is also observed in response to applied external forces. Strengthening of AJs involves the recruitment of new cadherins by myosin II (Kametani and Takeichi 2007) and the recruitment of vinculin by Myosin IV (Maddugoda et al. 2007) probably followed by its interaction withα-catenin. AJs mainly grow in the direction of the forces exerted on them. This anisotropic growth is likely to result in an anisotropic orientation of the underlying cytoskeleton which could influence directional cell behaviour.

Taken together, these observations show that cadherins can serve as mechanical anchors and as transducers that help cells migrate among others. Cadherin may also promote cell orientation, directional migration and pathfinding. The similarities between mechanical coupling at AJs and at focal adhesions suggest that the combination of interactions with cadherin-expressing cells and with the extracellular matrix can be easily integrated at the cellular level to lead to a common regulation of the cytoskeleton that will drive directed cell migration in a complex environment.

10.3 Adherens Junctions in Collective Cell Migration

During development, as well as during tissue renewal or wound healing in the adult, cells undergo collective movements in which cell–cell interactions play a crucial role. For instance, collective mesendoderm movement in Zebrafish (Arboleda-Estudillo et al. 2010), mammary duct formation (Ewald et al. 2008), or trachea development (Shaye et al. 2008) require cadherin-mediated adhesions. AJs are also involved in collective cell movements during cancer invasion (reviewed in (Friedl and Gilmour 2009)). More recently, *in vitro* wound healing assays have revealed that cadherin-mediated interactions regulate direction persistence and speed of migration (Camand et al. 2011).

The main functions of AJs during collective cell migration are to preserve cell–cell interactions and to promote the coordinated behaviour of cells. In the typical example of neural crest cells, which are collectively attracted toward a chemokine gradient, N-cadherin inhibits membrane protrusion at cell–cell contacts, whereas it promotes protrusion at the free edge of the cells (Theveneau et al. 2010). Cell–cell interactions mechanically link cells together and restrict their movement relative to each other, while simultaneously promoting cellular motility in regions devoid of cell–cell interactions.

10.3.1 Cadherin/Integrin Interplay and the Regulation of Cell Protrusion

During the migration of *Drosophila* border cells, E-cadherin-mediated lateral cell–cell interactions control frontal protrusion (Melani et al. 2008; Niewiadomska et al. 1999), suggesting that lateral interactions promote the coordinated protrusive activity at the leading edge of a cell group. During neural crest cell migration, N-cadherin is involved in Rac1 activation at the free edge of the cells (Theveneau et al. 2010). Similarly, during wound-induced astrocyte migration, N-cadherin is required for Cdc42 activation at the wound-edge of the cells (Camand et al. 2011). How AJs regulate intracellular signalling at distance in not clear, but seems to involve several distinct mechanisms. One mechanism relies on an intimate crosstalk between AJs and focal adhesions (Fig. 10.3b). The formation of focal adhesions at the leading edge of migrating cells is crucial for front edge cytoskeletal rearrangements and cell protrusion and migration. The presence of AJs appears to inhibit focal adhesions (Borghi et al. 2010; Dupin et al. 2009). The presence of AJs at the lateral and rear sides of migrating cells leads to the concentration of focal adhesions in regions devoid of cell–cell contacts (Camand et al. 2011). By influencing the localization of focal adhesions, AJs regulate the localization of integrin-mediated signalling pathways that are a determinant for the generation of a leading edge and for the establishment of cell migration.

Nectins are likely to participate in the crosstalk between cell–cell and cell-ECM adhesions, as Nectin-1 and nectin-3 inactivate αvβ3 integrins at cell–cell adhesion sites (Sakamoto et al. 2006). This inactivation, which involves the local decrease in the synthesis of PIP2 and the inhibition of talin binding to the integrin intracellular domain, inhibits cell motility (Sakamoto et al. 2008). Direct interaction with integrins has not been described in the case of cadherin but recent observations suggest that cadherins are directly required for the activation of leading edge signalling. The molecular mechanisms underlying the crosstalk between AJs and focal adhesions may also involve the mechanosensing properties of these two structures. Measurement of the forces exerted by cells on flexible substrates shows that, when cells are joined by AJs, the traction forces exerted on the substrate by each cell is unbalanced. This excess of traction forces is oriented perpendicularly to the region of cell–cell contacts and balances forces generated at AJs between neighbouring cells (Liu et al. 2010; Maruthamuthu et al. 2011). Vinculin might play a key role in balancing the forces between focal adhesions and AJs. Vinculin is recruited to stretched α-catenin in mature AJs. This may affect its localization to integrin-dependent structures and inhibit the maturation of focal adhesions. More generally, competition of integrin and cadherin complexes for common cytoplasmic molecules may be responsible for their mutual exclusion. Finally, vesicular traffic also links cadherin and integrin localization and functions. Key regulators of both AJs and focal adhesions, such as the small GTPase Rap1 (Nishimura and Kaibuchi 2007), are regulated by E-cadherin endocytosis (Balzac et al. 2005) and directly

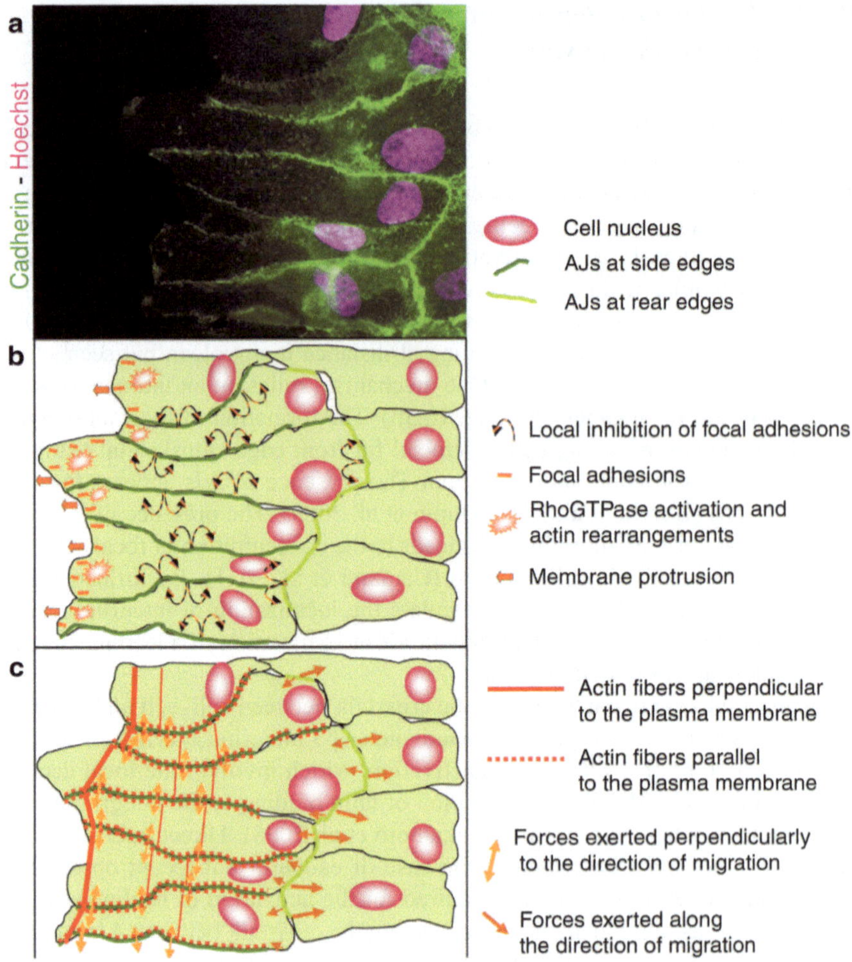

Fig. 10.3 Adherens junctions during collective migration. **a** Cadherin (*green*) and nucleus (*purple*) staining showing the localization of AJs between adjacent migrating astrocytes in an in vitro wound-healing assay. **b** During collective migration, AJs contribute to the coordination of the protrusive activity in front edge cells. AJs locally inhibit focal adhesions. Focal adhesions are excluded from the regions of cell–cell contacts and are concentrated at the cell leading edge, where they promote Rho GTPases activation and membrane protrusion. **c** Schematic representing the distribution of actin-myosin contractile forces in cells migrating in a collective manner. Wider arrows indicate stronger forces

interact with integrin complexes to control adhesion and migration (Tohyama et al. 2003). Rap1, for instance, is required for epithelial migration in *Drosophila* (Boettner and Van Aelst 2007) and for mesodermal movements during vertebrate gastrulation (Tsai et al. 2007).

10.3.2 AJs as Key Regulators of Cell Polarity

In addition to their impact on focal adhesion, AJs regulate the polarized organization of intracellular organelles. An anisotropic distribution of AJs results in the localization of the nucleus near cell–cell contacts and the orientation of the Golgi complex and the microtubule organizing center (MTOC) towards the cell–cell contact free edge (Desai et al. 2009; Dupin et al. 2009). The distribution of AJs at the cell periphery leads to the reorganization of actin microfilaments, microtubules and intermediate filaments (Dupin et al. 2009; Dupin et al. 2011). AJs possibly act through three different mechanisms.

First, AJs may directly influence MTOC and Golgi position by regulating microtubule dynamics and functions. This regulation is likely to involve microtubule interactions with cadherin/catenin complexes (Fig. 10.1B). Microtubule plus-ends are targeted to AJs (Shaw et al. 2007; Stehbens et al. 2006) and the microtubule associated motor dynein binds the β-catenin complex and localizes at cell–cell contacts (Ligon et al. 2001; Shaw et al. 2007). p120 catenin also contributes to microtubule interactions with AJs by recruiting a microtubule associated motor of the kinesin family (Chen et al. 2003) and by interacting with PLEKHA7 which binds the microtubule minus end-binding protein NEZHA (Meng et al. 2008). The polarization of the microtubule network and the positioning of the MTOC are paralleled by the orientation of the Golgi apparatus and of membrane trafficking towards the cell free edges (Fig. 10.2).

Second, the segregation of polarity protein complexes such as the Scribble complex (including Scrib, Discs large (Dlg) and Lethal giant larva (Lgl)), the Crumbs complex (including Crumbs, the Protein Associatedwith Lin Seven 1 (PALS1) and PALS1-Associated Tight Junctionprotein (PATJ)) or the Par complex (including Par-3, Par-6 and aPKC) may contribute to polarized membrane delivery (Macara 2004; Roh and Margolis 2003). In particular, Par-3 has recently been shown to interact with the dynein light intermediate chain 2. This interaction could influence the delivery of membrane vesicles by controlling both MT dynamics near cell–cell contacts and the global polarized organization of the microtubule network (Schmoranzer et al. 2009). The consequences of such a polarized intracellular organization are not completely understood but it is likely to favour membrane protrusion towards the non-adhesive cell region and thereby to facilitate directed migration.

Third, the interplay between AJs and cell interactions with the extracellular matrix through focal adhesions may contribute to the polarization of the actin flux from regions involved in cell-matrix interactions. Such effects appear to be reciprocal as integrin-dependent retrograde flow of actin fibres influences the organization of the intermediate filament network, which in turn pushes and moves the nucleus together with the MTOC close to AJs (Dupin et al. 2011) (Fig. 10.2).

10.3.3 Maintaining Cell–Cell Adhesion During Collective Cell Migration

During collective cell movement, adhesion between individual cells needs to be dynamically regulated and precisely coordinated to maintain cellular interactions while allowing cellular rearrangements within the group of co-migrating cells.

Through their link with the actin cytoskeleton, AJs provide connection between neighbouring cells. In addition to the actin fibers running underneath the contacting plasma membranes, cadherin/catenin complexes associate with actin bundles positioned perpendicularly to the plasma membrane. The actin bundles, which terminate in the cadherin-cell–cell junctions, are enriched in myosin II and arranged in antiparallel filaments that seem to link AJs from opposite cell sides (Miyake et al. 2006) (Fig. 10.3c). Myosin function is controlled by Cdc42 and its effector MRCK. The resulting increase in myosin activity at the edges of the group of cells generate inwardly directed contractile forces (Gaggioli et al. 2007; Maruthamuthu et al. 2011). In contrast, analysis of actomyosin contractility in collectively invading carcinoma cells revealed that cell–cell contacts located inside of the cell group displayed weaker actomyosin contractility. DDR1 (Discoidin Domain receptor 1) plays a key role in this process (Hidalgo-Carcedo et al. 2011). It interacts with the Par-3/Par-6 complex and controls RhoE localization at cell–cell contacts to inhibit actomyosin contractility.

The accumulation of contractile actin bundles at the limit of a cell group contributes to epithelial wound closure, dorsal closure in *Drosophila* (Jacinto et al. 2002; Martin and Parkhurst 2003) and migration of the hypodermis in *C. elegans* embryo (Costa et al. 1998). Additional tensile forces are applied along the axis of migration between leader cells and their followers (Fig. 10.3c). These forces predominately arise several rows behind the leading edge (Trepat et al. 2009). They do not seem to strongly support forward migration but they probably contribute to maintain the cohesion of cell sheet (Camand et al. 2011). Forces exerted from neighbouring cells are balanced in magnitude and orientation. Activation of actomyosin contraction by Rho in one cell of a pair causes a similar increase in junction-associated forces in both cells, indicating that the cell neighbour can sense and respond to pulling forces by regulating its own contractility (Liu et al. 2010). Myosin and vinculin are likely to play a key role in the mechanical coupling between adjacent cells (le Duc et al. 2010; Pasapera et al. 2010), but other proteins such as Zyxin have been shown to reinforce the interactions between adjacent migrating cells by connecting their actin cytoskeleton and by increasing junction stability (Sperry et al. 2010).

During morphogenesis, interacting cells migrate together but also reorganize and intercalate, allowing movements of convergent extension and remodelling of coherent tissues. Variations in size and strength of AJs and focal adhesions affect the distribution of forces throughout the cells. This spatial regulation of actomyosin contractility controls the coordinated cell movements that underlie convergent extension in *Drosophila* (Harris et al. 2009; Zallen 2007). Here, cell intercalation requires the control of adhesion strength by the tight regulation of AJ assembly and disassembly (Cox et al. 1996). Supracellular actomyosin cables potentially associated with AJs also contribute to the coordination of cell polarization and intercalation underlying convergence and extension movements during Xenopus gastrulation (Skoglund et al. 2008). Endocytic AJ regulation is also involved. In Xenopus, TGF-β induces convergent extension movements that involve the small GTPases dynamin and Rnd1 to control C-cadherin endocytosis (Jarrett et al. 2002; Ogata et al. 2007). Intracellular trafficking is similarly involved during tissue remodelling in the *Drosophila* wing and trachea (Classen et al. 2005; Shaye et al. 2008).

10.4 Establishment and Release of Contact Inhibition of Motility by AJs

Cell–cell contacts have been known for a long time to locally prevent cell migration, in particular in the context of epithelial sheets and more generally when two cells establish initial contacts. This so-called contact inhibition of migration, initially described by Abercrombie (Abercrombie 1970), can either lead to a complete arrest of cell migration or to a drastic change in the direction of migration away from the initial contact. Accumulating evidence suggests that AJs are involved in contact inhibition of cell migration.

10.4.1 Inhibition of Cell Protrusion at Cell-Cell Contacts

In keratinocytes, in migrating cells from *Drosophila* or *C. elegans* as well as in MDCK cells (McNeill et al. 1993; Vasioukhin et al. 2000; Vasioukhin and Fuchs 2001), prior to the formation of AJs, cadherins localize in actin-rich lamellipodial or filopodial protrusions which probe the environment for neighbouring cells. The association of VE-cadherin with myosin-X promotes cadherin trafficking to the filopodial tips (Almagro et al. 2010). VE cadherin accumulates in filopodia and contributes to the formation of early cell–cell contacts between neighbouring cells. As nectin- and then cadherin-mediated contacts start to form, lateral interactions between neighbouring cells increase (Takai et al. 2003; Togashi et al. 2006). The formation of AJs is associated with the local rearrangement of the actin cytoskeleton that extends the area of cell–cell contacts (Fig. 10.4). The extension of the contact region also involves the recruitment of vesicles that provide both membrane material and additional transmembrane cadherins necessary for the formation of new AJs (Etienne-Manneville 2011). Simultaneously, cadherin-mediated interactions locally inhibit membrane motility through the regulation of Rho GTPases. The small GTPase Rap1 (Pannekoek et al. 2009) and the Par complex (Par-3, Par-6 and aPKC) (Mishima et al. 2002) also contribute to contact inhibition. In addition, cadherins

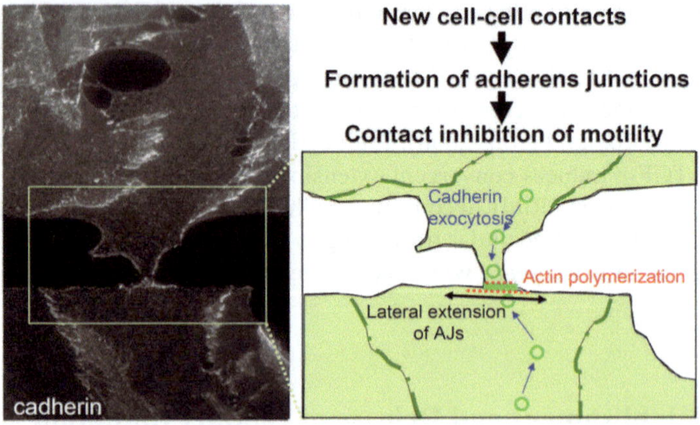

Fig. 10.4 Adherens junctions and contact inhibition of motility. *Left panel* cadherin staining showing nascent adherens junctions in contacting cells. *Right panel* cadherin and nectin engagement promotes the extension of the contact area via the regulation of actin rearrangements and cadherin vesicular traffic and local inhibition of cell motility

and nectins probably contribute to contact inhibition by preventing focal adhesion formation in the newly formed region of intercellular contact (see Sect. II.1)

The fact that destabilization or loss of cadherin-mediated junctions promotes cell separation and non-collective migration strongly supports the role AJs in contact inhibition of motility. During development, cells can detach from their neighbours to become motile and invade the surrounding tissues, as illustrated by a variety of cellular processes such as neural crest development, vascular sprouting, or wound healing. For instance, endothelial cells submitted to thrombin treatment tend to loosen their junctions and to migrate away from the endothelial wall to promote the formation of new vessels (Stepniak et al. 2009). This process has been mostly studied in epithelial cells, because a similar behaviour observed during carcinogenesis plays a crucial role in the formation of metastasis. During development and during cancer, epithelial cells can detach from their surrounding epithelium by undergoing epithelial-to-mesenchymal transition (EMT) (Baum et al. 2008). This results in the migration of isolated or grouped cells away from their original tissue. Variations in the adhesion strength also contribute to cell sorting and cell rearrangements within collectively migrating grouped cells. A wide variety of external signals including TGF-β, growth factors, Notch, have been involved. In all cases, loss of AJs at the apical side of the cells appears to be a critical step during EMT.

10.4.2 Cadherin Regulation and EMT in Development and Cancer

Misregulation of EMT can have dramatic consequences (Thiery and Sleeman 2006). Destabilization of AJs is critical for EMT both in physiological and pathological situations. Loss of AJs primarily results from a decrease in cadherin levels at the plasma membrane. Loss of E-cadherin in epithelial cells leads to the alteration of baso-apical polarity, to a motile morphology and to the formation of cell protrusions characteristic of EMT.

EMT is frequently associated with a decrease in E-cadherin expression. E-cadherin transcription is inhibited by the bHLH protein Snail (Barrallo-Gimeno and Nieto 2005) and by p38 MAPK (Mitogen-activated protein kinase) (Zohn et al. 2006). In many cases, downregulation of E-cadherin is associated with the increased expression of N-cadherin. During gastrulation, N-cadherin expression is increased in mesodermal cells and is required for their migration (Yang et al. 2008). This cadherin-switching is essential as both cadherins are required for normal development. Similarly, loss of E-cadherin in carcinoma cells is often coupled to an increased expression of N-cadherin, which increases cell motility. Loss of E-cadherin correlates with high tumor grades and poor prognosis (Perl et al. 1998) and up-regulation of N-cadherin is associated with increased cell motility (Nieman et al. 1999; Suyama et al. 2002).

A decrease in E-cadherin levels is also achieved by internalization, which is followed by recycling or lysosomal degradation of cadherin (Janda et al. 2006). Endocytosis of cadherins relies on cadherin interaction with p120-catenin, which prevents the recruitment of the clathrin coat. Phosphorylation of cadherin or p120-catenin impacts on p120-catenin binding and the amount of cadherin at the plasma membrane (Xiao et al. 2003; Xiao et al. 2005). Phosphorylation of E-cadherin by Src induces the dissociation of p120-catenin from E-cadherin to promote the binding of the cbl-like ubiquitin ligase Hakai (McLachlan et al. 2007). This results in E-cadherin ubiquitination followed by its internalization in clathrin-coated endosomes and by increased cell motility (Fujita et al. 2002; Palacios et al. 2005; Pece and Gutkind 2002). Activation of AJ-associated RTKs by their ligands is also likely to modulate, through phosphorylation of the cadherin complex, E-cadherin endocytosis as well as the strength of cadherin-mediated adhesion. Accordingly, FGF (Fibroblast Growth Factor) has been involved in mesodermal cell migration during gastrulation possibly through the regulation of cadherin endocytosis (Krens et al. 2008; Rottinger et al. 2008). For example, TGF-β signalling modulates E-cadherin endocytosis via the small GTPase Rnd1 in Xenopus (Ogata et al. 2007). Moreover, following VEGF stimulation, VE-cadherin is phosphorylated on S665 by Rac and Cdc42 effector PAK, which stimulates VE-cadherin internalization (Gavard and Gutkind 2006). VE-cadherin internalisation in clathrin-coated vesicles is induced by β-arrestin interaction with VE-cadherin cytoplasmic tail. The small GTPase Cdc42 has also been implicated in cadherin internalization. In the *Drosophila* pupal notum, CIP4 (Cdc42-interacting protein) associates with WASp and dynamin and contributes to E-cadherin endocytosis together with the WASp effector Arp2/3 (Leibfried et al. 2008).

Even when cadherin remains at the plasma membrane, stimulation of contractile force can sufficiently disrupt cell–cell adhesions to induce cell scattering, as upon stimulation by HGF (Hepatocyte Growth Factor) (de Rooij et al. 2005). The link between cadherins and the actin cytoskeleton can be modulated by several signalling pathways. EPB41L5, a protein of the band 4.1 superfamily, inhibits cadherin interaction with p120-catenin and destabilizes the cadherin-catenin complex to fa-

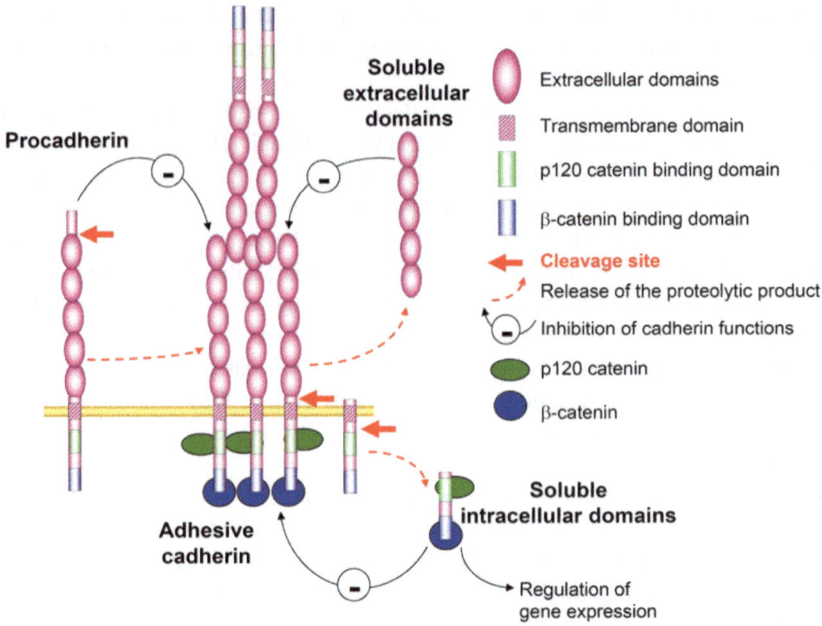

Fig. 10.5 Cadherin processing and shedding. Cadherins are initially produced as procadherins, which when expressed at the cell surface, inhibit the clustering of adhesive cadherins. Cleavage of the 130 amino terminal residues leads to the formation of adhesive cadherins which can interact both in *cis* and in *trans* with identical cadherins to form adherens junctions. Proteolytic cleavage near the transmembrane domain release soluble extracellular and intracellular fragments which both interfere with cadherin functions

vour its internalization (Hirano et al. 2008). The non-receptor tyrosine kinase Fer, which interacts with p120-catenin, stabilizes β-catenin-cadherin interactions (Xu et al. 2004). In general, phosphorylation of β-catenin affects its association with both cadherin and α-catenin (Coluccia et al. 2006; Coluccia et al. 2007; Daugherty and Gottardi 2007; Kajiguchi et al. 2007; Zinser et al. 2006).

Other changes that can affect the stability of cell–cell contacts and influence cell migration include expression of cadherin precursors and shedding of cadherins and nectin-1 by proteolytic processing, which all inhibit AJs (Fig. 10.5). Cadherin processing first occurs when the amino terminal prodomain of 130 amino acids is cleaved to generate the functional cadherin at the cell surface (Ozawa and Kemler 1990). The presence, at the plasma membrane, of a significant amount of the non-adhesive N-cadherin precursor may promote cell detachment and glioma cell migration (Maret et al. 2010). Additional proteolytic cleavage of surface cadherin releases the extracellular domain. This cadherin shedding has been observed in multiple developmental processes (McCusker et al. 2009; Seifert et al. 2009). Cadherin fragments are also found in the serum from cancer patients and cadherin cleavage has been observed in cultured cancer cells (De Wever et al. 2007). Several proteases can cleave cadherins near the transmembrane domain to release a 80kDa fragment corresponding to the entire cadherin ectodomain (Seifert et al. 2009). ADAM10 plays an essential role in E-cadherin and N-cadherin shedding (Maretzky et al. 2005; Reiss et al. 2005). The extracellular domain of E-cadherin is cleaved by MMP-3 (Xian et al. 2005). Interestingly MMP-3 is induced by N-cadherin signalling (Suyama et al. 2002). E-cadherin can also be cleaved by Kallikreins and Meprinβ (Huguenin et al. 2008; Klúcky et al. 2007). Cadherin cleavage has several consequences which all tend to weaken cell–cell interactions and promote cell migration. In addition to the loss of the adhesive functions of the cleaved protein, release of a soluble extracellular domain can perturb the interaction between full length cadherins. Altogether, these effects tend to weaken cell–cell adhesion and promote cell migration and invasion (Niessen et al. 2011). Moreover, cadherin shedding can also modify cadherin signalling either through the binding of the soluble fragment to surface cadherins or by the generation of membrane associated cytoplasmic domains. Intracellular proteases further cleave the cytoplasmic tail of cadherins to release soluble polypeptides with potential consequences on the regulation of gene expression (Marambaud et al. 2003; Shoval et al. 2007) (Fig. 10.5).

10.5 Summary and Future Perspectives

AJs can have distinct and apparently opposite effects on cell migration. Depending on the cell type and the cellular microenvironment, they can support single cell migration, coordinate collective cell migration or participate in contact inhibition of cell migration. Although the mechanisms that underlie each of these cellular responses are beginning to be understood, variations in the cell response to cadherin interactions still have to be investigated in detail. One of the most obvious hypotheses is that the nature of the cadherins involved in AJs may influence signalling cascades. This hypothesis could explain how a cadherin switch may dramatically modify the motility of epithelial cells. However, most cadherins seem to share similar intracellular partners and expression of one particular cadherin has different effects on cell behaviour depending on the cell type. For instance, whereas N-cadherin expression facilitates epithelial cell migration, it inhibits the motility of endothelial cells or astrocytes (Luo and Radice 2005); (Camand et al. 2011). The combination of particular cadherins with other transmembrane receptors is more likely to generate the large variety of signals. Receptors that could contribute are likely to fall in the growth factor receptor family or to be integrins involved in focal adhesions. The localization of cadherins in different domains of the plasma membrane may also play a significant role. In endothelial cells, for instance, VE-cadherin localizes at endothelial-endothelial AJs while N-cadherin is excluded from these contacts and is thus likely to interact with different extracellular and transmembrane partners. Another possibility is that quantitative differences in cadherin levels may induce different cellular responses, in particular during pathfinding or cell sorting (Takeichi 1990). An alternative parameter is the regulation of cadherin turnover in the cell, which may depend on cadherin-independent pathways. Cadherin turnover is likely to be involved in the control of cell migration versus inhibition of cell motility. It is tempting to speculate that high levels of surface cadherins would keep cells together while small amounts of cadherins quickly recycling between immature AJs would favour cell migration. Finally, the physiological importance of cadherin proteolytic processing remains to be clarified as it may strongly affect cadherin functions independently of the surface expression of the protein.

Acknowledgments I thank Florent Peglion for his immunofluorescence images shown in the figures. This work was supported by the Agence Nationale pour la recherche, the Institut Pasteur, Association pour la Recherche sur le Cancer, La Ligue contre le Cancer, and l'Institut National du Cancer. S.E.M. is a member of the EMBO YIP.

References

Abe K, Takeichi M (2008) EPLIN mediates linkage of the cadherin catenin complex to F-actin and stabilizes the circumferential actin belt. Proc Natl Acad Sci U S A 105:13–19

Abercrombie M (1970) Contact inhibition in tissue culture. In Vitro 6:128–142

Abraham S, Yeo M, Montero-Balaguer M, Paterson H, Dejana E, Marshall CJ, Mavria G (2009) VE-Cadherin-mediated cell-cell interaction suppresses sprouting via signaling to MLC2 phosphorylation. Curr Biol 19:668–674

Adams CL, Nelson WJ (1998) Cytomechanics of cadherin-mediated cell-cell adhesion. Curr Opin Cell Biol 10:572–577

Almagro S, Durmort C, Chervin-Petinot A, Heyraud S, Dubois M, Lambert O, Maillefaud C, Hewat E, Schaal JP, Huber P, Gulino-Debrac D (2010) The motor protein myosin-X transports VE-cadherin along filopodia to allow the formation of early endothelial cell-cell contacts. Mol Cell Biol 30:1703–1717

Anastasiadis PZ, Moon SY, Thoreson MA , Mariner DJ, Crawford HC, Zheng Y, Reynolds AB (2000) Inhibition of RhoA by p120 catenin. Nat Cell Biol 2:637–644

Arboleda-Estudillo Y, Krieg M, Stuhmer J, Licata NA, Muller DJ, Heisenberg CP (2010) Movement directionality in collective migration of germ layer progenitors. Curr Biol 20:161–169

Balzac F, Avolio M, Degani S, Kaverina I, Torti M, Silengo L, Small JV, Retta SF (2005) E-cadherin endocytosis regulates the activity of Rap1: a traffic light GTPase at the crossroads between cadherin and integrin function. J Cell Sci 118:4765–4783

Bard L, Boscher C, Lambert M, Mege RM, Choquet D, Thoumine O (2008) A molecular clutch between the actin flow and N-cadherin adhesions drives growth cone migration. J Neurosci 28:5879–5890

Barrallo-Gimeno A, Nieto MA (2005) The Snail genes as inducers of cell movement and survival: implications in development and cancer. Development 132:3151–3161

Baum B, Settleman J, Quinlan MP (2008) Transitions between epithelial and mesenchymal states in development and disease. Semin Cell Dev Biol 19:294–308

Boettner B, Van Aelst L (2007) The Rap GTPase activator Drosophila PDZ-GEF regulates cell shape in epithelial migration and morphogenesis. Mol Cell Biol 27:7966–7980

Borghi N, Lowndes M, Maruthamuthu V, Gardel ML, Nelson WJ (2010) Regulation of cell motile behavior by crosstalk between cadherin- and integrin-mediated adhesions. Proc Natl Acad Sci U S A 107(30):13199–13200. doi:10.1073/pnas.1008253107

Boscher C, Mege RM (2008) Cadherin-11 interacts with the FGF receptor and induces neurite outgrowth through associated downstream signalling. Cell Signal 20:1061–1072
Camand E, Peglion F, Osmani N et al (2011) N-cadherin expression level modulates integrin-mediated polarity and strongly impacts on the speed and directionality of glial cell migration J Cell Sci (In press)
Carmeliet P, Lampugnani MG, Moons L, Breviario F, Compernolle V, Bono F, Balconi G, Spagnuolo R, Oosthuyse B, Dewerchin M, Zanetti A, Angellilo A, Mattot V, Nuyens D, Lutgens E, Clotman F, de Ruiter MC, Gittenberger-de Groot A, Poelmann R, Lupu F, Herbert JM, Collen D, Dejana E (1999) Targeted deficiency or cytosolic truncation of the VE-cadherin gene in mice impairs VEGF-mediated endothelial survival and angiogenesis. Cell 98:147–157
Charrasse S, Meriane M, Comunale F, Blangy A, Gauthier-Rouviere C (2002) N-cadherin-dependent cell-cell contact regulates Rho GTPases and beta-catenin localization in mouse C2C12 myoblasts. J Cell Biol 158:953–965
Chen X, Kojima S, Borisy GG, Green KJ (2003) p120 catenin associates with kinesin and facilitates the transport of cadherin-catenin complexes to intercellular junctions. J Cell Biol 163:547–557
Chu YS, Thomas WA, Eder O, Pincet F, Perez E, Thiery JP, Dufour S (2004) Force measurements in E-cadherin-mediated cell doublets reveal rapid adhesion strengthened by actin cytoskeleton remodeling through Rac and Cdc42. J Cell Biol 167:1183–1194
Ciesiolka M, Delvaeye M, Van Imschoot G, Verschuere V, McCrea P, van Roy F, Vleminckx K (2004) p120 catenin is required for morphogenetic movements involved in the formation of the eyes and the craniofacial skeleton in Xenopus. J Cell Sci 117:4325–4339
Classen AK, Anderson KI, Marois E, Eaton S (2005) Hexagonal packing of Drosophila wing epithelial cells by the planar cell polarity pathway. Dev Cell 9:805–817
Coluccia AM, Benati D, Dekhil H, De Filippo A, Lan C, Gambacorti-Passerini C (2006) SKI-606 decreases growth and motility of colorectal cancer cells by preventing pp60(c-Src)-dependent tyrosine phosphorylation of beta-catenin and its nuclear signaling. Cancer Res 66:2279–2286
Coluccia AM, Vacca A, Dunach M, Mologni L, Redaelli S, Bustos VH, Benati D, Pinna LA, Gambacorti-Passerini C (2007) Bcr-Abl stabilizes beta-catenin in chronic myeloid leukemia through its tyrosine phosphorylation. Embo J 26:1456–1466
Costa M, Raich W, Agbunag C, Leung B, Hardin J, Priess JR (1998) A putative catenin-cadherin system mediates morphogenesis of the Caenorhabditis elegans embryo. J Cell Biol 141:297–308
Cox RT, Kirkpatrick C, Peifer M (1996) Armadillo is required for adherens junction assembly, cell polarity, and morphogenesis during Drosophila embryogenesis. J Cell Biol 134:133–148
Daugherty RL, Gottardi CJ (2007) Phospho-regulation of Beta-catenin adhesion and signaling functions. Physiology (Bethesda) 22:303–309
de Rooij J, Kerstens A, Danuser G, Schwartz MA, Waterman-Storer CM (2005) Integrin-dependent actomyosin contraction regulates epithelial cell scattering. J Cell Biol 171:153–164
De Wever O, Derycke L, Hendrix A, De Meerleer G, Godeau F, Depypere H, Bracke M (2007) Soluble cadherins as cancer biomarkers. Clin Exp Metastasis 24:685–697
Desai RA, Gao L, Raghavan S, Liu WF, Chen CS (2009) Cell polarity triggered by cell-cell adhesion via E-cadherin. J Cell Sci 122:905–911
Drees F, Pokutta S, Yamada S, Nelson WJ, Weis WI (2005) Alpha-catenin is a molecular switch that binds E-cadherin-beta-catenin and regulates actin-filament assembly. Cell 123:903–915
Dupin I, Camand E, Etienne-Manneville S (2009) Classical cadherins control nucleus and centrosome position and cell polarity. J Cell Biol 185:779–786
Dupin I, Sakamoto Y, Etienne-Manneville S (2011) Cytoplasmic intermediate filaments mediate actin-driven positioning of the nucleus. J Cell Sci 124:865–872
Etienne-Manneville S (2011) Control of polarized cell morphology and motility by adherens junctions. Semin Cell Dev Biol 22(8):850–857
Ewald AJ, Brenot A, Duong M, Chan BS, Werb Z (2008) Collective epithelial migration and cell rearrangements drive mammary branching morphogenesis. Dev Cell 14:570–581

Friedl P, Gilmour D (2009) Collective cell migration in morphogenesis, regeneration and cancer. Nat Rev Mol Cell Biol 10:445–457

Friedl P, Wolf K (2009) Proteolytic interstitial cell migration: a five-step process. Cancer Metastasis Rev 28:129–135

Fujita Y, Krause G, Scheffner M, Zechner D, Leddy HE, Behrens J, Sommer T, Birchmeier W (2002) Hakai, a c-Cbl-like protein, ubiquitinates and induces endocytosis of the E-cadherin complex. Nat Cell Biol 4:222–231

Gaggioli C, Hooper S, Hidalgo-Carcedo C, Grosse R, Marshall JF, Harrington K, Sahai E (2007) Fibroblast-led collective invasion of carcinoma cells with differing roles for RhoGTPases in leading and following cells. Nat Cell Biol 9:1392–1400

Ganz A, Lambert M, Saez A, Silberzan P, Buguin A, Mege RM, Ladoux B (2006) Traction forces exerted through N-cadherin contacts. Biol Cell 98:721–730

Gavard J, Gutkind JS (2006) VEGF controls endothelial-cell permeability by promoting the beta-arrestin-dependent endocytosis of VE-cadherin. Nat Cell Biol 8:1223–1234

Giannone G, Mege RM, Thoumine O (2009) Multi-level molecular clutches in motile cell processes. Trends Cell Biol 19:475–486

Harris TJ, Sawyer JK, Peifer M (2009) How the cytoskeleton helps build the embryonic body plan: models of morphogenesis from Drosophila. Curr Top Dev Biol 89:55–85

Harris TJ, Tepass U (2010) Adherens junctions: from molecules to morphogenesis. Nat Rev Mol Cell Biol 11:502–514

Hidalgo-Carcedo C, Hooper S, Chaudhry SI, Williamson P, Harrington K, Leitinger B, Sahai E (2011) Collective cell migration requires suppression of actomyosin at cell-cell contacts mediated by DDR1 and the cell polarity regulators Par3 and Par6. Nat Cell Biol 13:49–58

Hirano M, Hashimoto S, Yonemura S, Sabe H, Aizawa S (2008) EPB41L5 functions to post-transcriptionally regulate cadherin and integrin during epithelial-mesenchymal transition. J Cell Biol 182:1217–1230

Honig MG, Petersen GG, Rutishauser US, Camilli SJ (1998) In vitro studies of growth cone behavior support a role for fasciculation mediated by cell adhesion molecules in sensory axon guidance during development. Dev Biol 204:317–326

Hordijk PL, ten Klooster JP, van der Kammen RA, Michiels F, Oomen LC, Collard JG (1997) Inhibition of invasion of epithelial cells by Tiam1-Rac signaling. Science 278:1464–1466

Huguenin M, Muller EJ, Trachsel-Rosmann S, Oneda B, Ambort D, Sterchi EE, Lottaz D (2008) The metalloprotease meprinbeta processes E-cadherin and weakens intercellular adhesion. PLoS One 3:e2153

Iden S, Collard JG (2008) Crosstalk between small GTPases and polarity proteins in cell polarization. Nat Rev Mol Cell Biol 9:846–859

Iwai Y, Usui T, Hirano S, Steward R, Takeichi M, Uemura T (1997) Axon patterning requires DN-cadherin, a novel neuronal adhesion receptor, in the Drosophila embryonic CNS. Neuron 19:77–89

Jacinto A, Wood W, Woolner S, Hiley C, Turner L, Wilson C, Martinez-Arias A, Martin P (2002) Dynamic analysis of actin cable function during Drosophila dorsal closure. Curr Biol 12:1245–1250

Janda E, Nevolo M, Lehmann K, Downward J, Beug H, Grieco M (2006) Raf plus TGFbeta-dependent EMT is initiated by endocytosis and lysosomal degradation of E-cadherin. Oncogene 25:7117–7130

Jarrett O, Stow JL, Yap AS, Key B (2002) Dynamin-dependent endocytosis is necessary for convergent-extension movements in Xenopus animal cap explants. Int J Dev Biol 46:467–473

Jossin Y, Cooper JA (2011) Reelin, Rap1 and N-cadherin orient the migration of multipolar neurons in the developing neocortex. Nat Neurosci 14:697–703

Kajiguchi T, Chung EJ, Lee S, Stine A, Kiyoi H, Naoe T, Levis MJ, Neckers L, Trepel JB (2007) FLT3 regulates beta-catenin tyrosine phosphorylation, nuclear localization, and transcriptional activity in acute myeloid leukemia cells. Leukemia 21:2476–2484

Kametani Y, Takeichi M (2007) Basal-to-apical cadherin flow at cell junctions. Nat Cell Biol 9:92–98

Klucky B, Mueller R, Vogt I, Teurich S, Hartenstein B, Breuhahn K, Flechtenmacher C, Angel P, Hess J (2007) Kallikrein 6 induces E-cadherin shedding and promotes cell proliferation, migration, and invasion. Cancer Res 67:8198–8206

Kobielak A, Fuchs E (2004) Alpha-catenin: at the junction of intercellular adhesion and actin dynamics. Nat Rev Mol Cell Biol 5:614–625

Kobielak A, Pasolli HA, Fuchs E (2004) Mammalian formin-1 participates in adherens junctions and polymerization of linear actin cables. Nat Cell Biol 6:21–30

Kolkova K, Novitskaya V, Pedersen N, Berezin V, Bock E (2000) Neural cell adhesion molecule-stimulated neurite outgrowth depends on activation of protein kinase C and the Ras-mitogen-activated protein kinase pathway. J Neurosci 20:2238–2246

Kovacs EM, Ali RG, McCormack AJ, Yap AS (2002) E-cadherin homophilic ligation directly signals through Rac and phosphatidylinositol 3-kinase to regulate adhesive contacts. J Biol Chem 277:6708–6718

Krens SF, He S, Lamers GE, Meijer AH, Bakkers J, Schmidt T, Spaink HP, Snaar-Jagalska BE (2008) Distinct functions for ERK1 and ERK2 in cell migration processes during zebrafish gastrulation. Dev Biol 319:370–383

Ladoux B, Anon E, Lambert M, Rabodzey A, Hersen P, Buguin A, Silberzan P, Mege RM (2010) Strength dependence of cadherin-mediated adhesions. Biophys J 98:534–542

le Duc Q, Shi Q, Blonk I, Sonnenberg A, Wang N, Leckband D, de Rooij J (2010) Vinculin potentiates E-cadherin mechanosensing and is recruited to actin-anchored sites within adherens junctions in a myosin II-dependent manner. J Cell Biol 189:1107–1115

Leibfried A, Fricke R, Morgan MJ, Bogdan S, Bellaiche Y (2008) Drosophila Cip4 and WASp define a branch of the Cdc42-Par6-aPKC pathway regulating E-cadherin endocytosis. Curr Biol 18:1639–1648

Lele Z, Folchert A, Concha M, Rauch GJ, Geisler R, Rosa F, Wilson SW, Hammerschmidt M, Bally-Cuif L (2002) parachute/n-cadherin is required for morphogenesis and maintained integrity of the zebrafish neural tube. Development 129:3281–3294

Ligon LA, Karki S, Tokito M, Holzbaur EL (2001) Dynein binds to beta-catenin and may tether microtubules at adherens junctions. Nat Cell Biol 3:913–917

Liu Z, Tan JL, Cohen DM, Yang MT, Sniadecki NJ, Ruiz SA, Nelson CM, Chen CS (2010) Mechanical tugging force regulates the size of cell-cell junctions. Proc Natl Acad Sci U S A 107:9944–9949

Luo Y, Radice GL (2005) N-cadherin acts upstream of VE-cadherin in controlling vascular morphogenesis. J Cell Biol 169:29–34

Macara IG (2004) Par proteins: Partners in polarization. Curr Biol 14:R160–R162

Maddugoda MP, Crampton MS, Shewan AM, Yap AS (2007) Myosin VI and vinculin cooperate during the morphogenesis of cadherin cell cell contacts in mammalian epithelial cells. J Cell Biol 178:529–540

Marambaud P, Wen PH, Dutt A, Shioi J, Takashima A, Siman R, Robakis NK (2003) A CBP binding transcriptional repressor produced by the PS1/epsilon-cleavage of N-cadherin is inhibited by PS1 FAD mutations. Cell 114:635–645

Maret D, Gruzglin E, Sadr MS, Siu V, Shan W, Koch AW, Seidah NG, Del Maestro RF, Colman DR (2010) Surface expression of precursor N-cadherin promotes tumor cell invasion. Neoplasia 12:1066–1080

Maretzky T, Reiss K, Ludwig A, Buchholz J, Scholz F, Proksch E, de Strooper B, Hartmann D, Saftig P (2005) ADAM10 mediates E-cadherin shedding and regulates epithelial cell-cell adhesion, migration, and beta-catenin translocation. Proc Natl Acad Sci U S A 102:9182–9187

Martin P, Parkhurst SM (2003) Development. May the force be with you. Science 300:63–65

Maruthamuthu V, Aratyn-Schaus Y, Gardel ML (2010) Conserved F-actin dynamics and force transmission at cell adhesions. Curr Opin Cell Biol 22:583–588

Maruthamuthu V, Sabass B, Schwarz US, Gardel ML (2011) Cell-ECM traction force modulates endogenous tension at cell-cell contacts. Proc Natl Acad Sci U S A 108:4708–4713

McCusker C, Cousin H, Neuner R, Alfandari D (2009) Extracellular cleavage of cadherin-11 by ADAM metalloproteases is essential for Xenopus cranial neural crest cell migration. Mol Biol Cell 20:78–89

McLachlan RW, Kraemer A, Helwani FM, Kovacs EM, Yap AS (2007) E-cadherin adhesion activates c-Src signaling at cell-cell contacts. Mol Biol Cell 18:3214–3223

McNeill H, Ryan TA, Smith SJ, Nelson WJ (1993) Spatial and temporal dissection of immediate and early events following cadherin-mediated epithelial cell adhesion. J Cell Biol 120:1217–1226

Melani M, Simpson KJ, Brugge JS, Montell D (2008) Regulation of cell adhesion and collective cell migration by hindsight and its human homolog RREB1. Curr Biol 18:532–537

Meng W, Mushika Y, Ichii T, Takeichi M (2008) Anchorage of microtubule minus ends to adherens junctions regulates epithelial cell-cell contacts. Cell 135:948–959

Meng W, Takeichi M (2009) Adherens junction: molecular architecture and regulation. Cold Spring Harb Perspect Biol 1:a002899

Mishima A, Suzuki A, Enaka M, Hirose T, Mizuno K, Ohnishi T, Mohri H, Ishigatsubo Y, Ohno S (2002) Over-expression of PAR-3 suppresses contact-mediated inhibition of cell migration in MDCK cells. Genes Cells 7:581–596

Miyake Y, Inoue N, Nishimura K, Kinoshita N, Hosoya H, Yonemura S (2006) Actomyosin tension is required for correct recruitment of adherens junction components and zonula occludens formation. Exp Cell Res 312:1637–1650

Moh MC, Shen S (2009) The roles of cell adhesion molecules in tumor suppression and cell migration: a new paradox. Cell Adh Migr 3:334–336

Nakanishi H, Takai Y (2004) Roles of nectins in cell adhesion, migration and polarization. Biol Chem 385:885–892

Nieman MT, Prudoff RS, Johnson KR, Wheelock MJ (1999) N-cadherin promotes motility in human breast cancer cells regardless of their E-cadherin expression. J Cell Biol 147:631–644

Niessen CM, Gottardi CJ (2008) Molecular components of the adherens junction. Biochim Biophys Acta 1778:562–571

Niessen CM, Leckband D, Yap AS (2011) Tissue organization by cadherin adhesion molecules: dynamic molecular and cellular mechanisms of morphogenetic regulation. Physiol Rev 91:691–731

Niewiadomska P, Godt D, Tepass U (1999) DE-Cadherin is required for intercellular motility during Drosophila oogenesis. J Cell Biol 144:533–547

Nishimura T, Kaibuchi K (2007) Numb controls integrin endocytosis for directional cell migration with aPKC and PAR-3. Dev Cell 13:15–28

Oblander SA, Brady-Kalnay SM (2010) Distinct PTPmu-associated signaling molecules differentially regulate neurite outgrowth on E-, N-, and R-cadherin. Mol Cell Neurosci 44:78–93

Oda H, Takeichi M (2011) Evolution: structural and functional diversity of cadherin at the adherens junction. J Cell Biol 193:1137–1146

Ogata S, Morokuma J, Hayata T, Kolle G, Niehrs C, Ueno N, Cho KW (2007) TGF-beta signaling-mediated morphogenesis: modulation of cell adhesion via cadherin endocytosis. Genes Dev 21:1817–1831

Ozawa M, Kemler R (1990) Correct proteolytic cleavage is required for the cell adhesive function of uvomorulin. J Cell Biol 111:1645–1650

Palacios F, Tushir JS, Fujita Y, D'Souza-Schorey C (2005) Lysosomal targeting of E-cadherin: a unique mechanism for the down-regulation of cell-cell adhesion during epithelial to mesenchymal transitions. Mol Cell Biol 25:389–402

Pannekoek WJ, Kooistra MR, Zwartkruis FJ, Bos JL (2009) Cell-cell junction formation: the role of Rap1 and Rap1 guanine nucleotide exchange factors. Biochim Biophys Acta 1788:790–796

Pasapera AM, Schneider IC, Rericha E, Schlaepfer DD, Waterman CM (2010) Myosin II activity regulates vinculin recruitment to focal adhesions through FAK-mediated paxillin phosphorylation. J Cell Biol 188:877–890

Pece S, Gutkind JS (2000) Signaling from E-cadherins to the MAPK pathway by the recruitment and activation of epidermal growth factor receptors upon cell-cell contact formation. J Biol Chem 275:41227–41233

Pece S, Gutkind JS (2002) E-cadherin and Hakai: signalling, remodeling or destruction? Nat Cell Biol 4:E72–E74

Perez-Moreno M, Jamora C, Fuchs E (2003) Sticky business: orchestrating cellular signals at adherens junctions. Cell 112:535–548

Perez TD, Tamada M, Sheetz MP, Nelson WJ (2008) Immediate-early signaling induced by E-cadherin engagement and adhesion. J Biol Chem 283:5014–5022

Perl AK, Wilgenbus P, Dahl U, Semb H, Christofori G (1998) A causal role for E-cadherin in the transition from adenoma to carcinoma. Nature 392:190–193

Pokutta S, Drees F, Takai Y, Nelson WJ, Weis WI (2002) Biochemical and structural definition of the l-afadin- and actin-binding sites of alpha-catenin. J Biol Chem 277:18868–18874

Redies C (2000) Cadherins in the central nervous system. Prog Neurobiol 61:611–648

Reiss K, Maretzky T, Ludwig A, Tousseyn T, de Strooper B, Hartmann D, Saftig P (2005) ADAM10 cleavage of N-cadherin and regulation of cell-cell adhesion and beta-catenin nuclear signalling. Embo J 24:742–752

Ren G, Helwani FM, Verma S, McLachlan RW, Weed SA, Yap AS (2009) Cortactin is a functional target of E-cadherin-activated Src family kinases in MCF7 epithelial monolayers. J Biol Chem 284:18913–18922

Riehl R, Johnson K, Bradley R, Grunwald GB, Cornel E, Lilienbaum A, Holt CE (1996) Cadherin function is required for axon outgrowth in retinal ganglion cells in vivo. Neuron 17:837–848

Rimm DL, Koslov ER, Kebriaei P, Cianci CD, Morrow JS (1995) Alpha 1(E)-catenin is an actin-binding and -bundling protein mediating the attachment of F-actin to the membrane adhesion complex. Proc Natl Acad Sci U S A 92:8813–8817

Roh MH, Margolis B (2003) Composition and function of PDZ protein complexes during cell polarization. Am J Physiol Renal Physiol 285:F377–F387

Rorth P (2009) Collective cell migration. Annu Rev Cell Dev Biol 25:407–429

Rottinger E, Saudemont A, Duboc V, Besnardeau L, McClay D, Lepage T (2008) FGF signals guide migration of mesenchymal cells, control skeletal morphogenesis [corrected] and regulate gastrulation during sea urchin development. Development 135:353–365

Rudini N, Felici A, Giampietro C, Lampugnani M, Corada M, Swirsding K, Garre M, Liebner S, Letarte M, ten Dijke P, Dejana E (2008) VE-cadherin is a critical endothelial regulator of TGF-beta signalling. Embo J 27:993–1004

Sakamoto Y, Ogita H, Hirota T, Kawakatsu T, Fukuyama T, Yasumi M, Kanzaki N, Ozaki M, Takai Y (2006) Interaction of integrin alpha(v)beta3 with nectin. Implication in cross-talk between cell-matrix and cell-cell junctions. J Biol Chem 281:19631–19644

Sakamoto Y, Ogita H, Komura H, Takai Y (2008) Involvement of nectin in inactivation of integrin alpha(v)beta(3) after the establishment of cell-cell adhesion. J Biol Chem 283:496–505

Sawyer JK, Harris NJ, Slep KC, Gaul U, Peifer M (2009) The Drosophila afadin homologue Canoe regulates linkage of the actin cytoskeleton to adherens junctions during apical constriction. J Cell Biol 186:57–73

Schmoranzer J, Fawcett JP, Segura M, Tan S, Vallee RB, Pawson T, Gundersen GG (2009) Par3 and dynein associate to regulate local microtubule dynamics and centrosome orientation during migration. Curr Biol 19:1065–1074

Seifert K, Ibrahim H, Stodtmeister T, Winklbauer R, Niessen CM (2009) An adhesion-independent, aPKC-dependent function for cadherins in morphogenetic movements. J Cell Sci 122:2514–2523

Shaw RM, Fay AJ, Puthenveedu MA, von Zastrow M, Jan YN, Jan LY (2007) Microtubule plus-end-tracking proteins target gap junctions directly from the cell interior to adherens junctions. Cell 128:547–560

Shaye DD, Casanova J, Llimargas M (2008) Modulation of intracellular trafficking regulates cell intercalation in the Drosophila trachea. Nat Cell Biol 10:964–970

Shoval I, Ludwig A, Kalcheim C (2007) Antagonistic roles of full-length N-cadherin and its soluble BMP cleavage product in neural crest delamination. Development 134:491–501

Singleton PA, Dudek SM, Chiang ET, Garcia JG (2005) Regulation of sphingosine 1-phosphate-induced endothelial cytoskeletal rearrangement and barrier enhancement by S1P1 receptor, PI3 kinase, Tiam1/Rac1, and alpha-actinin. Faseb J 19:1646–1656

Skoglund P, Rolo A, Chen X, Gumbiner BM, Keller R (2008) Convergence and extension at gastrulation require a myosin IIB-dependent cortical actin network. Development 135:2435–2444

Smutny M, Yap AS (2010) Neighborly relations: cadherins and mechanotransduction. J Cell Biol 189:1075–1077

Sousa S, Cabanes D, El-Amraoui A, Petit C, Lecuit M, Cossart P (2004) Unconventional myosin VIIa and vezatin, two proteins crucial for Listeria entry into epithelial cells. J Cell Sci 117:2121–2130

Sperry RB, Bishop NH, Bramwell JJ, Brodeur MN, Carter MJ, Fowler BT, Lewis ZB, Maxfield SD, Staley DM, Vellinga RM, Hansen MD (2010) Zyxin controls migration in epithelial-mesenchymal transition by mediating actin-membrane linkages at cell-cell junctions. J Cell Physiol 222:612–624

Stehbens SJ, Paterson AD, Crampton MS, Shewan AM, Ferguson C, Akhmanova A, Parton RG, Yap AS (2006) Dynamic microtubules regulate the local concentration of E-cadherin at cell-cell contacts. J Cell Sci 119:1801–1811

Stepniak E, Radice GL, Vasioukhin V (2009) Adhesive and signaling functions of cadherins and catenins in vertebrate development. Cold Spring Harb Perspect Biol 1:a002949

Suyama K, Shapiro I, Guttman M, Hazan RB (2002) A signaling pathway leading to metastasis is controlled by N-cadherin and the FGF receptor. Cancer Cell 2:301–314

Tachibana K, Nakanishi H, Mandai K, Ozaki K, Ikeda W, Yamamoto Y, Nagafuchi A, Tsukita S, Takai Y (2000) Two cell adhesion molecules, nectin and cadherin, interact through their cytoplasmic domain-associated proteins. J Cell Biol 150:1161–1176

Taddei A, Giampietro C, Conti A, Orsenigo F, Breviario F, Pirazzoli V, Potente M, Daly C, Dimmeler S, Dejana E (2008) Endothelial adherens junctions control tight junctions by VE-cadherin-mediated upregulation of claudin-5. Nat Cell Biol 10:923–934

Takahashi K, Nakanishi H, Miyahara M, Mandai K, Satoh K, Satoh A, Nishioka H, Aoki J, Nomoto A, Mizoguchi A, Takai Y (1999) Nectin/PRR: an immunoglobulin-like cell adhesion molecule recruited to cadherin-based adherens junctions through interaction with Afadin, a PDZ domain-containing protein. J Cell Biol 145:539–549

Takai Y, Irie K, Shimizu K, Sakisaka T, Ikeda W (2003) Nectins and nectin-like molecules: roles in cell adhesion, migration, and polarization. Cancer Sci 94:655–667

Takeichi M (1988) The cadherins: cell-cell adhesion molecules controlling animal morphogenesis. Development 102:639–655

Takeichi M (1990) Cadherins: a molecular family important in selective cell-cell adhesion. Annu Rev Biochem 59:237–252

Theveneau E, Marchant L, Kuriyama S, Gull M, Moepps B, Parsons M, Mayor R (2010) Collective chemotaxis requires contact-dependent cell polarity. Dev Cell 19:39–53

Thiery JP, Sleeman JP (2006) Complex networks orchestrate epithelial-mesenchymal transitions. Nat Rev Mol Cell Biol 7:131–142

Togashi H, Miyoshi J, Honda T, Sakisaka T, Takai Y, Takeichi M (2006) Interneurite affinity is regulated by heterophilic nectin interactions in concert with the cadherin machinery. J Cell Biol 174:141–151

Tohyama Y, Katagiri K, Pardi R, Lu C, Springer TA, Kinashi T (2003) The critical cytoplasmic regions of the alphaL/beta2 integrin in Rap1-induced adhesion and migration. Mol Biol Cell 14:2570–2582

Trepat X, Wasserman MR, Angelini TE, Millet E, Weitz DA, Butler JP, Fredberg JJ (2009) Physical forces during collective cell migration. Nat Phys 5:426–430

Tsai IC, Amack JD, Gao ZH, Band V, Yost HJ, Virshup DM (2007) A Wnt-CKIvarepsilon-Rap1 pathway regulates gastrulation by modulating SIPA1L1, a Rap GTPase activating protein. Dev Cell 12:335–347

Vasioukhin V, Bauer C, Yin M, Fuchs E (2000) Directed actin polymerization is the driving force for epithelial cell-cell adhesion. Cell 100:209–219

Vasioukhin V, Fuchs E (2001) Actin dynamics and cell-cell adhesion in epithelia. Curr Opin Cell Biol 13:76–84

Weijer CJ (2009) Collective cell migration in development. J Cell Sci 122:3215–3223

Xian W, Schwertfeger KL, Vargo-Gogola T, Rosen JM (2005) Pleiotropic effects of FGFR1 on cell proliferation, survival, and migration in a 3D mammary epithelial cell model. J Cell Biol 171:663–673

Xiao K, Allison DF, Buckley KM, Kottke MD, Vincent PA, Faundez V, Kowalczyk AP (2003) Cellular levels of p120 catenin function as a set point for cadherin expression levels in microvascular endothelial cells. J Cell Biol 163:535–545

Xiao K, Garner J, Buckley KM, Vincent PA, Chiasson CM, Dejana E, Faundez V, Kowalczyk AP (2005) p120-Catenin regulates clathrin-dependent endocytosis of VE-cadherin. Mol Biol Cell 16:5141–5151

Xu G, Craig AW, Greer P, Miller M, Anastasiadis PZ, Lilien J, Balsamo J (2004) Continuous association of cadherin with beta-catenin requires the non-receptor tyrosine-kinase Fer. J Cell Sci 117:3207–3219

Yamada S, Nelson WJ (2007) Localized zones of Rho and Rac activities drive initiation and expansion of epithelial cell-cell adhesion. J Cell Biol 178:517–527

Yamada S, Pokutta S, Drees F, Weis WI, Nelson WJ (2005) Deconstructing the cadherin-catenin-actin complex. Cell 123:889–901

Yang X, Chrisman H, Weijer CJ (2008) PDGF signalling controls the migration of mesoderm cells during chick gastrulation by regulating N-cadherin expression. Development 135:3521–3530

Yonemura S, Wada Y, Watanabe T, Nagafuchi A, Shibata M (2010) alpha-Catenin as a tension transducer that induces adherens junction development. Nat Cell Biol 12:533–542

Yu X, Malenka RC (2003) Beta-catenin is critical for dendritic morphogenesis. Nat Neurosci 6:1169–1177

Zallen JA (2007) Planar polarity and tissue morphogenesis. Cell 129:1051–1063

Zhu H, Luo L (2004) Diverse functions of N-cadherin in dendritic and axonal terminal arborization of olfactory projection neurons. Neuron 42:63–75

Zinser GM, Leonis MA, Toney K, Pathrose P, Thobe M, Kader SA, Peace BE, Beauman SR, Collins MH, Waltz SE (2006) Mammary-specific Ron receptor overexpression induces highly metastatic mammary tumors associated with beta-catenin activation. Cancer Res 66:11967–11974

Zohn IE, Li Y, Skolnik EY, Anderson KV, Han J, Niswander L (2006) p38 and a p38-interacting protein are critical for downregulation of E-cadherin during mouse gastrulation. Cell 125:957–969

Chapter 11
Adherens Junctions and Cadherins in *Drosophila* Development

Annalisa Letizia and Marta Llimargas

Abstract *Drosophila* represents a paradigm for the analysis of the cellular, molecular and genetic mechanisms of development and is an ideal model system to study the contribution of Adherens Junctions (AJs) and their major components, cadherins, to morphogenesis. The combination of different techniques and approaches has allowed researchers to identify the requirements of these epithelial junctions *in vivo* in the context of a whole organism. The functional analysis of mutants for AJ core components, particularly for *Drosophila* DE-cadherin, has shown that AJs play critical roles in virtually all stages of development. For instance, AJs maintain tissue integrity while allowing the remodelling and homeostasis of many tissues. They control cell shape, contribute to cell polarity, facilitate cell–cell recognition during cell sorting, orient cell divisions, or regulate cell rearrangements, among other activities. Remarkably, these activities require a very fine control of the organisation and turnover of AJs during development. In addition, AJs engage in diverse and complex interactions with the cytoskeleton, signalling networks, intracellular trafficking machinery or polarity cues to perform these functions. Here, by summarising the requirements of AJs and cadherins during *Drosophila* morphogenesis, we illustrate the capital contribution of this model system to our knowledge of the mechanisms and biology of AJs.

11.1 Introduction

Evidence of the presence of AJs in *Drosophila* was documented many years ago (Eichenberger-Glinz 1979; Poodry and Schneiderman 1970). In 1994 Tepass and Hartenstein published a comprehensive ultrastructural analysis of the pattern and development of the junctions in *Drosophila*, including the AJs (Tepass and Hartenstein 1994b). In parallel to this systematic analysis, work from several labs identi-

M. Llimargas (✉) · A. Letizia
Institut de Biologia Molecular de Barcelona, CSIC, Parc Científic de Barcelona,
Baldiri Reixac 10–12, 08028 Barcelona, Spain
e-mail: mlcbmc@ibmb.csic.es

A. Letizia
e-mail: alebmc@ibmb.csic.es

T. Harris (ed.), *Adherens Junctions: From Molecular Mechanisms to Tissue Development and Disease,* Subcellular Biochemistry 60,
DOI 10.1007/978-94-007-4186-7_11, © Springer Science+Business Media Dordrecht 2012

fied for the first time in *Drosophila* the genes encoding the core components of AJs, namely Armadillo/β-catenin, α-Catenin and Shotgun/Drosophila DE-cadherin (McCrea et al. 1991; Oda et al. 1993, 1994; Peifer 1993; Peifer and Wieschaus 1990).

During the 1990s, researchers generated mutations in several components of AJs. The genetic analysis of these mutants provided some of the first functional evidence for the requirement of AJs during the development of a whole organism (Cox et al. 1996; Peifer et al. 1993; Tepass et al. 1996; Uemura et al. 1996). It was expected from the already known roles of AJs in other models that the loss of function of AJ components would generate strong defects in cell–cell adhesion and tissue integrity. Thus, it was somehow surprising to find only minor defects during embryonic development when AJ components were zygotically depleted. Further analysis indicated a strong maternal contribution of AJ components and revealed a critical specific requirement of AJs in those embryonic tissues engaged in dramatic morphogenetic events, like the ventral ectoderm, Malpighian tubules or tracheal system. By contrast, the maternal contribution of AJ components was sufficient to sustain cell-adhesion and integrity in more static embryonic tissues. In addition, the attempts to deplete completely the maternal contribution of AJ components indicated a strong requirement of AJs during oogenesis. Later, the use of transgenic lines containing different mutated forms of the AJ components helped to characterise and refine the specific tissue and temporal requirements of AJs in development (Pacquelet and Rorth 2005; Wang et al. 2004). In this chapter, we will describe the documented functions of AJs in a temporal order from gonad formation to adult tissues. We will pay special attention to the molecular and cellular mechanisms underlying these activities when they are known.

11.2 Gonad Formation and Organisation

Formation of the gonad starts very early in embryonic development and passes through several finely regulated steps in which cell adhesion plays a critical role. Primordial Germ Cells (PGCs) derive from the pole cells, and during gastrulation they form a tight, compact layer of cells that adhere to each other and to the underlying invaginating midgut primordium. The PGCs cluster is enveloped by the surrounding somatic tissue, which brings it into the embryo (Kunwar et al. 2008). This step is an active and regulated process that requires AJ stabilisation. In this case, AJ stabilisation depends on the concentration of H_2O_2, which is regulated by the activity of the peroxiredoxin Jafrac1 (DeGennaro et al. 2011). Once the PGCs are internalised they repolarise, disperse and transmigrate through the midgut epithelium. The repolarisation, which is controlled by G-protein coupled receptor signalling, requires downregulation and redistribution of DE-cadherin within the cell, which may also contribute to orient the transepithelial migration (Kunwar et al. 2008). After PGCs have migrated through the midgut tissue, they organise into two bilateral groups of cells that migrate towards the somatic gonadal precursors (SGPs). The SGPs arise initially as three clusters (4 in males) of cells positioned in parasegments

10–12 that later join to generate a single line of cells that will be contacted by the PGCs (Boyle et al. 1997; Boyle and DiNardo 1995). After contact, long cytoplasmic extensions of SGP cells ensheath each PGC (Jenkins et al. 2003), and finally the structure coalesces to form a compact spherical embryonic gonad that is located in parasegment 10 (Jenkins et al. 2003; Van Doren et al. 2003). Regulated levels of DE-cadherin accumulation play a critical role in allowing the compaction of the SGPs from the rest of the mesoderm to form the gonad, and to promote the ensheathment of PGCs by the SGPs. In these events, a cell sorting mechanism has been proposed (Jenkins et al. 2003; Van Doren et al. 2003). The activity of the transcription factor Eyes Absent and the zinc transporter protein Fear of intimacy trigger increased levels of DE-cadherin in the SGPs relative to the rest of mesoderm, which allows these precursor cells to maximise their contact and then to be sorted and become compacted (Jenkins et al. 2003; Mathews et al. 2006; Van Doren et al. 2003). In addition, higher accumulation of DE-cadherin between the SGPs and the contacting PGCs than between each group of somatic or germ cells would favour the ensheathment of the PGCs by the SGP. The Maf transcription factor Traffic Jam fine-tunes the levels of adhesion between germ and somatic cells to allow their correct intermingling (Li et al. 2003).

Cell adhesion and DE-cadherin again play a key role in the soma-germline cross-talk that ensures gonad organisation and gametogenesis during postembryonic gonad development (Fig. 11.1). DE-cadherin contributes to the recruitment of PGCs as Germline Stem Cells (GSCs) in the female ovaries (Song et al. 2002). Furthermore, increased levels of DE-cadherin and other AJ components are observed at the interface between stem cells (either GSCs or Somatic Stem Cells (SSCs)) and the soma to which they normally attach in both female ovaries and male testis. This increased adhesion (also indicated by the presence of AJs at the ultraestructural level) anchors the stem cells in their niche, allowing them to receive signals that maintain their stemness (Dansereau and Lasko 2008; Song and Xie 2002; Voog et al. 2008; Wang et al. 2006). During gametogenesis GSCs divide asymmetrically to render two daughter cells, one that inherits the AJs and remains in the niche to become a new GSC and another that breaks contact with the niche and initiates differentiation. The asymmetric cell division occurring in male and female GSCs is controlled by the orientation of the spindle, which in turn depends on the position of the AJs (Inaba et al. 2010; Xie 2008; Yamashita 2010; Yamashita et al. 2007) (for a more detailed description see Chap. 15).

11.3 Oogenesis

Drosophila ovaries are typically composed of 16–20 ovarioles that contain chains of oocytes at different stages of maturation. These are defined from stage 1, when they bud as cysts from the germarium (the tip of the ovariole containing the germline and somatic stem cell niche) to stage 14 (mature oocyte). When the oocyte-containing cysts proceed through the germarium they become surrounded by a population of

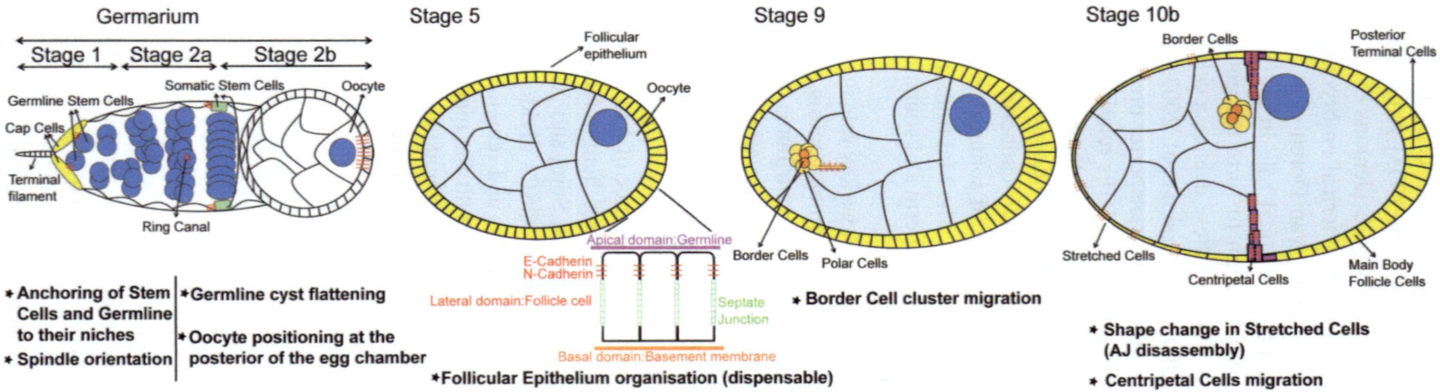

Fig. 11.1 AJs contribution to oocyte development. Diagrams of oocytes at different stages of maturation. AJs (symbolised by three red lines together in this and all figures) play critical and specific roles for the proper formation and development of the egg-chambers. AJs and DE-cadherin modulation are required both in the germline and in the somatic surrounding tissue throughout the whole process. The specific requirements for AJs are indicated at each step

somatic follicular cells and they are separated from contiguous egg chambers by somatic stalk cells. The female cystoblast initiates oogenesis by dividing four times with incomplete cytokinesis, which generates a germline cyst of 16 cells connected by cytoplasmic bridges known as ring canals. The stereotyped pattern of divisions generates two cells with 4 ring canals, known as pro-oocytes. One of these pro-oocytes will differentiate as the oocyte, which invariably positions at the posterior of the egg chamber, and will later proceed through meiosis. The remaining 15 germ cells undergo endoreplication and become nurse cells, which synthesise maternal products that will be dumped into the oocyte. In parallel, the somatic follicular epithelium becomes specified into different populations that perform specialised functions. Cell communication between the soma and germline coordinates the morphogenesis of both tissues (for reviews see Bastock and St Johnston 2008; Horne-Badovinac and Bilder 2005; Huynh and St Johnston 2004; Wu et al. 2008).

The positioning of the oocyte at the posterior of the germline cyst is a critical step to determine the anterio-posterior axis of the future embryo. Adhesion was shown to participate in this process (Fig. 11.1). In a first step, the homotypic adhesion between germline cells ensures that the pro-oocytes contact the follicular epithelium during the flattening of the germline cysts (Gonzalez-Reyes and St Johnston 1998). In a second step, increased accumulation of AJ components, independently at the interface between posterior follicular cells and at the already specified oocyte, anchors the oocyte at the posterior. This step is believed to rely on a sorting mechanism based on this differential heterotypic cell adhesion. The analysis of germline and somatic mosaics for *shotgun* (DE-cadherin) mutants indicated that the oocyte preferentially locates closest to those follicular cells with increased adhesion (Godt and Tepass 1998; Gonzalez-Reyes and St Johnston 1998). Furthermore, clonal analysis of *shotgun* and *armadillo* mutants revealed other germline-specific requirements during oogenesis, like the control of the number, shape and size of germ cells, the dumping of nurse cells contents into the oocyte, the cortical actin cytoskeleton organisation or the ring canal formation (Oda et al. 1997; Peifer et al. 1993; White et al. 1998), although the molecular mechanisms underlying these functions await further analysis.

The follicular cells that encapsulate the germline cyst form a polarised epithelium, the follicular epithelium (FE), with clear apical, lateral and basal domains, all of which provide polarising cues (Fig. 11.1). The basal domain contacts the basement membrane, whereas the AJs of the lateral domain establish adhesive contact between the follicular cells. In contrast to most epithelia in which the apical domain is free of contact, the apical domain of the FE faces and contacts the germline. AJs accumulate DE-cadherin throughout oogenesis, but in addition, *Drosophila* N-cadherin is also present in these AJs up to stage 10 of oogenesis. It was observed that the lack of AJs (due to the absence of *shotgun*, or *armadillo*) produced defects in the maintenance and organisation of the FE, however, it did not prevent the establishment of the integrity of this epithelium (Godt and Tepass 1998; Gonzalez-Reyes and St Johnston 1998; Tanentzapf et al. 2000). This surprising result was interpreted in the light of the particularities of the FE, and it was proposed that other mechanisms (probably the basal adhesion and the apical adhesion to the germline) might compensate for the lack of AJ-mediated cell adhesion in the formation of the FE (Tanentzapf et al. 2000).

By mid-oogenesis the FE differentiates into 5 different cell populations, namely, from anterior to posterior, the border, stretched, centripetal, main body and posterior terminal cells (Fig. 11.1). Each population undergoes particular morphogenetic events, in many of which AJs play a role. The border cells, which are recruited by the anterior polar cells, delaminate from the FE and migrate through the nurse cells to reach the oocyte. Border cells form long cytoplasmic extensions that accumulate high levels of DE-cadherin (Fulga and Rorth 2002). DE-cadherin-mediated adhesion both between border cells and to the germline migratory substratum is required to ensure efficient migration (Niewiadomska et al. 1999; Oda et al. 1997), rather than to maintain border cell cluster cohesiveness. It was shown that increased DE-cadherin accumulation in border cells is genetically controlled (Geisbrecht and Montell 2002; Niewiadomska et al. 1999) and a cytoplasmic domain of DE-cadherin promoting invasive border cell migration was identified (Pacquelet and Rorth 2005). FE cells are initially cuboidal, and most populations adopt a columnar shape, except the stretched cells, which become squamous, with expanded apical and basal domains and reduced lateral membranes. The stretched cells form a thin layer that covers the nurse cells while the main body cells cover the growing oocyte. The cell shape change in the stretched cells is accompanied by a gradual and ordered disassembly of AJs that dictates the orientation of the flattening and correlates with an accumulation of Myosin II. The Notch pathway and the transcription factor Hindsight regulate this process (Grammont 2007; Melani et al. 2008). The centripetal cells migrate inwards between the oocyte and the nurse cells, completely enclosing the oocyte at its anterior region. DE-cadherin accumulation is again required for proper migration of this follicular cell population (Niewiadomska et al. 1999; Oda et al. 1997), and again this increased accumulation is finely controlled at the genetic level (Hackney et al. 2007; Levine et al. 2010).

11.4 Adherens Junctions Biogenesis and Cellularisation

Cellularisation transforms the syncytial blastoderm into a cellular blastoderm formed by around 6000 individual columnar cells organised as an epithelium. During this process the embryo membrane invaginates synchronously between each nucleus positioned at the periphery of the blastoderm, forming the furrow canals at the front tip. When the furrow canals have passed the nuclei they expand laterally to form the basal membrane that separates the newly compartmentalised cell from the yolk. Membrane trafficking and remodelling, as well as microtubule and actin networks are required for these processes (for reviews see Harris et al. 2009; Lecuit 2004; Mazumdar and Mazumdar 2002). Cellularisation is concomitant to AJ assembly (Fig. 11.2), when two different and independent AJs structures form, a basal junction and apical spot AJs. Basal junctions form at early cellularisation stages just behind the tip of the furrow canals and their assembly requires the activity of Nullo

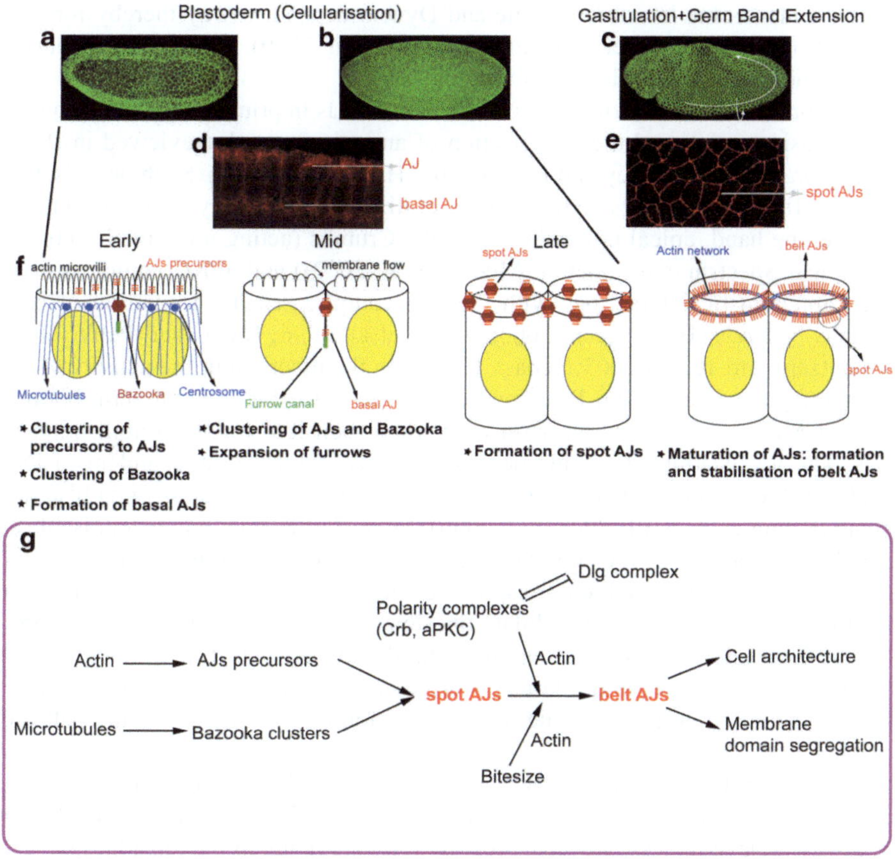

Fig. 11.2 AJs biogenesis and maturation. **a–c** Projections of confocal sections showing whole mount embryos in lateral views stained to visualise the cells at early **a** or late **b** cellularisation and at stage 6–7 **c**. **d,e** Close-up of embryos in **b** and **c**, respectively, showing DE-cadherin staining. **d** is a reconstruction of a Z-section that shows basal and apical AJs. **e** is a projection of XY sections showing belt AJs with recognisable foci that correspond to spot AJs. **f** Schematic representation of the formation and maturation of AJs at early, mid and late cellularisation and gastrulation. Each scheme shows a side view of two neighbouring cells attached by AJs. The different steps in the process are indicated. **g** Pathways and players involved in the assembly of spot AJs and in their maturation into belt AJs. Note the capital importance of the cytoskeleton and polarity cues in the process

protein. These structures, whose function is unclear, are transient and disappear by the end of cellularisation (Hunter and Wieschaus 2000; Sokac and Wieschaus 2008). On the other hand, the assembly of apical AJs starts by mid-cellularisation and is followed by a complex process of maturation as development proceeds.

AJ biogenesis and apico-basal polarity establishment are tightly coupled and interdependent. Precursors to AJs form first during early cellularisation at the actin-rich microvilli of the apical membrane. These apical puncta become trapped in clusters of Bazooka protein that organise in the apicolateral domain of the newly forming membrane in a microtubule and Dynein dependent way thereby forming spot AJs (Harris and Peifer 2004, 2005; McGill et al. 2009). At later stages, during gastrulation, spot AJs coalesce to form belt AJs.

The process of maturation and stabilisation of AJs in primary epithelia requires actomyosin tension and the organisation of an actin network (reviewed in Baum and Georgiou 2011; Cavey and Lecuit 2009; Harris et al. 2009). Such organisation of the actin network relies on the activity of different regulatory cues (Fig. 11.2g). On the one hand, apical polarising cues like Crumbs (acting most likely through the apical Spectrin cytoskeleton (Medina et al. 2002)) and aPKC complexes and their antagonistic interactions with the Dlg complex, ensure formation of continuous belt AJs (Bilder et al. 2003; Grawe et al. 1996; Harris and Peifer 2004; Tanentzapf and Tepass 2003; Tepass 1996). On the other hand, the activity of the synaptotagmin-like protein Bitesize, acting with the ERM-domain actin binding protein Moesin, organises an actin network responsible for the stabilisation of AJs. The localisation of apical cues and Bitesize, which is independent of DE-cadherin localisation, requires Bazooka and phosphatidylinositol (4,5)-bisphosphate (Krahn et al. 2010; Pilot et al. 2006). Interestingly, spot AJs recognisable as small microdomains or foci along the belt AJs have been observed at all embryonic stages. These are believed to correspond to sites of DE-cadherin trans-homophilic interactions that sustain intercellular adhesion. Two different types of actin populations associate with the AJs region: a small and very stable actin network that stabilises the spot AJs, and a larger, more dynamic population, which associates with α-Catenin, that prevents the lateral diffusion of spot AJs along the belt AJs (Cavey et al. 2008).

In spite of the correlation between cellularisation, AJ assembly, and epithelial polarity, formation of AJs is not required for cellularisation and initial establishment of polarity. Nevertheless, properly organised AJs are required for cell adhesion, cell architecture and membrane domain segregation at later stages (Cox et al. 1996; Harris and Peifer 2004).

11.5 Early Morphogenetic Movements: Gastrulation and Germ Band Extension

During gastrulation the future mesodermal cells undergo apical constriction and internalise at the ventral furrow (Leptin and Grunewald 1990; Sweeton et al. 1991). Pulsed contractions of an actomyosin network organised at the apical cortex of the ventral furrow cells drive their apical constriction in a ratchet-like mechanism (Dawes-Hoang et al. 2005; Martin et al. 2010; Sawyer et al. 2009). Experiments in which AJ levels are compromised (Dawes-Hoang et al. 2005; Martin et al. 2010) indicated that they are required to maintain actomyosin tension and to organise supracellular meshworks of actomyosin acting as anchor points that link actomyosin networks of adjacent cells. AJs are believed to transmit the forces generated in individual cells by actomyosin contraction to the whole tissue, thereby allowing the coordinated cell shape changes required for apical constriction (Martin et al. 2010).

The *Drosophila* Afadin protein Canoe acts as a link between AJs and the actin cytoskeleton to coordinate cell shape changes; in its absence apical constriction is not maintained (Sawyer et al. 2009).

At the onset of gastrulation AJs shift to a very apical position along the lateral cell membranes with respect to the neighboring neuroectodermal tissue, favoring apical constriction (Dawes-Hoang et al. 2005; Kolsch et al. 2007). Subsequently, during the process of internalisation of the mesodermal layer, a progressive loss of DE-cadherin and AJs is observed (Oda et al. 1998). This process parallels and promotes the epithelial to mesenchymal transition (EMT) reprogramming event that endows the cells with a migratory capacity (Wang et al. 2004). As expected, the genetic program controlling mesodermal development regulates specific changes at AJs. For instance, apical shift of AJs during the apical constriction phase is controlled by the transcriptional activator Twist, whereas the transcriptional repressor Snail represses DE-cadherin expression. In parallel to the loss of DE-cadherin, Twist activates the expression of DN-cadherin in mesodermal cells, although further experiments would be required to define the exact role of this cadherin molecule in EMT or mesodermal development (Oda et al. 1998; Iwai et al. 1997).

Germ band extension occurs in parallel to gastrulation and results in the elongation of the ectoderm along the anterio-posterior axis of the embryo. Early extension of the germ band occurs by two different mechanisms: cell intercalation and oriented cell shape changes in the ventrolateral ectodermal region. Cell intercalation is driven by actomyosin contractility, which results in the shrinking of particular cell contacts and the formation of new ones to promote the convergence and extension cell–cell rearrangements in the whole tissue. Actomyosin activity is anisotropic, and exhibits a planar polarised distribution which is key for correct intercalation (Bertet et al. 2004; Blankenship et al. 2006; Butler et al. 2009; Irvine and Wieschaus 1994; Zallen and Wieschaus 2004). Interestingly, DE-cadherin also shows a planar polarised distribution, which is complementary to the actomyosin one, and which depends on its spatially controlled Clathrin and Dynamin-mediated endocytosis (Levayer et al. 2011). DE-cadherin's planar polarised distribution orients the flows of actomyosin apical networks in such a way that they move toward the regions with lower DE-cadherin concentration, shrinking those junctions with increased actomyosin activity (Rauzi et al. 2010). At the same time the initial endocytosis of DE-cadherin also requires Myosin II (Levayer et al. 2011).

Taken together, these examples illustrate the importance of the interplay and feedback interactions between the contractile actomyosin network and AJs during morphogenesis, and at the same time they reflect the complex relationships between these two systems, which are modulated by tissue-specific and temporal factors.

11.6 Ectodermal Development

After the early events of gastrulation and germ band extension several organs and structures start to form and differentiate. The ectodermal layer gives rise to the epidermis, nervous system and internal organs, all of which need AJ activity.

The requirements for *shotgun* in the early embryonic epithelia depend on the degree of their morphogenetic activity, and clear differences between the ventral and the dorsal regions were described many years ago (Tepass et al. 1996; Uemura et al. 1996). The ventral neuroectoderm is morphogenetically more active due to the delamination of neuroblast precursors. The remaining cells stay at the surface and form the ventral epidermis. AJs disassemble and rapidly reform after neuroblast delamination to maintain AJ continuity and the epithelial integrity of the ectoderm. Thus, DE-cadherin regulation in the neuroectoderm is highly dynamic. This is achieved by the RhoGTPase Cdc42 and the Par complex, which finely tune the endocytosis of apical proteins controlling AJ stability. In particular, Cdc42 decreases the endocytosis rate of apical proteins from the membrane stabilising AJs. Interestingly, the activity of Cdc42 is not critical for the dorsal ectoderm since the levels of apical components and DE-cadherin are not affected there in Cdc42-compromised embryos (Harris and Tepass 2008). The Cdc42-Par complex also regulates DE-cadherin endocytosis in adult epithelial tissues like the dorsal thorax (Georgiou et al. 2008; Leibfried et al. 2008) or the retinal epithelium (Warner and Longmore 2009a, b), indicating that it could be a general mechanism to regulate DE-cadherin trafficking. AJs are also stabilised by the activity of Canoe in the ventral neuroectoderm but not in the dorsal ectoderm (Sawyer et al. 2009).

At later stages, the epidermal epithelium undergoes a major morphogenetic movement, dorsal closure, which encloses the dorsal part of the embryo (for reviews see (Harris et al. 2009; Heisenberg 2009; Jacinto et al. 2002b). Prior to dorsal closure the amnioserosa (AS) covers the dorsal part of the embryo. Different independent forces, i.e. apical constriction of AS cells, contractility of a supracellular actomyosin cable, zipping at the canthi, and an opposing force from the lateral epidermis, have been shown to drive dorsal closure. The constriction of AS cells helps the leading edge (LE) of the epidermal layer to approach. At the same time, the supracellular actin cable assembled at the LE acts as a purse-string helping the movement of the epidermal layer toward the midline. The epidermal cells at the LE emit filopodia and lamellipodia that grow over the AS to match and later interdigitate with the ones generated by the contralateral LE cells. The zippering pulls the cells of the contralateral epidermal layer together and at the same time forces the AS cells to move inside the embryo, where they undergo apoptosis (Blanchard et al. 2010; Franke et al. 2005; Gorfinkiel et al. 2009; Jacinto et al. 2000, 2002a, b; Kiehart et al. 2000; Solon et al. 2009). AJ-mediated adhesion plays a role in anchoring, modulating and transmitting the forces generated in the two different tissues involved in dorsal closure. Compromising either DE-cadherin or Armadillo activity leads to defects in both epidermal and AS cell behaviour and to detachments between the two layers that impair dorsal closure (Gorfinkiel and Arias 2007). Several signals regulate adhesion at the LE, like Src and Fer (Murray et al. 2006; Takahashi et al. 2005). Furthermore, AJ accumulation in the LE cells is dynamic: as closure proceeds AJs disassemble from the front of LE cells to concentrate in "actin-nucleating centres" at each edge of the contacts, and later they relocalise at the interface when the contralateral epidermal cells meet at the midline (Gorfinkiel and Arias 2007). Interestingly, this pattern correlates with that of Echinoid, a cell-adhesion molecule and AJ component (Wei et al. 2005), which also disappears from the leading edge of the dorsal-most epidermal cells due to a loss of Echinoid from the AS (Laplante and Nilson 2006; Lin et al. 2007). The planar polarity of Echinoid in LE cells is required for the assembly of the actomyosin supracellular cable (Laplante and Nilson 2011). Echinoid plays a similar role in other tissues, where expression borders (Echinoid+/Echinoid–) generate a contractile actomyosin structure that controls morphogenesis, like the formation of the dorsal appendages during oogenesis (Laplante and Nilson 2006). In addition, Echinoid regulates cell morphology during dorsal closure through its interaction with Myosin VI/Jaguar (Lin et al. 2007), which helps to maintain adhesion (Millo et al. 2004). Besides Echinoid, other known regulators and non-core components of AJs, generally involved in coordinating adhesion and cytoskeleton, are also required for efficient dorsal closure. This is the case for the Rap1 effector Canoe and for Polychaetoid (Boettner et al. 2003; Choi et al. 2011; Takahashi et al. 1998), p120ctn (Fox et al. 2005), the small GTPase Rho (Bloor and Kiehart 2002), and Diaphanous (Homem and Peifer 2008).

Besides these specific requirements in the embryonic epithelia, AJ integrity is also required during neurogenesis to allow the correct orientation in a horizontal plane of the mitotic spindles of neuroblast precursors. In embryos in which AJs function is compromised many neuroblasts lie at the surface of the embryo as a consequence of randomly oriented spindles and epithelial disruption (Harris and Peifer 2004; Lu et al. 2001; Wang et al. 2004).

11.7 Embryonic Organogenesis: Tubulogenesis

Most internal organs are composed of either simple tubes or a network of them and they are required to transport fluids or gases. Formation of most embryonic organs takes place from mid-embryogenesis onwards, and involves morphogenetic events like, for instance, migration, cell rearrangements or invagination. The requirements for AJ-based cell adhesion during the development of several tubular organs, like the Malpighian tubules (Uemura et al. 1996), salivary glands, heart or tracheal system (see below) illustrate the need to control and sustain cell adhesion for proper tubulogenesis (morphogenesis of branched tubular structures).

11.7.1 Tracheal System

The tracheal system is an excellent model for tubulogenesis. It arises from clusters of ectodermal cells, ten on each side of the embryo, that invaginate and undergo a complex process of tube formation. Tracheal morphogenesis occurs in the absence of cell proliferation and relies on processes of cell rearrangements, cell shape changes and directed migration of the tracheal cell group (collective cell migration). Tracheal development has been divided into consecutive steps of branching, branch

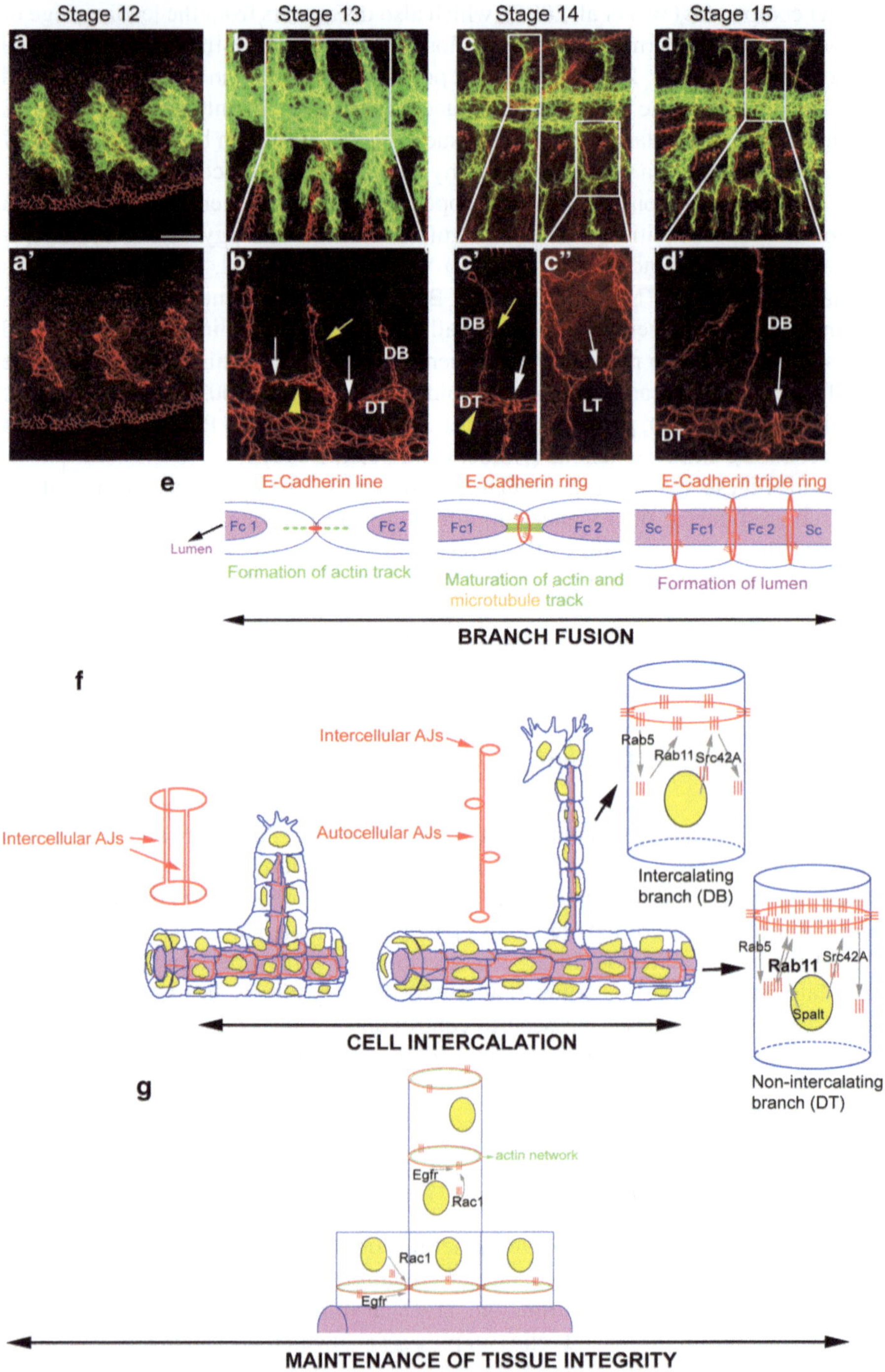
Stage 12
Stage 13
Stage 14
Stage 15
a
b
c
d
a'
b'
c'
c''
d'
DB
DT
LT
e
E-Cadherin line
E-Cadherin ring
E-Cadherin triple ring
Fc 1
Fc 2
Sc
Lumen
Formation of actin track
Maturation of actin and microtubule track
Formation of lumen
BRANCH FUSION
f
Intercellular AJs
Autocellular AJs
Rab5
Rab11
Src42A
Intercalating branch (DB)
Spalt
Non-intercalating branch (DT)
CELL INTERCALATION
g
actin network
Egfr
Rac1
MAINTENANCE OF TISSUE INTEGRITY

fusion and tube maturation that involve different genetic, molecular and cellular mechanisms (Ghabrial et al. 2003; Manning and Krasnow 1993; Samakovlis et al. 1996a). AJs perform specific and critical functions during the development of the trachea that we briefly describe below (Fig. 11.3).

11.7.1.1 Maintenance of Tracheal Epithelial Integrity

The morphogenesis of the tracheal tree involves a dramatic remodelling of the tissue, which requires sufficient plasticity to allow cell rearrangements, cell shape changes and cell movements. However, at the same time, cells forming the tracheal tubes must remain attached to one another to preserve the epithelial features and integrity. Accordingly, a balance between adhesion levels in the tracheal cells allows remodelling on the one hand, while maintaining epithelial continuity on the other. Therefore, a general decrease in DE-cadherin levels results in loss of tube continuity (Cela and Llimargas 2006; Chihara et al. 2003; Uemura et al. 1996). Conversely, increased levels of DE-cadherin correlate with tissue stiffness and impair cell rearrangements (Cela and Llimargas 2006; Chihara et al. 2003). Several factors fine-tune the balanced levels of DE-cadherin necessary to sustain cell adhesion and allow morphogenesis (Fig. 11.3g). The EGF receptor signalling pathway, which is active during tracheal development, ensures sufficient levels of DE-cadherin and the proper organisation of a cortical actin network required for tissue integrity. EGF receptor signalling posttrancriptionally controls DE-cadherin levels through the canonical MAPK pathway but independently of the transcription fac-

Fig 11.3 AJs in tracheal development. **a–d'** Projections of confocal sections showing lateral views of tracheal metameres of embryos at the indicated stages of development. Embryos carry a Src-GFP construct under the control of a tracheal enhancer and are stained with an antibody against GFP (in green, **a–d**) and with an DE-cadherin antibody (in red, **a–d'**). **b'–d'** are close-up of embryos in panels **b–d** respectively. *White arrows* indicate branch fusion structures like DE-cadherin lines or rings in different branches like the Dorsal Trunk (*DT*, **b'**) or Lateral Trunk (*LT*, **c''**), or formation of a triple ring (**c'**, **d'**) once the lumen has penetrated the fusion cells, which become doughnut-shaped. *Yellow arrows* (**b'**, **c'**) point to decreased levels of DE-cadherin in intercalating branches like the Dorsal Branch (*DB*) as compared to higher levels (*yellow arrowheads* in **b'**, **c'**) in non-intercalating branches like the *DT*. **e** Scheme of the fusion process showing two fusion cells (*Fc1*, *Fc2*) making contact and forming an DE-cadherin structure and a cytoskeleton track that helps to guide the lumen. After fusion the fusion cells become very compacted and remain in contact with the stalk cells (*Sc*) of their branches, thereby forming a triple ring structure. **f** Schematic representation of the cell intercalation process (adapted from Shaye et al. 2008). Cells originally positioned in pairs and attached by intercellular AJs reorganise to intercalate in a row where the lumen is sealed by an autocellular junction and the cells of the branch remain attached to one another by small rings of intercellular junctions. Intercalation occurs in most tracheal branches and blocked in others by the activity of Spalt, which regulates the balance of DE-cadherin intracellular trafficking. **g** Different factors regulate the general levels of DE-cadherin to allow a balance between tissue remodelling and maintenance of the integrity and continuity of the tracheal tissue

tor Pointed (Cela and Llimargas 2006). Similarly, the Rho GTPase protein Rac is required to control DE-cadherin levels by a posttranscriptional mechanism. In this case, Rac regulates the incorporation of newly synthesized DE-cadherin/α -Catenin complexes into the AJs (Chihara et al. 2003). These results highlight the strong requirement for DE-cadherin during tracheal development and are consistent with previous reports showing a specific need for DE-cadherin in morphogenetically active tissues (Tepass et al. 1996; Uemura et al. 1996).

11.7.1.2 Branch Fusion

The tracheal system arises as metameric units that later become interconnected, giving rise to a continuous structure that allows the flow of air. Specific branches of each tracheal metamere fuse with adjacent or contralateral branches in a highly stereotyped manner. Branch fusion is mediated by specialised cells at the tips of these branches, known as fusion cells (Samakovlis et al. 1996b; Tanaka-Matakatsu et al. 1996). These establish filopodia-mediated contacts through their basal domain and subsequently, at the contact point, a new DE-cadherin-containing apical domain forms, which is critical to fusion. DE-cadherin accumulates first in a line, initiating the formation of an actin and microtubule-based cytoskeleton track. Later, the line expands into a ring, allowing the passage of the lumen through the two cells (Oda and Tsukita 1999; Tanaka-Matakatsu et al. 1996) (Fig. 11.3e). A structure-function analysis of DE-cadherin identified two different regions in the protein with different abilities, the Armadillo-binding domain (and consequently Armadillo activity) is required for the initial formation of the cytoskeleton track, whereas a juxtamembrane domain is required for maturation and microtubule association to the track (Lee et al. 2003). The microtubule and actin cytoskeleton track is believed to play a key role in guiding and organising the penetration of the lumen through the fusion cells (Jiang et al. 2007; Kakihara et al. 2008; Lee et al. 2003; Lee and Kolodziej 2002; Tanaka et al. 2004). DE-cadherin is transcriptionally controlled in the fusion cells by Escargot, a transcription factor expressed and required for branch fusion (Samakovlis et al. 1996b; Tanaka-Matakatsu et al. 1996). This example emphasises the requirement for AJ regulation at different steps of branch fusion (Beitel and Krasnow 2000; Lee et al. 2003; Oda and Tsukita 1999; Tanaka-Matakatsu et al. 1996; Uemura et al. 1996).

11.7.1.3 Tracheal Cell Rearrangement

During branch extension tracheal cells undergo rearrangements and cell shape changes. A particular type of rearrangement, cell intercalation, occurs in most branches and leads to a transformation of AJs connecting different cells (intercellular AJs) into AJs sealing curled single tracheal cells (autocellular AJs). Intercalation is genetically controlled and specifically blocked by the activity of the transcription factor Spalt, which is expressed in particular branches (Ribeiro et al. 2004). Tracheal intercalation depends, at least in part, on the activity and accumulation of DE-cadherin (Choi et al. 2011; Jung et al. 2006; Shaye et al. 2008; Shindo et al. 2008) (Fig. 11.3f). Intercalating branches display low DE-cadherin levels and higher AJ turnover, whereas non-intercalating branches show high DE-cadherin levels and lower AJ turnover. Levels of DE-cadherin depend on a balance between Rab5-mediated endocytosis, which promotes intercalation, probably by loosening adhesion, and Rab11-mediated exocytosis, which prevents intercalation by strenghtening adhesion. Spalt regulates this intracellular trafficking of DE-cadherin by upregulating Rab11 (Shaye et al. 2008). On the other hand, Src42 A, which is activated in morphogenetically active tissues and localises to AJs, regulates AJ turnover in all tracheal tissue. Src42 A activation exerts a dual regulation on DE-cadherin levels by downregulating the protein and activating transcription. Thus, Src42 A facilitates the AJ remodelling required for cell rearrangements while helping to maintain tissue integrity (Shindo et al. 2008).

11.7.2 Other Embryonic Tubular Structures

The fine-tuned control of DE-cadherin levels also plays a role in the formation of other tubular structures. For instance, in salivary glands, DE-cadherin subcellular localization along the lateral membrane is dynamically controlled through endocytosis regulated by Pak1. This DE-cadherin subcellular localisation controls the shape and elongation of the apical domain of the salivary gland cells, which in turn regulate the expansion of the lumen (Pirraglia et al. 2010). DE-cadherin also plays a key role in the formation of the lumen of the heart, which utilises a specific mechanism. The two bilateral rows of cardioblasts meet at the dorsal midline region and establish two contact domains, leaving a lumen inside (Haag et al. 1999; Rugendorff et al. 1994). Interestingly, the luminal membrane is a non-adherent domain that displays basal features, whereas the junctional membrane containing AJs limits the luminal domain and seals the lumen. The generation of the lumen is controlled by Slit-Robo signaling, which exerts a repulsive activity preventing the formation of AJs in the luminal region and therefore maintaining membrane separation (Medioni et al. 2008; Santiago-Martinez et al. 2008).

11.8 Larval, Pupal and Adult Development

AJ-mediated adhesion also contributes to postembryonic development. The development and formation of several larval or pupal structures requires specific AJ activities.

11.8.1 Wing: Formation of Vein and Intervein

During pupal wing development DE-cadherin-mediated cell adhesion regulates cell shape changes, in particular the apical constriction of the presumptive vein cells required for vein lumen formation and morphogenesis. DE-cadherin accumulates at higher levels in the vein cells than in the intervein cells. The subcellular localisation of DE-cadherin also differs in these two cell types, as it localises apically at the AJs in vein cells while it accumulates basally in intervein cells. The upregulation of DE-cadherin and its subcellular localisation in vein cells depend on Ras signalling (O'Keefe et al. 2007).

11.8.2 Sensory Organ Formation

In the dorsal thorax of *Drosophila* pupae, each sensory organ derives from a single precursor cell, called pI, which undergoes several rounds of asymmetric cell division to generate different cell types (Gho et al. 1999). The pI divides along the A/P axis of the body to generate a posterior pIIa cell and an anterior pIIb cell. Distinct fates are conferred to the two daughter cells by the asymmetric distribution of Numb, a negative regulator of Notch signalling (Gho et al. 1999; Rhyu et al. 1994). The pIIb and the pIIa cells divide again and orient their mitotic spindles vertically and horizontally, respectively, along the epithelium (Gho and Schweisguth 1998; Roegiers et al. 2001). DE-cadherin, together with Armadillo, localises in a small cortical patch in the dividing pIIa cell, at the cell–cell contact with the pIIb cell. This patch is required to orient the division axis of the pIIa along the A/P axis and to asymmetrically localise proteins that regulate cell polarity such as Bazooka and Partner of Inscuteable (Le Borgne et al. 2002).

11.8.3 Axon Pathfinding

DE-cadherin is involved in the formation of the brain circuitry of the larvae that connects the brain compartments. DE-cadherin is required for the proliferation of neuroblasts that gives rise to the secondary lineage, for the correct placement of the secondary neurons in the cortex layer, and for axon patterning (Dumstrei et al. 2003). It is also important for the correct fasciculation, branching and trajectories of secondary axon tracts (Fung et al. 2009).

DN-cadherin, which is expressed in all neurons and their axons, is mainly involved in axon fasciculation and in the directionality of growth cone navigation in the embryo (Iwai et al. 1997). DN-cadherin is also required in later events of neuronal development, for instance for the formation and maintenance of synaptic connections (Hummel and Zipursky 2004; Iwai et al. 2002; Prakash et al. 2005), or for correct targeting and projections of axons and dendrites (Zhu and Luo 2004).

11.8.4 Eye Development

Cadherins play a capital role in *Drosophila* eye development and for this reason this organ has been extensively used as a model to analyse AJ regulation and function during the generation of highly organised patterns in epithelia (Fig. 11.4, see also Tepass and Harris 2007).

The *Drosophila* eye is a neuroepithelium composed of around 800 units called ommatidia, which organise in a hexagonal pattern at the level of the apical surface of the epithelium, near the AJs. It derives from a monolayered columnar epithelium, the eye-antenal imaginal disc of the larva, which remains proliferative and unpatterned until late in larval life, when a wave of differentiation generates the ommatidia. Mature ommatidia have a central core that consists of eight photoreceptor cells (PRCs) that accommodate the rhabdomeres, and four glial-like cone cells that lie above the PRCs. The core is enwrapped by two primary pigment cells. Surrounding each ommatidium there are additional pigment cells (secondary and tertiary) that arrange to generate the precise hexagonal pattern and to optically isolate individual ommatidia. Mechanosensory interommatidial bristles, which project their axons to the brain, occupy three vertices of the hexagon (Ready et al. 1976; Wolff and Ready 1993).

During larval development (Fig. 11.4a), ommatidial assembly starts posterior to an organising centre, the morphogenetic furrow. The PRCs form first, followed by the specification of the cone cells and, later, of the pigment cells. Clustering precedes PRC specification. The PRC precursors first organise in rosettes behind the morphogenetic furrow, a process accompanied by an increase in cell–cell adhesion. DE-cadherin and Armadillo levels increase between the cells of the clusters in relation with the surrounding epithelial cells. This regulation depends on Atonal and the Egfr pathway by two independent mechanisms: Atonal transcriptionally upregulates DE-cadherin while Egfr acts post-transcriptionally on AJs (Brown et al. 2006). These two mechanisms act in concert to promote adhesion between cells of the cluster, which are fated to form PRCs, while surrounding cells, which will divide again, become less tightly connected.

During PRC specification, the clusters rotate 90° toward the equator of the eye disc, independently of their undifferentiated, stationary neighbours called interommatidial precursor cells, from which the secondary and tertiary pigment cells will arise (Fiehler and Wolff 2007; Mlodzik 1999; Strutt and Strutt 1999; Wolff and Ready 1993). This movement is controlled by cadherin molecules. In particular, the rotation depends on a balance between DE-cadherin and two forms of DN-cadherin, encoded by two adjacent genes (Prakash et al. 2005; Yonekura et al. 2007). DE-cadherin promotes the movement of PRC clusters relative to the surrounding epithelial cells, while the DN-cadherins restrict the rate of rotation, counteracting DE-cadherin. Interestingly, DE-cadherin and DN-cadherins show a mutually exclusive localisation and are enriched in membranes that border different PRCs, in

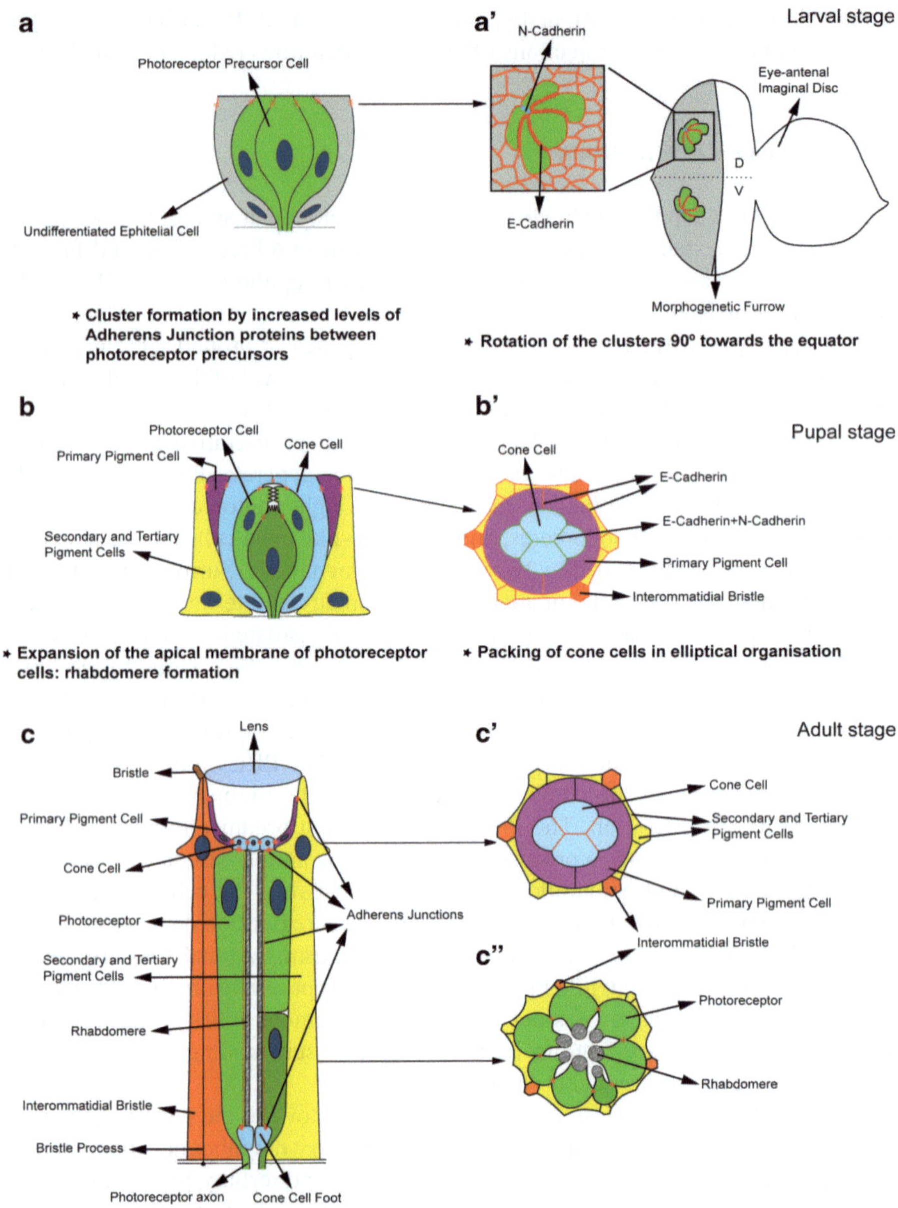

Fig. 11.4 AJs in eye development. Schematic representations of PRC clusters at different developmental stages (adapted from Tepass and Harris 2007). The specific requirements for AJs at each stage are indicated. **a–a'** Larval stages. A shows a longitudinal section of a cluster and **a'** an eye imaginal disc with two clusters depicted after rotation. *Red lines* represent AJs between cells. *Light blue lines* in **a'** represent DN-cadherin localisation between photoreceptors 3 and 5. D, dorsal; V, ventral. **b–b'** Longitudinal **b** or cross-section in the apical region of a PRC cluster. During the pupal stages, cone cells, primary, secondary and tertiary pigment cells are specified. Photoreceptors localise beneath the cone cells and start to elongate their apical membrane to form the rhabdomeres. **c–c''** Longitudinal **c** or cross-sections in the apical **c'** or basal **c''** region of a PRC cluster. In the adult retina, the rhabdomeres have elongated toward the basal membrane of the ommatidium

a complementary pattern. Armadillo is enriched in all of them, indicating that it can interact with the different cadherins. Moreover, genetic interactions with *armadillo* alleles suggests that DE-cadherin mediates rotation through mechanisms involving its cytoskeletal anchoring function (Mirkovic and Mlodzik 2006).

The differential expression of *Drosophila* cadherins is also crucial for the correct organisation of the cone cells (Fig. 11.4b). In the pupal eye DE-cadherin is expressed in all retinal cells, while DN-cadherin is restricted to the AJs of the cone–cone interface. This differential expression causes cone cells to minimise their surface contact with surrounding cells and to pack together in an elliptical form. When DN-cadherin is lost from cone cells or is mis-expressed in primary cells, the ellipse is lost (Hayashi and Carthew 2004).

Several other factors localising to AJs have also been shown to regulate cell movements within the cluster and the ommatidial patterning. This is the case of Polychaetoid (Seppa et al. 2008), Roughest (also called IrrC) and Hibris (Bao and Cagan 2005; Grzeschik and Knust 2005), and Canoe (Matsuo et al. 1999), whose activities are linked to cadherin function and often to cell-sorting mechanisms in retinal cells. Typically, loss of function of these genes gives rise to a rough eye, generally due to poorly arranged ommatidia. Here again, the activity of these factors emphasises the importance of fine-tuning the balance between adhesion molecules in tissues under strong morphogenetic stress.

During pupal development PRCs start to form rhabdomeres by the expansion of their apical membranes toward the basal side of the epithelium (Fig. 11.4b, c). The process requires the integrity of AJs, since mutations disrupting them impair rhabdomere extension (Izaddoost et al. 2002; Pellikka et al. 2002). Furthermore, DN-cadherin regulates the formation of the precise patterns of the PRC connections at this stage. In particular, it is essential in specific axons to select the correct synaptic partners in the lamina and in others to select the target layer in the medulla (Lee et al. 2001; Ting et al. 2005).

11.8.5 Midgut

In the adult midgut DE-cadherin is required for Notch signalling in the intestinal stem cells, participating in the correct specification and positioning of the different cell types of the intestine. After intestinal stem cell division, the two daughter cells remain attached by high levels of DE-cadherin to allow sufficient time for the intestinal stem cell and enteroblasts to interact via Delta-Notch binding (Maeda et al. 2008). Hence, Notch signalling induces cell differentiation before the enteroblasts separates from the stem cell and migrates apically (Ohlstein and Spradling 2007).

11.9 Summary and Future Perspectives

The contribution of AJs and cadherins to *Drosophila* development is a striking example of how these molecular complexes are not only integral to maintaining cell–cell adhesion and tissue integrity, but also regulate multiple aspects of morphogenesis.

The profound knowledge of *Drosophila* development, together with the simplicity of its genetics and the low functional redundancy makes it an ideal model in which to study AJs and cadherin function *in vivo.*

Remarkably, analysis of *Drosophila* development confirmed in vitro studies, showing that AJs are fundamental to cell structure, maintaining cell shape and cell adhesion, and also capable of instructing cell polarity. But what has become clear from studies *in vivo* in *Drosophila* is the enormous plasticity of this structure, that can be both passive, allowing cell movement while maintaining tissue architecture by loosening or increasing adhesion, but also instructive, such as in the case of cell rearrangements or segregation by cell sorting.

Furthermore, studies increasingly show that the formation and remodelling of AJs are under tight and precise control at different levels of regulation (e.g., transcriptional, post-transcriptional, intracellular trafficking, interactions with cytoskeleton and non-core components of AJs, etc.). Interestingly, this modulation differs even between different systems in *Drosophila.* Given this complexity, future investigations should focus on the analysis of the molecular mechanisms regulating AJs to understand how this relates to precise changes in cell behaviour and tissue dynamics.

Acknowledgments We would like to thank the members of Llimargas lab and K. Campbell for critical reading of the manuscript and R. Rycroft for corrections. Work in our lab is supported by funds from the Ministerio de Ciencia e Innovación (BFU2009-09041/BMC) and AGAUR-Generalitat de Catalunya (2009-SGR1333) to ML, and from Programme Consolider 2007 (CSD2007-00008) project. A.L. acknowledges a contract from the "Juan de la Cierva" programme. We apologize to those authors whose work we were unable to cite due to space restrictions.

References

Bao S, Cagan R (2005) Preferential adhesion mediated by Hibris and Roughest regulates morphogenesis and patterning in the Drosophila eye. Dev Cell 8:925–935

Bastock R, Johnston D St (2008) Drosophila oogenesis. Curr Biol 18:R1082–R1087

Baum B, Georgiou M (2011) Dynamics of adherens junctions in epithelial establishment, maintenance, and remodeling. J Cell Biol 192:907–917

Beitel GJ, Krasnow MA (2000) Genetic control of epithelial tube size in the Drosophila tracheal system. Development 127:3271–3282

Bertet C, Sulak L, Lecuit T (2004) Myosin-dependent junction remodelling controls planar cell intercalation and axis elongation. Nature 429:667–671

Bilder D, Schober M, Perrimon N (2003) Integrated activity of PDZ protein complexes regulates epithelial polarity. Nat Cell Biol 5:53–58

Blanchard GB, Murugesu S, Adams RJ, Martinez-Arias A, Gorfinkiel N (2010) Cytoskeletal dynamics and supracellular organisation of cell shape fluctuations during dorsal closure. Development 137:2743–2752

Blankenship JT, Backovic ST, Sanny JS, Weitz O, Zallen JA (2006) Multicellular rosette formation links planar cell polarity to tissue morphogenesis. Dev Cell 11:459–470

Bloor JW, Kiehart DP (2002) Drosophila RhoA regulates the cytoskeleton and cell-cell adhesion in the developing epidermis. Development 129:3173–3183

Boettner B, Harjes P, Ishimaru S, Heke M, Fan HQ, Qin Y, Van Aelst L, Gaul U (2003) The AF-6 homolog canoe acts as a Rap1 effector during dorsal closure of the Drosophila embryo. Genetics 165:159–169

Boyle M, DiNardo S (1995) Specification, migration and assembly of the somatic cells of the Drosophila gonad. Development 121:1815–1825

Boyle M, Bonini N, DiNardo S (1997) Expression and function of clift in the development of somatic gonadal precursors within the Drosophila mesoderm. Development 124:971–982

Brown KE, Baonza A, Freeman M (2006) Epithelial cell adhesion in the developing Drosophila retina is regulated by Atonal and the EGF receptor pathway. Dev Biol 300:710–721

Butler LC, Blanchard GB, Kabla AJ, Lawrence NJ, Welchman DP, Mahadevan L, Adams RJ, Sanson B (2009) Cell shape changes indicate a role for extrinsic tensile forces in Drosophila germ-band extension. Nat Cell Biol 11:859–864

Campos-Ortega JA, Hartenstein V (1997) The embryonic development of Drosophila melanogaster, 2nd edn. Springer-Verlag, Berlin

Cavey M, Lecuit T (2009) Molecular bases of cell-cell junctions stability and dynamics. Cold Spring Harb Perspect Biol 1:a002998

Cavey M, Rauzi M, Lenne PF, Lecuit T (2008) A two-tiered mechanism for stabilization and immobilization of E-cadherin. Nature 453:751–756

Cela C, Llimargas M (2006) Egfr is essential for maintaining epithelial integrity during tracheal remodelling in Drosophila. Development 133:3115–3125

Chihara T, Kato K, Taniguchi M, Ng J, Hayashi S (2003) Rac promotes epithelial cell rearrangement during tracheal tubulogenesis in Drosophila. Development 130:1419–1428

Choi W, Jung KC, Nelson KS, Bhat MA, Beitel GJ, Peifer M, Fanning AS (2011) The single Drosophila ZO-1 protein Polychaetoid regulates embryonic morphogenesis in coordination with Canoe/afadin and Enabled. Mol Biol Cell 22:2010–2030

Cox RT, Kirkpatrick C, Peifer M (1996) Armadillo is required for adherens junction assembly, cell polarity, and morphogenesis during Drosophila embryogenesis. J Cell Biol 134:133–148

Dansereau DA, Lasko P (2008) The development of germline stem cells in Drosophila. Methods Mol Biol 450:3–26

Dawes-Hoang RE, Parmar KM, Christiansen AE Phelps CB, Brand AH, Wieschaus EF (2005) folded gastrulation, cell shape change and the control of myosin localization. Development 132:4165–4178

DeGennaro M, Hurd TR, Siekhaus DE, Biteau B, Jasper H, Lehmann R (2011) Peroxiredoxin stabilization of DE-cadherin promotes primordial germ cell adhesion. Dev Cell 20:233–243

Doe CQ (1992) Molecular markers for identified neuroblasts and ganglion mother cells in the Drosophila central nervous system. Development 116:855–863

Dumstrei K, Wang F, Hartenstein V (2003) Role of DE-cadherin in neuroblast proliferation, neural morphogenesis, and axon tract formation in Drosophila larval brain development. J Neurosci 23:3325–3335

Eichenberger-Glinz S (1979) Intercellular junctions during development and in tissue cultures of Drosophila melanogaster: an electron-microscopic study. Wilhelm Roux's Arch Dev Biol 186:333–349

Fiehler RW, Wolff T (2007) Drosophila Myosin II, Zipper, is essential for ommatidial rotation. Dev Biol 310:348–362

Fox DT, Homem CC, Myster SH, Wang F, Bain EE, Peifer M (2005) Rho1 regulates Drosophila adherens junctions independently of p120ctn. Development 132:4819–4831

Franke JD, Montague RA, Kiehart DP (2005) Nonmuscle myosin II generates forces that transmit tension and drive contraction in multiple tissues during dorsal closure. Curr Biol 15:2208–2221

Fulga TA, Rorth P (2002) Invasive cell migration is initiated by guided growth of long cellular extensions. Nat Cell Biol 4:715–719

Fung S, Wang F, Spindler SR, Hartenstein V (2009) Drosophila E-cadherin and its binding partner Armadillo/ beta-catenin are required for axonal pathway choices in the developing larval brain. Dev Biol 332:371–382

Geisbrecht ER, Montell DJ (2002) Myosin VI is required for E-cadherin-mediated border cell migration. Nat Cell Biol 4:616–620
Georgiou M, Marinari E, Burden J, Baum B (2008) Cdc42, Par6, and aPKC regulate Arp2/3-mediated endocytosis to control local adherens junction stability. Curr Biol 18:1631–1638
Ghabrial A, Luschnig S, Metzstein MM, Krasnow MA (2003) Branching morphogenesis of the Drosophila tracheal system. Annu Rev Cell Dev Biol 19:623–647
Gho M, Schweisguth F (1998) Frizzled signalling controls orientation of asymmetric sense organ precursor cell divisions in Drosophila. Nature 393:178–181
Gho M, Bellaiche Y, Schweisguth F (1999) Revisiting the Drosophila microchaete lineage: a novel intrinsically asymmetric cell division generates a glial cell. Development 126:3573–3584
Godt D, Tepass U (1998) Drosophila oocyte localization is mediated by differential cadherin-based adhesion. Nature 395:387–391
Gonzalez-Reyes A, Johnston D St (1998) The Drosophila AP axis is polarised by the cadherin-mediated positioning of the oocyte. Development 125:3635–3644
Gorfinkiel N, Arias AM (2007) Requirements for adherens junction components in the interaction between epithelial tissues during dorsal closure in Drosophila. J Cell Sci 120:3289–3298
Gorfinkiel N, Blanchard GB, Adams RJ, Martinez Arias A (2009) Mechanical control of global cell behaviour during dorsal closure in Drosophila. Development 136:1889–1898
Grammont M (2007) Adherens junction remodeling by the Notch pathway in Drosophila melanogaster oogenesis. J Cell Biol 177:139–150
Grawe F, Wodarz A, Lee B, Knust E, Skaer H (1996) The Drosophila genes crumbs and stardust are involved in the biogenesis of adherens junctions. Development 122:951–959
Grzeschik NA, Knust E (2005) IrreC/rst-mediated cell sorting during Drosophila pupal eye development depends on proper localisation of DE-cadherin. Development 132:2035–2045
Haag TA, Haag NP, Lekven AC, Hartenstein V (1999) The role of cell adhesion molecules in Drosophila heart morphogenesis: faint sausage, shotgun/DE-cadherin, and laminin A are required for discrete stages in heart development. Dev Biol 208:56–69
Hackney JF, Pucci C, Naes E, Dobens L (2007) Ras signaling modulates activity of the ecdysone receptor EcR during cell migration in the Drosophila ovary. Dev Dyn 236:1213–1226
Harris KP, Tepass U (2008) Cdc42 and Par proteins stabilize dynamic adherens junctions in the Drosophila neuroectoderm through regulation of apical endocytosis. J Cell Biol 183:1129–1143
Harris TJ, Peifer M (2004) Adherens junction-dependent and -independent steps in the establishment of epithelial cell polarity in Drosophila. J Cell Biol 167:135–147
Harris TJ, Peifer M (2005) The positioning and segregation of apical cues during epithelial polarity establishment in Drosophila. J Cell Biol 170:813–823
Harris TJ, Sawyer JK, Peifer M (2009) How the cytoskeleton helps build the embryonic body plan: models of morphogenesis from Drosophila. Curr Top Dev Biol 89:55–85
Hartenstein V, Younossi-Hartenstein A, Lekven A (1994) Delamination and division in the Drosophila neurectoderm: spatiotemporal pattern, cytoskeletal dynamics, and common control by neurogenic and segment polarity genes. Dev Biol 165:480–499
Hayashi T, Carthew RW (2004) Surface mechanics mediate pattern formation in the developing retina. Nature 431:647–652
Heisenberg CP (2009) Dorsal closure in Drosophila: cells cannot get out of the tight spot. Bioessays 31:1284–1287
Homem CC, Peifer M (2008) Diaphanous regulates myosin and adherens junctions to control cell contractility and protrusive behavior during morphogenesis. Development 135:1005–1018
Horne-Badovinac S, Bilder D (2005) Mass transit: epithelial morphogenesis in the Drosophila egg chamber. Dev Dyn 232:559–574
Hummel T, Zipursky SL (2004) Afferent induction of olfactory glomeruli requires N-cadherin. Neuron. 42:77–88
Hunter C, Wieschaus E (2000) Regulated expression of nullo is required for the formation of distinct apical and basal adherens junctions in the Drosophila blastoderm. J Cell Biol 150:391–401

Huynh JR, Johnston D St (2004) The origin of asymmetry: early polarisation of the Drosophila germline cyst and oocyte. Curr Biol 14:R438–R449

Inaba M, Yuan H, Salzmann V, Fuller MT, Yamashita YM (2010) E-cadherin is required for centrosome and spindle orientation in Drosophila male germline stem cells. PLoS One 5:e12473

Irvine KD, Wieschaus E (1994) Cell intercalation during Drosophila germband extension and its regulation by pair-rule segmentation genes. Development 120:827–841

Iwai Y, Usui T, Hirano S, Steward R, Takeichi M, Uemura T (1997) Axon patterning requires DN-cadherin, a novel neuronal adhesion receptor, in the Drosophila embryonic CNS. Neuron. 19:77–89

Iwai Y, Hirota Y, Ozaki K, Okano H, Takeichi M, Uemura T (2002) DN-cadherin is required for spatial arrangement of nerve terminals and ultrastructural organization of synapses. Mol Cell Neurosci. 19:375–388

Izaddoost S, Nam SC, Bhat MA, Bellen HJ, Choi KW (2002) Drosophila Crumbs is a positional cue in photoreceptor adherens junctions and rhabdomeres. Nature 416:178–183

Jacinto A, Wood W, Balayo T, Turmaine M, Martinez-Arias A , Martin P (2000) Dynamic actin-based epithelial adhesion and cell matching during Drosophila dorsal closure. Curr Biol 10:1420–1426

Jacinto A, Wood W, Woolner S, Hiley C, Turner L, Wilson C, Martinez-Arias A, Martin P (2002a) Dynamic analysis of actin cable function during Drosophila dorsal closure. Curr Biol 12:1245–1250

Jacinto A, Woolner S, Martin P (2002b) Dynamic analysis of dorsal closure in Drosophila: from genetics to cell biology. Dev Cell 3:9–19

Jenkins AB, McCaffery JM, Van Doren M (2003) Drosophila E-cadherin is essential for proper germ cell-soma interaction during gonad morphogenesis. Development 130:4417–4426

Jiang L, Rogers SL, Crews ST (2007) The Drosophila Dead end Arf-like3 GTPase controls vesicle trafficking during tracheal fusion cell morphogenesis. Dev Biol 311:487–499

Jung AC, Ribeiro C, Michaut L, Certa U, Affolter M (2006) Polychaetoid/ZO-1 is required for cell specification and rearrangement during Drosophila tracheal morphogenesis. Curr Biol 16:1224–1231

Kakihara K, Shinmyozu K, Kato K, Wada H, Hayashi S (2008) Conversion of plasma membrane topology during epithelial tube connection requires Arf-like 3 small GTPase in Drosophila. Mech Dev 125:325–336

Kiehart DP, Galbraith CG, Edwards KA, Rickoll WL, Montague RA (2000) Multiple forces contribute to cell sheet morphogenesis for dorsal closure in Drosophila. J Cell Biol 149:471–490

Kolsch V, Seher T, Fernandez-Ballester GJ, Serrano L, Leptin M. (2007) Control of Drosophila gastrulation by apical localization of adherens junctions and RhoGEF2. Science 315:384–386

Krahn MP, Klopfenstein DR, Fischer N, Wodarz A (2010) Membrane targeting of Bazooka/PAR-3 is mediated by direct binding to phosphoinositide lipids. Curr Biol 20:636–642

Kunwar PS, Sano H, Renault AD, Barbosa V, Fuse N, Lehmann R (2008) Tre1 GPCR initiates germ cell transepithelial migration by regulating Drosophila melanogaster E-cadherin. J Cell Biol 183:157–168

Laplante C, Nilson LA (2006) Differential expression of the adhesion molecule Echinoid drives epithelial morphogenesis in Drosophila. Development 133:3255–3264

Laplante C, Nilson LA (2011) Asymmetric distribution of Echinoid defines the epidermal leading edge during Drosophila dorsal closure. J Cell Biol 192:335–348

Le Borgne R, Bellaiche Y, Schweisguth F (2002) Drosophila E-cadherin regulates the orientation of asymmetric cell division in the sensory organ lineage. Curr Biol 12:95–104

Lecuit T (2004) Junctions and vesicular trafficking during Drosophila cellularization. J Cell Sci 117:3427–3433

Lee CH, Herman T, Clandinin TR, Lee R, Zipursky SL (2001) N-cadherin regulates target specificity in the Drosophila visual system. Neuron 30:437–450

Lee M, Lee S, Zadeh AD, Kolodziej PA (2003) Distinct sites in E-cadherin regulate different steps in Drosophila tracheal tube fusion. Development 130:5989–5999

Lee S, Kolodziej PA (2002) The plakin Short Stop and the RhoA GTPase are required for E-cadherin-dependent apical surface remodeling during tracheal tube fusion. Development 129:1509–1520

Leibfried A, Fricke R, Morgan MJ, Bogdan S, Bellaiche Y (2008) Drosophila Cip4 and WASp define a branch of the Cdc42-Par6-aPKC pathway regulating E-cadherin endocytosis. Curr Biol 18:1639–1648

Leptin M, Grunewald B (1990) Cell shape changes during gastrulation in Drosophila. Development 110:73–84

Levayer R, Pelissier-Monier A, Lecuit T (2011) Spatial regulation of Dia and Myosin-II by RhoGEF2 controls initiation of E-cadherin endocytosis during epithelial morphogenesis. Nat Cell Biol 13:734

Levine B, Hackney JF, Bergen A, Dobens L, Truesdale 3rd A, Dobens L (2010) Opposing interactions between Drosophila cut and the C/EBP encoded by slow border cells direct apical constriction and epithelial invagination. Dev Biol 344:196–209

Li MA, Alls JD, Avancini RM, Koo K, Godt D (2003) The large Maf factor Traffic Jam controls gonad morphogenesis in Drosophila. Nat Cell Biol 5:994–1000

Lin HP, Chen HM, Wei SY, Chen LY, Chang LH, Sun YJ, Huang SY, Hsu JC (2007) Cell adhesion molecule Echinoid associates with unconventional myosin VI/Jaguar motor to regulate cell morphology during dorsal closure in Drosophila. Dev Biol 311:423–433

Llimargas M, Casanova J (2010) Apical constriction and invagination: a very self-reliant couple. Dev Biol 344:4–6

Lu B, Roegiers F, Jan LY, Jan YN (2001) Adherens junctions inhibit asymmetric division in the Drosophila epithelium. Nature 409:522–525

Maeda K, Takemura M, Umemori M, Adachi-Yamada T (2008) E-cadherin prolongs the moment for interaction between intestinal stem cell and its progenitor cell to ensure Notch signaling in adult Drosophila midgut. Genes Cells 13:1219–1227

Manning G, Krasnow MA (1993) Development of the *Drosophila* tracheal system. In: Bate M, Martínez-Arias A (eds) The development of Drosophila melanogaster, Vol I. Cold Spring Harbor Laboratory Press, Cold Spring Harbor, pp 609–685

Martin AC, Gelbart M, Fernandez-Gonzalez R, Kaschube M, Wieschaus EF (2010) Integration of contractile forces during tissue invagination. J Cell Biol 188:735–749

Mathews WR, Ong D, Milutinovich AB, Van Doren M (2006) Zinc transport activity of Fear of Intimacy is essential for proper gonad morphogenesis and DE-cadherin expression. Development 133:1143–1153

Matsuo T, Takahashi K, Suzuki E, Yamamoto D (1999) The Canoe protein is necessary in adherens junctions for development of ommatidial architecture in the Drosophila compound eye. Cell Tissue Res 298:397–404

Mazumdar A, Mazumdar M (2002) How one becomes many: blastoderm cellularization in Drosophila melanogaster. Bioessays 24:1012–1022

McCrea PD, Turck CW, Gumbiner B (1991) A homolog of the armadillo protein in Drosophila (plakoglobin) associated with E-cadherin. Science 254:1359–1361

McGill MA, McKinley RF, Harris TJ (2009) Independent cadherin-catenin and Bazooka clusters interact to assemble adherens junctions. J Cell Biol 185:787–796

Medina E, Williams J, Klipfell E, Zarnescu D, Thomas G, Le Bivic A (2002) Crumbs interacts with moesin and beta(Heavy)-spectrin in the apical membrane skeleton of Drosophila. J Cell Biol 158:941–951

Medioni C, Astier M, Zmojdzian M , Jagla K, Semeriva M (2008) Genetic control of cell morphogenesis during Drosophila melanogaster cardiac tube formation. J Cell Biol 182:249–261

Melani M, Simpson KJ, Brugge JS, Montell D (2008) Regulation of cell adhesion and collective cell migration by hindsight and its human homolog RREB1. Curr Biol 18:532–537

Millo H, Leaper K, Lazou V, Bownes M (2004) Myosin VI plays a role in cell-cell adhesion during epithelial morphogenesis. Mech Dev 121:1335–1351

Mirkovic I, Mlodzik M (2006) Cooperative activities of drosophila DE-cadherin and DN-cadherin regulate the cell motility process of ommatidial rotation. Development 133:3283–3293

Mlodzik M (1999) Planar polarity in the Drosophila eye: a multifaceted view of signaling specificity and cross-talk. EMBO J 18:6873–6879
Montell DJ (2003) Border-cell migration: the race is on. Nat Rev Mol Cell Biol 4:13–24
Montell DJ (2006) The social lives of migrating cells in Drosophila. Curr Opin Genet Dev 16:374–383
Murray MJ, Davidson CM, Hayward NM, Brand AH (2006) The Fes/Fer non-receptor tyrosine kinase cooperates with Src42 A to regulate dorsal closure in Drosophila. Development 133:3063–3073
Niewiadomska P, Godt D, Tepass U (1999) DE-Cadherin is required for intercellular motility during Drosophila oogenesis. J Cell Biol 144:533–547
O'Keefe DD, Prober DA, Moyle PS, Rickoll WL, Edgar BA (2007) Egfr/Ras signaling regulates DE-cadherin/Shotgun localization to control vein morphogenesis in the Drosophila wing. Dev Biol 311:25–39
Oda H, Tsukita S (1999) Dynamic features of adherens junctions during Drosophila embryonic epithelial morphogenesis revealed by a Dalpha-catenin-GFP fusion protein. Dev Genes Evol 209:218–225
Oda H, Uemura T, Shiomi K, Nagafuchi A, Tsukita S, Takeichi M (1993) Identification of a Drosophila homologue of alpha-catenin and its association with the armadillo protein. J Cell Biol 121:1133–1140
Oda H, Uemura T, Harada Y, Iwai Y, Takeichi M (1994) A Drosophila homolog of cadherin associated with armadillo and essential for embryonic cell-cell adhesion. Dev Biol 165:716–726
Oda H, Uemura T, Takeichi M (1997) Phenotypic analysis of null mutants for DE-cadherin and Armadillo in Drosophila ovaries reveals distinct aspects of their functions in cell adhesion and cytoskeletal organization. Genes Cells 2:29–40
Oda H, Tsukita S, Takeichi M (1998) Dynamic behavior of the cadherin-based cell-cell adhesion system during Drosophila gastrulation. Dev Biol 203:435–450
Ohlstein B, Spradling A (2007) Multipotent Drosophila intestinal stem cells specify daughter cell fates by differential notch signaling. Science 315:988–992
Pacquelet A, Rorth P (2005) Regulatory mechanisms required for DE-cadherin function in cell migration and other types of adhesion. J Cell Biol 170:803–812
Peifer M (1993) The product of the Drosophila segment polarity gene armadillo is part of a multiprotein complex resembling the vertebrate adherens junction. J Cell Sci 105(Pt 4):993–1000
Peifer M, Wieschaus E (1990) The segment polarity gene armadillo encodes a functionally modular protein that is the Drosophila homolog of human plakoglobin. Cell 63:1167–1176
Peifer M, Orsulic S, Sweeton D, Wieschaus E (1993) A role for the Drosophila segment polarity gene armadillo in cell adhesion and cytoskeletal integrity during oogenesis. Development 118:1191–207
Pellikka M, Tanentzapf G, Pinto M, Smith C, McGlade CJ, Ready DF, Tepass U (2002) Crumbs, the Drosophila homologue of human CRB1/RP12, is essential for photoreceptor morphogenesis. Nature 416:143–149
Pilot F, Philippe JM, Lemmers C, Lecuit T (2006) Spatial control of actin organization at adherens junctions by a synaptotagmin-like protein Btsz. Nature 442:580–584
Pirraglia C, Walters J, Myat MM (2010) Pak1 control of E-cadherin endocytosis regulates salivary gland lumen size and shape. Development 137:4177–4189
Poodry CA, Schneiderman HA (1970) The ultrastructure of the developing leg of Drosophila melanogaster. Wilhelm Roux's Arch Dev Biol 166:1–44
Prakash S, Caldwell JC, Eberl DF, Clandinin TR (2005) Drosophila N-cadherin mediates an attractive interaction between photoreceptor axons and their targets. Nat Neurosci 8:443–450
Rauzi M, Lenne PF, Lecuit T (2010) Planar polarized actomyosin contractile flows control epithelial junction remodelling. Nature 468:1110–1114
Ready DF, Hanson TE, Benzer S (1976) Development of the Drosophila retina, a neurocrystalline lattice. Dev Biol 53:217–240
Rhyu M.S, Jan LY, Jan YN (1994) Asymmetric distribution of numb protein during division of the sensory organ precursor cell confers distinct fates to daughter cells. Cell 76:477–491

Ribeiro C, Neumann M, Affolter M (2004) Genetic control of cell intercalation during tracheal morphogenesis in Drosophila. Curr Biol 14:2197–2207

Roegiers F, Younger-Shepherd S, Jan LY, Jan YN (2001) Two types of asymmetric divisions in the Drosophila sensory organ precursor cell lineage. Nat Cell Biol 3:58–67

Rugendorff A, Younossi-Hartenstein AY, Hartenstein V (1994) Embryonic origin and differentiation of the Drosophila heart. Roux's Arch Dev Biol 203

Samakovlis C, Hacohen N, Manning G, Sutherland DC, Guillemin K, Krasnow MA (1996a) Development of the Drosophila tracheal system occurs by a series of morphologically distinct but genetically coupled branching events. Development 122:1395–1407

Samakovlis C, Manning G, Steneberg P, Hacohen N, Cantera R, Krasnow MA (1996b) Genetic control of epithelial tube fusion during Drosophila tracheal development. Development 122:3531–3536

Santiago-Martinez E, Soplop NH, Patel R, Kramer SG (2008) Repulsion by Slit and Roundabout prevents Shotgun/E-cadherin-mediated cell adhesion during Drosophila heart tube lumen formation. J Cell Biol 182:241–248

Sawyer JK, Harris NJ, Slep KC, Gaul U, Peifer M (2009) The Drosophila afadin homologue Canoe regulates linkage of the actin cytoskeleton to adherens junctions during apical constriction. J Cell Biol 186:57–73

Sawyer JM, Harrell JR, Shemer G, Sullivan-Brown J, Roh-Johnson M, Goldstein B (2010) Apical constriction: a cell shape change that can drive morphogenesis. Dev Biol 341:5–19

Seppa MJ, Johnson RI, Bao S, Cagan RL (2008) Polychaetoid controls patterning by modulating adhesion in the Drosophila pupal retina. Dev Biol 318:1–16

Shaye DD, Casanova J, Llimargas M (2008) Modulation of intracellular trafficking regulates cell intercalation in the Drosophila trachea. Nat Cell Biol 10:964–970

Shindo M, Wada H, Kaido M, Tateno M, Aigaki T, Tsuda L, Hayashi S (2008) Dual function of Src in the maintenance of adherens junctions during tracheal epithelial morphogenesis. Development 135:1355–1364

Sokac AM, Wieschaus E (2008) Zygotically controlled F-actin establishes cortical compartments to stabilize furrows during Drosophila cellularization. J Cell Sci 121:1815–1824

Solon J, Kaya-Copur A, Colombelli J, Brunner D (2009) Pulsed forces timed by a ratchet-like mechanism drive directed tissue movement during dorsal closure. Cell 137:1331–1342

Song X, Xie T (2002) DE-cadherin-mediated cell adhesion is essential for maintaining somatic stem cells in the Drosophila ovary. Proc Natl Acad Sci U S A 99:14813–14818

Song X, Zhu CH, Doan C, Xie T (2002) Germline stem cells anchored by adherens junctions in the Drosophila ovary niches. Science 296:1855–1857

Strutt H, Strutt D (1999) Polarity determination in the Drosophila eye. Curr Opin Genet Dev 9:442–446

Sweeton D, Parks S, Costa M, Wieschaus E (1991) Gastrulation in Drosophila: the formation of the ventral furrow and posterior midgut invaginations. Development 112:775–789

Takahashi K, Matsuo T, Katsube T, Ueda R, Yamamoto D (1998) Direct binding between two PDZ domain proteins Canoe and ZO-1 and their roles in regulation of the jun N-terminal kinase pathway in Drosophila morphogenesis. Mech Dev 78:97–111

Takahashi M, Takahashi F, Ui-Tei K, Kojima T, Saigo K (2005) Requirements of genetic interactions between Src42 A, armadillo and shotgun, a gene encoding E-cadherin, for normal development in Drosophila. Development 132:2547–2559

Tanaka H, Takasu E, Aigaki T, Kato K, Hayashi S, Nose A (2004) Formin3 is required for assembly of the F-actin structure that mediates tracheal fusion in Drosophila. Dev Biol 274:413–425

Tanaka-Matakatsu M, Uemura T, Oda H, Takeichi M, Hayashi S (1996) Cadherin-mediated cell adhesion and cell motility in Drosophila trachea regulated by the transcription factor Escargot. Development 122:3697–3705

Tanentzapf G, Smith C, McGlade J, Tepass U (2000) Apical, lateral, and basal polarization cues contribute to the development of the follicular epithelium during Drosophila oogenesis. J Cell Biol 151:891–904

Tanentzapf G, Tepass U (2003) Interactions between the crumbs, lethal giant larvae and bazooka pathways in epithelial polarization. Nat Cell Biol 5:46–52

Tepass U (1996) Crumbs, a component of the apical membrane, is required for zonula adherens formation in primary epithelia of Drosophila. Dev Biol 177:217–225

Tepass U, Harris KP (2007) Adherens junctions in Drosophila retinal morphogenesis. Trends Cell Biol 17:26–35

Tepass U, Hartenstein V (1994a) Epithelium formation in the Drosophila midgut depends on the interaction of endoderm and mesoderm. Development 120:579–590

Tepass U, Hartenstein V (1994b) The development of cellular junctions in the Drosophila embryo. Dev Biol 161:563–596

Tepass U, Gruszynski-DeFeo E, Haag TA, Omatyar L, Torok T, Hartenstein V (1996) shotgun encodes Drosophila E-cadherin and is preferentially required during cell rearrangement in the neurectoderm and other morphogenetically active epithelia. Genes Dev 10:672–685

Tepass U, Godt D, Winklbauer R (2002) Cell sorting in animal development: signalling and adhesive mechanisms in the formation of tissue boundaries. Curr Opin Genet Dev 12:572–582

Ting CY, Yonekura S, Chung P, Hsu SN, Robertson HM, Chiba A, Lee CH (2005) Drosophila N-cadherin functions in the first stage of the two-stage layer-selection process of R7 photoreceptor afferents. Development 132:953–963

Uemura T, Oda H, Kraut R, Hayashi S, Kotaoka Y, Takeichi M (1996) Zygotic Drosophila E-cadherin expression is required for processes of dynamic epithelial cell rearrangement in the Drosophila embryo. Genes Dev 10:659–671

Van Doren M, Mathews WR, Samuels M, Moore LA, Broihier HT, Lehmann R (2003) fear of intimacy encodes a novel transmembrane protein required for gonad morphogenesis in Drosophila. Development 130:2355–2364

Voog J, D'Alterio C, Jones DL (2008) Multipotent somatic stem cells contribute to the stem cell niche in the Drosophila testis. Nature 454:1132–1136

Wang F, Dumstrei K, Haag T, Hartenstein V (2004) The role of DE-cadherin during cellularization, germ layer formation and early neurogenesis in the Drosophila embryo. Dev Biol 270:350–363

Wang H, Singh SR, Zheng Z, Oh SW, Chen X, Edwards K, Hou SX (2006) Rap-GEF signaling controls stem cell anchoring to their niche through regulating DE-cadherin-mediated cell adhesion in the Drosophila testis. Dev Cell 10:117–126

Warner SJ, Longmore GD (2009a) Cdc42 antagonizes Rho1 activity at adherens junctions to limit epithelial cell apical tension. J Cell Biol 187:119–133

Warner SJ, Longmore GD (2009b) Distinct functions for Rho1 in maintaining adherens junctions and apical tension in remodeling epithelia. J Cell Biol 185:1111–1125

Wei SY, Escudero LM, Yu F, Chang LH, Chen LY, Ho YH, Lin CM, Chou CS, Chia W, Modolell J, Hsu JC (2005) Echinoid is a component of adherens junctions that cooperates with DE-Cadherin to mediate cell adhesion. Dev Cell 8:493–504

White P, Aberle H, Vincent JP (1998) Signaling and adhesion activities of mammalian beta-catenin and plakoglobin in Drosophila. J Cell Biol 140:183–195

Wolff T, Ready DF (1993) Pattern formation in the Drosophila retina. In: The development of Drosophila melanogaster. Cold Spring Harbor Laboratory Press. Vol. 2:1277–1325

Wu X, Tanwar PS, Raftery LA (2008) Drosophila follicle cells: morphogenesis in an eggshell. Semin Cell Dev Biol 19:271–282

Xie T (2008) Germline stem cell niches

Yamashita YM (2010) Cell adhesion in regulation of asymmetric stem cell division. Curr Opin Cell Biol 22:605–610

Yamashita YM, Mahowald AP, Perlin JR, Fuller MT (2007) Asymmetric inheritance of mother versus daughter centrosome in stem cell division. Science 315:518–521

Yonekura S, Xu L, Ting CY, Lee CH (2007) Adhesive but not signaling activity of Drosophila N-cadherin is essential for target selection of photoreceptor afferents. Dev Biol 304:759–770

Zallen JA, Wieschaus E (2004) Patterned gene expression directs bipolar planar polarity in Drosophila. Dev Cell 6:343–355

Zhu H, Luo L (2004) Diverse functions of N-cadherin in dendritic and axonal terminal arborization of olfactory projection neurons. Neuron 42:63–75

Chapter 12
Adherens Junctions in *C. elegans* Embryonic Morphogenesis

Stephen T. Armenti and Jeremy Nance

Abstract *Caenorhabditis elegans* provides a simplified, *in vivo* model system in which to study adherens junctions (AJs) and their role in morphogenesis. The core AJ components—HMR-1/E-cadherin, HMP-2/β-catenin and HMP-1/α-catenin—were initially identified through genetic screens for mutants with body axis elongation defects. In early embryos, AJ proteins are found at sites of contact between blastomeres, and in epithelial cells AJ proteins localize to the multifaceted apical junction (*Ce*AJ)—a single structure that combines the adhesive and barrier functions of vertebrate adherens and tight junctions. The apically localized polarity proteins PAR-3 and PAR-6 mediate formation and maturation of junctions, while the basolaterally localized regulator LET-413/Scribble ensures that junctions remain apically positioned. AJs promote robust adhesion between epithelial cells and provide mechanical resistance for the physical strains of morphogenesis. However, in contrast to vertebrates, *C. elegans* AJ proteins are not essential for general cell adhesion or for epithelial cell polarization. A combination of conserved and novel proteins localizes to the *Ce*AJ and works together with AJ proteins to mediate adhesion.

12.1 Introduction

The relative simplicity of *Caenorhabditis elegans*, combined with a deep understanding of its development and numerous tools for genetic and cell biological analysis, has made it a rich model system for the study of morphogenesis. Homologues of many vertebrate junction proteins and their regulators are found in *C. elegans* and are encoded by single genes rather than gene families. Combining genetic tools such as feeding RNAi and chemically induced mutations with live imaging and immunohistochemistry, researchers are uncovering the roles of both conserved and novel morphogenesis genes.

J. Nance (✉)
Department of Cell Biology, NYU School of Medicine, 10016 New York, NY, USA
e-mail: jeremy.nance@med.nyu.edu

S. T. Armenti · J. Nance
Helen L. and Martin S. Kimmel Center for Biology and Medicine at the Skirball Institute for Biomolecular Medicine, NYU School of Medicine, 10016 New York, NY, USA

T. Harris (ed.), *Adherens Junctions: From Molecular Mechanisms to Tissue Development and Disease,* Subcellular Biochemistry 60,
DOI 10.1007/978-94-007-4186-7_12, © Springer Science+Business Media Dordrecht 2012

Genes encoding the core AJ components—*hmr-1/E-cadherin, hmp-2/β-catenin* and *hmp-1/α-catenin*—were identified in genetic screens for embryos with morphological defects (Costa et al. 1998). It was shown that these genes are important for proper morphogenesis of the epidermis (Costa et al. 1998; Raich et al. 1999), whose movements and shape changes are responsible for converting the elliptical embryo into its final worm-like shape (Priess and Hirsh 1986; Sulston et al. 1983). This chapter focuses on how AJ proteins, as well as their regulators and downstream effectors, contribute to cell adhesion and morphogenesis in the *C. elegans* embryo. We begin the chapter with a description of the core *C. elegans* AJ proteins, highlighting similarities and differences with their mammalian homologues. We next provide an overview of the major morphogenetic events that shape the *C. elegans* embryo, and describe the contribution of AJ proteins to these events. Finally, we describe how junctions assemble and mature, and introduce the regulatory proteins that influence AJ placement, stability and activity.

12.2 Core Components of *C. elegans* AJs: The Cadherin-Catenin Complex

C. elegans contains single genes encoding the major AJ proteins E-cadherin (*hmr-1*), β-catenin (*hmp-2*), and α-catenin (*hmp-1*). All three genes were originally identified in a mutant screen for embryos with defects in epidermal morphogenesis, and their names reflect the abnormal shapes that mutant embryos form (*hmr*=*Ham*mer*h*ead, *hmp*=*Hump*back) (Costa et al. 1998). One of the more remarkable features of these genes is that putative null mutations have a relatively mild effect on general cell adhesion and apicobasal polarity but severe effects on morphogenesis, even when both maternal and zygotic contributions are removed. Along these lines, while many of the molecular properties of HMR-1/E-cadherin, HMP-2/β-catenin, and HMP-1/α-catenin appear to be similar to those of their vertebrate homologues, the worm proteins have a few distinct features that we highlight below.

12.2.1 HMR-1/E-Cadherin

hmr-1 is the sole gene in *C. elegans* encoding a classic cadherin (Costa et al. 1998), although there are approximately a dozen additional genes that can encode proteins with cadherin repeats and a transmembrane domain (Hill et al. 2001). Through the use of alternative promoters, *hmr-1* produces two distinct isoforms that differ in the length and composition of the extracellular domain (Broadbent and Pettitt 2002). The shorter HMR-1a isoform is more homologous in organization to mammalian E-cadherin, while the longer HMR-1b isoform contains additional cadherin repeats and is more similar to mammalian N-cadherin. HMR-1a is found in AJs (HMR-1b expression has only been detected in the nervous system), and for simplicity,

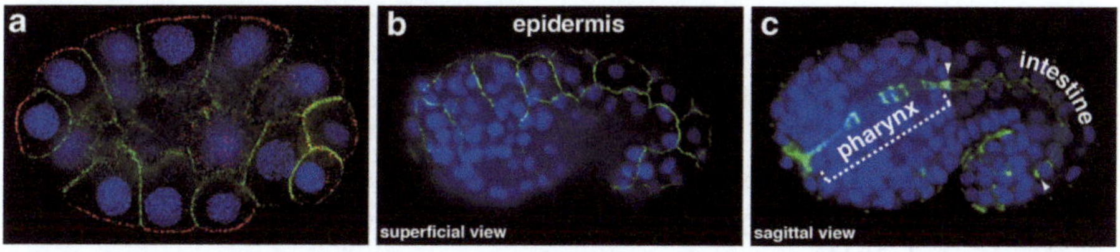

Fig. 12.1 HMR-1/E-cadherin localization in blastomeres and epithelial cells. All panels show immunostained embryos; DNA is stained with DAPI (*blue*). **a** HMR-1 (*green*) in early embryos localizes to sites of contact between blastomeres (a 26-cell stage embryo is shown). Contact-free surfaces are marked with PAR-3 (*red*). **b–c** After epithelial cells begin to form during mid-embryogenesis, HMR-1 localizes to AJs. **b** *Superficial view* showing HMR-1 at junctions between epidermal cells. **c** *Sagittal view* of the central region of an embryo showing HMR-1 at junctions in pharyngeal and intestinal epithelial cells, which form a tube comprising the digestive tract; region of intestinal cells is bounded by arrowheads

hereafter we refer to HMR-1a as HMR-1 (Broadbent and Pettitt 2002; Costa et al. 1998).

The extracellular domain of HMR-1 contains three cadherin repeats, as well as EGF and Laminin G domains that are found in other invertebrate classic cadherins (Costa et al. 1998; Hill et al. 2001). Although it is widely assumed that cadherins participate in homotypic binding, differences with vertebrate classic cadherins in the composition and organization of the extracellular domain make it unclear how, or even whether, HMR-1 extracellular domains interact with one another (Shapiro and Weis 2009). However, studies in *Drosophila* suggest that another related invertebrate cadherin, Shotgun/E-cadherin, can undergo homophilic binding (Oda et al. 1994). The HMR-1 cytoplasmic tail has been shown to interact with HMP-2/β-catenin as well as JAC-1, the sole p120-catenin homologue found in worms (Kwiatkowski et al. 2010; Pettitt et al. 2003). HMR-1 expression begins prior to the formation of AJs. In early embryos, maternally supplied HMR-1 localizes uniformly at contacts between each blastomere, which lack cell-cell junctions (Fig. 12.1a) (Costa et al. 1998; Nance and Priess 2002). During later embryogenesis, when epithelial tissues and organs begin to develop, zygotically expressed HMR-1 is found in epithelial cells and is enriched at AJs (Fig. 12.1b, c) (Costa et al. 1998; Sulston et al. 1983). As described below, HMR-1 contributes to cell adhesion and morphogenesis during both of these stages of development.

12.2.2 HMP-2/β-Catenin

Rather uniquely, worms have parceled the functions of their β-catenins into separate signaling and junctional proteins (Korswagen et al. 2000). HMP-2 is the only β-catenin known to localize to blastomere contacts and AJs, and similar to its vertebrate counterpart binds directly to the cytoplasmic tail of HMR-1/E-cadherin as well as to HMP-1/α-catenin (Costa et al. 1998; Kwiatkowski et al. 2010; Pettitt et al. 2003). HMP-2 colocalizes with HMR-1/E-cadherin both at contacts between blastomeres and at AJs in epithelial cells (Costa et al. 1998). Loss of HMR-1/E-cadherin causes HMP-2 to redistribute to the cytoplasm (Costa et al. 1998). Although the function of HMP-2 in epithelial cell adhesion is well established (described below), recent evidence has shown that HMP-2 can also contribute to Wnt/Wingless signaling in early embryonic cell fate specification (Putzke and Rothman 2010; Sumiyoshi et al. 2011).

12.2.3 HMP-1/α-Catenin

HMP-1/α-catenin contains a β-catenin-binding domain and an F-actin-binding domain that have been shown to be operative *in vitro* (Costa et al. 1998; Kwiatkowski et al. 2010). However, unlike vertebrate αE-catenin, recombinant HMP-1 dimers cannot be detected *in vitro*, and the ability of HMP-1 to bind actin appears to require additional proteins (Kwiatkowski et al. 2010). Nonetheless, both the β-catenin-binding domain and F-actin-binding domain are required for HMP-1 to function *in vivo*, suggesting that HMP-1 does indeed provide a bridge between HMR-1/E-cadherin and F-actin (Kwiatkowski et al. 2010). HMP-1 colocalizes with HMP-2/β-catenin and HMR-1/E-cadherin, and depends on both proteins for its recruitment to blastomere cell contacts and to epithelial cell AJs (Costa et al. 1998).

12.3 Embryonic Morphogenesis

The *C. elegans* embryo undergoes two major morphogenetic rearrangements prior to hatching, and AJ proteins contribute to both events. The first morphogenetic event is gastrulation, when cells fated to become endoderm, mesoderm, and germ cells ingress from the surface of the embryo into the interior. The second event is epidermal morphogenesis, when the epidermis wraps around the embryo's surface then squeezes the embryo to elongate it into a worm-like shape. Forces driving gastrulation and epidermal morphogenesis arise from cell shape changes, and both events require cells to generate nascent cell contacts.

12.3.1 Gastrulation

Gastrulation begins 90 min after the first cleavage (26-cell stage) when the daughters of the E endoderm precursor (Ea and Ep) ingress from the surface of the embryo into the interior (Fig. 12.3a) (Sulston et al. 1983). Gastrulation continues over the next few hours with the ingression of mesodermal precursors and primordial germ cells (Nance and Priess 2002). In order for ingression to commence, Ea and

Ep accumulate non-muscle myosin (NMY-2) on their apical, contact-free surfaces, causing this surface to constrict and helping to move these cells into the interior of the embryo (Lee and Goldstein 2003; Nance and Priess 2002). At this stage of development, embryonic cells do not show characteristics of epithelial cells such as electron-dense intercellular junctions or asymmetrically positioned organelles, and HMR-1/E-cadherin, HMP-2/β-catenin, and HMP-1/α-catenin localize uniformly to all sites of cell contact (Costa et al. 1998; Grana et al. 2010; Nance and Priess 2002; Priess and Hirsh 1986). Removing HMR-1/E-cadherin from early embryos does not globally disrupt cell adhesion or gastrulation—a somewhat surprising finding given that there are no other classic cadherins that could contribute redundantly (Costa et al. 1998; Grana et al. 2010). This paradox was partially resolved when it was shown that SAX-7/L1CAM functions redundantly with HMR-1/E-cadherin in early embryonic cell adhesion and gastrulation (Grana et al. 2010); in embryos lacking both HMR-1/E-cadherin and SAX-7/L1CAM, cell adhesion is compromised and the E daughters fail to ingress (Grana et al. 2010). The adhesive role of HMR-1 appears to be at least partially independent of HMP-2/β-catenin and HMP-1/α-catenin, as *sax-7* mutant embryos lacking HMR-1/E-cadherin show more severe cell adhesion defects than do *sax-7* mutants lacking either HMP-1/α-catenin or HMP-2/β-catenin. It is not known whether HMR-1/E-cadherin has a specific role in promoting gastrulation (e.g., creating tissue-specific differences in adhesion), or whether gastrulation movements themselves simply require robust cell adhesion. Interestingly, although clear adhesion defects arise upon simultaneous loss of HMR-1/E-cadherin and SAX-7/L1CAM, embryonic cells still maintain a basal level of adhesion. Therefore, additional proteins that promote the adhesion of early embryonic cells remain to be identified.

12.3.2 Epidermal Morphogenesis

AJs do not begin to form until the middle stages of embryogenesis, when most cell divisions have ceased and epithelial tissues begin to form (~300 min after the first cell division) (Leung et al. 1999; Podbilewicz and White 1994; Priess and Hirsh 1986; Sulston et al. 1983). Epithelial tissues fall into two major classes: the internal epithelia that comprise the digestive tract (pharynx and intestine) and an external epithelium that surrounds the embryo (epidermis). Cells in both epithelial classes undergo dramatic shape changes as the embryo elongates, and junctions must be created and remodeled to ensure appropriate cell adhesion. Below, we describe the important role of AJs in morphogenesis of the epidermis, whose directed movements and cell shape changes drastically alter the shape of the embryo.

Ventral enclosure: Epidermal cells are born in a monolayer on the dorsal surface during mid-embryogenesis. In order to encase the embryo, epidermal cells undergo a rapid and dramatic migration called ventral enclosure (Fig. 12.2) (Sulston et al. 1983). Ventral enclosure begins soon after epidermal cells differentiate to form an epithelial sheet (Podbilewicz and White 1994; Sulston et al. 1983; Williams-Masson

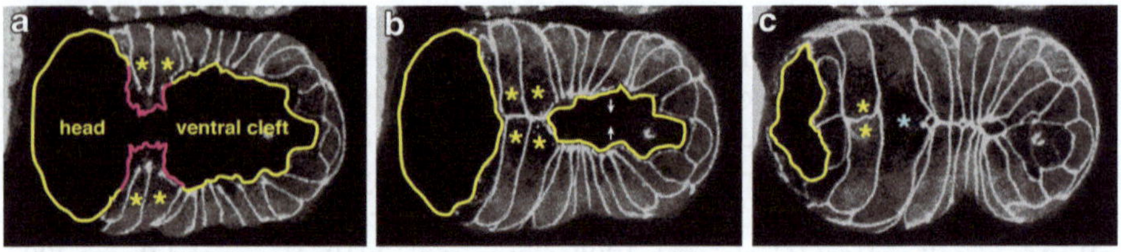

Fig. 12.2 Ventral Enclosure. **a** Embryo near the start of ventral enclosure expressing DLG-1-GFP to mark junctions, *ventral* view. Leading cells are marked by asterisks and filopodia are depicted by magenta tracings. The *ventral cleft* and future *head* region are indicated. **b** Middle of ventral enclosure. Leading cells have generated nascent contacts with contralateral cells. The *ventral cleft* is closing as pocket cells begin to come together. **c** End of ventral enclosure. The *ventral cleft* has closed and the posterior pair of leading cells has fused (cyan asterisk), abolishing the initial contact. Images reproduced with permission from Chisholm and Hardin 2005

et al. 1997). AJs develop at the apicolateral interface of each epidermal cell and connect neighboring epidermal cells together (Costa et al. 1998; Priess and Hirsh 1986). Just before ventral enclosure, epidermal cells are arranged in bilaterally symmetric rows that are aligned with the anterior-posterior axis: two dorsal rows (which intercalate to form a single row), two lateral rows, and two ventral rows that are born with a free edge that does not contact other epidermal cells and lacks junctions (Priess and Hirsh 1986; Sulston et al. 1983; Williams-Masson et al. 1997). The two anterior-most pairs of ventral epidermal cells, called leading cells, extend actin-rich filopodia and begin to migrate, pulling the entire epidermal sheet ventrally (Raich et al. 1999; Williams-Masson et al. 1997). Once arriving at the ventral surface, filopodia from contralateral pairs of leading cells meet, and the cells seal together at the ventral midline and form new junctions (Raich et al. 1999; Williams-Masson et al. 1997). Subsequently, many of the remaining ventral epidermal cells (called pocket cells) seal the ventral cleft in a process that has been described as a 'purse-string' mechanism, culminating with the formation of new junctions between contralateral cell pairs (Raich et al. 1999; Williams-Masson et al. 1997). Embryos mutant for AJ components can show defects in forming stable junctions between contralateral pairs of epidermal cells that join and seal at the ventral surface during ventral enclosure (Raich et al. 1999), preventing proper cell adhesion and causing the epidermis to contract dorsally.

In comparison to cultured mammalian cells (Adams et al. 1998), *C. elegans* epidermal cells form new junctions very rapidly upon contact, suggesting that there is a highly mobile pool of AJ proteins that are primed to be delivered to cell contacts (Raich et al. 1999). As in mammalian cells, oriented actin filaments present within filopodia at sites of nascent contact could facilitate the linkage of F-actin to new junctions (Raich et al. 1999; Vasioukhin et al. 2000; Williams-Masson et al. 1997).

Upon the completion of ventral enclosure, an epithelial monolayer surrounds the embryo, and junctions join each cell in the monolayer to its neighbors. Many of the epidermal cells ultimately fuse together, abolish cell junctions and form large, multinucleate syncytia (see Fig. 12.2c) (Podbilewicz and White 1994). Fusion

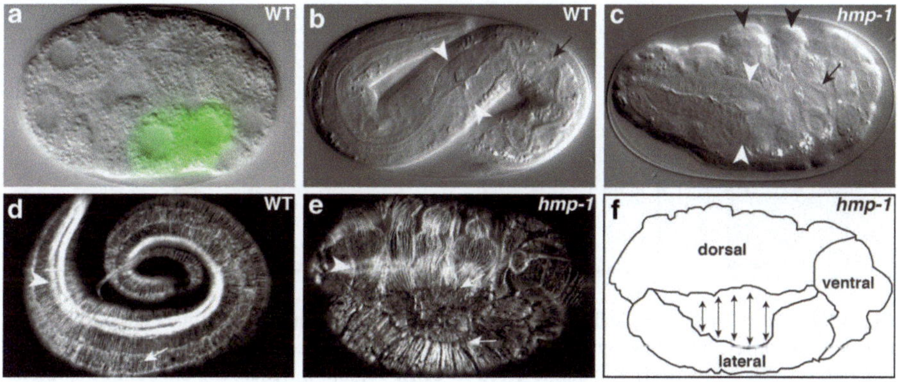

Fig. 12.3 Gastrulation and epidermal morphogenesis. **a** Early embryo at the onset of gastrulation. The E daughters (labeled *green* by *end-1::GFP* transgene) have flattened their apical surfaces and are beginning to ingress into the embryo. **b** Fully elongated wild-type embryo. In **b** and **c**, white arrowheads indicate pharyngeal bulb and arrows mark the intestinal lumen. **c** *hmp-1* mutant embryo displaying the Humpback (Hmp) phenotype, with characteristic dorsal epidermal bulges (black arrowheads). Note shortened pharynx and intestine due to elongation failure. **d** Fully elongated wild-type embryo stained with phalloidin to visualize F-actin. Arrow indicates parallel bundles of circumferential actin filaments, arrowhead indicates actin filaments in underlying muscle tissue. **e** Phalloidin-stained *hmp-1* mutant embryo. Circumferential actin bundles between *dorsal* and *lateral* epidermal cells have detached (indicated by arrows). Arrowhead indicates intact underlying muscle actin filaments. **f** Schematic diagram of embryo in E showing points of separation between *dorsal* and *lateral* epidermal cells. Images in panels B-E ©1998 Rockefeller University Press; originally published in *J. Cell Biol.* 141:297–308; Panel **f** was redrawn from a similar panel in Costa et al. 1998

is triggered when adjacent epidermal cells express the fusogenic protein EFF-1 at their surfaces (the related protein AFF-1 mediates fusion of other cell types) (Mohler et al. 2002; Sapir et al. 2007; Shemer et al. 2004). A fusion pore appears in an apical zone at or near the AJ, and expands laterally and basally (Mohler et al. 1998). Membrane vesiculation and junction dissolution occurs as the fusion pore expands to encompass the length of the contact between EFF-1-expressing cells. *eff-1* mutant worms, which cannot fuse their epidermal cells, are viable but have severe defects in body morphogenesis (Mohler et al. 2002). The AJ proteins do not seem to be required for fusion, as *hmp-1* mutants undergo dorsal epidermal fusions normally (Costa et al. 1998).

Elongation: The second phase of epidermal morphogenesis is elongation. During elongation, the epidermis squeezes the internal cells of the embryo, causing the entire embryo to elongate four-fold and adopt a worm-like shape (Fig. 12.3b). Elongation requires asymmetric changes in epidermal cell shape, which occur in the absence of cell division (Sulston et al. 1983). Each epidermal cell shortens along its circumferential (radial) axis and lengthens along its anterior-posterior (AP) axis (Priess and Hirsh 1986). These shape changes are promoted by the contraction of parallel bundles of circumferentially oriented apical microfilaments that anchor to AJs at the border of the cell (Fig. 12.3d) (Priess and Hirsh 1986). In *hmr-1, hmp-2,*

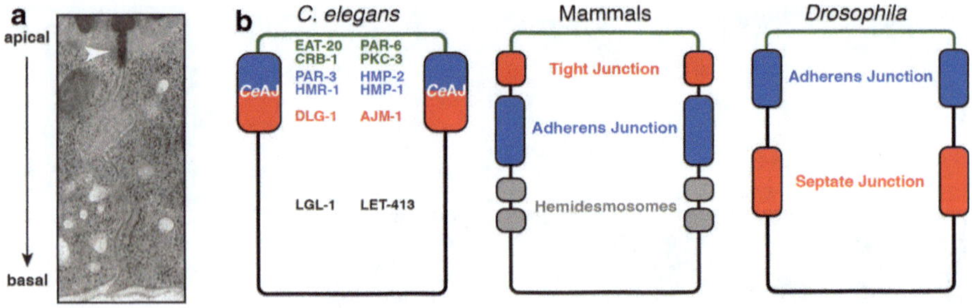

Fig. 12.4 The *C. elegans* apical junction and analogous junctions in mammals and *Drosophila*. **a** Transmission electron micrograph of intestinal epithelial cells showing the *C. elegans* apical junction (*Ce*AJ, arrowhead) as a single electron-dense region. **b** Schematic diagram of epithelial domains and junction structures in *C. elegans*, mammals and *Drosophila*. Major polarity and junction proteins and their localization pattern in mature *C. elegans* epithelia are depicted. Functionally analogous regions in mammals and *Drosophila* are shown in common colors. Panel **a** was reproduced with permission from Muller and Bossinger 2003

and *hmp-1* mutant embryos, circumferential actin bundles detach from AJs, particularly between dorsal and lateral epidermal cells, causing epidermal bulges to form on the dorsal surface (Fig. 12.3c, e, f) (Costa et al. 1998). The regulation of circumferential microfilament contractions during elongation, which is beyond the focus of this chapter, has been the topic of several recent reviews (Chisholm and Hardin 2005; Zhang et al. 2010).

12.4 Molecular Organization of *C. elegans* Junctions

In contrast to mammals and *Drosophila*, which have functionally and physically distinct adherens and tight/septate junctions, *C. elegans* epithelial cells contain a single junction that executes both adhesive and barrier functions (Fig. 12.4). The *C. elegans* apical junction (or *Ce*AJ) is positioned near the apicolateral interface. Despite its organization in electron micrographs as a single, electron-dense structure (Leung et al. 1999; Priess and Hirsh 1986), the *Ce*AJ contains subdomains that are enriched with different subsets of junction proteins (Table 12.1). AJ proteins (HMR-1/E-cadherin, HMP-2/β-catenin, and HMP-1/α-catenin) are found in the apical-most region of the CeAJ (McMahon et al. 2001; Segbert et al. 2004), as are the ZO-1 homologue ZOO-1 and the BCMP1/claudin protein VAB-9 (Lockwood et al. 2008a; Simske et al. 2003). By contrast, the Discs large homolog DLG-1 and its novel binding partner AJM-1 are found in more basal regions of the CeAJ (Bossinger et al. 2001; Firestein and Rongo 2001; Koppen et al. 2001; McMahon et al. 2001). As we discuss below, junctions assemble in a step-wise fashion that begins with the formation of junction protein puncta, and follows with their apical accumulation and maturation into belt-like junctions that surround each cell. In addition to *Ce*AJs, epidermal cells contain unique junctions called hemidesmosomes, which

Table 12.1 *C. elegans* junction and polarity proteins

Protein	Type of protein	References
PAR-6	PAR-6 protein: PDZ and CRIB domain	(Hung and Kemphues 1999; Watts et al. 1996)
PKC-3	Atypical protein kinase C	(Tabuse et al. 1998)
CRB-1	Crumbs: transmembrane protein	(Bossinger et al. 2001)
EAT-20	Crumbs-like protein: transmembrane protein	(Shibata et al. 2000)
PAR-3	PAR-3 protein: Multi-PDZ domain	(Cheng et al. 1995; Etemad-Moghadam et al. 1995; Kemphues et al. 1988)
HMR-1	E-cadherin	(Costa et al. 1998)
HMP-2	β-catenin	(Costa et al. 1998)
HMP-1	α-catenin	(Costa et al. 1998)
DLG-1	Discs large: MAGUK protein	(Bossinger et al. 2001; Firestein and Rongo 2001; Koppen et al. 2001; McMahon et al. 2001)
AJM-1	Novel coiled-coil protein	(Koppen et al. 2001)
LET-413	Scribble (LAP protein)	(Legouis et al. 2000)
LGL-1	Lethal (2) giant larvae: WD-40 repeat protein	(Beatty et al. 2010; Fichelson et al. 2010)

form a bridge that connects each epidermal cell to overlying cuticle and underlying muscle (Ding et al. 2004; Labouesse 2006; Zhang and Labouesse 2010).

12.5 PAR Proteins and Junction Assembly

Formation of the *Ce*AJ requires the function of several conserved polarity determinants, including the PAR polarity proteins. PAR proteins were originally identified for their role in polarizing the one-celled *C. elegans* embryo (zygote) along its anterior-posterior axis (Etemad-Moghadam et al. 1995; Hung and Kemphues 1999; Kemphues et al. 1988; Tabuse et al. 1998; Watts et al. 1996). PAR-3 (a multi-PDZ domain scaffolding protein), PAR-6 (a PDZ and CRIB-domain protein), and PKC-3 (atypical protein kinase C, aPKC) can physically interact and localize to the anterior of the one-celled embryo, establishing a spatially localized signaling center that polarizes the cell (Nance and Zallen 2011; St Johnston and Ahringer 2010). The PAR proteins function together with the Rho GTPase CDC-42 to polarize the one-celled embryo (Chen et al. 1993; Gotta et al. 2001; Kay and Hunter 2001). Subsequent studies in many species have revealed that homologues of PAR-3, PAR-6, PKC-3/aPKC and CDC-42 are essential for numerous cell polarization events, including polarization of epithelial cells (Goldstein and Macara 2007; St Johnston and Ahringer 2010). PAR-3, PAR-6, and PKC-3/aPKC are also found in *C. elegans* epithelial cells and develop asymmetric localizations during polarization (CDC-42 distribution has not been determined at this stage) (Bossinger et al. 2001; Leung et al. 1999; McMahon et al. 2001). The functions of PAR-3 and PAR-6 in epithelia were determined relatively recently, when genetic tools became available to circumvent their earlier requirement in polarization of the one-cell embryo (Achilleos et al. 2010; Aono et al. 2004; Totong et al. 2007). These studies have revealed that PAR-3 and PAR-6 function sequentially to regulate epithelial polarization and junction maturation, respectively.

12.5.1 PAR-3

The cellular role of PAR-3 in epithelial cells has been examined most extensively in the embryonic intestine. The intestine is a 20-cell tube consisting of nine rings of cells along its length; most rings contain a pair of cells that connect to each other, and to cells in neighboring rings, through apical junctions (Leung et al. 1999). Intestinal cells form during embryogenesis when mesenchymal-like intestinal precursors polarize along their apicobasal axis and assemble junctions. During the very initial stages of polarization, scattered puncta of PAR-3 appear at sites of contact between intestinal precursor cells (Achilleos et al. 2010). PAR-3 puncta also contain AJ proteins (HMR-1/E-cadherin, HMP-2/β-catenin, HMP-1/α-catenin) and other PAR proteins (PAR-6 and PKC-3/aPKC) (Fig. 12.5a) (Achilleos et al. 2010). Soon after forming, PAR-3 puncta move asymmetrically along the cell cortex to the future apical surface (Fig. 12.5b). In embryos lacking PAR-3, puncta are not detected and junction and polarity proteins mislocalize (Fig. 12.5c) (Achilleos et al. 2010). These findings suggest that PAR-3 polarizes intestinal cells by concentrating junction and polarity proteins into puncta, which then move apically and are enriched at the site of future junction formation. This scaffolding function of PAR-3 appears to be conserved, as the *Drosophila* PAR-3 homologue Bazooka has been shown to concentrate and trap clusters of E-cadherin at AJs (McGill et al. 2009).

It is largely unknown what functions upstream of PAR-3 to ensure its apical localization during polarization. Mutations in the kinesin-like protein ZEN-4 prevent polarization and apical PAR-3 recruitment in one lineage of epithelial cells—the arcade cells (which connect the pharynx to the mouth) (Portereiko et al. 2004). The asymmetric localization of PAR-3 by motor proteins has been documented in several other types of polarized cells, including dynein in *Drosophila* epithelial cells (Harris and Peifer 2005) and kinesin in cilia (Fan et al. 2004), raising the possibility that ZEN-4 shuttles PAR-3 to the apical surface. However, *zen-4* mutants polarize other epithelial lineages normally, suggesting that such a mechanism of PAR-3 localization would be specialized.

RNAi experiments have shown that PAR-3 is also required for the apical localization of junction proteins and F-actin in spermathecal epithelial cells, which form during late larval stages (Aono et al. 2004). By contrast, embryonic epidermal cells do not require PAR-3 to polarize and assemble apical junctions, even though PAR-3 is expressed in these cells (Achilleos et al. 2010). An important difference between epidermal cells and tube-forming internal epithelia such as the intestine and the

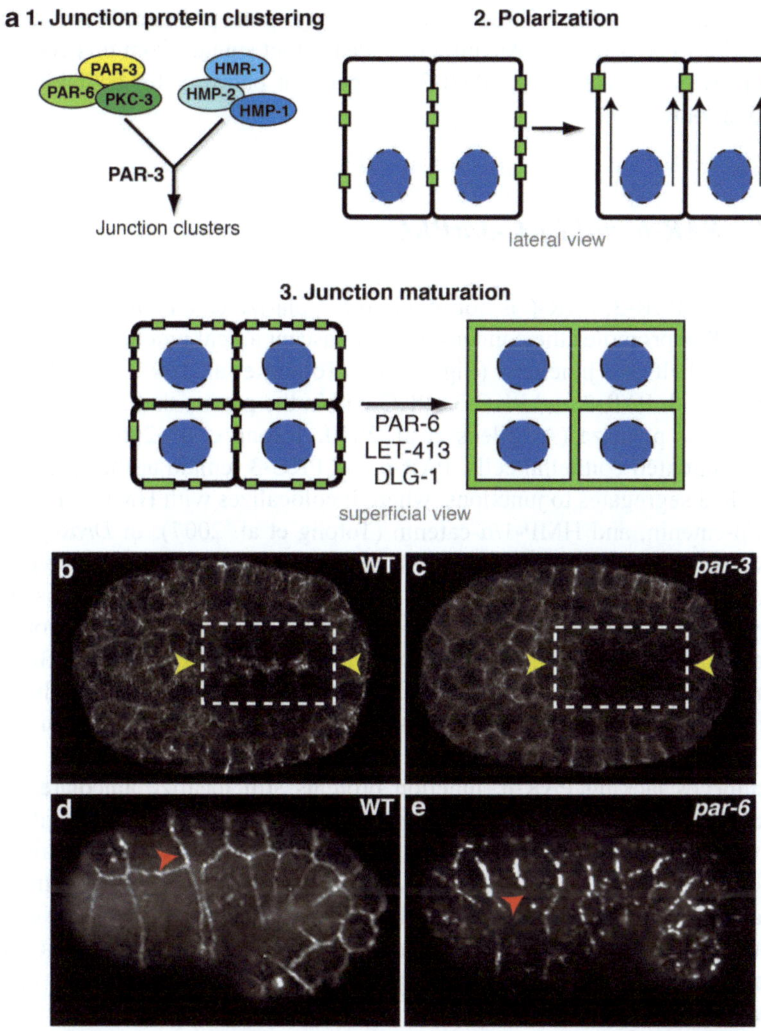

Fig. 12.5 AJ formation. **a** Junction formation involves distinct steps: (**1**) PAR-3 is required for clustering AJ proteins early during polarization. PAR proteins and AJ proteins are all present in puncta. (**2**) Epithelial cells polarize and junction clusters are recruited apically to the site of future junction formation. (**3**) Apically localized puncta mature into belt-like junctions through the function of PAR-6, LET-413, and DLG-1. **b**, **c** Polarization of junction clusters in WT **b** and embryos lacking both maternal and zygotic PAR-3 **c**. HMR-1-GFP is shown. In each panel, the box represents the polarizing intestine. *Left* and *right* rows of intestinal cells show mirror symmetry, and the future apical surface between them is indicated by yellow arrowheads. **d**, **e** Junction maturation in wild-type **d** and embryos lacking both maternal and zygotic PAR-6 **e**. Red arrowheads indicate continuous junctions in wild-type **d** and fragmented junctions in *par-6* mutants. Junctions are stained with DLG-1, and the epidermis is shown. Panels B-E reproduced with permission from Achilleos et al. 2010

spermatheca, where PAR-3 is required for polarity, is the presence of a contact-free surface prior to polarization. An attractive idea is that signals from this contact-free surface provide an alternative, PAR-3-independent pathway for achieving apico-basal polarization.

12.5.2 PAR-6 and PKC-3/aPKC

In contrast to PAR-3, PAR-6 is not required to polarize *C. elegans* epithelial cells. Rather, PAR-6 promotes the condensation of nascent apical junction protein puncta into mature, belt-like junctions (Fig. 12.5c) (Totong et al. 2007). Initially, PAR-6 colocalizes with PAR-3 and PKC-3/aPKC within the puncta that form as intestinal epithelial cells polarize (Achilleos et al. 2010; Totong et al. 2007). However, in fully differentiated epithelial cells, PAR-6 and PKC-3 remain at the apical surface while PAR-3 segregates to junctions, where it colocalizes with HMR-1/E-cadherin, HMP-2/β-catenin, and HMP-1/α-catenin (Totong et al. 2007). In *Drosophila* epithelial cells, an analogous relocation of Bazooka/PAR-3 was shown to involve Par-6, aPKC/PKC-3, and the apical transmembrane polarity protein Crumbs (CRB-1 in *C. elegans*) (Morais-De-Sa et al. 2010). However, while Crumbs proteins are important for both mammalian and *Drosophila* epithelial polarization (Bulgakova and Knust 2009), RNAi co-depletion of CRB-1 and EAT-20 (which has homology to Crumbs in the cytoplasmic tail) does not affect epithelial polarization in *C. elegans* (Bossinger et al. 2001; Shibata et al. 2000).

In embryos lacking PAR-6, junction proteins still localize apicolaterally, but fail to coalesce into mature, belt-like junctions (Fig. 12.5e, f). Consequently, *par-6* mutant embryos arrest at the beginning of elongation and develop ruptures within the epidermis (Totong et al. 2007). It is not yet known how PAR-6 controls junction maturation, although studies in other systems suggest that it is likely to do so through PKC-3/aPKC (St Johnston and Ahringer 2010), with which it interacts physically (Aceto et al. 2006). However, the role of PKC-3 in *C. elegans* epithelial cells has not yet been determined.

12.6 Basolateral Polarity Regulators and Junction Maturation and Maintenance

Studies of epithelial polarization and cell proliferation in *Drosophila* have identified a group of basolaterally localized polarity regulators defined by the genes *scribble*, *discs large*, and *lethal (2) giant larvae* (Bilder et al. 2000; Bilder and Perrimon 2000; Woods et al. 1996). *C. elegans* contains a single homologue of each of these genes, and *let-413/scribble* and *dlg-1/discs large* have been shown to be important for maintaining polarity and forming junctions (Fig. 12.5a) (Bossinger et al. 2001; Firestein and Rongo 2001; Koppen et al. 2001; Legouis et al. 2000; McMahon et al. 2001). However, neither gene appears to affect cell proliferation in *C. elegans*, which develops with a largely invariant cell lineage. A *C. elegans* homolog of *lethal (2) giant larvae* (*lgl-1*) has been identified and is expressed in embryonic epithelia, although *lgl-1* mutants have no defects in junction formation or epithelial polarity (Beatty et al. 2010; Fichelson et al. 2010).

let-413, which was identified in a chromosomal deficiency screen for genes important for junction morphology (Legouis et al. 2000), encodes a leucine-rich repeat (LRR) and PDZ domain protein (LAP protein) homologous to *Drosophila* Scribble (Bilder and Perrimon 2000). Like Scribble and the human orthologue, SCRIB (Navarro et al. 2005), LET-413 is restricted to basolateral surfaces of epithelial cells (Bilder and Perrimon 2000; Navarro et al. 2005). During epithelial polarization, loss of LET-413 results in a delayed initial apical compaction of junction proteins (Koppen et al. 2001; McMahon et al. 2001). After compaction, apical proteins expand basolaterally in *let-413* mutant embryos and junctions fail to mature into belt-like junctions (Legouis et al. 2000; McMahon et al. 2001). These defects cause *let-413* mutant embryos to arrest during the beginning stages of elongation (Legouis et al. 2000).

dlg-1, a homologue of *Drosophila discs large*, encodes a MAGUK family scaffolding protein that contains three PDZ domains, one L27 domain, an SH3 domain and a (GuK) guanylate kinase domain (Bossinger et al. 2001; Firestein and Rongo 2001; Koppen et al. 2001; McMahon et al. 2001). DLG-1 is expressed just after epithelial cells begin to differentiate and localizes to the basal region of the *Ce*AJ. Junctions fail to mature in *dlg-1* mutant embryos, though not as severely as in *let-413* embryos, and embryos arrest during elongation. Through its L27 domain, DLG-1 is able to multimerize and bind to AJM-1, a coiled-coil domain protein that has no clear orthologues in *Drosophila* or mammals (Koppen et al. 2001; Lockwood et al. 2008b). *dlg-1* and *ajm-1* mutants have similar phenotypes, although *dlg-1* embryos arrest at an earlier stage. The colocalization of DLG-1 and AJM-1, ability of the two proteins to interact, and similarity of phenotypes, all suggest that DLG-1 and AJM-1 function in a common pathway to regulate junction maturation.

dlg-1 and *ajm-1* mutants have only minor defects in apicobasal polarity maintenance compared to *let-413* mutant embryos, and apical proteins show only a modest expansion into the basolateral domain (Koppen et al. 2001; McMahon et al. 2001). In contrast to *Drosophila* Discs large, which colocalizes with Scribble and Lethal (2) giant larvae (Bilder et al. 2000; Bilder and Perrimon 2000), DLG-1/Discs large localizes to a distinct domain immediately basal to AJs, while LET-413/Scribble extends basolaterally. In *dlg-1* and *ajm-1* mutants, large vacuoles appear in the posterior of the embryo, suggesting these genes may have a role in maintaining a permeability barrier (Firestein and Rongo 2001; Koppen et al. 2001; McMahon et al. 2001). Consistent with this idea, both *dlg-1* and *ajm-1* mutants have defects in formation of the apical, electron-dense structure that corresponds to the *Ce*AJ. Since *Drosophila* Discs large is required for septate junction formation (Bilder et al. 2003), these findings suggest a functional conservation between Discs large orthologues in both species.

The molecular roles of LET-413, DLG-1, and AJM-1 in junction maturation and positioning have not yet been elucidated. However, the fragmented junctions of *let-413(RNAi)* or *dlg-1(RNAi)* embryos can be partially rescued by depletion of the inositol 5-phosphatase homolog IPP-5 (Pilipiuk et al. 2009). This family of enzymes regulates levels of inositol triphosphate [IP_3] (Bui and Sternberg 2002), which can trigger calcium efflux and signaling within the cell. *ipp-5* mutants also rescue sterile phenotypes caused by RNAi depletion of PAR-3 in larval stages (which disrupts junction formation in reproductive tract epithelial cells), suggesting that calcium signaling may be a more general regulator of junction formation (Aono et al. 2004).

12.7 Modifiers of AJs

12.7.1 JAC-1/p120-catenin

p120-catenins bind to the cytoplasmic tail of classic cadherins and can regulate cadherin clustering and downstream signaling events (Ireton et al. 2002; Thoreson et al. 2000; Yap et al. 1998). *C. elegans* contains a single gene, *jac-1*, that encodes a p120-catenin (Pettitt et al. 2003). JAC-1/p120-catenin includes 10 armadillo-like repeats, which in other p120 catenins have been shown to bind to a juxtamembrane region in the cadherin cytoplasmic tail (Yap et al. 1998). JAC-1/p120-catenin also contains four N-terminal fibronectin-like repeats and a putative PDZ-binding domain, whose functional role is currently unknown (Pettitt et al. 2003).

JAC-1/p120-catenin co-localizes with the cadherin-catenin complex during elongation, depends on HMR-1/E-cadherin for localization, and binds to the HMR-1/E-cadherin cytoplasmic tail (Pettitt et al. 2003). In contrast to mammalian systems, where p120-catenins are required to stabilize and inhibit endocytosis of classic cadherins (Davis et al. 2003; Ireton et al. 2002), loss of JAC-1/p120-catenin causes only mild defects in HMR-1/E-cadherin localization and does not disrupt epidermal morphogenesis. However, JAC-1/p120-catenin depletion greatly enhances the epidermal morphogenesis defects caused by a weak mutation in *hmp-1*, suggesting that JAC-1/p120-catenin has a regulatory role in promoting AJ function (Pettitt et al. 2003). This enhancement is at least partially caused by defects in the organization of circumferential actin filament bundles, which become irregularly dense and separate from AJs during elongation, similar to null mutants in *hmr-1*, *hmp-1* and *hmp-2* mutants (Costa et al. 1998; Pettitt et al. 2003). Although mammalian p120-catenins function by modulating Rho GTPase signaling at junctions (Anastasiadis 2007), the molecular mechanism of JAC-1/p120-catenin junction regulation has not yet been determined.

12.7.2 F-BAR Proteins SRGP-1 and TOCA-1/2

SRGP-1 is the lone *C. elegans* ortholog of Slit-Robo GTPase activating protein (srGAP). SRGP-1 contains both an F-BAR domain responsible for inducing membrane curvature and a Rho GTPase activation (GAP) domain that can inhibit Rho

GTPase signaling (Frost et al. 2009; Guerrier et al. 2009; Zaidel-Bar et al. 2010). Originally, srGAP proteins were found to be downstream effectors of Slit-Robo in axon guidance (Wong et al. 2001). In *C. elegans* epithelia, SRGP-1 localizes to *Ce*AJs, although it can do so independently of HMR-1 (Zaidel-Bar et al. 2010). Loss of SRGP-1 results in slowed ventral enclosure and enhances the embryonic lethality of a weak *hmp-1* mutant. Overexpression of SRGP-1 induces membrane undulation at junctions, suggesting that local regulation of membrane dynamics likely contributes to AJ function.

F-BAR proteins of the TOCA family are conserved regulators of CDC-42 and N-WASP-dependent actin polymerization (Ho et al. 2004; Takano et al. 2008). The *C. elegans* TOCA proteins, TOCA-1 and 2, have been shown to regulate endocytosis in oocytes and localize to junctions in epidermal cells (Giuliani et al. 2009). In *Drosophila*, TOCA homologue Cip4 is required for proper endocytosis of E-cadherin at junctions (Leibfried et al. 2008). Similarly, *toca-1; toca-2* mutants show a significant increase in AJM-1 at junctions in epidermal cells (Giuliani et al. 2009). Loss of TOCA-1 and TOCA-2 causes internal tissues to extrude prior to ventral enclosure (*G*ut on the *ex*terior, or Gex phenotype), suggesting that in mutant embryos, a decrease in junction protein recycling significantly alters adhesiveness between epithelial cells (Giuliani et al. 2009). The Gex phenotype is also seen in *C. elegans* mutants for branched actin regulators (Bernadskaya et al. 2011; Patel et al. 2008; Soto et al. 2002; Withee et al. 2004), suggesting that a link between actin organization and trafficking of junction proteins is critical for morphogenesis.

12.8 Relatives of Vertebrate Tight Junction Proteins

12.8.1 VAB-9/BCMP1

Homologues or structural relatives of several vertebrate tight junction proteins localize to *Ce*AJs and help to ensure proper cell adhesion. VAB-9 is a four-pass transmembrane protein, structurally related to the claudin family of tight junction molecules and most similar to human brain cell membrane protein 1 (BCMP1) (Christophe-Hobertus et al. 2001; Simske et al. 2003). VAB-9 colocalizes with AJ proteins in the *Ce*AJ, and depends on HMR-1 (but not HMP-2 or HMP-1) for its initial recruitment to junctions (Simske et al. 2003). *vab-9* mutant embryos have defects in the organization of circumferential actin filaments within the epidermis and develop dorsal bulges during elongation (Simske et al. 2003), although most mutant embryos complete elongation and hatch into misshapen larvae. However, *vab-9* mutations enhance the elongation and adhesion defects of embryos lacking AJM-1 or DLG-1, as well as embryos with reduced levels of AJ proteins (Simske et al. 2003). Based on these interactions, it is likely that *vab-9* functions downstream of *hmr-1* to help mediate proper adhesion between epithelial cells.

12.8.2 ZOO-1/ZO-1

A likely downstream effector of VAB-9 is ZOO-1, the sole *C. elegans* homologue of vertebrate tight junction proteins in the zonula occludens (ZO) family (Lockwood et al. 2008a). ZOO-1 contains 3 PDZ domains, a *S*rc *h*omology 3 (SH3) domain and a *gu*anylate *k*inase (GuK) domain, and colocalizes with AJ proteins and VAB-9. ZOO-1 depends on both HMR-1 and VAB-9 for its junctional localization, but does not require HMP-2 or HMP-1. Like *vab-9* mutants, *zoo-1(RNAi)* embryos have reduced levels of actin at junctions, elongate slowly, develop bulges within the epidermis, and a small percentage of embryos rupture. Loss of *zoo-1* function significantly enhances the lethality of weak mutations in *hmp-1* and *hmp-2*, but does not enhance *vab-9* null mutants (Lockwood et al. 2008a), suggesting that it functions downstream of *vab-9* to mediate junctional actin organization and adhesion.

12.8.3 MAGI-1

Membrane-associated guanylate kinases (MAGUKs) proteins were originally described in vertebrate tight junctions and function to scaffold protein complexes at the cytoplasmic side of intercellular junctions (Funke et al. 2005). These proteins contain a PDZ domain, SH3 domain and GuK domain. Classical MAGUK family proteins found in *C. elegans* epithelial cells include DLG-1/Discs large and ZOO-1/ZO-1. MAGI-1, a homologue of the mammalian MAGUK MAGI-1, had previously been shown to function in the *C. elegans* nervous system (Stetak et al. 2009). In embryonic epithelia, MAGI-1 localizes apically and, unlike other modifiers of *C. elegans* AJs, MAGI-1 localization is largely independent of both the AJ complex and DLG-1 and AJM-1 (Stetak and Hajnal 2011). Although localization does not depend on other *Ce*AJ components, *magi-1* enhances both *dlg-1/ajm-1* and *hmr-1/hmp-1/hmp-2* mutants. Additionally, loss of MAGI-1 results in an increased overlap between AJ proteins and the DLG-1/AJM-1 domain within the *Ce*AJ (Stetak and Hajnal 2011). These data indicate that MAGI-1 may have an important regulatory role in segregating functional domains within the *Ce*AJ, and it will be important to learn how MAGI-1 functions.

12.9 Future Perspectives

The ease of genetic screening in *C. elegans* has allowed for the identification of many proteins that contribute to the formation, function, and regulation of AJs, providing an outstanding opportunity to learn how AJs function *in vivo*. However, many important questions remain. For example, a key difference between *C. elegans* and other systems is that the cadherin-catenin complex plays a rather minor role in cell adhesion. A remaining challenge will be to identify the proteins that function redundantly with AJ proteins to promote adhesion between epithelial cells and blastomeres, such as SAX-7 in early embryos, and to learn whether they work together with cadherins and catenins. It will also be important to determine whether genes that affect a similar process, such as *let-413* and *par-6* in junction maturation, function in a common pathway or have distinct molecular targets. Answering these and related questions will require a detailed understanding of the molecular function of junction and polarity proteins, and learning how they interface with the cadherin–catenin complex.

References

Aceto D, Beers M, Kemphues KJ (2006) Interaction of PAR-6 with CDC-42 is required for maintenance but not establishment of PAR asymmetry in *C. elegans*. Dev Biol 299:386–97

Achilleos A, Wehman AM, Nance J (2010) PAR-3 mediates the initial clustering and apical localization of junction and polarity proteins during *C. elegans* intestinal epithelial cell polarization. Development 137:1833–42

Adams CL, Chen YT, Smith SJ, Nelson WJ (1998) Mechanisms of epithelial cell-cell adhesion and cell compaction revealed by high-resolution tracking of E-cadherin-green fluorescent protein. J Cell Biol 142:1105–19

Anastasiadis PZ (2007) p120-ctn: a nexus for contextual signaling via Rho GTPases. Biochim et Biophys Acta 1773:34–46

Aono S, Legouis R, Hoose WA, Kemphues KJ (2004) PAR-3 is required for epithelial cell polarity in the distal spermatheca of *C. elegans*. Development 131:2865–74

Beatty A, Morton D, Kemphues K (2010) The *C. elegans* homolog of *Drosophila lethal giant larvae* functions redundantly with PAR-2 to maintain polarity in the early embryo. Development 137:3995–4004

Bernadskaya YY, Patel FB, Hsu HT, Soto MC. (2011) Arp2/3 promotes junction formation and maintenance in the *Caenorhabditis elegans* intestine by regulating membrane association of apical proteins. Mol Biol Cell 22:2886–99

Bilder D, Li M, Perrimon N (2000) Cooperative regulation of cell polarity and growth by *Drosophila* tumor suppressors. Science 289:113–6

Bilder D, Perrimon N (2000) Localization of apical epithelial determinants by the basolateral PDZ protein Scribble. Nature 403:676–80

Bilder D, Schober M, Perrimon N (2003) Integrated activity of PDZ protein complexes regulates epithelial polarity. Nat Cell Biol 5:53–58

Bossinger O, Klebes A, Segbert C, Theres C, Knust E (2001) Zonula adherens formation in *Caenorhabditis elegans* requires *dlg-1*, the homologue of the *Drosophila* gene *discs large*. Dev Biol 230:29–42

Broadbent ID, Pettitt J (2002) The *C. elegans hmr-1* gene can encode a neuronal classic cadherin involved in the regulation of axon fasciculation. Curr Biol CB 12:59–63

Bui YK, Sternberg PW (2002) *Caenorhabditis elegans* inositol 5-phosphatase homolog negatively regulates inositol 1,4,5-triphosphate signaling in ovulation. Mol Biol Cell 13:1641–51

Bulgakova NA, Knust E (2009) The Crumbs complex: from epithelial-cell polarity to retinal degeneration. J Cell Sci 122:2587–2596

Chen W, Lim HH, Lim L (1993) The CDC42 homologue from *Caenorhabditis elegans*. Complementation of yeast mutation. J Biol Chem 268:13280–5

Cheng NN, Kirby CM, Kemphues KJ (1995) Control of cleavage spindle orientation in *Caenorhabditis elegans*: the role of the genes *par-2* and *par-3*. Genetics 139:549–59

Chisholm AD, Hardin J (2005) Epidermal morphogenesis. WormBook: the online review of *C. elegans* Biology 1–22

Christophe-Hobertus C, Szpirer C, Guyon R, Christophe D (2001) Identification of the gene encoding brain cell membrane protein 1 (BCMP1), a putative four-transmembrane protein distantly related to the peripheral myelin protein 22/epithelial membrane proteins and the claudins. BMC Genomics 2:3

Costa M, Raich W, Agbunag C, Leung B, Hardin J, Priess JR (1998) A putative catenin-cadherin system mediates morphogenesis of the *Caenorhabditis elegans* embryo. J Cell Biol 141:297–308

Davis MA, Ireton RC, Reynolds AB (2003) A core function for p120-catenin in cadherin turnover. J Cell Biol 163:525–34

Ding M, Woo WM, Chisholm AD (2004) The cytoskeleton and epidermal morphogenesis in *C. elegans*. Exp Cell Res 301:84–90

Etemad-Moghadam B, Guo S, Kemphues KJ (1995) Asymmetrically distributed PAR-3 protein contributes to cell polarity and spindle alignment in early *C. elegans* embryos. Cell 83:743–52

Fan S, Hurd TW, Liu CJ, Straight SW, Weimbs T, Hurd EA, Domino SE, Margolis B (2004) Polarity proteins control ciliogenesis via kinesin motor interactions. Curr Biol 14:1451–61

Fichelson P, Jagut M, Lepanse S, Lepesant J-A, Huynh J-R (2010) *lethal giant larvae* is required with the *par* genes for the early polarization of the *Drosophila* oocyte. Development 137:815–824

Firestein BL, Rongo C (2001) DLG-1 is a MAGUK similar to SAP97 and is required for adherens junction formation. Mol Biol Cell 12:3465–75

Frost A, Unger VM, De Camilli P (2009) The BAR domain superfamily: membrane-molding macromolecules. Cell 137:191–6

Funke L, Dakoji S, Bredt DS (2005) Membrane-associated guanylate kinases regulate adhesion and plasticity at cell junctions. Annu Rev Biochem 74:219–45

Giuliani C, Troglio F, Bai Z, Patel FB, Zucconi A, Malabarba MG, Disanza A, Stradal TB, Cassata G, Confalonieri S, Hardin JD, Soto MC, Grant BD, Scita G (2009) Requirements for F-BAR proteins TOCA-1 and TOCA-2 in actin dynamics and membrane trafficking during *Caenorhabditis elegans* oocyte growth and embryonic epidermal morphogenesis. PLoS Genet 5 e1000675

Goldstein B, Macara IG (2007) The PAR proteins: fundamental players in animal cell polarization. Dev Cell 13:609–22

Gotta M, Abraham MC, Ahringer J (2001) CDC-42 controls early cell polarity and spindle orientation in *C. elegans*. Curr Biol 11:482–8

Grana TM, Cox EA, Lynch AM, Hardin J (2010) SAX-7/L1CAM and HMR-1/cadherin function redundantly in blastomere compaction and non-muscle myosin accumulation during *Caenorhabditis elegans* gastrulation. Dev Biol 344:731–744

Guerrier S, Coutinho-Budd J, Sassa T, Gresset A, Jordan NV, Chen K, Jin W-L, Frost A, Polleux F (2009) The F-BAR domain of srGAP2 induces membrane protrusions required for neuronal migration and morphogenesis. Cell 138:990–1004

Harris TJ, Peifer M (2005) The positioning and segregation of apical cues during epithelial polarity establishment in *Drosophila*. J Cell Biol 170:813–23

Hill E, Broadbent ID, Chothia C, Pettitt J (2001) Cadherin superfamily proteins in *Caenorhabditis elegans* and *Drosophila melanogaster*. J Mol Biol 305:1011–24

Ho HY, Rohatgi R, Lebensohn AM, Le M, Li J, Gygi SP, Kirschner MW (2004) Toca-1 mediates Cdc42-dependent actin nucleation by activating the N-WASP-WIP complex. Cell 118:203–16

Hung TJ, Kemphues KJ (1999) PAR-6 is a conserved PDZ domain-containing protein that colocalizes with PAR-3 in *Caenorhabditis elegans* embryos. Development 126:127–35

Ireton RC, Davis MA, van Hengel J, Mariner DJ, Barnes K, Thoreson MA, Anastasiadis PZ, Matrisian L, Bundy LM, Sealy L, Gilbert B, van Roy F, Reynolds AB (2002) A novel role for p120 catenin in E-cadherin function. J Cell Biol 159:465–76

Kay AJ, Hunter CP (2001) CDC-42 regulates PAR protein localization and function to control cellular and embryonic polarity in *C. elegans*. Curr Biol 11:474–81

Kemphues KJ, Priess JR, Morton DG, Cheng NS (1988) Identification of genes required for cytoplasmic localization in early *C. elegans* embryos. Cell 52:311–20

Koppen M, Simske JS, Sims PA, Firestein BL, Hall DH, Radice AD, Rongo C, Hardin JD (2001) Cooperative regulation of AJM-1 controls junctional integrity in *Caenorhabditis elegans* epithelia. Nat Cell Biol 3:983–991

Korswagen HC, Herman MA, Clevers HC (2000) Distinct beta-catenins mediate adhesion and signalling functions in *C. elegans*. Nature 406:527–32

Kwiatkowski AV, Maiden SL, Pokutta S, Choi HJ, Benjamin JM, Lynch AM, Nelson WJ, Weis WI, Hardin J (2010) *In vitro* and *in vivo* reconstitution of the cadherin-catenin-actin complex from *Caenorhabditis elegans*. Proc Natl Acad Sci U S A 107:14591–6

Labouesse M (2006) Epithelial junctions and attachments. WormBook: the online review of *C. elegans*. Biology 13:1–21

Lee JY, Goldstein B (2003) Mechanisms of cell positioning during *C. elegans* gastrulation. Development 130:307–20

Legouis R, Gansmuller A, Sookhareea S, Bosher JM, Baillie DL, Labouesse M (2000) LET-413 is a basolateral protein required for the assembly of adherens junctions in *Caenorhabditis elegans*. Nat Cell Biol 2:415–22

Leibfried A, Fricke R, Morgan MJ, Bogdan S, Bellaiche Y (2008) *Drosophila* Cip4 and WASp define a branch of the Cdc42-Par6-aPKC pathway regulating E-Cadherin endocytosis. Curr Biol 18:1639–1648

Leung B, Hermann GJ, Priess JR (1999) Organogenesis of the *Caenorhabditis elegans* intestine. Dev Biol 216:114–34

Lockwood C, Zaidel-Bar R, Hardin J (2008a) The *C. elegans* zonula occludens ortholog cooperates with the cadherin complex to recruit actin during morphogenesis. Curr Biol 18:1333–7

Lockwood CA, Lynch AM, Hardin J (2008b) Dynamic analysis identifies novel roles for DLG-1 subdomains in AJM-1 recruitment and LET-413-dependent apical focusing. J Cell Sci 121:1477–1487

McGill MA, McKinley RF, Harris TJ (2009) Independent cadherin-catenin and Bazooka clusters interact to assemble adherens junctions. J Cell Biol 185:787–96

McMahon L, Legouis R, Vonesch JL, Labouesse M (2001) Assembly of *C. elegans* apical junctions involves positioning and compaction by LET-413 and protein aggregation by the MAGUK protein DLG-1. J Cell Sci 114:2265–77

Mohler WA, Shemer G, del Campo JJ, Valansi C, Opoku-Serebuoh E, Scranton V, Assaf N, White JG, Podbilewicz B (2002) The type I membrane protein EFF-1 is essential for developmental cell fusion. Dev Cell 2:355–62

Mohler WA, Simske JS, Williams-Masson EM, Hardin JD, White JG (1998) Dynamics and ultrastructure of developmental cell fusions in the *Caenorhabditis elegans* hypodermis. Curr Biol 8:1087–90

Morais-De-Sa E, Mirouse V, Johnston DS (2010) aPKC phosphorylation of Bazooka defines the apical/lateral border in *Drosophila* epithelial cells. Cell 141:509–523

Muller HJ, Bossinger O (2003) Molecular networks controlling epithelial cell polarity in development. Mech Dev 120:1231–1256

Nance J, Priess JR (2002) Cell polarity and gastrulation in *C. elegans*. Development 129:387–97

Nance J, Zallen JA (2011) Elaborating polarity: PAR proteins and the cytoskeleton. Development 138:799–809

Navarro C, Nola S, Audebert S, Santoni MJ, Arsanto JP, Ginestier C, Marchetto S, Jacquemier J, Isnardon D, Le Bivic A, Birnbaum D, Borg JP (2005) Junctional recruitment of mammalian Scribble relies on E-cadherin engagement. Oncogene 24:4330–9

Oda H, Uemura T, Harada Y, Iwai Y, Takeichi M (1994) A *Drosophila* homolog of cadherin associated with Armadillo and essential for embryonic cell-cell adhesion. Dev Biol 165:716–26

Patel FB, Bernadskaya YY, Chen E, Jobanputra A, Pooladi Z, Freeman KL, Gally C, Mohler WA, Soto MC (2008) The WAVE/SCAR complex promotes polarized cell movements and actin enrichment in epithelia during *C. elegans* embryogenesis. Dev Biol 324:297–309

Pettitt J, Cox EA, Broadbent ID, Flett A, Hardin J (2003) The *Caenorhabditis elegans* p120 catenin homologue, JAC-1, modulates cadherin-catenin function during epidermal morphogenesis. J Cell Biol 162:15–22

Pilipiuk J, Lefebvre C, Wiesenfahrt T, Legouis R, Bossinger O (2009) Increased IP3/Ca^{2+} signaling compensates depletion of LET-413/DLG-1 in *C. elegans* epithelial junction assembly. Dev Biol 327:34–47

Podbilewicz B, White JG (1994) Cell fusions in the developing epithelia of *C. elegans*. Dev Biol 161:408–24

Portereiko MF, Saam J, Mango SE (2004) ZEN-4/MKLP1 is required to polarize the foregut epithelium. Curr Biol 14:932–41

Priess JR, Hirsh DI (1986) *Caenorhabditis elegans* morphogenesis: the role of the cytoskeleton in elongation of the embryo. Dev Biol 117:156–73

Putzke AP, Rothman JH (2010) Repression of Wnt signaling by a Fer-type nonreceptor tyrosine kinase. Proc Natl Acad Sci U S A 107:16154–9

Raich WB, Agbunag C, Hardin J (1999) Rapid epithelial-sheet sealing in the *Caenorhabditis elegans* embryo requires cadherin-dependent filopodial priming. Curr Biol 9:1139–46

Sapir A, Choi J, Leikina E, Avinoam O, Valansi C, Chernomordik LV, Newman AP, Podbilewicz B (2007) AFF-1, a FOS-1-regulated fusogen, mediates fusion of the anchor cell in *C. elegans*. Dev Cell 12:683–98

Segbert C, Johnson K, Theres C, van Fürden D, Bossinger O (2004) Molecular and functional analysis of apical junction formation in the gut epithelium of *Caenorhabditis elegans*. Dev Biol 266:17–26

Shapiro L, Weis WI (2009) Structure and biochemistry of cadherins and catenins. Cold Spring Harb Perspect Biol 1:a003053

Shemer G, Suissa M, Kolotuev I, Nguyen KC, Hall DH, Podbilewicz B (2004) EFF-1 is sufficient to initiate and execute tissue-specific cell fusion in *C. elegans*. Curr Biol 14:1587–91

Shibata Y, Fujii T, Dent JA, Fujisawa H, Takagi S (2000) EAT-20, a novel transmembrane protein with EGF motifs, is required for efficient feeding in *Caenorhabditis elegans*. Genetics 154:635–46

Simske JS, Köppen M, Sims P, Hodgkin J, Yonkof A, Hardin J (2003) The cell junction protein VAB-9 regulates adhesion and epidermal morphology in *C. elegans*. Nat Cell Biol 5:619–25

Soto MC, Qadota H, Kasuya K, Inoue M, Tsuboi D, Mello CC, Kaibuchi K (2002) The GEX-2 and GEX-3 proteins are required for tissue morphogenesis and cell migrations in *C. elegans*. Genes Dev 16:620–32

St Johnston D, Ahringer J (2010) Cell polarity in eggs and epithelia: parallels and diversity. Cell 141:757–74

Stetak A, Hajnal A (2011) The *C. elegans* MAGI-1 protein is a novel component of cell junctions that is required for junctional compartmentalization. Dev Biol 350:24–31

Stetak A, Horndli F, Maricq AV, Van Den Heuvel S, Hajnal A (2009) Neuron-specific regulation of associative learning and memory by MAGI-1 in *C. elegans*. PloS One 4:e6019

Sulston JE, Schierenberg E, White JG, Thomson JN (1983) The embryonic cell lineage of the nematode *Caenorhabditis elegans*. Dev Biol 100:64–119

Sumiyoshi E, Takahashi S, Obata H, Sugimoto A, Kohara Y (2011) The beta-catenin HMP-2 functions downstream of src in parallel with the Wnt pathway in early embryogenesis of *C. elegans*. Dev Biol 355:302–12

Tabuse Y, Izumi Y, Piano F, Kemphues KJ, Miwa J, Ohno S (1998) Atypical protein kinase C cooperates with PAR-3 to establish embryonic polarity in *Caenorhabditis elegans*. Development 125:3607–14

Takano K, Toyooka K, Suetsugu S (2008) EFC/F-BAR proteins and the N-WASP-WIP complex induce membrane curvature-dependent actin polymerization. EMBO J 27:2817–28

Thoreson MA, Anastasiadis PZ, Daniel JM, Ireton RC, Wheelock MJ, Johnson KR, Hummingbird DK, Reynolds AB (2000) Selective uncoupling of p120(ctn) from E-cadherin disrupts strong adhesion. J Cell Biol 148:189–202

Totong R, Achilleos A, Nance J (2007) PAR-6 is required for junction formation but not apicobasal polarization in *C. elegans* embryonic epithelial cells. Development 134:1259–68

Vasioukhin V, Bauer C, Yin M, Fuchs E (2000) Directed actin polymerization is the driving force for epithelial cell-cell adhesion. Cell 100:209–19

Watts JL, Etemad-Moghadam B, Guo S, Boyd L, Draper BW, Mello CC, Priess JR, Kemphues KJ (1996) par-6, a gene involved in the establishment of asymmetry in early *C. elegans* embryos, mediates the asymmetric localization of PAR-3. Development 122:3133–40

Williams-Masson EM, Malik AN, Hardin J (1997) An actin-mediated two-step mechanism is required for ventral enclosure of the *C. elegans* hypodermis. Development 124:2889–901

Withee J, Galligan B, Hawkins N, Garriga G (2004) *Caenorhabditis elegans* WASP and Ena/VASP proteins play compensatory roles in morphogenesis and neuronal cell migration. Genetics 167:1165–76

Wong K, Ren XR, Huang YZ, Xie Y, Liu G, Saito H, Tang H, Wen L, Brady-Kalnay SM, Mei L, Wu JY, Xiong WC, Rao Y (2001) Signal transduction in neuronal migration: roles of GTPase activating proteins and the small GTPase Cdc42 in the Slit-Robo pathway. Cell 107:209–21

Woods DF, Hough C, Peel D, Callaini G, Bryant PJ (1996) Dlg protein is required for junction structure, cell polarity, and proliferation control in *Drosophila* epithelia. J Cell Biol 134:1469–82

Yap AS, Niessen CM, Gumbiner BM (1998) The juxtamembrane region of the cadherin cytoplasmic tail supports lateral clustering, adhesive strengthening, and interaction with p120ctn. J Cell Biol 141:779–89

Zaidel-Bar R, Joyce MJ, Lynch AM, Witte K, Audhya A, Hardin J (2010) The F-BAR domain of SRGP-1 facilitates cell-cell adhesion during *C. elegans* morphogenesis. J Cell Biol 191:761–769

Zhang H, Gally C, Labouesse M (2010) Tissue morphogenesis: how multiple cells cooperate to generate a tissue. Curr Opin Cell Biol 22:575–82

Zhang H, Labouesse M (2010) The making of hemidesmosome structures *in vivo*. Devel Dyn 239:1465–76

Chapter 13
Cadherin Function During Xenopus Gastrulation

Rudolf Winklbauer

Abstract Xenopus gastrulation consists of the orderly deformation of a single, multilayered cell sheet that resembles a multilayered epithelium, and flexible cell–cell adhesion has to provide tissue cohesion while allowing for cell rearrangements that drive gastrulation. A few classic cadherins are expressed in the Xenopus early embryo. The prominent C-cadherin is essential for the cohesion of the animal part of the gastrula including ectoderm and chordamesoderm, and it contributes to the adhesion of endoderm and anterior mesoderm in the vegetal moiety. The cadherin/catenin complex is expressed in a graded pattern which is stable during early development. Regional differences in cell adhesion conform to the graded cadherin/catenin expression pattern. However, although the cadherin/catenin pattern seems to be actively maintained, and cadherin function is modulated to reinforce differential adhesiveness, it is not clear how regional differences in tissue cohesion affect gastrulation. Manipulating cadherin expression or function does not induce cell sorting or boundary formation in the embryo. Moreover, known boundary formation mechanisms in the gastrula are based on active cell repulsion. Cell rearrangement is also compatible with variable tissue cohesion. Thus, identifying roles for differential adhesion in the Xenopus gastrula remains a challenge.

13.1 The Xenopus Gastrula: Structure and Movements of a Cohesive Tissue

The Xenopus early embryo forms a single, continuous, cohesive tissue which is locally deformed during gastrulation. For example, it folds back on itself during involution, stretches in the course of epiboly, or narrows and lengthens during convergent extension (Fig. 13.1). Thus, flexible tissue cohesion which ensures tissue integrity yet permits cell movement must be an essential property of the gastrula.

R. Winklbauer (✉)
Department of Cell and Systems Biology, University of Toronto, Toronto, Canada
e-mail: r.winklbauer@utoronto.ca

T. Harris (ed.), *Adherens Junctions: From Molecular Mechanisms to Tissue Development and Disease,* Subcellular Biochemistry 60,
DOI 10.1007/978-94-007-4186-7_13, © Springer Science+Business Media Dordrecht 2012

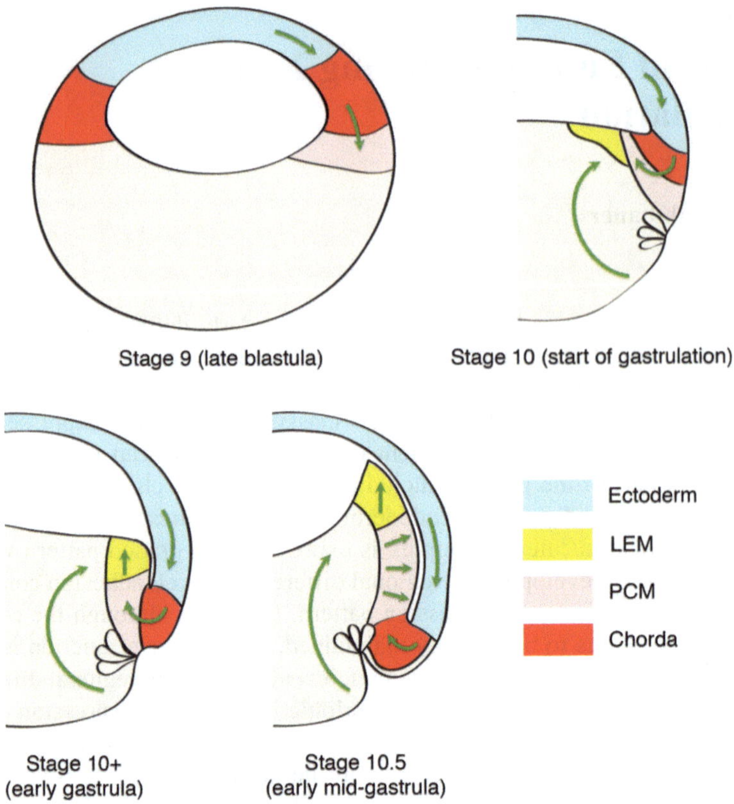

Fig. 13.1 Major gastrula regions and gastrulation movements. Whole blastula or dorsal half of gastrula is depicted for different stages, and *arrows* indicate directions of main movements during early gastrulation (see text for details). *LEM* leading edge mesoderm; *PCM* prechordal mesoderm; *Chorda* chordamesoderm. Vegetal cell mass, beige

The large egg of Xenopus is transformed by a series of rapid holoblastic cleavage divisions into a coherent mass of cells, or blastomeres. Inside the mass, a blastocoel cavity develops. Blastomeres divide either in the plane of the surface, or perpendicular to it, thus generating a multilayered blastocoel wall (Figs. 13.1, 13.2a, b). In the surface layer of the blastula thus formed, cells are polarized apico-basally. Their apical membrane is derived from the egg membrane, whereas basolateral membranes, like the membranes of the deep blastomeres, are newly formed during cleavage. The new membranes contain cadherins and catenins, and mediate cell–cell adhesion (Fig. 13.2a, b). In contrast, the egg-derived apical membrane is non-adhesive. Between apical and basolateral membrane domains, junctional complexes consisting of tight junctions, adherens junctions and desmosomes develop during cleavage (Byers and Armstrong 1986; Regen and Steinhardt 1986; Roberts

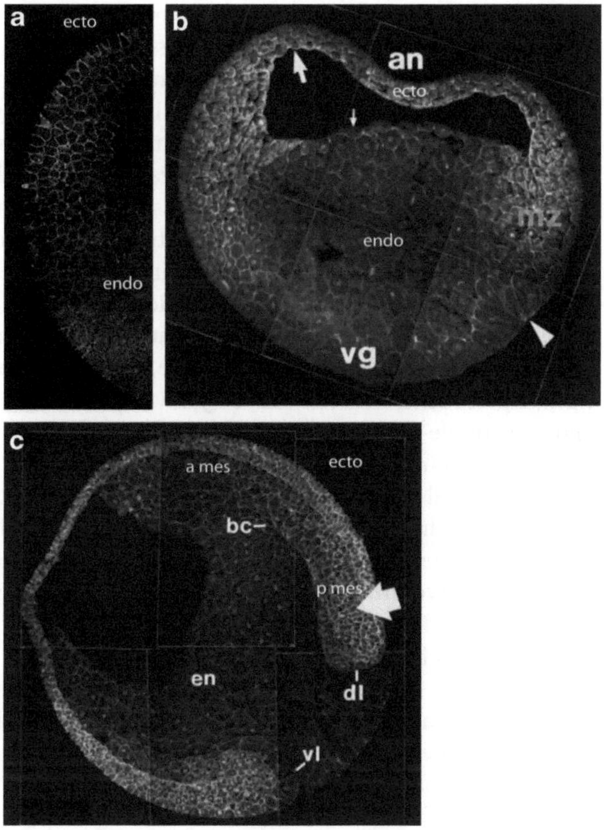

Fig. 13.2 The cadherin/catenin complex in the early Xenopus embryo. **a** Late blastula, α-catenin antibody staining of the ventral side, sagittal section. Modified after Schneider et al. (1993). **b, c** β-catenin antibody staining of sagittal sections of the early **b** and late **c** gastrula. Modified after Fagotto and Gumbiner (1994). *an* animal pole; *vg* vegetal pole; *ecto* ectoderm; *endo* endoderm; *mz* marginal zone mesoderm; *a mes* anterior mesoderm; *p mes* posterior mesoderm; *vl, dl* ventral and dorsal blastopore lip; *arrowhead* in **b**, bc in **c**: bottle cells of blastopore; *arrows* in **b**, free basolateral surfaces of ectoderm and endoderm; *arrow* in **c**, blastopore lip

et al. 1992; Müller and Hausen 1995; Chalmers et al. 2003; Müller 2001; Strauss et al. 2006).

Due to its apico-basolateral polarity and its junctional complexes, the outer layer of cells is often considered to be an epithelium that covers the inner, 'mesenchymal' layers. However, cells in all layers express the same cadherins (see below) and epithelial intermediate filaments (Klymkowsky et al. 1992). Moreover, apical and deep layers are not separated by a basal lamina. Thus, the blastocoel wall may as well be regarded as a multilayered epithelium (Shook and Keller 2003).

The blastula wall is multilayered in all regions, but it varies in thickness (Figs. 13.1, 13.2). Its thin, small-celled animal part forms the blastocoel roof. The vegetal part consists of a compact mass of large, yolk-rich cells, and of an equatorial ring of intermediate-sized cells, the marginal zone (Nieuwkoop and Florschutz 1950; Nakatsuji 1975; Keller 1976; Keller and Schoenwolf 1977). The marginal zone contains the prospective mesoderm, the vegetal cell mass will become included in the endoderm, and most of the blastocoel roof will form ectoderm (Fig. 13.1) (Keller 1975, 1976; Moody 1987; Dale and Slack 1987).

During gastrulation, the prospective mesoderm and endoderm are moved to the interior of the embryo, and different mesodermal and endodermal tissue precursors are positioned along the future dorsoventral and anteroposterior axes. In these gastrulation movements, three-dimensional active cell rearrangement, or 'intercellular migration' (Gumbiner 2005)—the movement of cells across the surface of neighbouring cells—plays a fundamental role. In contrast, epithelial shape change is of minor importance. Its most prominent expression is the establishment of the blastopore by bottle cell formation at the onset of gastrulation: epithelial cells at the vegetal boundary of the dorsal marginal zone constrict their apical surfaces to generate the blastoporal indentation (Fig. 13.2b, arrowhead) (Keller 1978, 1981; Hardin and Keller 1988).

Mesoderm and endoderm internalization are driven by the cooperation of two intercellular migration processes. Above the blastopore, in the so-called blastopore lip, the mesoderm moves inward by involution (Figs. 13.1, 13.2c) (Keller 1981), and this tissue-autonomous movement is associated cell rearrangement (Winklbauer and Schurfeld 1999; Ibrahim and Winklbauer 2001). Below the blastopore, the endodermal vegetal cell mass internalizes by vegetal rotation. Cells surge animally toward the blastocoel floor, and outward (Winklbauer and Schurfeld 1999), thus moving the peripheral blastocoel floor against the ectodermal blastocoel roof (Fig. 13.1). Involution and vegetal rotation together generate a vortex pattern of movement, with involution contributing a downward and inward flow of material, and vegetal rotation a complementary upward and outward movement of cells (Fig. 13.1).

Once internalized, several processes contribute to the animally directed movement of mesoderm and endoderm. Most anteriorly, vegetal cell mass is constantly added at the leading edge by a continuation of vegetal rotation (Bauer et al. 1994; Ibrahim and Winklbauer 2001), and cells migrate across the ectoderm (Fig. 13.1) (Nakatsuji 1975; Keller and Schoenwolf 1977; Winklbauer 1990; Winklbauer and Nagel 1991; Davidson et al. 2002; Nagel et al. 2004). Posterior to it, the dorsal prechordal mesoderm spreads on the overlying ectoderm by radial cell intercalation (Fig. 13.1), transforming the thick multilayered tissue into a single-layered sheet (Damm and Winklbauer 2011). Still further posterior, the somitic and chordamesoderm cells rearrange by medio-lateral and radial intercalation to narrow and lengthen the tissue in the process of convergent extension (Keller and Danilchik 1988; Keller and Tibbetts 1989; Wilson et al. 1989; Wilson and Keller 1991; Shih and Keller 1992; Lane and Keller 1997; for review see Keller 2002; Keller et al. 2000, 2003; Wallingford et al. 2002). In the ectoderm, a similar convergent extension process elongates the prospective central nervous system (Keller et al. 1992; Elul and Keller 2000). Eventually, the blastopore closes below the vegetal cell mass. Inside the embryo, the archenteron inflates dorsally (Nieuwkoop and Florschutz 1950; Keller 1981; Ewald et al. 2004). As mesoderm and endoderm are removed from the surface of the embryo, the ectoderm spreads to cover the embryo in the process of epiboly (Fig. 13.1). The respective increase in area is due to the passive stretching of the outer, epithelial layer, and to a rearrangement of inner cells which reduces the number of cell layers (Keller 1978, 1980; Marsden and DeSimone 2001; Luu et al. 2011).

13.2 Classic Cadherins and Catenins in the Early Xenopus Embryo

13.2.1 Cadherin Isoforms

Gastrula tissue cohesion is essentially mediated by cadherin-based cell–cell adhesion. The vertebrate classic cadherins contain five extracellular cadherin repeats (EC domains) and a conserved cytoplasmic domain which binds the armadillo-repeat proteins β-catenin, plakoglobin, and p120-catenins, and other factors. β-catenin interacts with α-catenin to dynamically link cadherins to the actin cytoskeleton, whereas p120-catenins regulate cadherin stability at the cell membrane (Gumbiner 2005; Halbleib and Nelson 2006; Nelson 2008; Pokutta and Weiss 2007; Harris and Tepass 2010; Niessen et al. 2011).

In Xenopus, a limited number of cadherin isoforms are expressed in the early embryo. At cleavage and blastula stages, maternal cadherins support cell adhesion. EP- and C-cadherin (Ginsberg et al. 1991; Choi et al. 1990) are allelic variants, as are XB- and U-cadherin, respectively (Herzberg et al. 1991; Angres et al. 1991). EP/C- and XB/U-cadherins in turn are pseudoalleles generated in Xenopus laevis during a tetraploidization event (Kuhl and Wedlich 1996), and they are closely related to mammalian P-cadherin (Gallin 1998). These cadherins are expressed on the basolateral membranes of all blastomeres (Angres et al. 1991; Herzberg et al. 1991; Levi et al. 1991), with XB/U-cadherin being a minor component (Müller et al. 1994). They continue to be expressed after the onset of zygotic transcription at the mid-blastula stage, and throughout gastrulation. The type II classic cadherin-11 is maternally expressed at a low level, and at increased levels in the animal and marginal parts of the gastrula (Hadeball et al 1998). XmN-cadherin is also expressed maternally, but is soon downregulated (Hojyo et al. 1998). E-cadherin appears at mid-gastrula stages in the ectoderm, preferentially in its outer, epithelial layer (Choi and Gumbiner 1989; Angres et al. 1991; Schneider et al. 1993; Nandadasa et al. 2009), and N-cadherin is expressed on prospective neural plate cells in the dorsal ectoderm of late gastrulae (Detrick et al. 1990; Nandadasa et al. 2009).

13.2.2 Catenins

In the Xenopus embryo, β- and α-catenin are maternally expressed. β-catenin constitutively binds to cadherins, and in the gastrula most of it is associated with EP/C- and XB/U-cadherin (DeMarais and Moon 1992; Schneider et al. 1993; Fagotto and Gumbiner 1994). When these cadherins are overexpressed or depleted, β-catenin membrane density changes accordingly (Kurth et al. 1999; Ninomiya et al. 2012). Whereas β-catenin colocalizes with cadherin in cell–cell contacts, but also on exposed cell membranes (Fig. 13.2b, arrow), α-catenin is restricted to areas of contact (Fig. 13.2a) (Schneider et al. 1993; Kurth et al. 1999). Apparently, α-catenin is not a constitutive component of the cadherin/catenin complex, but correlated with actual cell contact formation. Inhibiting α-catenin function diminishes blastomere adhesion (Kofron et al. 1997; Sehgal et al. 1997). The inhibition of β-catenin has been studied in the context of canonical Wnt signaling, but not of adhesion, in the Xenopus embryo.

Plakoglobin, which is closely related to β-catenin, associates with desmosomal cadherins, but is also found in adherens junctions (Chitaev et al. 1998). In Xenopus, it is maternally expressed (Fouquet et al. 1992; Kofron et al. 1997) and forms puncta at cell membranes, but these puncta do not colocalize with overexpressed cadherin (Kurth et al. 1999). In plakoglobin-depleted embryos, cell adhesion is reduced in the vegetal cell mass and gastrulation is delayed (Kofron et al. 1997, 2002).

Members of the p120-catenin subfamily of armadillo-repeat proteins bind to the cytoplasmic domain of classic cadherins and attenuate cadherin endocytosis (Reynolds and Carnahan 2004). They potentially activate Rac, inhibit RhoA (Noren et al. 2000; Grosheva et al. 2001; Yanagisawa and Anastiasiadis 2006), and control the lateral clustering of cadherins (Yap et al. 1998). In the Xenopus early embryo, p120-catenin, ARVCF, and δ-catenin are expressed. Their depletion or overexpression diminishes or increases cadherin protein levels, respectively (Fang et al. 2004; Tao et al. 2007; McCrea and Park 2007; Gu et al. 2009). Knockdown of δ-catenin was shown not only to reduce cadherin expression, but also ectoderm cell adhesion and the rate of blastopore closure, in partial redundancy with p120-catenin and through the activation of RhoA (Gu et al. 2009). Overexpression of membrane-tethered ARVCF in gastrula ectoderm cells diminished adhesion independently of cadherin expression, probably through an inhibition of Rac1 (Reintsch et al. 2008).

13.2.3 Subcellular Localization of the Cadherin/Catenin Complex

Cadherins and associated catenins are present in all basolateral membranes of the blastula and gastrula, but are usually absent from apical membranes (Fig. 13.2). Thus, the deep cells of the blastocoel wall are uniformly decorated with cadherins, whereas the epithelial cells at the embryo surface are polarized apico-basolaterally (Angres et al. 1991; Fagotto and Gumbiner 1994; Schneider et al. 1993; Kurth et al. 1999). At the end of gastrulation, however, apical expression of cadherins is observed in certain regions. C-cadherin becomes apically localized in the future epidermis of the ventral ectoderm, whereas in neural ectoderm, N-cadherin shows apical expression in addition to its basolateral localization (Nandadasa et al. 2007). An apical localization of β-catenin at the very end of gastrulation is consistent with this redistribution of cadherins (Fagotto and Gumbiner 1994). The function of apical cadherin in these regions is not known.

Although transmission electron microscopy shows that sub-apical junctional complexes develop already in the blastula, this is not reflected in a corresponding localization of cadherins or catenins. β-catenin distribution is rather uniform at blastula stages. The first indication of sub-apical concentrations is seen in bottle cells at gastrulation (Fig. 13.2b, arrowhead), and in ectoderm-derived epidermis after gastrulation (Fagotto and Gumbiner 1994). Also, no increased sub-apical cadherin expression has been reported for any of the cadherins during gastrulation. A possible explanation for this discrepancy could be that cadherin/β-catenin expression is initially high in all membrane domains, which to some degree masks the junctional cadherin accumulation. Reduction of cadherin density at later stages could then reveal the presence of junctions. Consistent with this possibility, the less densely spaced α-catenin (see below) shows a distinct sub-apical enrichment suggestive of adherens junctions at the blastula stage (Schneider et al. 1993).

Cadherins often occur as clusters on cell membranes, which appear in the light microscope as fine puncta (e.g. Angres et al 1996; Cavey et al. 2008). Such cadherin puncta are apparently not a striking feature in the Xenopus gastrula, as they are usually not noted in respective descriptions of cadherin localization. However, Tao et al. (2007) describe C-cadherin puncta for the late blastula, and a finely punctate pattern is seen for membrane β-catenin at high magnification (Kurth et al. 1999). Thus, it seems likely that cadherin and β-catenin form puncta in the Xenopus early embryo, but due to their high density they are usually not resolved. In contrast to this, α-catenin forms well-defined puncta at blastula and gastrula stages (Schneider et al. 1993). This suggests that only a subset of cadherin puncta contain α-catenin. Since small gaps are present between cells (Johnson 1970; Nakatsuji 1976), perhaps only cadherin engaged in *trans*-binding attracts α-catenin.

13.2.4 Temporal and Spatial Expression of the Cadherin/Catenin Complex

In the gastrula, cadherin and β-catenin densities increase from a relatively high level in the ectoderm to their strongest expression in the dorsal chordamesoderm, become lower again in anterior mesoderm, and are weakest in the endodermal vegetal mass (Fig. 13.2b, c) (Fagotto and Gumbiner 1994; Angres et al. 1991). The densities change in a graded fashion between regions except where tissues form new contacts in the course of gastrulation. Thus, as the anterior mesoderm internalizes and advances toward the animal pole, the contrast between less intensely stained anterior mesoderm and strongly labeled ectoderm becomes apparent (Fig. 13.2c) (Fagotto and Gumbiner 1994; Ogata et al. 2007). In the late gastrula, β-catenin staining increases in the epithelial layer of the ectoderm, probably reflecting the onset of E-cadherin expression (Fagotto and Gumbiner 1994). A similar overall pattern is seen with α-catenin staining (Fig. 13.2a) (Schneider et al. 1993).

Cadherin/catenin intensity differences corresponding to the gastrula pattern are observed already at early cleavage stages. XB/U-cadherin, β-catenin and α-catenin are strongly expressed in marginal zone blastomere, at slightly lower levels animally, and weakly vegetally (Herzberg et al. 1991; Schneider et al. 1993; Fagotto and Gumbiner 1994). As this pattern is set up before the onset of zygotic transcription, it apparently has to be established in the egg cytoplasm under maternal control. Maternal cadherin and catenin proteins are indeed prelocalized in the egg (Herzberg et al. 1991; Ginsberg et al. 1991; Schneider et al. 1993). During gastrulation, this basic cadherin/catenin pattern is distorted by tissue rearrangements, but is otherwise not very dynamic (i.e. the deformation of the single, multilayered tissue which forms the Xenopus gastrula occurs without much change in local cadherin/catenin membrane expression). This maintenance of the expression pattern is not trivial, given the extensive changes during early development, like the transition from maternal control to zygotic transcription, or the induction and patterning of mesodermal and neural tissues.

It is not well understood how the overall cadherin/catenin pattern is actively maintained, although a few components of its regulation are known. The best-characterized modulator of cadherin membrane expression in the Xenopus gastrula is the FLRT3/Rnd1 pathway. It involves the interaction of a fibronectin-leucine-rich transmembrane (FLRT) protein with a small GTPase-like factor, Rnd1 in the anterior mesoderm (Ogata et al. 2007). FLRTs contain a series of leucine-rich repeats and a fibronectin type III domain in their extracellular part, and ectopic expression of cytoplasmically truncated FLRT3 in the embryo is sufficient to drive cell sorting (Karaulanov et al. 2006). Rnd1 differs from other Rho-like GTPases by lacking intrinsic GTPase activity, and its overexpression disrupts cell adhesion in the Xenopus embryo (Wunnenberg-Stapleton et al. 1999). In the gastrula, FLRT3 and Rnd1 are expressed in the involuted anterior mesoderm. The proteins interact physically, and FLRT3 acts upstream of Rnd1 to upregulate dynamin-mediated endocytosis of C-cadherin. This leads to the apposition of tissues with low (anterior mesoderm) and high levels (ectoderm) of membrane-expressed cadherin (Fig. 13.2c). Knockdown of FLRT or Rnd increases cadherin density in the mesoderm and leads to gastrulation defects (Ogata et al. 2007). Similar modulators of cadherin membrane expression could exist for other regions to locally control the cadherin/catenin pattern in a mosaic fashion.

At a global level, ARVCF, δ-catenin and p120-catenin affect cadherin expression of gastrula cells (Fang et al. 2004; Tao et al. 2007; Gu et al. 2009). Also, two G-protein-coupled receptors, the LPA receptor and a receptor for an unknown ligand, Xflop, are expressed in the early embryo, and overexpression of either receptor increases cadherin membrane density in the late blastula, whereas depletion diminishes it (Tao et al. 2007). However, a role for any of these factors in determining the endogenous pattern of cadherin/catenin complex expression has not yet been demonstrated.

13.3 Cadherin Function in the Xenopus Gastrula

13.3.1 Cadherins and Basic Gastrula Tissue Cohesion

Although the cadherin/catenin complex is expressed in a steep animal-to-vegetal gradient, it could still provide basic tissue cohesion in all regions. The maternal cadherins are indeed essential for cell adhesion in the blastula (Angres et al. 1991;

Müller et al. 1994; Heasman et al. 1994; Kurth et al. 1999). Blocking the maternal expression of EP/C-cadherin, or of both EP/C- and XB/U-cadherin, by antisense oligonucleotide injection into oocytes diminishes the adhesion of inner blastula cells while leaving the epithelial layer intact (Heasman et al. 1994; Kurth et al. 1999).

Injection of antisense oligonucleotides into the fertilized egg affects zygotic expression of C-cadherin, and consequently reduces cell adhesion in the gastrula (Ninomiya et al. 2012). Due to the lower cadherin density in the vegetal regions, one might expect these to be more sensitive to cadherin knockdown. However, the effect is strongest in the ectoderm and chordamesoderm where complete dissociation of the tissue can occur, while it is weak in anterior mesoderm and endoderm, where it is only detected as reductions of *in vitro* cell adhesion or tissue surface tension (Ninomiya et al. 2012). This differential sensitivity develops at the onset of gastrulation. In the blastula, both animal and vegetal regions are affected by the knockdown of maternal cadherin (Heasman et al. 1994), but in the early gastrula, vegetal cells become able to adhere normally while the animally located ectoderm cells are still dissociated (Kurth et al. 1999). Thus, cadherins and in particular C-cadherin are essential for the cohesion of gastrula tissues which are derived from the animal part of the egg, such as ectoderm and chordamesoderm. In vegetally derived regions such as anterior mesoderm or endoderm, C-cadherin contributes to adhesion, but other factors seem to be more important.

13.3.2 Cadherins and Regional Differences of Adhesion

Differences in adhesiveness between tissues have been claimed to be important determinants of morphogenesis and embryo structure (Steinberg 1970; Foty and Steinberg 2005; Krieg et al. 2008). Quantitative differences in cadherin expression are sufficient to drive cell sorting *in vitro* (Steinberg and Takeichi 1994; Duguay et al. 2003), and the expression of different cadherins is associated with the positioning of cell populations *in vivo* (Price et al. 2002). The strength of cell adhesion in the amphibian embryo has been estimated based on concepts borrowed from fluid mechanics. In general, mesenchymal cell aggregates exhibit liquid-like behavior, and surface tension has been used to explain phenomena such as cell sorting or the mutual engulfment of tissues (Steinberg 1970; Foty et al. 1996; Krieg et al. 2008; Schotz et al. 2008). By definition, surface tension is equal to half the work required, per unit area, to separate a liquid into two parts, and is thus a measure of cohesion; it can be used to probe cell–cell adhesion in intact tissues. In a more qualitative assay, cell sorting can be treated analogously to the demixing of immiscible fluids (Graner 1993; Beysen et al. 2000), and can be used to infer differences in adhesion between cell populations. As in fluids, the more cohesive phase of a mixture usually sorts to the center, the less cohesive fraction to the periphery.

Tissue surface tension differs between regions of the Xenopus gastrula. It is lowest in the endodermal vegetal cell mass, intermediate in the mesoderm, and equally high or higher in the ectoderm (Luu et al. 2011). In another frog species, Rana, surface tension values show a similar regional pattern (Davis et al. 1997). Differences between regions are also seen in cell sorting experiments. Animal, vegetal and marginal zone cells sort from each other after mixing at the earliest blastula stages when such experiments are feasible (Turner et al. 1989). Later, differences between germ layers are observed. When ectoderm is experimentally induced to form mesoderm or endoderm, cadherin-dependent cell adhesiveness changes accordingly (Brieher and Gumbiner 1994), and the cell types sort from each other and from ectodermal cells (Kuroda et al. 1999). Differences exist also within germ layers. Cells from anterior and posterior mesoderm sort out (Ninomiya et al. 2004), and when ectoderm cells are induced by different doses of the TGF-β like factor, activin, to form prechordal mesoderm, anterior and posterior chordamesoderm, and then mixed with uninduced ectoderm, all four cell types segregate from each other (Ninomiya and Winklbauer 2008). Apparently, a pattern of differential adhesiveness is established in early development. Generally, tissue cohesiveness increases from the vegetal endoderm to the mesoderm and ectoderm.

This pattern of differential adhesion approximately fits the graded quantitative differences in cadherin expression along the animal-vegetal axis. In Xenopus, moderate knockdown of C-cadherin can reduce surface tension in the ectoderm to an endodermal level, implying that increased cell adhesion in the ectoderm is at least partially due to the high levels of cadherin expression in this tissue (Ninomiya et al. 2012). However, region-specific expression of different cadherins as a mechanism to establish differently adhesive cell populations is rare in the Xenopus gastrula. E-cadherin (Choi and Gumbiner 1989; Angres et al. 1991) appears only in the late gastrula, in the epithelial layer of the ectoderm. Expression of cytoplasmically truncated E-cadherin affects the integrity of the epithelial layer, and this defect can be rescued by full-length E-cadherin, but not C-cadherin (Levine et al. 1994; Nandadasa et al. 2009). However, it is not known whether this specific function of E-cadherin in the epithelial layer leads to adhesion differences between this layer and the deep ectoderm cells.

Further evidence for a role of cadherins in establishing regional differences of adhesiveness comes from studies that show a corresponding regional modulation not of cadherin expression, but of cadherin activity. Cdc42 is expressed in an animal-to-vegetal gradient similar to the cadherin/catenin complex (Choi and Han 2002), and is possibly linked to this complex by IQGAPs. IQGAPs contain binding sites for Cdc42 and Rac, actin, cadherin and β-catenin, and regulate the cytoskeleton and cell adhesion (Kuroda et al. 1998; Fukata et al. 1999). Two isoforms, IQGAP1 and IQGAP2, are expressed in the Xenopus embryo (Yamashiro et al. 2003). IQGAP1 overexpression activates Cdc42 (Sokol et al. 2001), and when levels of IQGAP1 and the nuclear-localized IQGAP2 are diminished simultaneously, β-catenin and IQGAP1 localization at cell contacts are reduced and adhesion is decreased (Yamashiro et al. 2007). This suggests a positive regulation of adhesion by Cdc42 and IQGAP, potentially reinforcing adhesion differences due to graded cadherin expression.

In the mesoderm, paraxial protocadherin (PAPC) interacts with the above-mentioned FLRT/Rnd pathway to downregulate adhesiveness independently of the

FLRT-induced reduction of C-cadherin expression. PAPC is initially expressed in the dorsal anterior mesoderm, under the control of activin and Wnt signaling, and the transcription factor Xlim1 (Kim et al. 1998; Hukriede et al. 2003; Medina et al. 2004; Schambony and Wedlich 2007). In the course of gastrulation, expression spreads laterally and ventrally, ceases in the dorsal mesoderm, and increases in the somitic mesoderm (Kim et al. 1998; Hukriede et al. 2003). PAPC seems not to act as an adhesion molecule (Chen and Gumbiner 2006). However, its ectopic expression promotes cell sorting, and expression of its extracellular domain decreases C-cadherin dependent adhesion while the knockdown of PAPC increases the adhesiveness of mesoderm cells (Kim et al. 1998; Chen and Gumbiner 2006). For this modulation of C-cadherin dependent cell adhesion, PAPC forms a complex with C-cadherin, FLRT3, Rnd1, and the netrin receptors Unc5B and Unc5D (Karaulanov et al. 2009; Chen et al. 2009). Interactions within the complex seem to be complicated. PAPC and FLRT3 are able to reduce adhesion independently, but when both are present, PAPC counteracts FLRT3 activity and limits the strong dissociating effect of FLRT3 to a physiological level (Chen et al. 2009).

Non-protein membrane constituents could also modulate cadherin-dependent adhesion. Neutral glycolipids are present in an animal to vegetal gradient in the blastula and gastrula, and antibodies against them interfere with the reaggregation of animal, but not vegetal blastomeres (Turner et al. 1992). Ganglioside GM1 is detected in the animal hemisphere and marginal zone, whereas sulfated galactosylceramide is present in the vegetal blastomeres (Kubo et al. 1995). Adhesion between blastomeres depends also on blood-group B trisaccharide-bearing molecules including glyco-sphingolipids and GPI-anchored glycoproteins (Nomura et al. 1998). Glycolipids and glycoproteins can colocalize with cadherins and promote cell adhesion in a cooperative fashion (Adachi et al. 2008), consistent with the cadherin/catenin complex and the glycolipids/glycoproteins both being essential for cell–cell adhesion. However, Ca^{++}-dependent cell adhesion can also be mediated by direct carbohyrate-carbohydrate interaction (Bucior and Burger 2004), and in animal blastomeres of Xenopus, neutral glycolipids and cadherins seem to act additively (Heasman et al. 1994).

13.3.3 The Function of Cadherin-Dependent Adhesion Differences

Gastrula regions differ in cell–cell adhesion, and the modulation of cadherin expression and function at least contributes to these differences. This raises the question of the functional relevance of differential adhesion in the Xenopus gastrula. A potential role for differential adhesion is cell sorting. However, the various tissues of the Xenopus embryo are formed by the partitioning of a common cell mass, and not from dispersed cells that become individually specified and would then need to sort out.

A related role for differential adhesion is the formation of a sorting boundary. However, the available evidence suggests that cadherin-dependent adhesion differences are not able to generate such boundaries in the Xenopus embryo. Casual observations had indicated that localized C-cadherin overexpression or inhibition by dominant-negative constructs do not induce the sorting of respective gastrula cell clones into domains with distinct sorting boundaries; instead, normal dispersal of injected cells was observed (Lee and Gumbiner 1995; Broders and Thiery 1995; Kurth et al. 1999; Reintsch et al. 2005). A further analysis of this unexpected effect showed that modulations of cadherin expression that change cell adhesion and drive the sorting of dispersed cells *in vitro*, do not lead to similar sorting in the embryo. This lack of *in vivo* sorting was observed for ectoderm cells which rearrange passively during epiboly, but also for actively intercalating chordamesoderm cells (Ninomiya et al. 2012). Together, these findings argue against a role for cadherin-dependent differential adhesion in boundary formation.

The best-studied example of boundary formation in the Xenopus gastrula is the establishment of Brachet's cleft at the boundary between ectoderm and mesoderm (Figs. 13.1, 13.2c). Consistent with the above conclusion, it is not based on differential adhesion but on active, localized cell–cell repulsion at the boundary. In the embryo, mesoderm and ectoderm remain separated despite the fact that no physical obstacle, such as a basal lamina, prevents their fusion. The same cadherins are expressed on either side of Brachet's cleft, but expression levels differ, which would be consistent with differential adhesion-based sorting. However, Brachet's cleft forms even if cadherin density is rendered similar on both sides (Ogata et al. 2007). Moreover, PAPC function is required for tissue separation (Hukriede et al. 2003; Medina et al. 2004), but its ability to reduce the strength of C-cadherin mediated adhesion is not important for this role (Chen and Gumbiner 2006). Tissue separation requires signaling of PAPC through its cytoplasmic effector, ankyrin repeat domain protein 5 (xANR5) (Chung et al. 2007), whereas C-cadherin function is modulated by the extracellular domain of PAPC (Chen and Gumbiner 2006). Together, the available evidence suggests that tissue separation at Brachet's cleft is not due to a difference in cadherin-mediated adhesion. However, overexpression of either EP/C- or XB/U-cadherin in the mesoderm interferes with separation (Wacker et al. 2000). Apparently, cadherin-mediated adhesion must somehow be overcome to allow for the separation of ectoderm and mesoderm, and the respective mechanism can be overwhelmed by an excess of cadherin.

Such a mechanism for separation at Brachet's cleft has been identified. Various EphB receptors and ephrinB ligands are expressed in the mesoderm and ectoderm of the Xenopus gastrula, and EphB forward signalling across the ectoderm-mesoderm boundary leads to cycles of cell–cell repulsion at the boundary (Rohani et al. 2011). When cells come into contact, ephrinB ligands can bind to the EphB receptors and induce signaling, which leads to cell detachment. Upon detachment, signaling ceases, which in turn allows for the reattachment of cells, and a further repetition of the cycle. This agrees with the notion that a mechanism independent of differential adhesion establishes Brachet's cleft. Its existence complicates the interpretation of sorting experiments, as factors other than classical differential adhesion have to be taken into account.

13.3.4 Cadherin Adhesion and Intercellular Migration

During intercellular migration, cells use each other as substrata for translocation, and one can expect that cadherins play a major role in such processes. To study the role of cadherins in Xenopus gastrula intercellular migration, extracellularly or cytoplasmically truncated constructs have previously been used which were thought to act as dominant negatives. The isolated cytoplasmic domain of cadherins is thought to compete with endogenous cadherins for factors such as catenins (Kintner 1992). Expression of the membrane-anchored extracellular domain of cadherins also diminishes adhesion. Moderate doses reduce tissue surface tension (Kalantarian et al. 2009; Ninomiya et al. submitted), and high doses cause disintegration of the ectoderm (Lee and Gumbiner 1995), although the construct still mediates cell binding to immobilized cadherin (Brieher et al. 1996; Seifert et al. 2009). Also, the effect of this construct on adhesion *in vivo* may, in part, be related to its apparent ability to regulate the interaction of aPKC with the endogenous cadherins (Seifert et al. 2009).

In Xenopus, expression of the C-cadherin extracellular domain impedes mesoderm involution, blastopore closure and convergent extension, and a similar phenotype is obtained with an analogous XB-cadherin construct (Lee and Gumbiner 1995; Kuhl et al. 1996; Kalantarian et al. 2009; Seifert et al. 2009). Expression of the cytoplasmic domain also leads to an arrest of gastrulation and a loss of tissue integrity (Lee and Gumbiner 1995). When expressed in a small region of the gastrula only, involution and convergent extension are compromised when the affected cells are included in the chordamesoderm and adjacent ectoderm. If present in the anterior mesoderm, gastrulation movements are almost normal (Broders and Thiery 1995), in agreement with the finding that this region is less dependent on cadherin function.

The interpretation of these experiments is difficult due to the use of cadherin constructs with ill-defined effects on cell adhesion. Recently, the knockdown of C-cadherin by morpholino antisense oligonucleotides was used to examine this question (Ninomiya et al. 2012). As mentioned above, a moderate knockdown affected tissue cohesion, cell adhesion to cadherin substratum, and cell sorting *in vitro*, but neither did it prompt cell sorting in the embryo, nor did it interfere with gastrulation movements. In particular, mediolateral cell intercalation in the chordamesoderm progressed and almost normal larvae formed despite the fact that morpholino injected cells were round and only loosely attached to each other. Only at high doses, when chordamesoderm and ectoderm tissues began to dissociate, were cell intercalation and consequently convergent extension inhibited. Movement of the less affected anterior mesoderm was still normal. Together, these findings indicate that gastrulation is strikingly insensitive to variations in tissue cohesion. Movements are blocked when tissue integrity is disrupted, but above a critical threshold of adhesive strength, embryos complete gastrulation. Interestingly, however, increasing C-cadherin affinity with an activating antibody attenuates convergent extension movements (Zhong et al. 1999).

13.4 Conclusion

Xenopus gastrulation consists essentially of the ordered deformation of a single, multilayered cell sheet that resembles a multilayered epithelium. Consequently, cell–cell adhesion has to be sufficiently firm to guarantee tissue cohesion, yet flexible enough to allow for the various forms of cell rearrangement that drive gastrula morphogenesis. A small number of classic cadherins are expressed in the Xenopus early embryo, most prominently among them the P-cadherin related C-cadherin. C-cadherin is essential for the cohesion of the animal part of the gastrula including ectoderm and chordamesoderm, and it contributes to adhesion in the vegetal moiety which harbors the endoderm and the anterior mesoderm. The cadherin/catenin complex is expressed in a graded pattern which is surprisingly stable during early development from early cleavage to the late gastrula. The region-specific expression of different cadherins plays a minor role before the late gastrula stage.

Regional differences in cell–cell adhesion and hence tissue cohesion roughly conform to the graded cadherin/catenin expression pattern. However, although evidence suggests that the cadherin/catenin pattern is actively maintained, and that cadherin function is modulated to reinforce the anticipated differential adhesiveness, it is not clear how the observed differences in tissue cohesion between gastrula regions affect gastrulation. Manipulating cadherin expression or function does not lead to cell sorting and boundary formation in the embryo, and the best-known boundary formation mechanism in the gastrula is based on localized cell–cell repulsion. Cell rearrangement seems also compatible with a wide range of tissue cohesion values, further drawing the function of adhesion differences into question. Thus, identifying definitive roles for differential adhesion in the Xenopus gastrula remains a major challenge.

Acknowledgements Research of the author is funded by the Canadian Institutes of Health Research (CIHR). I thank Olivia Luu and Erich Damm for preparing the figures.

References

Adachi T, Sato C, Kishi Y, Totani K, Murata T, Usui T, Kitajima K (2008) Membrane microdomains from early gastrula embryos of medaka, Oryzias latipes, are a platform of E-cadherin- and carbohydrate-mediated cell-cell interaction during epiboly. Glycoconj J 26:285–299

Angres B, Müller AHJ, Kellermann J, Hausen P (1991) Differential expression of two cadherins in Xenopus laevis. Development 111:8229–8244

Angres B, Barth A, Nelson WJ (1996) Mechanism for transition from initial to stable cell--cell adhesion: kinetic analysis of E-cadherin-mediated adhesion using a quantitative adhesion assay. J Cell Biol 134:549–557

Bauer DV, Huang S, Moody SA (1994) The cleavage stage origin of Spemann's organizer: analysis of the movements of blastomere clones before and during gastrulation. Development 120:1179–1189

Beysen DA, Forgacs G, Glazier JA (2000) Cell sorting is analogous to phase ordering in fluids. Proc Natl Acad Sci U S A 97:9467–9471

Brieher WM, Gumbiner BM (1994) Regulation of C-cadherin function during activin induced morphogenesis of Xenopus animal caps. J Cell Biol 126:519–527

Brieher WM, Yap AS, Gumbiner BM (1996) Lateral dimerization is required for the homophilic binding activity of C-cadherin. J Cell Biol 135:487–496

Broders F, Thiery J-P (1995) Contribution of cadherins to directional cell migration and histogenesis in Xenopus embryos. Cell Adh Comm 3:419–440

Bucior I, Burger MM (2004) Carbohydrate-carbohydrate interactions in cell recognition. Curr Opin Struct Biol 14:631–637

Byers TJ, Armstrong PB (1986) Membrane protein redistribution during Xenopus first cleavage. J Cell Biol 102:2176–2184

Cavey M, Rauzi M, Lenne PF, Lecuit T (2008) A two-tiered mechanism for stabilization and immobilization of E-cadherin. Nature 453:751–756

Chalmers AD, Strauss B, Papalopulu N (2003) Oriented cell divisions asymmetrically segregate aPKC and generate cell fate diversity in the early Xenopus embryo. Development 130:2657–2668

Chen X, Gumbiner BM (2006) Paraxial protocadherin mediates cell sorting and tissue morphogenesis by regulating C-cadherin adhesion activity. J Cell Biol 174:301–313

Chen X, Koh E, Gumbiner BM (2009) A protocadherin-cadherin-FLRT3 complex controls cell adhesion and morphogenesis. PLoS One 4(12):e8411

Chitaev NA, Averbakh AZ, Troyanovsky RB, Troyanovsky SM (1998) Molecular organization of the desmoglein-plakoglobin complex. J Cell Sci 111:1941–1949

Choi YS, Gumbiner B (1989) Expression of cell adhesion molecule E-cadherin in Xenopus embryos begins at gastrulation and predominates in the ectoderm. J Cell Biol 108:2449–2458

Choi YS, Seghal R, McCrea P, Gumbiner B (1990) A cadherin-like protein in eggs and cleaving embryos of Xenopus laevis is expressed in oocytes in response to progesterone. J Cell Biol 110:1575–1582

Choi S-C, Han J-K (2002) Xenopus Cdc42 regulates convergent extension movements during gastrulation through Wnt/Ca^{2+} signaling pathway. Dev Biol 244:342–357

Chung HA, Yamamoto TS, Ueno N (2007) ANR5, an FGF target gene product, regulates gastrulation in Xenopus. Curr Biol 17:932–939

Dale L, Slack JMW (1987) Fate map for the 32-cell stage of Xenopus laevis. Development 99:527–551

Damm EW, Winklbauer R (2011) PDGF-A controls mesoderm cell orientation and radial intercalation during Xenopus gastrulation. Development 138:565–575

Davidson LA, Hoffstrom BG, Keller R, DeSimone DW (2002) Mesendoderm extension and mantle closure in Xenopus laevis gastrulation: combined roles for integrin aa, fibronectin, and tissue geometry. Dev Biol 242:109–129

Davis GS, Phillips HM, Steinberg MS (1997) Germ-layer surface tensions and "tissue affinities" in Rana pipiens gastrulae: Quantitative measurements. Dev Biol 192:630–644

DeMarais AA, Moon RT (1992) The armadillo homologs β-catenin and plakoglobin are differentially expressed during early development of Xenopus laevis. Dev Biol 153:337–346

Detrick RJ, Dickey D, Kintern CR (1990) The effects of N-cadherin misexpression on morphogenesis in Xenopus embryos. Neuron 4:493–506

Duguay D, Foty RA, Steinberg MS (2003) Cadherin-mediated cell adhesion and tissue segregation: qualitative and quantitative determinants. Dev Biol 253:309–323

Elul T, Keller R (2000) Monopolar protrusive activity: a new morphogenetic cell behavior in the neural plate dependent on vertical interactions with the mesoderm in Xenopus. Dev Biol 224:3–19

Ewald AJ, Peyrot SM, Tyszka JM, Fraser SE, Wallingford JB (2004) Regional requirements for Dishevelled signaling during Xenopus gastrulation: separable effects on blastopore closure, mesendoderm internalization and archenteron formation. Development 131:6195–6209

Fagotto F, Gumbiner BM (1994) Beta-catenin localization during Xenopus embryogenesis: accumulation at tissue and somite boundaries. Development 120:3667–3679

Fang X, Ji H, Kim SW, Park JI, Vaught TG, Anastasiadis PZ, Ciesiolka M, McCrea PD (2004) Vertebrate development requires ARVCF and p120 catenins and their interplay with RhoA and Rac. J Cell Biol 165:87–98

Foty RA, Pfleger CM, Forgacs G, Steinberg MS (1996) Surface tensions of embryonic tissues predict their mutual envelopment behavior. Development 122:1611–1620

Foty RA, Steinberg MS (2005) The differential adhesion hypothesis: a direct evaluation. Dev Biol 278:255–263

Fouquet B, Zimbelmann R, Franke WW (1992) Identification of plakoglobin in oocytes and early embryos of Xenopus laevis: maternal expression of a gene encoding a junctional plaque protein. Differentiation 51:187–194

Fukata M, Nakagawa M, Kuroda S, Kaibuchi K (1999) Cell adhesion and Rho small GTPases. J Cell Sci 112:4491–4500

Gallin WJ (1998) Evolution of the "classical" cadherin family of cell adhesion molecules in vertebrates. Mol Biol Evol 15:1099–1107

Ginsberg D, DeSimone D, Geiger B (1991) Expression of a novel cadherin (EP-cadherin) in unfertilized eggs and early Xenopus embryos. Development 111:315–325

Graner F (1993) Can surface adhesion drive cell-rearrangement? Part I: Biological cell-sorting. J theor Biol 164:445–476

Grosheva I, Shtutman M, Elbaum M, Bershadsky AD (2001) P120 catenin affects cell motility via modulation of activity of Rho-family GTPases: a link between cell-cell contact formation and regulation of cell locomotion. J Cell Sci 114:695–707

Gu D, Sater AK, Ji H, Cho K, Clark M, Stratton SA, Barton MC, Lu Q, McCrea PD (2009) Xenopus δ-catenin is essential in early embryogenesis and is functionally linked to cadherins and small GTPases. J Cell Sci 122:4049–4061

Gumbiner BM (2005) Regulation of cadherin-mediated adhesion in morphogenesis. Nature Rev Mol Cell Biol 6:622–634

Hadeball B, Borchers A, Wedlich D (1998) Xenopus cadherin-11 expression requires the Wg/Wnt signal. Mech Dev 72:101–113.

Halbleib JM, Nelson WJ (2006) Cadherins in development: cell adhesion, sorting, and tissue morphogenesis. Genes Dev 20:3199–3214

Hardin J, Keller R (1988) The behavior and function of bottle cells during gastrulation of Xenopus laevis. Development 103:211–230

Harris TJC, Tepass U (2010) Adherens junctions: from molecules to morphogenesis. Nat Rev Mol Cell Biol 11:502–514

Heasman J, Ginsberg D, Geiger B, Goldstone K, Pratt T, Yoshida-Noro C, Wylie C (1994) A functional test for maternally inherited cadherin in Xenopus shows its importance in cell adhesion at the blastula stage. Development 120:49–57

Herzberg F, Wildermuth V, Wedlich D (1991) Expression of XB-cad, a novel cadherin, during oogenesis and early development of Xenopus. Mech Dev 35:33–42

Hojyo T, Tooi O, Tashiro K, Shiokawa K (1998) Exogastrula formation in Xenopus laevis embryos depleted with maternal XmN-cadherin mRNA by antisense S-oligo DNA. Biochem Biophys Res Commun 242:170–175

Hukriede NA, Tsang TE, Habas R, Khoo PL, Steiner K, Weeks DL, Tam PP, Dawid IB (2003) Conserved requirement of Lim1 function for cell movements during gastrulation. Dev Cell 4:83–94

Ibrahim H, Winklbauer R (2001) Mechanisms of mesendoderm internalization in the Xenopus gastrula: Lessons from the ventral side. Dev Biol 240:108–122

Johnson KE (1970) The role of changes in cell contact behavior in amphibian gastrulation. J exp Zool 175:391–428

Kalantarian A, Ninomiya H, Saad SM, David R, Winklbauer R, Neumann AW (2009) Axisymmetric drop shape analysis for estimating the surface tension of cell aggregates by centrifugation. Biophys J 96:1606–1616

Karaulanov EE, Bottcher RT, Niehrs C (2006) A role for fibronectin-leucin-rich transmembrane cell-surface proteins in homotypic cell adhesion. EMBO Rep 7:283–290

Karaulanov EE, Bottcher RT, Stannek P, Wu W, Rau M, Ogata S, Cho KWY, Niehrs C (2009) Unc5B interacts with FLRT3 and Rnd1 to modulate cell adhesion in Xenopus embryos. PLoS One 4(5):e5742
Keller RE (1975) Vital dye mapping of the gastrula and neurula of Xenopus laevis. I. Prospective areas and morphogenetic movements of the superficial layer. Dev Biol 42:222–241
Keller RE (1976) Vital dye mapping of the gastrula and neurula of Xenopus laevis. II. Prospective areas and morphogenetic movements of the deep layer. Dev Biol 51:118–137
Keller RE (1978) Time-lapse cinemicrographic analysis of superficial cell behavior during and prior to gastrulation in Xenopus laevis. J Morphol 157:223–248
Keller RE (1980) The cellular basis of epiboly: an SEM study of deep-cell rearrangement during gastrulation in Xenopus laevis. J Embryol exp Morph 60:201–234
Keller R (1981) An experimental analysis of the role of bottle cells and the deep marginal zone in gastrulation of Xenopus laevis. J exp Zool 216:81–101
Keller R (2002) Shaping the vertebrate body plan by polarized embryonic cell movements. Science 298:1950–1954
Keller R, Danilchik M (1988) Regional expression, pattern and timing of convergence and extension during gastrulation of Xenopus laevis. Dev 103:193–209
Keller RE, Schoenwolf GC (1977) An SEM study of cellular morphology, contact, and arrangement, as related to gastrulation in Xenopus laevis. W. Roux's Arch 182:165–186
Keller R, Tibbetts P (1989) Mediolateral cell intercalation in the dorsal, axial mesoderm of Xenopus laevis. Dev Biol 131:539–549
Keller R, Shih J, Sater A (1992) The cellular basis of the convergence and extension of the Xenopus neural plate. Dev Dyn 193:199–217
Keller R, Davidson R, Edlund A, Elul T, Ezin M, Shook D, Skoglund P (2000) Mechanisms of convergence and extension by cell intercalation. Phil Trans R Soc Lond B 355:897–922
Keller R, Davidson LA, Shook DR (2003) How we are shaped: the biomechanics of gastrulation. Different 71:171–205
Kim S-H, Yamamoto A, Bouwmeester T, Agius E, De Robertis EM (1998) The role of Paraxial Protocadherin in selective adhesion and cell movements of the mesoderm during Xenopus gastrulation. Development 125:4681–4691
Kintner CR (1992) Regulation of embryonic cell adhesion by the cadherin cytoplasmic domain. Cell 69:225–236
Klymkowsky MW, Shook DR, Maynell LA (1992) Evidence that the deep keratin filament systems of the Xenopus embryo act to ensure normal gastrulation. Proc Natl Acad Sci U S A 89:8736–8740
Kofron M, Spagnuolo A, Klymkowsky M, Wylie C, Heasman J (1997) The roles of maternal α-catenin and plakoglobin in the early Xenopus embryo. Development 124:1553–1560
Kofron M, Heasman J, Lang SA, Wylie CC (2002) Plakoglobin is required for maintenance of the cortical actin skeleton in early Xenopus embryos and for cdc42-mediated wound healing. J Cell Biol 158:695–708
Krieg M, Arboleda-Estudillo Y, Puech P-H, Kafer J, Graner F, Müller DJ, Heisenberg C-P (2008) Tensile forces govern germ-layer organization in zebrafish. Nat Cell Biol 10:429–436
Kubo H, Kotani M, Ozawa H, Kawashima I, Tai T, Suzuki A (1995) Differential distribution of ganglioside GM1 and sulfatide during the development of Xenopus embryos. Develop Growth Differ 37:243–255
Kuhl M, Finnemann S, Binder O, Wedlich D (1996) Dominant-negative expression of a cytoplasmatically deleted mutant of XB/U-cadherin disturbs mesoderm migration during gastrulation in Xenopus laevis. Mech Dev 54:71–82
Kuhl M, Wedlich D (1996) Xenopus cadherins: sorting out types and functions in embryogenesis. Dev Dyn 207:121–134
Kuroda S, Fukata M, Nakagawa M, Fuji K, Nakamura T (1998) Role of IQGAP1, a target of small GTPases Cdc42 and Rac1, in regulation of E-cadherin-mediated cell-cell adhesion. Science 281:832–835

Kuroda H, Sakumoto H, Kinoshita K, Asashima M (1999) Changes in the adhesive properties of dissociated and reaggregated Xenopus laevis embryo cells. Dev Growth Differ 41:283–291

Kurth T, Fesenko IV, Schneider S, Munchberg FE, Joos TE, Spieker TP, Hausen P (1999) Immunocytochemical studies of the interactions of cadherins and catenins in the early Xenopus embryo. Dev Dyn 215:155–169

Lane MC, Keller R (1997) Microtubule disruption reveals that Spemann's organizer is subdivided into two domains by the vegetal alignment zone. Development 124:895–906

Lee C-H, Gumbiner B (1995) Disruption of gastrulation movements in Xenopus by a dominant-negative mutant for C-cadherin. Dev Biol 171:363–373

Levi G, Ginsberg D, Girault J-M, Sabanay I, Thiery J-P, Geiger B (1991) EP-cadherin in muscles and epithelia of Xenopus laevis embryos. Development 113:1335–1344

Levine E, Lee CH, Kintner C, Gumbiner BM (1994) Selective disruption of E-cadherin function in early Xenopus embryos by a dominant negative mutant. Development 120:901–909

Luu O, David R, Ninomiya H, Winklbauer R (2011) Large-scale mechanical properties of Xenopus embryonic epithelium. Proc natl Acad Sci U S A 108:4000–4005

Marsden M, DeSimone DW (2001) Regulation of cell polarity, radial intercalation and epiboly in Xenopus: novel roles for integrin and fibronectin. Development 128:3635–3647

McCrea PD, Park JI (2007) Developmental functions of the P120-catenin sub-family. Biochim Biophys Acta 1773:17–33

Medina A, Swain RK, Kuerner K-M, Steinbeisser H (2004) Xenopus paraxial protocadherin has signaling functions and is involved in tissue separation. EMBO J 23:3249–3258

Moody SA (1987) Fates of the blastomeres of the 32-cell-stage Xenopus embryo. Dev Biol 122:300–319

Müller HA (2001) Of mice, frogs and flies: generation of membrane asymmetries in early development. Dev Growth Differ 43:327–342

Müller H-AJ, Hausen P (1995) Epithelial cell polarity in early Xenopus development. Dev Dyn 202:405–420

Müller H-AJ, Kuhl M, Finnemann S, Schneider S, Van Der Poel SZ, Hausen P, Wedlich D (1994) Xenopus cadherins: the maternal pool comprises distinguishable members of the family. Mech Dev 47:213–223

Nagel M, Tahinci E, Symes K, Winklbauer R (2004) Guidance of mesoderm cell migration in the Xenopus gastrula requires PDGF signaling. Development 131:2727–2736

Nakatsuji N (1975) Studies on the gastrulation of amphibian embryos: cell movement during gastrulation in Xenopus laevis embryos. W Roux Arch 178:1–14

Nakatsuji N (1976) Studies on the gastrulation of amphibian embryos: ultrastructure of the migrating cells of anurans. W Roux's Arch 180:229–240

Nandadasa S, Tao Q, Menon NR, Heasman J, Wylie C (2009) N- and E-cadherins in Xenopus are specifically required in the neural and non-neural ectoderm, respectively, for F-actin assembly and morphogenetic movements. Development 136:1327–1338

Nelson WJ (2008) Regulation of cell-cell adhesion by the cadherin-catenin complex. Biochem Soc Trans 36:149–155

Niessen CM, Leckband D, Yap AS (2011) Tissue organization by cadherin adhesion molecules: dynamic molecular and cellular mechanisms of morphogenetic regulation. Physiol Rev 91:691–731

Nieuwkoop PD, Florschutz PA (1950) Quelques caracteres speciaux de la gastrulation et de la neurulation de l'oeuf de Xenopus laevis, Daud. Et de quelques autres Anoures. 1ere partie. Etude descriptive. Arch Biol 61:113–150

Ninomiya H, Winklbauer R (2008) Epithelial coating controls mesenchymal shape change through tissue-positioning effects and reduction of surface-minimizing tension. Nat Cell Biol 10:61–69

Ninomiya H, Elinson RP, Winklbauer R (2004) Antero-posterior tissue polarity links mesoderm convergent extension to axial patterning. Nature 430:364–367

Ninomiya H, David R, Damm EW, Fagotto F, Niessen C, Winklbauer R (2012) Cadherin-dependent differential cell adhesion in Xenopus causes cell sorting in vitro, but not in the embryo. J Cell Sci in press (Epub ahead of print Feb 10, 2012)

Nomura KH, Kobayashi R, Hirabayashi Y, Fujisue-Sakai M, Mizuguchi S, Nomura K (1998) Involvement of blood-group-B-active trisaccharides in Ca^{2+}-dependent cell-cell adhesion in the Xenopus blastula. Dev Genes Evol 208:9–18

Noren NK, Liu BP, Burridge K, Kreft B (2000) P120 catenin regulates the actin cytoskeleton via Rho family GTPases. J Cell Biol 150:567–580

Ogata S, Morokuma J, Hayata T, Kolle G, Niehrs C, Ueno N, Cho KWY (2007) TGF-beta signaling-mediated morphogenesis: modulation of cell adhesion via cadherin endocytosis. Genes Dev 21:1817–1831

Pokutta S, Weis WI (2007) Structure and mechanism of cadherins and catenins in cell-cell contacts. Ann Rev Cell Dev Biol 23:237–261

Price SR, De Marco Garcia NV, Ranscht B, Jessell TM (2002) Regulation of motor neuron pool sorting by differential expression of type II cadherins. Cell 109:205–216

Regen CM, Steinhardt RA (1986) Global properties of the Xenopus blastula are mediated by a high-resistance epithelial seal. Dev Biol 113:147–154

Reintsch WE, Habring-Mueller A, Wang RW, Schohl A, Fagotto F (2005) β-catenin controls cell sorting at the notochord-somite boundary independently of cadherin-mediated adhesion. J Cell Biol 170:675–686

Reintsch WE, Mandato CA, McCrea PD, Fagotto F (2008) Inhibition of cell adhesion by xARVCF indicates a regulatory function at the plasma membrane. Dev Dyn 237:2328–2341

Reynolds AB, Carnahan RH (2004) Regulation of cadherin stability and turnover by p120ctn: implications in disease and cancer. Semin Cell Dev Biol 15:657–663

Roberts SJ, Leaf DS, Moore H-P, Gerhart JC (1992) The establishment of polarized membrane traffic in Xenopus laevis embryos. J Cell Biol 118:1359–1369

Rohani N, Canty L, Luu O, Fagotto F, Winklbauer R (2011) EphrinB/EphB signaling controls embryonic germ layer separation by contact-induced cell detachment. PLOS Biol 9(3):e1000597

Schambony A, Wedlich D (2007) Wnt-5a/Ror2 regulate expression of XPAPC through an alternative noncanonical signaling pathway. Dev Cell 12:779–792

Schneider S, Herrenknecht K, Butz S, Kemler R, Hausen P (1993) Catenins in Xenopus embryogenesis and their relation to the cadherin-mediated cell-cell adhesion system. Development 118:629–640

Schotz E-M, Burdine RD, Julicher F, Steinberg MS, Heisenberg C-P, Foty RA (2008) Quantitative differences in tissue surface tension influence zebrafish germ layer positioning. HFSP J 2:42–56

Sehgal RNM, Gumbiner BM, Reichardt LF (1997) Antagonism of cell adhesion by an α-catenin mutant, and of the Wnt signaling pathway by α-catenin in Xenopus embryos. J Cell Biol 139:1033–1046

Seifert K, Ibrahim H, Stodtmeister T, Winklbauer R, Niessen CM (2009) An adhesion-independent, aPKC-dependent function for cadherins in morphogenetic movements. J Cell Sci 122:2514–2523

Shih J, Keller R (1992) Cell motility driving mediolateral intercalation in explants of Xenopus laevis. Development 116:901–914

Shook D, Keller R (2003) Mechanisms, mechanics and function of epithelial-mesenchymal transitions in early development. Mech Dev 120:1351–1383

Sokol SY, Li Z, Sacks DB (2001) The effect of IQGAP1 on Xenopus embryonic ectoderm requires Cdc42. J Biol Chem 276:48425–48430

Steinberg MS (1970) Does differential adhesion govern self-assembly processes in histogenesis? Equilibrium configurations and the emergence of a hierarchy among populations of embryonic cells. J exp Zool 173:395–434

Steinberg MS, Takeichi M (1994) Experimental specification of cell sorting, tissue spreading, and specific spatial patterning by quantitative differences in cadherin expression. Proc Natl Acad Sci U S A 91:206–209

Strauss B, Adams RJ, Papalopulu N (2006) A default mechanism of spindle orientation based on cell shape is sufficient to generate cell fate diversity in polarised Xenopus blastomeres. Development 133:3883–3893

Tao Q, Nandadasa S, McCrea PD, Heasman J, Wylie C (2007) G-protein-coupled signals control cortical actin assembly by controlling cadherin expression in the early Xenopus embryo. Development 134:2651–2661

Turner A, Snape AM, Wylie CC, Heasman J (1989) Regional identity is established before gastrulation in the Xenopus embryo. J exp Zool 251:245–252

Turner AP, Brown D, Heasman J, Cook GM, Evans J, Vickers L, Wylie CC (1992) Involvement of a neutral glycolipid in differential cell adhesion in the Xenopus blastula. EMBO J 11:3845–3855

Wacker S, Grimm K, Joos T, Winklbauer R (2000) Development and control of tissue separation at gastrulation in Xenopus. Dev Biol 224:428–439

Wallingford JB, Fraser SE, Harland RM (2002) Convergent extension: the molecular control of polarized cell movement during embryonic development. Dev Cell 2:695–706

Wilson PA, Oster G, Keller R (1989) Cell rearrangement and segmentation in Xenopus: direct observation of cultured explants. Development 105:155–166

Wilson PA, Keller R (1991) Cell rearrangement during gastrulation of Xenopus: direct observation of cultured explants. Development 112:289–300

Winklbauer R (1990) Mesodermal cell migration during Xenopus gastrulation. Dev Biol 142:155–168

Winklbauer R, Nagel M (1991) Directional mesoderm cell migration in the Xenopus gastrula. Dev Biol 148:573–589

Winklbauer R, Schürfeld M (1999) Vegetal rotation, a new gastrulation movement involved in the internalization of the mesoderm and endoderm in Xenopus. Development 126:3703–3713

Wunnenberg-Stapleton K, Blitz IL, Hashimoto C, Cho KW (1999) Involvement of the small GTPases XrhoA and XRnd1 in cell adhesion and head formation in early Xenopus development. Development 126:5339–5351

Yamashiro S, Noguchi T, Mabuchi I (2003) Localization of two IQGAPs in cultured cells and early embryos of Xenopus laevis. Cell Motil Cytoskelet 55:36–50

Yamashiro S, Abe H, Mabuchi I (2007) IQGAP2 is required for the cadherin-mediated cell-to-cell adhesion in Xenopus laevis embryos. Dev Biol 308:485–493

Yanagisawa M, Anastasiadis PZ (2006) P120 catenin is essential for mesenchymal cadherin-mediated regulation of cell motility and invasiveness. J Cell Biol 174:1087–1096

Yap AS, Niessen CM, Gumbiner BM (1998) The juxtamembrane region of the cadherin cytoplasmic tail supports lateral clustering, adhesive strengthening, and interaction with p120ctn. J Cell Biol 141:779–789

Zhong Y, Brieher WM, Gumbiner BM (1999) Analysis of C-cadherin regulation during tissue morphogenesis with an activating antibody. J Cell Biol 144:351–359

Chapter 14
Adherens Junctions in Mammalian Development, Homeostasis and Disease: Lessons from Mice

Barbara Boggetti and Carien M. Niessen

Abstract Mice have proven to be a particularly powerful model to study molecular mechanisms of development and disease. The reason for this is the close evolutionary relationship between rodents and humans, similarities in physiological mechanisms in mice and human, and the large number of techniques available to study gene functions in mice. A large number of mice mutations, either germ line, conditional or inducible, have been generated in the past years for adherens junctions components, and the number is still increasing. In this review we will discuss mice models that have contributed to understanding the developmental and physiological role of adherens junctions and their components in mammals and have revealed novel mechanistic aspects of how adherens junctions regulate morphogenesis and tissue homeostasis.

14.1 Introduction to Adherens Junctions in Mammals

As the name implies adherens junctions (AJs) are structures in between cells that mediate intercellular adhesion (Niessen and Gottardi 2008). The most well known example is the *zonula adherens* (ZA), which connects to the actin cytoskeleton. The ZA was first identified as the intermediate structure of a tripartite complex called the apical junctional complex in polarized simple epithelial cells of the intestine (Farquhar and Palade 1963). This complex also consists of the more basally localized desmosomes, another adhesive junction connected to the intermediate filament system, and the apically found tight junctions, which provide tissues with a paracellular size and ion diffusion barrier. Ultrastructurally discernible AJs are

C. M. Niessen (✉) · B. Boggetti
Department of Dermatology, Center for Molecular Medicine, University of Cologne, Room 4A.05, Robert Kochstrasse 21, 50931 Cologne, Germany
e-mail: Carien.niessen@uni-koeln.de

Center for Molecular Medicine Cologne, University of Cologne, Cologne, Germany

C. M. Niessen
Cologne Excellence Cluster on Cellular Stress Responses in Aging associated Diseases (CECAD), University of Cologne, Cologne, Germany

T. Harris (ed.), *Adherens Junctions: From Molecular Mechanisms to Tissue Development and Disease,* Subcellular Biochemistry 60,
DOI 10.1007/978-94-007-4186-7_14, © Springer Science+Business Media Dordrecht 2012

also observed in various other tissues, such as the intercalated disc/fascia adherens of cardiac myocytes, paranodal loops/intracellular junctions of Schwann cells, between Sertoli cells and spermatids in the testis and punctae adherens that border synapses of neuronal and neuromuscular junctions. More diffuse AJs, sometimes referred to as spot AJs, are also found outside of the tripartite complex in both epithelial and non-epithelial cells, often showing a more discontinuous or spot-like pattern.

The importance of AJs for tissue morphogenesis and homeostasis and how they regulate their adhesive and signaling function has initially come mostly from lower organisms and *in vitro* cell culture that either use overexpression of mutant proteins, knockdowns using small RNAs or hairpin RNAs to study AJ function. Even though such studies have revealed a wealth of insight into AJs and their function, it is important to realize that mutant proteins may have different or broader effects than loss of these proteins. In addition, although knockdown can be very effective with protein reduction of over 90%, often its efficiency is less leaving substantial residual protein. Inactivation of genes through homologous recombination technology in mice allows one to ask the physiological significance of the protein in question in the context of the organism. Earlier studies using germ line inactivation of different AJ components showed that, not unexpectedly, many of these components are essential during early mammalian development (Stepniak et al. 2009). However, early embryonic lethality prevented examination of their roles at later developmental stages and in the adult. Conditional knockouts (cKOs) have alleviated this issue. The development of the Cre-LoxP technique in combination with the identification of increasing numbers of tissue specific promoters as well as increasingly sophisticated mouse techniques has provided a wealth of information in recent years on the role of different AJ components in tissue formation, homeostasis as well as in disease (Table 14.1). This chapter will provide an overview of the *in vivo* roles of AJs and their components, the cadherin/catenin complex and the nectin/afadin complex, in mammalian biology during development, tissue homeostasis and how they contribute to disease. We will not discuss in detail the core components of AJs since this has been done in other chapters of this volume and elsewhere (Harris and Tepass 2010; Oda and Takeichi 2011; Yonemura 2011; Niessen and Gottardi 2008) and also we will not discuss *in vitro* studies unless they provide direct insight into the phenotypes of mice transgenic for AJ components.

We will focus on the *in vivo* roles of the most ubiquitously expressed cadherins, the type I classical cadherins, N-cadherin, E-cadherin and P-cadherin, as well as on the most predominant cadherin of endothelial cells, VE-cadherin, and their associated catenins. Whereas other type II cadherins will be mentioned whenever relevant, their *in vivo* function will not be discussed in detail. For the analysis of germ line and conditional knockout mice, it is important to realize that in mice as well as humans most tissues express more than one cadherin or nectin and several homologues with the capacity to bind classical cadherins exist for each of the catenins and may thus compensate for the loss of one specific component. In particular, plakoglobin is closely related to β-catenin. Although in many epithelial tissues it is found at desmosomes, where through desmoplakin binding it links desmosomal

Table 14.1 Role of different AJ components in tissue formation

Gene	Tissue/Time	Cre line	Phenotype	Reference
E-cadherin (CDH1)	Oocyte	ZP3-Cre	Adhesion of blastomeres only at late morula stage followed by normal development	(De Vries et al. 2004)
	Zygote		Lethal at E4; fail to form a trophoblast epithelium	(Larue et al. 1994)
			Lethal at E4; fail to form a trophoblast epithelium	(Riethmacher et al. 1995)
	Zygote (gene replacement, N-cadherin into E-cadherin locus + maternal E-cad knockout)	ZP3-Cre	Lethal at E3.5–4.5; rescue of initial blastomeres adhesion defects but not of the defective trophoblast epithelium	(Kan et al. 2007)
	Zygote (gene replacement, N-cadherin into E-cadherin locus); small intestine and colon/E12.5	Villin-Cre	Lethal at P14–21; increased cell proliferation, formation of polyps	(Libusova et al. 2010)
	Epidermis/E11.5	K14-Cre	Progressive hyperproliferation and loss of hair follicles, impaired differentiation	(Tinkle et al. 2004)
			Lethal at P0; epidermal water barrier defects	(Tunggal et al. 2005)
	+ P-cadherin knockdown Epidermis/E11.5	K14-Cre	Lethal at P0; loss of epidermal integrity, increased apoptosis, defective epidermal barrier	(Tinkle et al. 2008)
	Epidermis/P15	Krox20-Cre	Progressive hair loss, dermal fibrosis, altered epidermal differentiation	(Young et al. 2003)
	Myelinated nerve fibers/	Krox20-Cre	Loss of AJs in the outer mesaxon, unaffected demyelination	(Young et al. 2002)
	Liver/E15	Alfp-Cre	No phenotype	(Battle et al. 2006)
	Mammary gland	MMTV-Cre	Impaired differentiation and apoptosis of alveolar epithelial cells	(Boussadia et al. 2002)
	Developing lens placode/E9	Lens-Cre	Microphthalmia, severe iris hyperplasia, vacuolization within the fibre cell region, lens epithelial cell deterioration	(Pontoriero et al. 2009)
	Thyroid gland/E14.5–E15	Tg-Cre	Small and irregularly shaped thyroid follicle lumens, unaffected cell-cell junctions	(Cali et al. 2007)

Table 14.1 (continued)

Gene	Tissue/Time	Cre line	Phenotype	Reference
N-cadherin (CDH2)	Zygote		Lethal at E10; defective somites and early heart development	(Radice et al. 1997b)
	Zygote + transgene expression; cardiomyocytes/E7.5	αMHC, βMHC	Lethal at E12; partial rescue of cardiac defects by either N- or E-cadherin	(Luo et al. 2001)
	Chimeric embryos (wt/N-cadherin$^{-/-}$)		Segregation of N-cadherin$^{-/-}$ cells from wt cells, formation of N-cadherin$^{-/-}$ aggregates	(Kostetskii et al. 2001)
	Zygote +/–		Increased susceptibility to ventricular arrhythmias	(Li et al. 2008)
	Endothelial cells/E7.5	Tie2-Cre	Lethal at mid-gestation; vascular defects	(Luo and Radice 2005)
	Cerebral cortex and hippocampus/E10.5	D6-Cre	Severe disorganization of intra-cortical structures	(Kadowaki et al. 2007)
	CNS, neural crest stem cells/E8.5	Wnt1-Cre	Lethal at E12.5–E13.5; severe cardiovascular defects	(Luo et al. 2006)
	Cardiomyocytes/E7.5	αMHC-Cre	Lethal at mid-gestation; malformed heart, delayed embryonic development	(Piven et al. 2011)
	Developing lens placode/E9	Lens-Cre	Microphthalmia, severe iris hyperplasia, vacuolization within the fibre cell region, lens epithelial cell deterioration	(Pontoriero et al. 2009)
E-cadherin (CDH1) + N-cadherin (CDH2)	Developing lens placode/E9	Lens-Cre	Defective lens vesicle separation, impaired lens epithelial cell adhesion and survival	(Pontoriero et al. 2009)
P-cadherin (CDH3)	Zygote		Precocious differentiation of the mammary gland, focal alveolar hyperplasia and ductal dysplasia in aged animals	(Radice et al. 1997a)
VE-cadherin (CDH5)	Zygote		Lethal at E9.5; impaired remodeling and maturation of vascular plexi, endothelial apoptosis	(Carmeliet et al. 1999)
			Lethal at E11.5; severe vascular morphogenesis defects	(Gory-Fauré et al. 1999)

Table 14.1 (continued)

Gene	Tissue/Time	Cre line	Phenotype	Reference
αE-catenin	Zygote		Lethal at E4; defective trophoblast epithelium	(Torres et al. 1997)
	Epidermis/E11.5	K14-Cre	Lethal at P0; blocked hair follicle development, loss of epidermal integrity and defective morphogenesis, hyperproliferation	(Vasioukhin et al. 2001)
	CNS/E10.5	Nestin-Cre	Lethal at P14–21; massive cortical hyperplasia, loss of polarity	(Lien et al. 2006)
	Bulge region of the hair follicle/P2	GFAP-Cre	Defective hair follicle morphogenesis, skin lesions, inflammation and keratoacanthoma squamous cell carcinoma	(Silvis et al. 2011)
	Mammary epithelium (embryogenesis)/adult	MMTV-Cre	Defective alveolar structures, alveolar differentiation and mammary function, increased apoptosis	(Nemade et al. 2004)
	Mammary epithelium (second half of pregnancy)/adult	WAP-Cre	Defective alveolar structures, alveolar differentiation and mammary function, increased apoptosis	(Nemade et al. 2004)
	Cardiomyocytes/E7.5	αMHC-Cre	No phenotype	(Piven et al. 2011)
	Ventricular cardiac myocytes/E9	MLC2v-Cre	Progressive cardiomyopathy, defects in the right ventricle, severe disorganization in intercalated disc structure	(Sheikh et al. 2006)
αN-catenin	Zygote		Impaired dendritic spine morphogenesis in hippocampal neurons and related synaptic functions	(Togashi et al. 2002)
			Disrupted positioning of Purkinje cells in the cerebellum and lamination defects in the hippocampus	(Park et al. 2002)
			Defects in axon migration and nuclear organization	(Uemura and Takeichi 2006)
β-catenin	Oocyte	ZP3-Cre	Dissociation of 2-cell stage blastomeres followed by rescued adhesion from paternal transcripts	(De Vries et al. 2004)
	Zygote		Lethal at E7.5; failure of primitive streak formation	(Haegel et al. 1995)
			Lethal between E7.5 and 9.5; block in anterior-posterior axis formation	(Huelsken et al. 2000)

Table 14.1 (continued)

Gene	Tissue/Time	Cre line	Phenotype	Reference
	Primitive streak/E7.5	T-Cre	Mesoderm formation and somitogenesis defects	(Aulehla et al. 2008; Dunty et al. 2008)
	CNS/E8.5	Nes8-Cre	Failure of anterior neural tube closure, dorsal-ventral fate shift before neurogenesis	(Backman et al. 2005)
	CNS/E9.5	Nes11-Cre	Defective cell-cell adhesion	(Backman et al. 2005)
	Cerebellun	Nestin-Cre	Defective midbrain and cerebellum	(Schüller and Rowitch, 2007)
	CNS, neural crest stem cells/ E8.5	Wnt1-Cre	Lethal before P0; brain malformation and failure of craniofacial development; defective cardiac outflow tract	(Brault et al. 2001; Kioussi et al. 2002)
			Lack of melanocytes and dorsal root ganglia	(Hari et al. 2002)
	CNS/E8.75	Foxg1-Cre	Lethal at P0; lack of forebrain and anterior facial structures	(Junghans et al. 2005)
	CNS/E10	Brn4-Cre	Decreased proliferation and increased apoptosis in spinal cord and brain	(Zechner et al. 2003)
	Apical ectodermal ridge/E10	Brn4-Cre	Severe hindimb malformation	(Soshnikova et al. 2003)
	Cerebral cortex and hippocampus/E10.5	D6-Cre	Disrupted radial migration of neurons, decreased cortical proliferation	(Machon et al. 2003)
	Cerebral cortex and hippocampus/P5	POMC-Cre	Defects in dendritic morphology	(Gao et al. 2007)
	Telencephalon	Emx 1-Cre	Cortical malformation, increased seizure susceptibility	(Campos et al. 2004)
	Myotome	HSA-Cre	Defects in motoneuron terminals of neuromuscular junctions	(Li et al. 2008)
	Lens surface ectoderm/E9.5	Lens-Cre	Defects in lens morphogenesis, ectopic lentoid bodies	(Smith et al. 2005)
	Lens ectoderm, partially in retina	LR-Cre		(Miller et al. 2006)
	Developing retina/E9	Six3-Cre	Loss of arrangement of retinal progenitors cells, defects in neurons migration	(Fu et al. 2006)
	Otic precursors/E8.25	Pax2-Cre	Lethal at E10.5; reduced otic placodes	(Ohyama et al. 2006)

Table 14.1 (continued)

Gene	Tissue/Time	Cre line	Phenotype	Reference
	Embryonic endoderm/E5.5	K19-Cre	Lethal at E10.5; lack of node, somites and notochord; multiple hearts	(Lickert et al. 2002)
	Osteoblast progenitors/E14.5	Col1a-Cre	Low bone mass and high bone resorption	(Glass et al. 2005)
	Osteo-chondroprogenitors/E14.5	Col2a-Cre	Lethal at P0; craniofacial deformities, short limbs	(Akiyama et al. 2004)
			Lethal at P0; skeletal defects, short limbs	(Guo et al. 2004)
	Osteoblast progenitors/E14.5	Osx1-GFP-Cre	Osteoblasts convert to chondrocyte fate	(Rodda and McMahon 2006)
	Hematopoeitic precursors	Vav-Cre	Hematopoeitic stem cells deficient in long-term growth and maintenance; reduced development of leukemia	(Zhao et al. 2007)
	Hepatocytes	Albumin-Cre	Decreased mass of liver recovered after hepatectomy, defective ammonia detoxyfication	(Sekine et al. 2007; Sekine et al. 2006; Tan et al. 2006)
	Foregut and liver endoderm	Foxa3-Cre	Lethal at E17; underdeveloped liver, defects in hepatoblast expansion and maturation	(Tan et al. 2008)
	Pancreatic and duodenum epithelium/E8.5	Pdx1-Cre	Acute edematous pancreatitis perinatally, lack of acinar cells	(Dessimoz et al. 2005; Wells et al. 2007) (Murtaugh et al. 2005)
	Rathke's pouch progenitors/E9.0	Pitx1-Cre	Smaller gland, loss of Pit 1 expression	(Olson et al. 2006)
	Sublineages of the pituitary/E13.5	Pit1-Cre	No effect	(Olson et al. 2006)
	Müllerian duct mesenchyme/E13.5	MisrII-Cre	Switch from myogenesis to adipogenesis	(Arango et al. 2005)
	Müllerian duct mesenchyme/E15.5	Amhr2-Cre	Defects in oviduct differentiation, decrease cell proliferation	(Deutscher and Hung-Chang Yao 2007)
	Sertoli cells	AMH-Cre	No effect	(Chang et al. 2008)

Table 14.1 (continued)

Gene	Tissue/Time	Cre line	Phenotype	Reference
	Limb and head mesenchyme/E9.5	Prx1-Cre	Lethal at P0; early osteoblast differentiation arrest, limb truncations	(Hill et al. 2005, 2006)
	Limb ectoderm/E9.5	Msx2-Cre	Defected formation and maintenance of the apical ectodermal ridge	(Barrow et al. 2003)
	Epidermis/E11.5	K14-Cre	Impaired follicle morphogenesis	(Huelsken et al. 2001)
	Central dermomyotome/E9.5	En 1-Cre	Loss of dorsal dermal progenitors and dermis	(Atit et al. 2006; Ohtola et al. 2008)
	Lateral plate mesoderm/E8.5	HoxB6	Lack of ventral dermis	(Ohtola et al. 2008)
	Mesoderm progenitors/E7.5	T-Cre	Lack of primitive streak, posterior mesoderm and obvious somites	(Dunty et al. 2008)
	Endothelial cells/E7.5	Tie2-Cre	Lethal between E11.5–13.5; altered vascular patterning, irregular vascular lumen, lack of heart cushion formation	(Cattelino et al. 2003; Liebner et al. 2004)
	Lung epithelium/E13.5	SP-C-Cre	Lethal at P0; pulmonary malformations, undeveloped peripheral airways	(Mucenski et al. 2003)
	Mesodermal lineages/E12.5–13.5	Dermo1-Cre	Lethal at P0; impaired lung growth	(Yin et al. 2008)
			Lethal at between E13.5–14.5; multiple mesenchymally-related defects	(De Langhe et al. 2008)
			Lethal before E11.5; synovial joint fusions, abnormal differentiation of osteoblasts and chondrocytes	(Day et al. 2005)
	Kidney (ureteric bud pithelium)/E11.5	Hoxb7-Cre	Lethal at P1; defective renal branching morphogenesis, bilateral renal aplasia and dysplasia	(Bridgewater et al. 2008)
			Lethal at P2; bilateral renal aplasia and dysplasia	(Marose et al. 2008)
	Kidney (renal vesicles mesenchymal progenitor cells)/E12.5	Six2GFP-Cre	Reduced kidney structure	(Park et al. 2007)
	Heart (myocardial precursors)/E8	Nkx2.5-Cre	Lethal at E12.5; reduced ventricular size, thinner ventricular wall	(Kwon et al. 2007)

Table 14.1 (continued)

Gene	Tissue/Time	Cre line	Phenotype	Reference
	Heart (cardiovascular progenitors, pharyngeal arch mesoderm)/E8	Isl1-Cre	Failure to form the right ventricle, hypoplasia of pharyngeal arches	(Kwon et al. 2007)
			Lethal at E13; cardiac and pharyngeal arch artery defects	(Lin et al. 2007)
	Heart (proepicardum)/E9.5	GATA5-Cre	Lethal between E15.5-birth; impaired coronary artery formation and epicardial development	(Zamora et al. 2007)
	Heart (vascular smooth muscle cells, cardiac myocytes)/E9.5	SM22α-Cre	Lethal between E10.5–11.5; severely hypoplastic right ventricle	(Cohen et al. 2007)
	Heart (FHF and SHF)/E6.5	MesP1-Cre	Blocked cardiac looping and right ventricle formation	(Klaus et al. 2007)
	Heart (SHF)/E8 -E9.5	Mef2c-Cre	Right ventricle and interventricular myocardial defects	(Ai et al. 2007)
	Cardiomyocytes/E7.5	αMHC-Cre	Partial lethality due to functional redundancy with plakoglobin	(Piven et al. 2011)
	Cardiomyocytes/5 week-old	αMHC-MCM	No basal phenotype but reduced hypertrophy following TAC	(Chen et al. 2006)
	Cardiomyocytes/3–4 month-old	αMHC-MCM	No basal phenotype due to compensatory role of plakoglobin	(Zhou et al. 2007)
γ-catenin (plakoglobin)	Zygote		Lethal at E9.5; cardiac rupture, skin blistering, desmosomes absence in the heart	(Bierkamp et al. 1996)
			Lethal between E12 and E16; defects in heart function, absence of desmosomes in intercalated discs	(Ruiz et al. 1996)
	Zygote +/–		Postnatal ARVC after aging and training	(Kirchhof et al. 2006)
	Cardiomyocytes/E7.5	αMHC-Cre	Sudden death at 4.5 months old; ARVC	(Li et al. 2011)
δ-catenin	Zygote		Severely impaired cognitive function, progressive loss of dendrites and spines	(Israely et al. 2004; Matter et al. 2009)
p120-catenin	Epidermis/E11.5	K14-Cre	Epidermal hyperplasia and chronic subcutaneous inflammation; cell-cell adhesion and ameloblast morphology defects	(Bartlett et al. 2010; Perez-Moreno et al. 2006, 2008)

Table 14.1 (continued)

Gene	Tissue/Time	Cre line	Phenotype	Reference
	Epithelia of the salivary gland/E14	MMTV-Cre	Blocked acinar differentiation, intraepithelial neoplasia, defects in cell adhesion, polarity and epithelial morphology	(Davis and Reynolds 2006)
	Dorsal forebrain/E9	Emx1-Cre	Reduced spine and synapse densities	(Elia et al. 2006)
	Small intestine and colon/E12.5	Villin-Cre	Lethal at P21; cell-cell adhesion defects, inflammation	(Smalley-Freed et al. 2010)
	Squamous oral cavity, esophagus, and forestomach/2 month-old	L2-Cre	Tumor development: preneoplastic and neoplastic lesions in the oral cavity, esophagus, and squamous forestomach	(Stairs et al. 2011)
	Endothelial cells/E7.5	Tie2-Cre	Lethal between E11.5-birth; decreased microvascular density, haemorrhages	(Oas et al. 2010)
Vinculin	Zygote		Lethal at E10.5; neural tube and heart defects	(Xu et al. 1998)
	Zygote+/-		Cardiac dysfunction, increased mortality after TAC	(Zemljic-Harpf et al. 2004)
	Ventricular cardiac myocytes/E9	MLC2v-Cre	Sudden death by ventricular tachycardia	(Zemljic-Harpf et al. 2007)
Nectin 1	Zygote		Microphthalmia; defects in terminal differentiation of the epidermis; disrupted checkerboard-like pattern of the auditory epithelia; abnormal mossy-fibre trajectories in the hippocampus, reduced number of puncta adherentia junctions at hippocampal synapses	(Inagaki et al. 2005; Wakamatsu et al. 2007) (Honda et al. 2006; Togashi et al. 2011)
			Defective enamel formation, reduced size and number of desmosomes (between ameloblasts and cells of the stratum intermedium)	(Barron et al. 2008)
Nectin 2	Zygote		Aberrant spermatid morphologenesis, male-specific infertility; cardiac dysfunction in response to pressure overload	(Bouchard et al. 2000) (Satomi-Kobayashi et al. 2009) (Ozaki-Kuroda et al. 2002)
			Male-specific infertility, malformations of the spermatozoa	(Mueller et al. 2003)

Table 14.1 (continued)

Gene	Tissue/Time	Cre line	Phenotype	Reference
Nectin 3	Zygote		Microphthalmia, male-specific infertility, disrupted checkerboard-like pattern of the auditory epithelia, abnormal mossy-fibre trajectories in the hippocampus, reduced number of puncta adherentia junctions at hippocampal synapses	(Honda et al. 2006; Inagaki et al. 2005; Togashi et al. 2011)
Nectin 1+3	Zygote		Abormal ameloblast function and crown shape development, reduced desmosomes between ameloblasts and cells of the stratum intermedium	(Yoshida et al. 2010)
Necl1	Zygote		Delayed axonal myelination in the optic nerve and spinal cord	(Park et al. 2008)
Necl2	Zygote		Male-specific infertility, oligo-astheno-teratozoospermia; delayed wound healing	(Fujita et al. 2006; Giangreco et al. 2009)
Necl5	Zygote		Decreased colon tumours after DMH/DSS treatment	(Abe et al. 2009)
mXinα	Zygote		Cardiac hypertrophy, cardiomyopathy	(Gustafson-Wagner et al. 2007)
Afadin	Zygote		Lethal at E10.5; disorganized ectoderm, impaired mesoderm migration	(Zhadanov et al. 1999)
			Lethal at E10.5; impaired organization of ectoderm, defective mesoderm migration, no somites, disorganized AJs and TJs in the ectoderm	(Ikeda et al. 1999)
	Small intestine and colon/E12.5	Villin-Cre	Increased paracellular permeability	(Tanaka-Okamoto et al. 2011)
	Forebrain/postnatal	Camk2a-Cre	Perforated synapses	(Majima et al. 2009)
VASP	Dominant negative TD-GFP	K14-Cre	Blistered epidermis, highly perturbed cell adhesion	(Vasioukhin et al. 2000)

ARVC Arrhythmogenic right ventricular cardiomyopathy; *DMH* dimethylhydrazine (DMH) and/or dextran sodium sulphate (DSS); *TAC* transverse aortic constriction; *FHF* First heart field; *SHF* Second heart field

cadherins with the intermediate filament system, in other tissues, e.g. the heart, it can be found at AJs (Dejana et al. 2009; Li and Radice 2010). The mammalian α-catenin family consists of three members: αE-catenin, which is ubiqioutisly expressed, αN-catenin, which is exclusively found in the brain, and αT-catenin (Smith et al. 2011; Maiden and Hardin 2011). Lastly, several members of the p120 subfamily of armadillo proteins, such as δ-catenin at synapses, bind classical cadherins and are found at AJs (Carnahan et al. 2011; McCrea and Park 2007; Pieters et al. 2012).

14.2 Adherens Junctions in Development: New Insights from Mice

14.2.1 Early Development

In the preimplantation mouse embryo, loosely attached blastomeres increase their intercellular contacts starting at the late 8-cells stage. This leads to compaction and the subsequent formation of the first epithelium in development, the trophectoderm (Fleming et al. 1993; Fleming and Johnson 1988), which surrounds the inner cell mass cells (Fig. 14.1a). This process is accompanied by polarization and changes in cell shape. Inactivation of different AJ components has shown a crucial role for the E-cadherin/catenin complex in these early processes. For example embryos with zygotic deletion of either E-cadherin or αE-catenin fail to form the throphectoderm even though compaction still occurs (Larue et al. 1994; Riethmacher et al. 1995; Torres et al. 1997). Activity at compaction is likely due to maternal expression of E-cadherin/catenins since inactivation of maternal E-cadherin in oocytes results in two-cell stage blastomeres that no longer adhere properly (De Vries et al. 2004). Interestingly, insertion of the N-cadherin cDNA into the E-cadherin locus does rescue blastomere adhesion defects but is unable to restore throphectoderm formation (Kan et al. 2007), indicating a specific role for E-cadherin in the formation of this epithelium that is independent of maintaining cell-cell adhesion.

Zygotic inactivation of β-catenin does not mimic either E-cadherin$^{-/-}$ or αE-catenin$^{-/-}$ mice. Instead these mice die during gastrulation at around E7.5 due to a failure to form the anterior posterior axis, likely due to the absence of Wnt/β-catenin signaling in the establishment of this axis (Huelsken et al. 2000). AJs are still observed because plakoglobin, a closely related β-catenin homologue predominantly found at desmosomes, can substitute for β-catenin its adhesive function (Huelsken et al. 2000), thus explaining initial normal development of these mice. Nevertheless, the ectoderm starts to dissociate at E7 in these mice, indicative of cell adhesion defects (Haegel et al. 1995) and suggesting that plakoglobin is not completely sufficient to replace β-catenin.

Loss of individual nectins is not embryonic lethal (Bouchard et al. 2000; Inagaki et al. 2005; Mueller et al. 2003; Yoshida et al. 2010). However, since afadin knockout mice die at around E10.5 with cell adhesion and polarity defects (Ikeda et al.

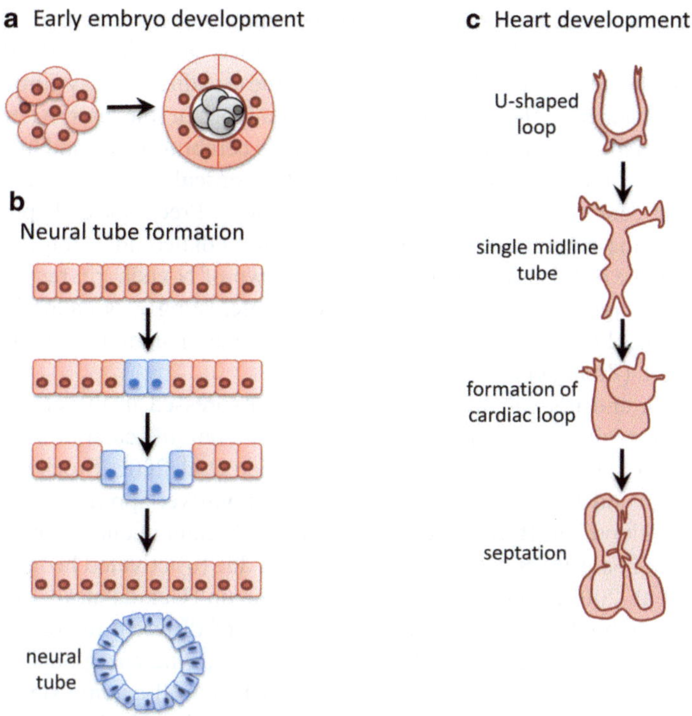

Fig. 14.1 Adherens junctions in early development. **a** Cell adhesion mediated by the cadherin-catenin system is required during early development for the transition from compact morula to blastocyst. **b** The cadherin-catenin complex regulates cells segregation and neurulation. **c** The cadherin-catenin complex regulates the primitive heart tube changes in shape, from an elongated cylinder to a looping structure

1999; Zhadanov et al. 1999), it is likely that loss of individual nectins is either not important due to the presence of other nectins or its function is compensated by one of the other nectin family members.

14.2.2 Neuronal Development

The cadherin-catenin adhesion complex has been implicated in the regulation of different aspects of neuronal development, e.g. the establishment and maintenance of the neuroepithelium, neurite extension as well as polarity, proliferation and migration of neuronal cells. During neurulation, a gradual switch from E- to N-cadherin expression occurs with almost all cells exclusively expressing N-cadherin upon neural tube closure (Fig. 14.1b). This cadherin-switching is thought to drive segregation of neuro-epithelial cells away from the overlying ectoderm (Hatta et al. 1987). It was thus quite surprising that inactivation of N-cadherin in mice only resulted in mild changes in neural tube shape with no overt defects in neural tube closure or organization (Radice et al. 1997b). Since cadherin-6 and -6B are also expressed in the neural tube (Inoue et al. 1998), these cadherins may compensate for the loss of N-cadherin. N-cadherin does play an important role later in neurogenesis since conditional inactivation of N-cadherin in the cerebral cortex and hippocampus of the mouse results in highly disorganized intra-cortical structures due to the loss of AJs in the cerebral cortex (Kadowaki et al. 2007). Precursor cells provide their own niche in the cerebral cortex and N-cadherin mediated adhesion is crucial for the maintenance of this niche as *in vivo* knockdown of N-cadherin results in differentiation and migration of neuronal stem cells. Interestingly, N-cadherin exerts its effect through Akt- dependent maintenance of β-catenin signaling that counteracts differentiation (Zhang et al. 2010).

αE-catenin and αN-catenin are differentially expressed in the central nervous system, αN-catenin is present throughout the brain, and αE-catenin is restricted to the ventricular zone in neural progenitor cells (Lien et al. 2006; Stocker and Chenn 2006; Uemura and Takeichi 2006), and they serve non-overlapping functions during neuronal development. Brain specific deletion of αE-catenin during embryogenesis leads to loss of polarity and extensive hyperplasia due to activated hedgehog signaling (Lien et al. 2006). In contrast, loss of αN-catenin (Uchida et al. 1994), results in cerebellar hypoplasia and incorrect positioning of different neuronal cell types in the hippocampus (Park et al. 2002). Interestingly, whereas global loss of αE-catenin during brain development did not obviously alter β-catenin signaling activity (Lien et al. 2008), focal loss of αE-catenin in cerebral cortical precursors induces premature differentiation accompanied by reduced β-catenin signaling (Stocker and Chenn 2006), similar to what was observed upon N-cadherin down-regulation. Thus, N-cadherin-αE-catenin intercellular contacts in the ventricular zone are necessary to maintain progenitor cell identity.

Specific loss- and gain-of-function mutations of β-catenin in different areas of the brain have shown that the Wnt signaling function of β-catenin regulates cell fate, cortical proliferation and anterior-posterior patterning of the brain and spinal cord (reviewed in Grigoryan et al. 2008). These results thus suggest that, as in other tissues, in most regions of the brain plakoglobin replaces β-catenin in its adhesive AJ function. However, mice with a deletion of β-catenin in the forebrain lack forehead structures and anterior facial structures associated with loss of AJs and apoptosis. Using Wnt/β-catenin reporter mice no obvious β-catenin signals could be detected during early telencephalon development. These results thus indicate that during telencephalon development AJs provide essential survival signals (Backman et al. 2005; Junghans et al. 2005).

AJs also are essential for the proper formation and function of synapses. Loss of αN-catenin impaired dendritic spine morphogenesis and synaptic function in hippocampal neurons (Togashi et al. 2002), in addition to defective axonal migration (Uemura and Takeichi 2006). Two other components of the AJ which regulate cell adhesion during the development of the nervous system are p120ctn and δ-catenin, an armadillo protein that belongs to the p120ctn subfamily. Indeed, mice lacking these proteins show severe defects in synaptogenesis (Elia et al. 2006; Israely et al.

2004; Matter et al. 2009). Noteworthy, the nectins and their binding partner afadin cooperate with the cadherin-catenin system during synaptogenesis. In both the nectin-1 and nectin-3 mutants, hippocampal synapses present a reduced number of *puncta adherentia* junctions and impairments of the mossy fiber tract (Honda et al. 2006) and conditional mutation of afadin in the forebrain leads to defective morphogenesis of synapses (Majima et al. 2009).

14.2.3 Epithelial Organ Development

Mice mutant for AJ components have provided indispensable experimental models to investigate the molecular mechanisms underlying epithelial organ development.

E-cadherin is indispensable for the formation and maintenance of polarized simple epithelia in lower organisms. Tissue specific overexpression or inactivation of E-cadherin or other cadherin/catenin complex members enabled the examination of tissue specific functions of AJ complex components in epithelial organogenesis in a cell autonomous manner. The intestine is one of the most rapidly renewing tissues of mammals with stem cells residing in the crypt that drive proliferation. Enterocytes then migrate along the crypt-villus axis to be sloughed off through programmed cell death at the villus tip. Intestinal overexpression of E-cadherin, the predominant intestinal cadherin, in the crypts and villi slowed migration, inhibited proliferation and induced apoptosis in the crypts (Hermiston et al. 1996). In contrast, expression of a dominant negative N-cadherin in post-mitotic enterocytes disrupted AJs and differentiation, enhanced migration but also induced precocious apoptosis (Hermiston and Gordon 1995a). Expressing this dominant negative along the whole crypt-villus axis, including stem cells, additionally caused an inflammatory bowel disease-like phenotype characterized by inflammation, altered epithelial architecture and in the long run adenoma formation (Hermiston and Gordon 1995b), indicating a specific role for AJs in the regulation of inflammation and growth in the proliferative/stem cell compartment. Later, similar phenotypes were observed when either E-cadherin or p120ctn were inactivated along the whole crypt/villi axis (Schneider et al. 2010; Smalley-Freed et al. 2010), thus showing that these phenotypes are indeed a direct consequence of loss of E-cadherin/catenin complex, the predominant cadherin complex in the intestine. Moreover, they indicate an important role for E-cadherin in the formation of adhesive AJs and tissue integrity. It was therefore surprising that inactivation of E-cadherin in either mammary glands (Boussadia et al. 2002), thyroid gland (Cali et al. 2007) or the epidermis (Tinkle et al. 2004; Tunggal et al. 2005), a stratified epithelium, did not obviously alter intercellular adhesion and AJs. Since these epithelia express other cadherins, loss of E-cadherin may not be sufficient to disturb AJs and thus intercellular adhesion.

As in intestine, the E-cadherin/catenin complex does promote cell survival in the mammary gland, since increased apoptosis was the predominant phenotype observed upon tissue specific loss of E-cadherin or αE-catenin (Boussadia et al. 2002; Nemade et al. 2004). Inactivation of E-cadherin in the thyroid gland also resulted in smaller glands although alterations in survival were not directly tested. E-cadherin may promote cell survival through its association with PI3-kinase, which regulates Akt mediated survival signaling (Cali et al. 2007). Together, these results indicate a specific role for E-cadherin in the regulation of apoptosis and differentiation. In contrast, in the epidermis increased apoptosis was only observed when inactivation was combined with *in vivo* knockdown of the other epidermal cadherin, P-cadherin (Tinkle et al. 2008).

14.2.4 Early Heart and Vascular Development

During mouse development the primitive heart tube changes in shape from an elongated cylinder to a looping structure (Fig. 14.1c). N-cadherin, the only classical cadherin expressed in cardiac muscle cells, is a key determinant of this developmental process. Germ line deletion or cardiac muscle cell specific deletion of N-cadherin disturbs myocyte adhesion resulting in deformation of the heart tube and early embryonic lethality (Piven et al. 2011; Radice et al. 1997b). Interestingly, the cardiac phenotype can be rescued by muscle specific expression of not only N-cadherin but also E-cadherin, showing that E-cadherin can functionally substitute for N-cadherin during cardiogenesis (Luo et al. 2001).

N-cadherin is also essential during vasculogenesis, as mice lacking N-cadherin in the endothelial compartment die at mid-gestation due to severe vascular defects (Luo and Radice 2005). Very interestingly, these mice display a very similar phenotype as that of VE-cadherin germ line knockouts (Carmeliet et al. 1999; Gory-Fauré et al. 1999). Both knockouts fail to develop a proper vascular plexus. This was surprising since VE-cadherin localizes to endothelial AJs and excludes N-cadherin from these junctions (Navarro et al. 1998). N-cadherin shows a more diffuse distribution and was thought to be more relevant for adhesion of endothelial cells with either pericytes or smooth muscle cells in more mature vessels. Closer examination revealed that VE-cadherin protein expression was strongly reduced in the absence of N-cadherin, whereas loss of VE-cadherin did not affect N-cadherin levels. This indicates that N-cadherin regulates vasculogenesis through posttranscriptional control of VE-cadherin expression (Luo and Radice 2005).

As in other tissues, p120ctn regulates cell surface levels of VE-cadherin by inhibiting endocytosis (Xiao et al. 2005). *In vivo*, p120ctn regulates VE-cadherin and N-cadherin stability and is indeed required for vasculogenesis in mice since loss of p120ctn in the endothelial compartment leads to severe defects in vascular patterning and morphogenesis and compromised vessel integrity resulting in hemorrhages and death at E11.5 (Oas et al. 2010). In addition, these mice also showed a strongly reduced association of endothelial cells with pericytes, which is likely explained by reduced N-cadherin levels on endothelial cells that are necessary for adhesion to pericytes. Endothelial β-catenin mutants are also embryonic lethal with similar phenotypes as other AJ components (Cattelino et al. 2003). Interestingly, in these mice, loss of β-catenin is not compensated by plakoglobin, as is observed during e.g. early development or in the epidermis. Instead, in β-catenin$^{-/-}$ endothelia, cell-cell adhesion strength is reduced, α-catenin at contacts is decreased, while desmoplakin expression and membrane localization is increased. This suggests that upon loss of β-catenin endothelial cells shift from α-catenin based AJs to *complexus adhaerentes*, in which VE-cadherin binds to plakoglobin and in turn recruits desmoplakin that connects to vimentin intermediate filaments. Next to its essential role in endothelial adhesion strength, endothelial Wnt/β-catenin signaling also regulates heart cushion formation by promoting TGFβ-dependent endocardial epithelial mesenchymal transformation (EMT), which does not occur in the absence of endothelial β-catenin (Liebner et al. 2004).

14.3 *In Vivo* Regulation of Other Intercellular Junctions

Initial contact mediated by adherens junction components has been linked not only to the establishment of polarity but also to the formation of other intercellular junctions such as desmosomes, gap and tight junctions. Below we will discuss if and how *in vivo* loss of AJs components affect the formation and/or stability of desmosomes and tight junctions

14.3.1 In Vivo *Interaction Between AJs and Desmosomes*

In light of the strong *in vitro* evidence that especially E-cadherin regulates desmosome formation (Gumbiner et al. 1988; Lewis et al. 1994), it was thus rather surprising that *in vivo* inactivation of E-cadherin in e.g. mammary gland (Boussadia et al. 2002), thyroid follicle cells (Cali et al. 2007) or in the epidermis (Tinkle et al. 2004; Tunggal et al. 2005; Young et al. 2002) in which desmosomes are crucial for tissue integrity, did not cause any obvious impairment in cell junctions or desmosome ultrastructure. Since these tissues express other cadherins this provided a potential explanation for how desmosomes were assembled. Similarly, lack of the AJ components p120ctn or β-catenin did not affect either the number or structure of desmosomes (Huelsken et al. 2001; Perez-Moreno et al. 2006). Although epidermal loss of p120ctn results in destabilization, E-cadherin cell surface expression is likely high enough since AJs and desmosomes can still be observed. As during development, plakoglobin replaces β-catenin in the AJs of β-catenin negative epidermis and therefore no obvious defects in intercellular junctions are observed. Desmosome formation is perturbed when both classical cadherins are downregulated in the epidermis. Deletion of E-cadherin in combination with short hairpin mediated downregulation of P-cadherin, the two main epidermal classical cadherins, either *in vivo* or *in vitro*, strongly impaired intercellular cohesion and desmosome formation (Michels et al. 2009; Tinkle et al. 2008). This could be rescued by expressing either E-cadherin or

P-cadherin, indicating that the levels but not cadherin specificity was important for the formation of desmosomes (Michels et al. 2009).

In addition, inactivation of N-cadherin, the main cadherin in the adult heart, resulted in the disassembly of the intercalated disc, a structure that encompasses AJs, gap junctions and desmosomes (Kostetskii et al. 2005), and destabilized gap junction function (Li et al. 2005). Together, these results indicate classical cadherins can regulate the formation of other junctions *in vivo* and that total classical cadherin levels but not cadherin specificity are important.

Why are desmosomal cadherins unable to engage in productive adhesive interactions to form stable desmosomes in the absence of classical cadherins? Although the mechanism is unclear, several findings provide interesting clues. First, desmosome formation cannot be rescued by a chimeric IL2 receptor extracellular and transmembrane domain-cadherin cytoplasmic domain protein (Michels et al. 2009), previously shown to rescue E-cadherin dependent growth and migration inhibition (Gottardi et al. 2001; Wong and Gumbiner 2003). This thus suggests that classical cadherin adhesive activity is required for desmosome formation. Moreover, epidermal deletion of α-catenin also strongly reduces intercellular contacts and the number and size of desmosomes (Vasioukhin et al. 2000, 2001). Interestingly, E-cadherin-β-catenin complexes are recruited to sites of cell-cell contacts at the cell surface of the mutants but these contacts do not mature and no reorganization of actin into a cortical network is observed. Taken together, these data suggest that low levels of adhesive classical cadherins are sufficient to initiate desmosome assembly and suggest that α-catenin is necessary for the maturation of desmosomes by a mechanism that may involve actin cytoskeleton reorganization and/or the connection with the nectin-afadin complex. In this respect, it is worth mentioning that desmosomes are also reduced in size and number between ameloblasts and cells of the stratum intermedium in the enamel organ of the nectin 1 knockout mouse (Barron et al. 2008).

14.3.2 Contribution to in Vivo *Tight Junctional Barrier Formation and Function in Epithelia and Endothelia*

In epithelial and endothelial cells, tight junctions act as barriers that prevent the paracellular passage of soluble molecules in an ion and size selective manner. As for desmosomes, *in vitro* data indicate a role for E-cadherin in the formation of tight junctions (Gumbiner et al. 1988). During early mouse development E-cadherin is indeed essential for the formation of tight junctions in the trophectoderm (reviewed in Fleming et al. 2000). Since E-cadherin regulates trophectoderm differentiation and compaction, both necessary events for tight junction assembly in this epithelium, this may, however, be a rather indirect effect. Inactivation of E-cadherin either alone or in combination with knocked down P-cadherin in the epidermis results in impaired tight junctional function (Michels et al. 2009; Tinkle et al. 2008; Tunggal et al. 2005). Similarly, loss of either E-cadherin or p120ctn impairs intestinal barrier function (Schneider et al. 2010; Smalley-Freed et al. 2010). In the epidermal E-cadherin cKO mice, differentiation and growth are unaltered thus uncoupling impaired tight junction function from differentiation and growth (Tunggal et al. 2005) and suggesting a more direct role for E-cadherin in the regulation of tight junctions.

Since ZO-1 can directly bind to α-catenin (Itoh et al. 1997) and accumulates very early in cell-cell contacts upon the initiation of cadherin adhesion, one proposed model was that cadherins recruit ZO-1 to intercellular contacts and thereby initiate tight junctions that upon maturation separate out. Since ZO-1 is still recruited to sites of cell-cell contacts in the E-cadherin cKO tissue (Tunggal et al. 2005), this rules out a simple recruitment model.

Interestingly, intercellular contact recruitment of other tight junctional components, such as occludin or claudin1, is also not impaired (Michels and Niessen, unpublished observations), suggesting that E-cadherin regulates a late step in the assembly of functional tight junctions. This may occur through the regulation of activity of the cell polarity protein atypical protein kinase C (aPKC) and the small GTPase Rac. Staining for phospho-aPKC, indicative of aPKC activation, and Rac revealed a loss of intercellular membrane staining in E-cadherin negative epidermis (Tunggal et al. 2005). More importantly, inhibition of aPKC, through either pharmacological inhibitors, or through expression of dominant-negative constructs impaired *in vitro* tight junction transepithelial resistance function but did not disturb tight junction protein localization, similar to E-cadherin$^{-/-}$ keratinocytes. In support of these findings are the results on epidermal inactivation in mice of the neurofibromatosis type 2 tumor suppressor Merlin, a protein closely related to Ezrin, Radixin, and Moesin, which regulates the establishment of stable AJs in several types of cultured mammalian cells (Lallemand et al. 2003). These mice die perinatally due to water barrier defects, and in keratinocytes, aPKC is no longer targeted to the membrane (Gladden et al. 2010). These authors show that Merlin binds simultaneously to α-catenin and to the polarity protein Par-3, an aPKC binding partner. Based on these data, one model would be that Merlin links the Par-3/aPKC polarity complex to AJs to regulate the activity of the small GTPase Rac and thereby promoting tight junctional function. However, the observation that Par-3 membrane localization is not altered upon loss of E-cadherin (Tunggal et al. 2005) suggests that Par-3 is upstream of the cadherin in this model, similar to what has been observed in *Drosophila* (Harris and Peifer 2005).

Since either E- or P-cadherin re-expression can rescue tight junctional function in E-cadherin$^{-/-}$ keratinocytes (Michels and Niessen, unpublished results), tight junction function, like desmosome assembly, depends on levels but not specificity of classical cadherins. This may also explain why inactivation of E-cadherin in the mammary epithelium (Boussadia et al. 2002), in the thyroid (Cali et al. 2007) or liver (Battle et al. 2006) did not obviously impair tight junctions.

Several studies illustrate that VE-cadherin expression and organization at AJs is also a crucial determinant for vascular barrier function. VE-cadherin-specific antibodies injected *in vivo* induce the release of the cadherin from AJs, its diffuse distribution on the cell surface and then impaired vascular barrier function (Corada et al. 1999). Moreover, activation of the VEGF receptor-2 promotes endocytosis of VE-cadherin in a β-arrestin2-dependent manner, thereby increasing vascular permeability (Gavard and Gutkind 2006). VE-cadherin may also regulate vascular tight junction barrier function more indirectly by controlling the expression of claudin-5, one important barrier-promoting claudin in endothelia (Nitta et al. 2003). In VE-cadherin$^{-/-}$ cells, or cells not engaged in intercellular contacts, β-catenin forms a complex with the transcription factors Tcf4 and FoxO1, which bind and represses the claudin-5 promoter. Engagement of VE-cadherin results in recruitment of β-catenin to cell-cell contacts and, at the same time activates PI(3) kinase-Akt signaling. Activated Akt then phosphorylates FoxO1 resulting in its translocation from the nucleus to the cytoplasm and thus releasing its repressive function on claudin-5 expression (Taddei et al. 2008).

14.4 Regulation of Growth and Inflammation and Their Role in Carcinogenesis

It is by now well established that loss of several AJ components, and especially of E-cadherin, is a key step in epithelial tumorigenesis and observed in a wide range of tumors. Not only is the reduced E-cadherin expression a hallmark of snail/slug mediated EMT that promotes migration and invasion behavior of cancer cells, but also altered E-cadherin function is associated with tumor initiation as mutations in human E-cadherin are found in familiar gastric cancer and sporadic lobular breast cancer (reviewed in Berx and van Roy 2009). Moreover, down-regulation or mislocalization of p120ctn serves to classify different human tumor types in many tissues, including the skin (Reynolds and Roczniak-Ferguson 2004; Thoreson and Reynolds 2002; van Hengel and van Roy 2007).

Mouse models have been instrumental in showing the causal relation between loss of E-cadherin/catenin complex and carcinogenesis. For example, while loss of E-cadherin in mammary gland, epidermis or stomach is not sufficient to initiate tumors, additional loss of the tumor suppressor p53 not only results in more invasive tumors but also increases tumor burden, indicating that E-cadherin is both a tumor and invasion suppressor (Derksen et al. 2006, 2011; Shimada et al. 2011). In addition, mice heterozygous for E-cadherin and deficient for mismatch DNA repair form endometrial tumors that show loss of E-cadherin heterozygosity and develop more aggressive lymphomas, indicating that E-cadherin also plays a role in hematological malignancies (Kovtun et al. 2011). In fact, inactivation of αE-catenin or p120ctn in the epidermis or salivary gland is sufficient for hyperplasia (Davis and Reynolds 2006; Vasioukhin et al. 2001).

How loss of cadherin-dependent AJs drives carcinogenesis is less clear. Initially it was assumed that loss of adhesion is the primary driver at least for increased migration and invasion. However, *in vitro* studies using cancer cell lines revealed that growth and invasion inhibitory properties of E-cadherin were independent of adhesion (Gottardi et al. 2001; Wong and Gumbiner 2003), raising questions of how alterations in cadherin/catenins drive tumor growth and invasion. Several recent *in vivo* studies in mice have now revealed important mechanistic links between the cadherin/catenin complex and the regulation of not only growth and survival but also inflammation, which are considered key contributors to cancer initiation and progression (reviewed in Grivennikov et al. 2010). For example, in p53$^{-/-}$ mouse models for human invasive lobular breast cancer, loss of E-cadherin induces the translocation of p120ctn to the cytosol and this promotes anchorage independent survival in a Rho/Rock dependent manner (Schackmann et al. 2011).

Epidermal inactivation of either αE-catenin or p120ctn induces inflammation in the underlying dermis. Grafting of αE-catenin cKO skin onto the backs of thymus-defective mice (*Nude*) induced inflammation and the formation of tumor-like nodules that after 70 days resemble human grade III squamous cell carcinomas (Kobielak and Fuchs 2006) with atypical cells over time gaining a more fibroblastic-like morphology that invade the underlying basement membrane. Transcriptional profiling of E18.5 control and αE-catenin cKO epidermis revealed early up-regulation of not only growth factor signaling but also of different Nf-κB targets, many of which are associated with a wound healing and inflammatory response. Such an Nf-κB signature is also a feature of human squamous cell carcinomas (Kobielak and Fuchs 2006). A more recent study showed that mice with conditional ablation of αE-catenin in the hair follicle stem cell compartment show local inflammation and within these areas epidermal tumors develop that resemble human keratoacanthoma squamous cell carcinoma (Silvis et al. 2011). Notably, mice in which both αE-catenin and p53 are inactivated in the hair follicle stem cell compartment do not present inflammation and develop multifocal tumors that grow faster than those originated from αE-catenin$^{-/-}$ cells only (Silvis et al. 2011). These results indicate that p53 mediates the inflammatory response downstream of αE-catenin and that loss of p53 allows αE-catenin$^{-/-}$ cells to escape apoptosis thus developing more tumors in a shorter time.

Another pathway through which αE-catenin may exert its tumor suppressive function is through regulation of the transcription factor Yap, which is negatively regulated by the Hippo signaling pathway involved in contact inhibition, organ size and tumor suppression. Loss of αE-catenin results in constitutive nuclear localization of Yap independent of cell density. Yap directly interacts with αE-catenin in a 14-3-3 dependent manner resulting in its translocation out of the nucleus when cells are allowed to form intercellular contacts and reach confluency (Schlegelmilch et al. 2011; Silvis et al. 2011). In line with these findings, gain and loss of function analysis identified Yap1 as a key regulator of epidermal progenitor cell proliferation and active Yap1 promotes skin tumor formation (Schlegelmilch et al. 2011). Taken together these studies suggest that αE-catenin functions as a tumor suppressor at two levels: it inhibits cell autonomous Nf-κB signaling thereby suppressing cancer promoting skin inflammation and it negatively regulates Yap1 transcriptional activity to suppress growth.

Mice in which p120ctn is inactivated in the epidermis display hyperplasia, chronic inflammation and abnormal mitosis with loss of hair and body fat (Fig. 14.2; Perez-Moreno et al. 2006). p120ctn has previously been identified as a negative regulator of the small GTPase Rho (Anastasiadis et al. 2000; Noren et al. 2000). Rho was indeed

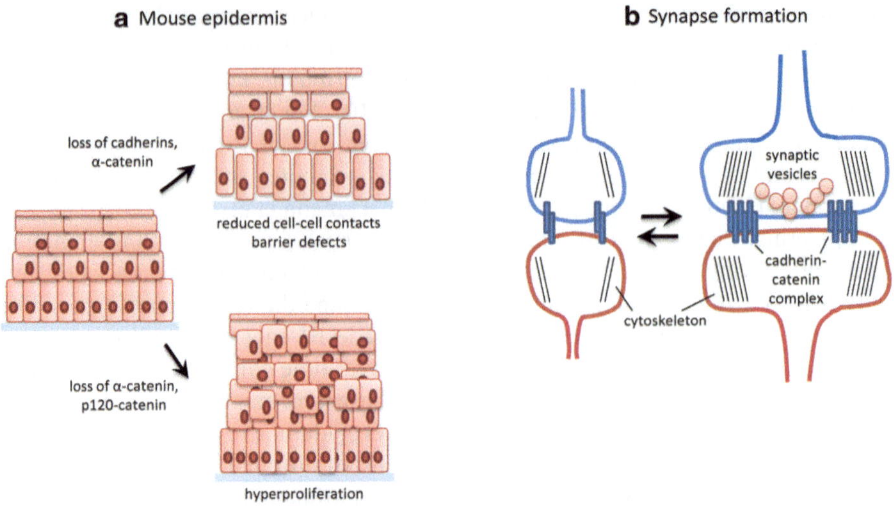

Fig. 14.2 Adherens junctions in organ development. **a** AJs are essential for tissue cohesion and barrier function in the epidermis. **b** Synapse formation is regulated by cell adhesion molecules. Presynaptic and postsynaptic compartments engage in contacts mediated by the cadherin-catenin system, which progressively promotes the expansion of the dendritic spine head and the maturation of the synapse

activated upon *in vivo* epidermal p120ctn loss and this is sufficient to drive mitotic defects. Moreover, chronic inflammation was associated with increased Nf-κB signaling in the epidermis. Blocking Nf-κB signaling inhibits epidermal hyperproliferation indicating that inflammation drives the formation of skin neoplasias (Perez-Moreno et al. 2008). These data thus suggest that p120ctn functions as a tumor suppressor. More definitive proof that loss of p120ctn promotes tumor development was provided recently using mice with a conditional deletion of p120ctn in the squamous oral cavity, esophagus, and forestomach (Stairs et al. 2011). These tissues display severe dysplasia that develop into invasive squamous cancer associated with increased Nf-κB signaling and immune cell infiltration, resembling human oral and esophageal squamous cell cancers. In summary, loss-of-function experiments revealed critical roles for E-cadherin, αE-catenin and p120ctn not only in cell adhesion but also in the regulation of cell proliferation, survival and inflammation that, upon loss or altered localization of these components, may all contribute to carcinogenesis.

14.5 Transgenic Mice as Models for Human Diseases Associated with Adherens Junctions

Mutations in several AJ complex components are associated with human disease. Transgenic mice generated for these components may serve as important models for these diseases. They will allow study of the underlying mechanisms and can address if these mutations contribute in a cell autonomous versus non-autonomous manner.

14.5.1 Ectodermal Dysplasias

Mutations in P-cadherin are associated with two very related human diseases named hypotrichosis with juvenile macular dystrophy (HJMD syndrome) and ectodermal dysplasia, ectrodactyly, and macular dystrophy (EEM syndrome). HJMD and EEM patients display short and sparse hair and progressive degeneration of the central retina, leading to blindness between the first and third decades (Sprecher et al. 2001). Additionally, the EEM syndrome is associated with limb defects like ectrodactyly and syndactyly, hypodontia and oligodontia, enamel hypoplasia and widely spaced teeth (Kjaer et al. 2005). However, the reasons for this phenotypic range need still to be determined. Although P-cadherin is widely expressed in the mouse including in epidermal structures (Hirai et al. 1989), retinal pigment epithelium (Nose and Takeichi 1986) and limbs (Kjaer et al. 2005) germ line deletion in mice did not reveal any obvious phenotypes in these organs but instead showed precocious differentiation of the mammary gland (Radice et al. 1997a). As discussed previously E- and P-cadherin can compensate for each other at least with respect to tissue structure and integrity in the mouse epidermis, thus providing one explanation. At least for the skin phenotype, epidermal E-cadherin knockout/P-cadherin knockdown mice may serve as models. Loss of P-cadherin in a different genetic setting than the mouse strain used by Radice et al. (1997a) may also result in a stronger phenotype.

P-cadherin transcription is regulated by p63, a key regulator of epidermal development and maintenance. In humans, heterozygous mutations in the DNA-binding domain of p63 can cause ectrodactyly, ectodermal dysplasia, and cleft lip (EEC syndrome) (reviewed in Koster 2010), a syndrome remarkably similar to EEM. It has been reported that various isoforms of p63 bind to two distinct regions of the human P-cadherin promoter thereby regulating its transcription (Shimomura et al. 2008). Interestingly, one of these p63-binding regions is not conserved in the mouse P-cadherin gene (Yang et al. 2006).

Nectins have also been implicated in human ectodermal dysplasia syndromes. By positional cloning, nectin-1 was found responsible for the autosomal recessive cleft lip/palate-ectodermal dysplasia syndromes, Zlotogora-Ogur syndrome and Margarita Island ectodermal dysplasia in human, which are characterized by cleft lip/palate, syndactyly, mental retardation, and ectodermal dysplasia (Sozen et al. 2001; Suzuki et al. 2000). In addition, a recent report identified Nectin4 mutations in ectodermal-dysplasia-syndactyly syndrome (EDSS1 (Jelani et al. 2011)). Until now, no knockout for Nectin4 has been reported and nectin1 mutant mice have no obvious ectodermal dysplasia phenotype perhaps due to expression of other nectins. However, double knockouts may serve as models to study the role of nectins in ectodermal dysplasias.

14.5.2 Deafness

A recent report identified roles for nectin-1 and nectin-3 in the development of the checkerboard-like pattern of the auditory epithelia (Togashi et al. 2011). The cochlea of mouse nectin-1 knockouts, and more severely for nectin-3 knockouts, displays a disorganized arrangement of both inner and outer hair cells in the organ of Corti. Here, supporting cells are no longer uniformly intercalated between hair cells that aberrantly attach to each other disrupting the highly organized pattern of the auditory epithelia. The reason for these phenotypic alterations seems to relate to the differential adhesive strengths between different nectins, with the heterophilic interaction of nectin 1–3 being stronger than that of nectin 2–3, which is in turn stronger than all the homophilic associations. The absence of either nectin-1 or nectin-3 disrupts this strict hierarchy required for proper cell-cell sorting and adhesion, and results in a disorganized pattern formation. It would be interesting to examine if this disorganization affects hearing abilities of the nectin-1 and nectin-3 knockout mice.

14.5.3 Nervous System

Synapses of the central nervous system are asymmetric, intercellular junctions that mediate neuronal transmission. Defects in synapse development and function may lead to neurodevelopmental disorders like mental retardation and autism, and could also play a role in neurodegenerative disorders, like Alzheimer's disease (reviewed in McAllister 2007). Mouse models for different AJ components have revealed important roles for these proteins in synaptogenesis and plasticity through both cell-adhesion-dependent and -independent mechanisms (Table 14.1). These transgenic mice may serve as important models for understanding human disease since mutations in several AJ components have been associated with different cognitive disorders. δ-catenin is mutated in the cri-du-chat syndrome, which is characterized by severe mental retardation (Cerruti Mainardi 1996). Additionally, δ-catenin associates with presenilin-1, the most common mutated protein in familial Alzheimer's disease (Zhou et al. 1997). Mice mutants for δ-catenin present severe deficits in memory, learning and synaptic plasticity, pointing to a specific role of δ-catenin in experience-dependent synaptic modifications (Israely et al. 2004). Moreover, further studies revealed that these mice encounter a progressive dendritic retraction, reduced spine density and stability, and declined cortical responsiveness, demonstrating that δ-catenin is essential for the maintenance of neuronal structure and function *in vivo* (Matter et al. 2009). The δ-catenin and other brain specific cadherin mutant mice may thus serve as models for these human diseases.

14.5.4 Heart Disease

The organization of the intercalated disc of the myocardium is indispensable to maintain mechanical and electrical function of the heart (Li and Radice 2010) since it couples structural integrity and mechanical strength through desmosomes and AJs to electrical communication through the channel forming gap junctions. Indeed, mutations in different intercalated disc components, including plakoglobin, are associated with cardiac arrhythmia in humans (Li and Radice 2010). These mutations are likely causal since loss of plakoglobin in the adult heart results in cardiac dysfunction that resembles the heart phenotype in these human patients (Li et al. 2011). Although loss of nectin-2 does not cause spontaneous heart dysfunction, likely due to compensation by nectin-4, chronic pressure overload resulted in intercalated disc disassembly accompanied by cardiac dysfunction (Satomi-Kobayashi et al. 2009). Inducible tissue specific inactivation of N-cadherin in the myocardium showed that N-cadherin is also essential for the maintenance of intercalated discs (Kostetskii et al. 2005). More importantly, these mice suffer from impaired heart function with spontaneous ventricular tachyarrhythmia likely causing their sudden cardiac death (Li et al. 2005). This is likely due to reduced formation of gap junction-mediated conductivity since the expression of connexin43, the major channel forming gap junction protein in myocardium, is strongly reduced in these mice, a feature commonly observed in diseased myocardium (Akar and Tomaselli 2005; Kanno and Saffitz 2001). Interestingly, N-cadherin heterozygous or N-cadherin/connexin43 compound heterozygous mice showed increased susceptibility to ventricular arrhythmias accompanied by reduced connexin43 protein expression (Li et al. 2008). This suggests that in humans N-cadherin haploin sufficiency could be a predisposition to increased risk of cardiac arrhythmias and thus that these mouse models serve as useful tools to study human heart disease.

14.6 Concluding Statement

Initial studies with transgenic mice either deficient or overexpressing cadherins revealed essential roles for many of these components during mammalian development, as was perhaps expected in light of the many-fold *in vivo* findings in lower organisms and *in vitro* cell culture data. The use of the Cre-LoxP system in combination with inducible systems have allowed tissue specific and temporal inactivation of adherens junction components. These models have provided novel and sometimes surprising insights into how dysfunctional AJs contribute not only to organ development but also disturb mammalian tissue homeostasis resulting in disease. Such mice, together with newly generated temporal and tissue specific knockouts, will likely provide crucial models for human disease and will likely help to not only decipher the role of dysfunctional AJ components in disease but also will serve to test novel therapeutic avenues to treat these diseases.

References

Abe A, Fukui H, Fujii S, Kono T, Mukawa K, Yoshitake N, Sekikawa A, Ichikawa K, Tomita S, Yamagishi H, Imai Y, Shinoda M, Ishizaki H, Tanaka-Okamoto M, Kubota K, Miyoshi J, Takai Y, Fujimori T (2009) Role of Necl-5 in the pathophysiology of colorectal lesions induced by dimethylhydrazine and/or dextran sodium sulphate. J Pathol 217:42–53

Ai D, Fu X, Wang J, Lu M, Chen L, Baldini A, Klein W, Martin J (2007) Canonical Wnt signaling functions in second heart field to promote right ventricular growth. Proc Natl Acad Sci U S A 104:9319–9324

Akar F, Tomaselli G (2005) Conduction abnormalities in nonischemic dilated cardiomyopathy: basic mechanisms and arrhythmic consequences. Trends Cardiovasc Med 15:259–264

Akiyama H, Lyons J, Mori-Akiyama Y, Yang X, Zhang R, Zhang Z, Deng J, Taketo M, Nakamura T, Behringer R, McCrea P, de Crombrugghe B (2004) Interactions between Sox9 and beta-catenin control chondrocyte differentiation. Genes Dev 18:1072–1087

Anastasiadis PZ, Moon SY, Thoreson MA, Mariner DJ, Crawford HC, Zheng Y, Reynolds AB (2000) Inhibition of RhoA by p120 catenin. Nat Cell Biol 2:637–644

Arango N, Szotek P, Manganaro T, Oliva E, Donahoe P, Teixeira J (2005) Conditional deletion of beta-catenin in the mesenchyme of the developing mouse uterus results in a switch to adipogenesis in the myometrium. Dev Biol 288:276–283

Atit R, Sgaier S, Mohamed O, Taketo M, Dufort D, Joyner A, Niswander L, Conlon R (2006) β-catenin activation is necessary and sufficient to specify the dorsal dermal fate in the mouse. Dev Biol 296:164–176

Aulehla A, Wiegraebe W, Baubet V, Wahl M, Deng C, Taketo M, Lewandoski M, Pourquié O (2008) A beta-catenin gradient links the clock and wavefront systems in mouse embryo segmentation. Nat Cell Biol 10:186–193

Backman M, Machon O, Mygland L, van den Bout C, Zhong W, Taketo M, Krauss S (2005) Effects of canonical Wnt signaling on dorso-ventral specification of the mouse telencephalon. Dev Biol 279:155–168

Barron M, Brookes S, Draper C, Garrod D, Kirkham J, Shore R, Dixon M (2008) The cell adhesion molecule nectin-1 is critical for normal enamel formation in mice. Hum Mol Genet 17:3509–3520

Barrow J, Thomas K, Boussadia-Zahui O, Moore R, Kemler R, Capecchi M, McMahon A (2003) Ectodermal Wnt3/beta-catenin signaling is required for the establishment and maintenance of the apical ectodermal ridge. Genes Dev 17:394–409

Bartlett J, Dobeck J, Tye C, Perez-Moreno M, Stokes N, Reynolds A, Fuchs E, Skobe Z (2010) Targeted p120-catenin ablation disrupts dental enamel development. PLoS One 5:e12703

Battle M, Konopka G, Parviz F, Gaggl A, Yang C, Sladek F, Duncan S (2006) Hepatocyte nuclear factor 4alpha orchestrates expression of cell adhesion proteins during the epithelial transformation of the developing liver. Proc Natl Acad Sci U S A 103:8419–8424

Berx G, van Roy F (2009) Involvement of members of the cadherin superfamily in cancer. Cold Spring Harb Perspect Biol 1:a003129

Bierkamp C, Mclaughlin K, Schwarz H, Huber O, Kemler R (1996) Embryonic heart and skin defects in mice lacking plakoglobin. Dev Biol 180:780–785

Bouchard M, Dong Y, McDermott BJ, Lam D, Brown K, Shelanski M, Bellvé A, Racaniello V (2000) Defects in nuclear and cytoskeletal morphology and mitochondrial localization in spermatozoa of mice lacking nectin-2, a component of cell-cell adherens junctions. Mol Cell Biol 20:2865–2873

Boussadia O, Kutsch S, Hierholzer A, Delmas V, Kemler R (2002) E-cadherin is a survival factor for the lactating mouse mammary gland. Mech Dev 115:53–62

Brault V, Moore R, Kutsch S, Ishibashi M, Rowitch D, McMahon A, Sommer L, Boussadia O, Kemler R (2001) Inactivation of the beta-catenin gene by Wnt1-Cre-mediated deletion results in dramatic brain malformation and failure of craniofacial development. Development 128:1253–1264

Bridgewater D, Cox B, Cain J, Lau A, Athaide V, Gill P, Kuure S, Sainio K, Rosenblum N (2008) Canonical WNT/beta-catenin signaling is required for ureteric branching. Dev Biol 317:83–94

Cali G, Zannini M, Rubini P, Tacchetti C, D'Andrea B, Affuso A, Wintermantel T, Boussadia O, Terracciano D, Silberschmidt D, Amendola E, De Felice M, Schutz G, Kemler R, Di Lauro R, Nitsch L (2007) Conditional inactivation of the E-cadherin gene in thyroid follicular cells affects gland development but does not impair junction formation. Endocrinology 148:2737–2746

Campos V, Du M, Li Y (2004) Increased seizure susceptibility and cortical malformation in beta-catenin mutant mice. Biochem Biophys Res Commun 320:606–614

Carmeliet P, Lampugnani M, Moons L, Breviario F, Compernolle V, Bono F, Balconi G, Spagnuolo R, Oosthuyse B, Dewerchin M, Zanetti A, Angellilo A, Mattot V, Nuyens D, Lutgens E, Clotman F, de Ruiter M, Gittenberger-de Groot A, Poelmann R, Lupu F, Herbert J, Collen D, Dejana E (1999) Targeted deficiency or cytosolic truncation of the VE-cadherin gene in mice impairs VEGF-mediated endothelial survival and angiogenesis. Cell 98:147–157

Carnahan R, Rokas A, Gaucher E, Reynolds A (2011) The molecular evolution of the p120-catenin subfamily and its functional associations. PLoS One 5:e15747

Cattelino A, Liebner S, Gallini R, Zanetti A, Balconi G, Corsi A, Bianco P, Wolburg H, Moore R, Oreda B, Kemler R, Dejana E (2003) The conditional inactivation of the beta-catenin gene in endothelial cells causes a defective vascular pattern and increased vascular fragility. J Cell Biol 162:1111–1122

Cerruti Mainardi P (1996) Cri du Chat syndrome. Orphanet J Rare Dis 1:33

Chang H, Gao F, Guillou F, Taketo M, Huff V, Behringer R (2008) Wt1 negatively regulates beta-catenin signaling during testis development. Development 135:1875–1885

Chen X, Shevtsov S, Hsich E, Cui L, Haq S, Aronovitz M, Kerkelä R, Molkentin J, Liao R, Salomon R, Patten R, Force T (2006) The beta-catenin/T-cell factor/lymphocyte enhancer factor signaling pathway is required for normal and stress-induced cardiac hypertrophy. Mol Cell Biol 26:4462–4473

Cohen E, Wang Z, Lepore J, Lu M, Taketo M, Epstein D, Morrisey E (2007) Wnt/beta-catenin signaling promotes expansion of Isl-1-positive cardiac progenitor cells through regulation of FGF signaling. J Clin Invest 117:1794–1804

Corada M, Mariotti M, Thurston G, Smith K, Kunkel R, Brockhaus M, Lampugnani M, Martin-Padura I, Stoppacciaro A, Ruco L, Macdonald D, Ward P, Dejana E (1999) Vascular endothelial-cadherin is an important determinant of microvascular integrity *in vivo*. Proc Natl Acad Sci U S A 96:9815–9820

Davis M, Reynolds A (2006) Blocked acinar development, E-cadherin reduction, and intraepithelial neoplasia upon ablation of p120-catenin in the mouse salivary gland. Dev Cell 10:21–31

Day T, Guo X, Garrett-Beal L, Yang Y (2005) Wnt/beta-catenin signaling in mesenchymal progenitors controls osteoblast and chondrocyte differentiation during vertebrate skeletogenesis. Dev Cell 8:739–750

De Langhe S, Carraro G, Tefft D, Li C, Xu X, Chai Y, Minoo P, Hajihosseini M, Drouin J, Kaartinen V, Bellusci S (2008) Formation and differentiation of multiple mesenchymal lineages during lung development is regulated by beta-catenin signaling. PLoS One 3:e1516

De Vries W, Evsikov A, Haac B, Fancher K, Holbrook A, Kemler R, Solter D, Knowles B (2004) Maternal beta-catenin and E-cadherin in mouse development. Development 131:4435–4445

Dejana E, Tournier-Lasserve E, Weinstein BM (2009) The control of vascular integrity by endothelial cell junctions: molecular basis and pathological implications. Dev Cell 16:209–221

Derksen PW, Liu X, Saridin F, van der Gulden H, Zevenhoven J, Evers B, van Beijnum JR, Griffioen AW, Vink J, Krimpenfort P, Peterse JL, Cardiff RD, Berns A, Jonkers J (2006) Somatic inactivation of E-cadherin and p53 in mice leads to metastatic lobular mammary carcinoma through induction of anoikis resistance and angiogenesis. Cancer Cell 10:437–449

Derksen PW, Braumuller TM, van der Burg E, Hornsveld M, Mesman E, Wesseling J, Krimpenfort P, Jonkers J (2011) Mammary-specific inactivation of E-cadherin and p53 impairs functional gland development and leads to pleomorphic invasive lobular carcinoma in mice. Dis Model Mech 4:347–358

Dessimoz J, Bonnard C, Huelsken J, Grapin-Botton A (2005) Pancreas-specific deletion of beta-catenin reveals Wnt-dependent and Wnt-independent functions during development. Curr Biol 15:1677–1683

Deutscher E, Hung-Chang Yao H (2007) Essential roles of mesenchyme-derived beta-catenin in mouse Müllerian duct morphogenesis. Dev Biol 307:227–236

Dunty WJ, Biris K, Chalamalasetty R, Taketo M, Lewandoski M, Yamaguchi T (2008) Wnt3a/beta-catenin signaling controls posterior body development by coordinating mesoderm formation and segmentation. Development 135:85–94

Elia L, Yamamoto M, Zang K, Reichardt L (2006) p120 catenin regulates dendritic spine and synapse development through Rho-family GTPases and cadherins. Neuron 51:43–56

Farquhar MG, Palade GE (1963) Junctional complexes in various epithelia. J Cell Biol 17:375–412
Fleming T, Johnson M (1988) From egg to epithelium. Annu Rev Cell Biol 4:459–485
Fleming T, Javed Q, Collins J, Hay M (1993) Biogenesis of structural intercellular junctions during cleavage in the mouse embryo. J Cell Sci Suppl 17:119–125
Fleming TP, Papenbrock T, Fesenko I, Hausen P, Sheth B (2000) Assembly of tight junctions during early vertebrate development. Semin Cell Dev Biol 11:291–299
Fu X, Sun H, Klein W, Mu X (2006) Beta-catenin is essential for lamination but not neurogenesis in mouse retinal development. Dev Biol 299:424–437
Fujita E, Kouroku Y, Ozeki S, Tanabe Y, Toyama Y, Maekawa M, Kojima N, Senoo H, Toshimori K, Momoi T (2006) Oligo-astheno-teratozoospermia in mice lacking RA175/TSLC1/SynCAM/IGSF4A, a cell adhesion molecule in the immunoglobulin superfamily. Mol Cell Biol 26:718–726
Gao X, Arlotta P, Macklis J, Chen J (2007) Conditional knock-out of beta-catenin in postnatal-born dentate gyrus granule neurons results in dendritic malformation. J Neurosci 27:14317–14325
Gavard J, Gutkind J (2006) VEGF controls endothelial-cell permeability by promoting the β-arrestin-dependent endocytosis of VE-cadherin. Nat Cell Biol 8:1223–1234
Giangreco A, Jensen K, Takai Y, Miyoshi J, Watt F (2009) Necl2 regulates epidermal adhesion and wound repair. Development 136:3505–3514
Gladden A, Hebert A, Schneeberger E, McClatchey A (2010) The NF2 tumor suppressor, Merlin, regulates epidermal development through the establishment of a junctional polarity complex. Dev Cell 19:727–739
Glass D, Bialek P, Ahn J, Starbuch M, Patel M, Clevers H, Taketo M, Long F, McMahon A, Lang R et al (2005) Canonical Wnt signaling in differentiated osteoblasts controls osteoclast differentiation. Dev Cell 8:751–764
Gory-Fauré S, Prandini M, Pointu H, Roullot V, Pignot-Paintrand I, Vernet M, Huber P (1999) Role of vascular endothelial-cadherin in vascular morphogenesis. Development 126:2093–2102
Gottardi CJ, Wong E, Gumbiner BM (2001) E-cadherin suppresses cellular transformation by inhibiting beta-catenin signaling in an adhesion-independent manner. J Cell Biol 153:1049–1060
Grigoryan T, Wend P, Klaus A, Birchmeier W (2008) Deciphering the function of canonical Wnt signals in development and disease: conditional loss- and gain-of-function mutations of b-catenin in mice. Genes Dev 22:2308–2341
Grivennikov S, Greten F, Karin M (2010) Immunity, inflammation, and cancer. Cell 140:883–899
Gumbiner B, Stevenson B, Grimaldi A (1988) The role of the cell adhesion molecule uvomorulin in the formation and maintenance of the epithelial junctional complex. J Cell Biol 107:1575–1587
Guo X, Day T, Jiang X, Garrett-Beal L, Topol L, Yang Y (2004) Wnt/beta-catenin signaling is sufficient and necessary for synovial joint formation. Genes Dev 18:2404–2417
Gustafson-Wagner E, Sinn H, Chen Y, Wang D, Reiter R, Lin J, Yang B, Williamson R, Chen J, Lin C, Lin J (2007) Loss of mXinalpha, an intercalated disk protein, results in cardiac hypertrophy and cardiomyopathy with conduction defects. Am J Physiol Heart Circ Physiol 293:H2680–2692
Haegel H, Larue L, Ohsugi M, Fedorov L, Herrenknecht K, Kemler R (1995) Lack of beta-catenin affects mouse development at gastrulation. Development 121:3529–3537
Hari L, Brault V, Kléber M, Lee H, Ille F, Leimeroth R, Paratore C, Suter U, Kemler R, Sommer L (2002) Lineage-specific requirements of beta-catenin in neural crest development. J Cell Biol 159:867–880
Harris TJ, Peifer M (2005) The positioning and segregation of apical cues during epithelial polarity establishment in Drosophila. J Cell Biol 170:813–823
Harris TJ, Tepass U (2010) Adherens junctions: from molecules to morphogenesis. Nat Rev Mol Cell Biol 11:502–514
Hatta K, Takagi S, Fujisawa H, Takeichi M (1987) Spatial and temporal expression pattern of N-cadherin cell adhesion molecules correlated with morphogenetic processes of chicken embryos. Dev Biol 120:215–227
Hermiston ML, Gordon JI (1995a) In vivo analysis of cadherin function in the mouse intestinal epithelium: essential roles in adhesion, maintenance of differentiation, and regulation of programmed cell death. J Cell Biol 129:489–506

Hermiston ML, Gordon JI (1995b) Inflammatory bowel disease and adenomas in mice expressing a dominant negative N-cadherin. Science 270:1203–1207

Hermiston ML, Wong MH, Gordon JI (1996) Forced expression of E-cadherin in the mouse intestinal epithelium slows cell migration and provides evidence for nonautonomous regulation of cell fate in a self-renewing system. Genes Dev 10:985–996

Hill T, Später D, Taketo M, Birchmeier W, Hartmann C (2005) Canonical Wnt/beta-catenin signaling prevents osteoblasts from differentiating into chondrocytes. Dev Cell 8:727–738

Hill T, Taketo M, Birchmeier W, Hartmann C (2006) Multiple roles of mesenchymal beta-catenin during murine limb patterning. Development 133:1219–1229

Hirai Y, Nose A, Kobayashi S, Takeichi M (1989) Expression and role of E- and P-cadherin adhesion molecules in embryonic histogenesis II. Skin morphogenesis. Development 105:271–277

Honda T, Sakisaka T, Yamada T, Kumazawa N, Hoshino T, Kajita M, Kayahara T, Ishizaki H, Tanaka-Okamoto M, Mizoguchi A, Manabe T, Miyoshi J, Takai Y (2006) Involvement of nectins in the formation of puncta adherentia junctions and the mossy fiber trajectory in the mouse hippocampus. Mol Cell Neurosci 31:315–325

Huelsken J, Vogel R, Brinkmann V, Erdmann B, Birchmeier C, Birchmeier W (2000) Requirement for beta-catenin in anterior-posterior axis formation in mice. J Cell Biol 148:567–578

Huelsken J, Vogel R, Erdmann B, Cotsarelis G, Birchmeier W (2001) beta-Catenin controls hair follicle morphogenesis and stem cell differentiation in the skin. Cell 105:533–545

Ikeda W, Nakanishi H, Miyoshi J, Mandai K, Ishizaki H, Tanaka M, Togawa A, Takahashi K, Nishioka H, Yoshida H, Mizoguchi A, Nishikawa S, Takai Y (1999) Afadin: a key molecule essential for structural organization of cell-cell junctions of polarized epithelia during embryogenesis. J Cell Biol 146:1117–1132

Inagaki M, Irie K, Ishizaki H, Tanaka-Okamoto M, Morimoto K, Inoue E, Ohtsuka T, Miyoshi J, Takai Y (2005) Roles of cell-adhesion molecules nectin 1 and nectin 3 in ciliary body development. Development 132:1525–1537

Inoue T, Tanaka T, Suzuki S, Takeichi M (1998) Cadherin-6 in the developing mouse brain: expression along restricted connection systems and synaptic localization suggest a potential role in neuronal circuitry. Dev Dyn 211:338–351

Israely I, Costa R, Xie C, Silva A, Kosik K, Liu X (2004) Deletion of the neuron-specific protein delta-catenin leads to severe cognitive and synaptic dysfunction. Curr Biol 14:1657–1663

Itoh M, Nagafuchi A, Moroi S, Tsukita S (1997) Involvement of ZO-1 in cadherin-based cell adhesion through its direct binding to alpha catenin and actin filaments. J Cell Biol 138:181–192

Jelani M, Chishti MS, Ahmad W (2011) Mutation in PVRL4 gene encoding nectin-4 underlies ectodermal-dysplasia-syndactyly syndrome (EDSS1). J Hum Genet 56:352–357

Junghans D, Hack I, Frotscher M, Taylor V, Kemler R (2005) Beta-catenin-mediated cell-adhesion is vital for embryonic forebrain development. Dev Dyn 233:528–539

Kadowaki M, Nakamura S, Machon O, Krauss S, Radice G, Takeichi M (2007) N-cadherin mediates cortical organization in the mouse brain. Dev Biol 304:22–33

Kan N, Stemmler M, Junghans D, Kanzler B, de Vries W, Dominis M, Kemler R (2007) Gene replacement reveals a specific role for E-cadherin in the formation of a functional trophectoderm. Development 134:31–41

Kanno S, Saffitz J (2001) The role of myocardial gap junctions in electrical conduction and arrhythmogenesis. Cardiovasc Pathol 10:169–177

Kioussi C, Briata P, Baek S, Rose D, Hamblet N, Herman T, Ohgi K, Lin C, Gleiberman A, Wang J, Brault V, Ruiz-Lozano P, Nguyen H, Kemler R, Glass C, Wynshaw-Boris A, Rosenfeld M (2002) Identification of a Wnt/Dvl/beta-Catenin → Pitx2 pathway mediating cell-type-specific proliferation during development. Cell 111:673–685

Kirchhof P, Fabritz L, Zwiener M, Witt H, Schafers M, Zellerhoff S, Paul M, Athai T, Hiller K, Baba H, Breithardt G, Ruiz P, Wichter T, Levkau B (2006) Age- and training-dependent development of arrhythmogenic right ventricular cardiomyopathy in heterozygous plakoglobin-deficient mice. Circulation 114:1799–1806

Kjaer K, Hansen L, Schwabe G, Marques-de-Faria A, Eiberg H, Mundlos S, Tommerup N, Rosenberg T (2005) Distinct CDH3 mutations cause ectodermal dysplasia, ectrodactyly, macular dystrophy (EEM syndrome). J Med Genet 42:292–298

Klaus A, Saga Y, Taketo M, Tzahor E, Birchmeier W (2007) Distinct roles of Wnt/beta-catenin and Bmp signaling during early cardiogenesis. Proc Natl Acad Sci U S A 104:18531–18536
Kobielak A, Fuchs E (2006) Links between alpha-catenin, NF-kB, and squamous cell carcinoma in skin. Proc Natl Acad Sci U S A 103:2322–2327
Koster M (2010) p63 in skin development and ectodermal dysplasias. J Invest Derm 130:2352–2358
Kostetskii I, Moore R, Kemler R, Radice G (2001) Differential adhesion leads to segregation and exclusion of N-cadherin-deficient cells in chimeric embryos. Dev Biol 234:72–79
Kostetskii I, Li J, Xiong Y, Zhou R, Ferrari VA, Patel VV, Molkentin JD, Radice GL (2005) Induced deletion of the N-cadherin gene in the heart leads to dissolution of the intercalated disc structure. Circ Res 96:346–354
Kovtun IV, Harris KJ, Jatoi A, Jevremovic D (2011) Increased incidence of endometrioid tumors caused by aberrations in E-cadherin promoter of mismatch repair-deficient mice. Carcinogenesis 32:1085–1092
Kwon C, Arnold J, Hsiao E, Taketo M, Conklin B, Srivastava D (2007) Canonical Wnt signaling is a positive regulator of mammalian cardiac progenitors. Proc Natl Acad Sci U S A 104:10894–10899
Lallemand D, Curto M, Saotome I, Giovannini M, McClatchey AI (2003) NF2 deficiency promotes tumorigenesis and metastasis by destabilizing adherens junctions. Genes Dev 17:1090–1100
Larue L, Ohsugi M, Hirchenhain J, Kemler R (1994) E-cadherin null mutant embryos fail to form a trophectoderm epithelium. Proc Natl Acad Sci U S A 91:8263–8267
Lewis JE, Jensen PJ, Wheelock MJ (1994) Cadherin function is required for human keratinocytes to assemble desmosomes and stratify in response to calcium. J Invest Dermatol 102:870–877
Li J, Radice GL (2010) A new perspective on intercalated disc organization: implications for heart disease. Dermatol Res Pract 2010:207835
Li J, Patel VV, Kostetskii I, Xiong Y, Chu AF, Jacobson JT, Yu C, Morley GE, Molkentin JD, Radice GL (2005) Cardiac-specific loss of N-cadherin leads to alteration in connexins with conduction slowing and arrhythmogenesis. Circ Res 97:474–481
Li J, Levin MD, Xiong Y, Petrenko N, Patel VV, Radice GL (2008) N-cadherin haploinsufficiency affects cardiac gap junctions and arrhythmic susceptibility. J Mol Cell Cardiol 44:597–606
Li X, Dong X, Luo S, Zhang B, Lee D, Ting A, Neiswender H, Kim C, Carpenter-Hyland E, Gao T, Xiong W, Mei L (2008) Retrograde regulation of motoneuron differentiation by muscle beta-catenin. Nat Neurosci 11:262–268
Li D, Liu Y, Maruyama M, Zhu W, Chen H, Zhang W, Reuter S, Lin S, Haneline L, Field L, Chen P, Shou W (2011) Restrictive loss of plakoglobin in cardiomyocytes leads to arrhythmogenic cardiomyopathy. Hum Mol Genet 20:4582–4596
Libusova L, Stemmler M, Hierholzer A, Schwarz E, Kemler R (2010) N-cadherin can structurally substitute for E-cadherin during intestinal development but leads to polyp formation. Development 137:2297–2305
Lickert H, Kutsch S, Kanzler B, Tamai Y, Taketo M, Kemler R (2002) Formation of multiple hearts in mice following deletion of beta-catenin in the embryonic endoderm. Dev Cell 3:171–181
Liebner S, Cattelino A, Gallini R, Rudini N, Iurlaro M, Piccolo S, Dejana E (2004) Beta-catenin is required for endothelial-mesenchymal transformation during heart cushion development in the mouse. Cell Biol 166:359–367
Lien W, Klezovitch O, Fernandez T, Delrow J, Vasioukhin V (2006) alphaE-catenin controls cerebral cortical size by regulating the hedgehog signaling pathway. Science 311:1609–1612
Lien WH, Klezovitch O, Null M, Vasioukhin V (2008) alphaE-catenin is not a significant regulator of beta-catenin signaling in the developing mammalian brain. J Cell Sci 121:1357–1362
Lin L, Cui L, Zhou W, Dufort D, Zhang X, Cai C, Bu L, Yang L, Martin J, Kemler R, Rosenfeld M, Chen J, Evans S (2007) Beta-catenin directly regulates Islet1 expression in cardiovascular progenitors and is required for multiple aspects of cardiogenesis. Proc Natl Acad Sci U S A 104:9313–9318
Luo Y, Radice G (2005) N-cadherin acts upstream of VE-cadherin in controlling vascular morphogenesis. J Cell Biol 169:29–34

Luo Y, Ferreira-Cornwell M, Baldwin H, Kostetskii I, Lenox J, Lieberman M, Radice G (2001) Rescuing the N-cadherin knockout by cardiac-specific expression of N- or E-cadherin. Development 128:459–469

Luo Y, High F, Epstein J, Radice G (2006) N-cadherin is required for neural crest remodeling of the cardiac outflow tract. Dev Biol 299:517–528

Machon O, van den Bout C, Backman M, Kemler R, Krauss S (2003) Role of beta-catenin in the developing cortical and hippocampal neuroepithelium. Neuroscience 122:129–143

Maiden SL, Hardin J (2011) The secret life of alpha-catenin: moonlighting in morphogenesis. J Cell Biol 195:543–552

Majima T, Ogita H, Yamada T, Amano H, Togashi H, Sakisaka T, Tanaka-Okamoto M, Ishizaki H, Miyoshi J, Takai Y (2009) Involvement of afadin in the formation and remodeling of synapses in the hippocampus. Biochem Biophys Res Commun 385:539–544

Marose T, Merkel C, McMahon A, Carroll T (2008) Beta-catenin is necessary to keep cells of ureteric bud/Wolffian duct epithelium in a precursor state. Dev Biol 314:112–126

Matter C, Pribadi M, Liu X, Trachtenberg J (2009) Delta-catenin is required for the maintenance of neural structure and function in mature cortex in vivo. Neuron 64:320–327

McAllister A (2007) Dynamic aspects of CNS synapse formation. Annu Rev Neurosci 30:425–450

McCrea PD, Park JI (2007) Developmental functions of the P120-catenin sub-family. Biochim Biophys Acta 1773:17–33

Michels C, Buchta T, Bloch W, Krieg T, Niessen C (2009) Classical cadherins regulate desmosome formation. J Invest Derm 129:2072–2075

Miller L, Smith A, Taketo M, Lang R (2006) Optic cup and facial petterning defects in ocular ectoderm β-catenin gain-of-function mice. BMC Dev Biol 6:14

Mucenski M, Wert S, Nation J, Loudy D, Huelsken J, Birchmeier W, Morrisey E, Whitsett J (2003) beta-Catenin is required for specification of proximal/distal cell fate during lung morphogenesis. J Biol Chem 278:40231–40238

Mueller S, Rosenquist T, Takai Y, Bronson R, Wimmer E (2003) Loss of nectin-2 at Sertoli-spermatid junctions leads to male infertility and correlates with severe spermatozoan head and midpiece malformation, impaired binding to the zona pellucida, and oocyte penetration. Biol Reprod 69:1330–1340

Murtaugh L, Law A, Dor Y, Melton D (2005) Beta-catenin is essential for pancreatic acinar but not islet development. Development 132:4663–4674

Navarro P, Ruco L, Dejana E (1998) Differential localization of VE- and N-cadherins in human endothelial cells: VE-cadherin competes with N-cadherin for junctional localization. J Cell Biol 140:1475–1484

Nemade R, Bierie B, Nozawa M, Bry C, Smith G, Vasioukhin V, Fuchs E, Hennighausen L (2004) Biogenesis and function of mouse mammary epithelium depends on the presence of functional alpha-catenin. Mech Dev 121:91–99

Niessen CM, Gottardi CJ (2008) Molecular components of the adherens junction. Biochim Biophys Acta 1778:562–571

Nitta T, Hata M, Gotoh S, Seo Y, Sasaki H, Hashimoto N, Furuse M, Tsukita S (2003) Size-selective loosening of the blood-brain barrier in claudin-5-deficient mice. J Cell Biol 161:653–660

Noren NK, Liu BP, Burridge K, Kreft B (2000) p120 catenin regulates the actin cytoskeleton via Rho family GTPases. J Cell Biol 150:567–580

Nose A, Takeichi M (1986) A novel cadherin cell adhesion molecule: its expression patterns associated with implantation and organogenesis of mouse embryos. J Cell Biol 103:2649–2658

Oas R, Xiao K, Summers S, Wittich K, Chiasson C, Martin W, Grossniklaus H, Vincent P, Reynolds A, Kowalczyk A (2010) p120-Catenin is required for mouse vascular development. Circ Res 106:941–951

Oda H, Takeichi M (2011) Evolution: structural and functional diversity of cadherin at the adherens junction. J Cell Biol 193:1137–1146

Ohtola J, Myers J, Akhtar-Zaidi B, Zuzindlak D, Sandesara P, Yeh K, Mackem S, Atit R (2008) beta-Catenin has sequential roles in the survival and specification of ventral dermis. Development 135:2321–2329

Ohyama T, Mohamed O, Taketo M, Dufort D, Groves A (2006) Wnt signals mediate a fate decision between otic placode and epidermis. Development 133:865–875
Olson L, Tollkuhn J, Scafoglio C, Krones A, Zhang J, Ohgi K, Wu W, Taketo M, Kemler R, Grosschedl R, Rose D, Li X, Rosenfeld M (2006) Homeodomain-mediated beta-catenin-dependent switching events dictate cell-lineage determination. Cell 125:593–605
Ozaki-Kuroda K, Nakanishi H, Ohta H, Tanaka H, Kurihara H, Mueller S, Irie K, Ikeda W, Sakai T, Wimmer E, Nishimune Y, Takai Y (2002) Nectin couples cell-cell adhesion and the actin scaffold at heterotypic testicular junctions. Curr Biol 12:1145–1150
Park C, Finger J, Cooper J, Ackerman S (2002) The cerebellar deficient folia (cdf) gene acts intrinsically in Purkinje cell migrations. Genesis 32:32–41
Park J, Valerius M, McMahon A (2007) Wnt/beta-catenin signaling regulates nephron induction during mouse kidney development. Development 134:2533–2539
Park J, Liu B, Chen T, Li H, Hu X, Gao J, Zhu Y, Zhu Q, Qiang B, Yuan J, Peng X, Qiu M (2008) Disruption of Nectin-like 1 cell adhesion molecule leads to delayed axonal myelination in the CNS. J Neurosci 28:12815–12819
Perez-Moreno M, Davis M, Wong E, Pasolli H, Reynolds A, Fuchs E (2006) p120-catenin mediates inflammatory responses in the skin. Cell 124:631–644
Perez-Moreno M, Song W, Pasolli H, Williams S, Fuchs E (2008) Loss of p120 catenin and links to mitotic alterations, inflammation, and skin cancer. Proc Natl Acad Sci U S A 105:15399–15404
Pieters T, van Hengel J, van Roy F (2012) Functions of p120ctn in development and disease. Front Biosci 17:760–783
Piven O, Kostetskii I, Macewicz L, Kolomiets Y, Radice G, Lukash L (2011) Requirement for N-cadherin-catenin complex in heart development. Exp Biol Med (Maywood) 236:816–822
Pontoriero G, Smith A, Miller L, Radice G, West-Mays J, Lang R (2009) Co-operative roles for E-cadherin and N-cadherin during lens vesicle separation and lens epithelial cell survival. Dev Biol 326:403–417
Radice G, Ferreira-Cornwell M, Robinson S, Rayburn H, Chodosh L, Takeichi M, Hynes R (1997a) Precocious mammary gland development in P-cadherin-deficient mice. J Cell Biol 139:1025–1032
Radice G, Rayburn H, Matsunami H, Knudsen K, Takeichi M, Hynes R (1997b) Developmental defects in mouse embryos lacking N-cadherin. Dev Biol 181:64–78
Reynolds A, Roczniak-Ferguson A (2004) Emerging roles for p120-catenin in cell adhesion and cancer. Oncogene 23:7947–7956
Riethmacher D, Brinkmann V, Birchmeier C (1995) A targeted mutation in the mouse E-cadherin gene results in defective preimplantation development. Proc Natl Acad Sci U S A 92:855–859
Rodda S, McMahon A (2006) Distinct roles for Hedgehog and canonical Wnt signaling in specification, differentiation and maintenance of osteoblast progenitors. Development 133:3231–3244
Ruiz P, Brinkmann V, Ledermann B, Behrend M, Grund C, Thalhammer C, Vogel F, Birchmeier C, Gtinthert U, Franke W, Birchmeier W (1996) Targeted mutation of plakoglobin in mice reveals essential functions of desmosomes in the embryonic heart. J Cell Biol 135:215–225
Satomi-Kobayashi S, Ueyama T, Mueller S, Toh R, Masano T, Sakoda T, Rikitake Y, Miyoshi J, Matsubara H, Oh H, Kawashima S, Hirata K, Takai Y (2009) Deficiency of nectin-2 leads to cardiac fibrosis and dysfunction under chronic pressure overload. Hypertension 54:825–831
Schackmann RC, van Amersfoort M, Haarhuis JH, Vlug EJ, Halim VA, Roodhart JM, Vermaat JS, Voest EE, van der Groep P, van Diest PJ, Jonkers J, Derksen PW (2011) Cytosolic p120-catenin regulates growth of metastatic lobular carcinoma through Rock1-mediated anoikis resistance. J Clin Invest 121:3176–3188
Schlegelmilch K, Mohseni M, Kirak O, Pruszak J, Rodriguez JR, Zhou D, Kreger BT, Vasioukhin V, Avruch J, Brummelkamp TR, Camargo FD (2011) Yap1 acts downstream of alpha-catenin to control epidermal proliferation. Cell 144:782–795
Schneider MR, Dahlhoff M, Horst D, Hirschi B, Trulzsch K, Muller-Hocker J, Vogelmann R, Allgauer M, Gerhard M, Steininger S, Wolf E, Kolligs FT (2010) A key role for E-cadherin in intestinal homeostasis and Paneth cell maturation. PLoS One 5:e14325

Schüller U, Rowitch D (2007) Beta-catenin function is required for cerebellar morphogenesis. Brain Res 1140:161–169

Sekine S, Lan B, Bedolli M, Feng S, Hebrok M (2006) Liver-specific loss of beta-catenin blocks glutamine synthesis pathway activity and cytochrome p450 expression in mice. Hepatology 43:817–825

Sekine S, Gutiérrez P, Lan B, Feng S, Hebrok M (2007) Liver-specific loss of beta-catenin results in delayed hepatocyte proliferation after partial hepatectomy. Hepatology 45:361–368

Sheikh F, Chen Y, Liang X, Hirschy A, Stenbit A, Gu Y, Dalton N, Yajima T, Lu Y, Knowlton K, Peterson K, Perriard J, Chen J (2006) alpha-E-catenin inactivation disrupts the cardiomyocyte adherens junction, resulting in cardiomyopathy and susceptibility to wall rupture. Circulation 114:1046–1055

Shimada S, Mimata A, Sekine M, Mogushi K, Akiyama Y, Fukamachi H, Jonkers J, Tanaka H, Eishi Y, Yuasa Y (2011) Synergistic tumour suppressor activity of E-cadherin and p53 in a conditional mouse model for metastatic diffuse-type gastric cancer. Gut 61:344–353

Shimomura Y, Wajid M, Shapiro Lea (2008) P-cadherin is a p63 target gene with a crucial role in the developing human limb bud and hair follicle. Development 135:743–753

Silvis M, Kreger B, Lien W, Klezovitch O, Rudakova G, Camargo F, Lantz D, Seykora J, Vasioukhin V (2011) α-catenin is a tumor suppressor that controls cell accumulation by regulating the localization and activity of the transcriptional coactivator Yap1. Sci Signal 4:ra33

Smalley-Freed W, Efimov A, Burnett P, Short S, Davis M, Gumucio D, Washington M, Coffey R, Reynolds A (2010) p120-catenin is essential for maintenance of barrier function and intestinal homeostasis in mice. J Clin Invest 120:1824–1835

Smith A, Miller L, Song N, Taketo M, Lang R (2005) The duality of beta-catenin function: a requirement in lens morphogenesis and signaling suppression of lens fate in periocular ectoderm. Dev Biol 285:477–489

Smith JD, Meehan MH, Crean J, McCann A (2011) Alpha T-catenin (CTNNA3): a gene in the hand is worth two in the nest. Cell Mol Life Sci 68:2493–2498

Soshnikova N, Zechner D, Huelsken J, Mishina Y, Behringer R, Taketo M, Crenshaw Er, Birchmeier W (2003) Genetic interaction between Wnt/beta-catenin and BMP receptor signaling during formation of the AER and the dorsal-ventral axis in the limb. Genes Dev 17:1963–1968

Sozen M, Suzuki K, Tolarova M, Bustos T, Fernandez Iglesias J, Spritz R (2001) Mutation of PVRL1 is associated with sporadic, non-syndromic cleft lip/palate in northern Venezuela. Nat Genet 29:141–142

Sprecher E, Bergman R, Richard G, Lurie R, Shalev S, Petronius D, Shalata A, Anbinder Y, Leibu R, Perlman I, Cohen N, Szargel R (2001) Hypotrichosis with juvenile macular dystrophy is caused by a mutation in CDH3, encoding P-cadherin. Nat Genet 29:134–136

Stairs D, Bayne L, Rhoades B, Vega M, Waldron T, Kalabis J, Klein-Szanto A, Lee J, Katz J, Diehl J, Reynolds A, Vonderheide R, Rustgi A (2011) Deletion of p120-catenin results in a tumor microenvironment with inflammation and cancer that establishes it as a tumor suppressor gene. Cancer Cell 19:470–483

Stepniak E, Radice GL, Vasioukhin V (2009) Adhesive and signaling functions of cadherins and catenins in vertebrate development. Cold Spring Harb Perspect Biol 1:a002949

Stocker AM, Chenn A (2006) Differential expression of alpha-E-catenin and alpha-N-catenin in the developing cerebral cortex. Brain Res 1073–1074:151–158

Suzuki K, Hu D, Bustos T, Zlotogora J, Richieri-Costa A, Helms J, Spritz R (2000) Mutations of PVRL1, encoding a cell-cell adhesion molecule/herpesvirus receptor, in cleft lip/palate-ectodermal dysplasia. Nat Genet 25:427–430

Taddei A, Giampietro C, Conti A, Orsenigo F, Breviario F, Pirazzoli V, Potente M, Daly C, Dimmeler S, Dejana E (2008) Endothelial adherens junctions control tight junctions by VE-cadherin-mediated upregulation of claudin-5. Nat Cell Biol 10:923–934

Tan X, Behari J, Cieply B, Michalopoulos G, Monga S (2006) Conditional deletion of beta-catenin reveals its role in liver growth and regeneration. Gastroenterology 131:1561–1572

Tan X, Yuan Y, Zeng G, Apte U, Thompson M, Cieply B, Stolz D, Michalopoulos G, Kaestner K, Monga S (2008) Beta-catenin deletion in hepatoblasts disrupts hepatic morphogenesis and survival during mouse development. Hepatology 47:1667–1679

Tanaka-Okamoto M, Hori K, Ishizaki H, Itoh Y, Onishi S, Yonemura S, Takai Y, Miyoshi J (2011) Involvement of afadin in barrier function and homeostasis of mouse intestinal epithelia. J Cell Sci 124:2231–2240

Thoreson M, Reynolds A (2002) Altered expression of the catenin p120 in human cancer: implications for tumor progression. Differentiation 70:583–589

Tinkle C, Lechler T, Pasolli H, Fuchs E (2004) Conditional targeting of E-cadherin in skin: insights into hyperproliferative and degenerative responses. Proc Natl Acad Sci U S A 101:552–557

Tinkle C, Pasolli H, Stokes N, Fuchs E (2008) New insights into cadherin function in epidermal sheet formation and maintenance of tissue integrity. Proc Natl Acad Sci U S A 105:15405–15410

Togashi H, Abe K, Mizoguchi A, Takaoka K, Chisaka O, Takeichi M (2002) Cadherin regulates dendritic spine morphogenesis. Neuron 35:77–89

Togashi H, Kominami K, Waseda M, Komura H, Miyoshi J, Takeichi M, Takai Y (2011) Nectins establish a checkerboard-like cellular pattern in the auditory epithelium. Science 333:1144–1147

Torres M, Stoykova A, Huber O, Chowdhury K, Bonaldo P, Mansouri A, Butz S, Kemler R, Gruss P (1997) An alpha-E-catenin gene trap mutation defines its function in preimplantation development. Proc Natl Acad Sci U S A 94:901–906

Tunggal J, Helfrich I, Schmitz A, Schwarz H, Gunzel D, Fromm M, Kemler R, Krieg T, Niessen C (2005) E-cadherin is essential for in vivo epidermal barrier function by regulating tight junctions. EMBO J 24:1146–1156

Uchida N, Shimamura K, Miyatani S, Copeland N, Gilbert D, Jenkins N, Takeichi M (1994) Mouse alpha N-catenin: two isoforms, specific expression in the nervous system, and chromosomal localization of the gene. Dev Biol 163:75–85

Uemura M, Takeichi M (2006) Alpha N-catenin deficiency causes defects in axon migration and nuclear organization in restricted regions of the mouse brain. Dev Dyn 235:2559–2566

van Hengel J, van Roy F (2007) Diverse functions of p120ctn in tumors. Biochim Biophys Acta 1773:78–88

Vasioukhin V, Bauer C, Yin M, Fuchs E (2000) Directed actin polymerization is the driving force for epithelial cell--cell adhesion. Cell 100:209–219

Vasioukhin V, Bauer C, Degenstein L, Wise B, Fuchs E (2001) Hyperproliferation and defects in epithelial polarity upon conditional ablation of alpha-catenin in skin. Cell 104:605–617

Wakamatsu K, Ogita H, Okabe N, Irie K, Tanaka-Okamoto M, Ishizaki H, Ishida-Yamamoto A, Takai Y (2007) Up-regulation of loricrin expression by cell adhesion molecule nectin-1 through Rap1-ERK signaling in keratinocytes. J Biol Chem 282:18173–18181

Wells J, Esni F, Boivin G, Aronow B, Stuart W, Combs C, Sklenka A, Leach S, Lowy A (2007) Wnt/beta-catenin signaling is required for development of the exocrine pancreas. BMC Dev Biol 7:4

Wong AS, Gumbiner BM (2003) Adhesion-independent mechanism for suppression of tumor cell invasion by E-cadherin. J Cell Biol 161:1191–1203

Xiao K, Garner J, Buckley K, Vincent P, Chiasson C, Dejana E, Faundez V, Kowalczyk A (2005) p120-Catenin regulates clathrin-dependent endocytosis of VE-cadherin. Mol Biol Cell 16:5141–5151

Xu W, Baribault H, Adamson E (1998) Vinculin knockout results in heart and brain defects during embryonic development. Development 125:327–337

Yang A, Zhu Z, Kapranov P, McKeon F, Church G, Gingeras T, Struhl K (2006) Relationships between p63 binding, DNA sequence, transcription activity, and biological function in human cells. Mol Cell 24:593–602

Yin Y, White A, Huh S, Hilton M, Kanazawa H, Long F, Ornitz D (2008) An FGF-WNT gene regulatory network controls lung mesenchyme development. Dev Biol 319:426–436

Yonemura S (2011) Cadherin-actin interactions at adherens junctions. Curr Opin Cell Biol 23:515–522

Yoshida T, Miyoshi J, Takai Y, Thesleff I (2010) Cooperation of nectin-1 and nectin-3 is required for normal ameloblast function and crown shape development in mouse teeth. Dev Dyn 239:2558–2569

Young P, Boussadia O, Berger P, Leone D, Charnay P, Kemler R, Suter U (2002) E-cadherin is required for the correct formation of autotypic adherens junctions of the outer mesaxon but not for the integrity of myelinated fibers of peripheral nerves. Mol Cell Neurosci 21:341–351

Young P, Boussadia O, Halfter H, Grose R, Berger P, Leone D, Robenek H, Charnay P, Kemler R, Suter U (2003) E-cadherin controls adherens junctions in the epidermis and the renewal of hair follicles. EMBO J 22:5723–5733

Zamora M, Männer J, Ruiz-Lozano P (2007) Epicardium-derived progenitor cells require beta-catenin for coronary artery formation. Proc Natl Acad Sci U S A 104:18109–18114

Zechner D, Fujita Y, Hülsken J, Müller T, Walther I, Taketo M, Crenshaw Er, Birchmeier W, Birchmeier C (2003) beta-Catenin signals regulate cell growth and the balance between progenitor cell expansion and differentiation in the nervous system. Dev Biol 258:406–418

Zemljic-Harpf A, Ponrartana S, Avalos R, Jordan M, Roos K, Dalton N, Phan V, Adamson E, Ross R (2004) Heterozygous inactivation of the vinculin gene predisposes to stress-induced cardiomyopathy. Am J Pathol 165:1033–1044

Zemljic-Harpf A, Miller J, Henderson S, Wright A, Manso A, Elsherif L, Dalton N, Thor A, Perkins G, McCulloch A, Ross R (2007) Cardiac-myocyte-specific excision of the vinculin gene disrupts cellular junctions, causing sudden death or dilated cardiomyopathy. Mol Cell Biol 27:7522–7537

Zhadanov A, Provance DJ, Speer C, Coffin J, Goss D, Blixt J, Reichert C, Mercer J (1999) Absence of the tight junctional protein AF-6 disrupts epithelial cell-cell junctions and cell polarity during mouse development. Curr Biol 9:880–888

Zhang J, Woodhead GJ, Swaminathan SK, Noles SR, McQuinn ER, Pisarek AJ, Stocker AM, Mutch CA, Funatsu N, Chenn A (2010) Cortical neural precursors inhibit their own differentiation via N-cadherin maintenance of beta-catenin signaling. Dev Cell 18:472–479

Zhao C, Blum J, Chen A, Kwon H, Jung S, Cook J, Lagoo A, Reya T (2007) Loss of beta-catenin impairs the renewal of normal and CML stem cells in vivo. Cancer Cell 12:528–541

Zhou J, Liyanage U, Medina M, Ho C, Simmons A, Lovett M, Kosik K (1997) Presenilin 1 interaction in the brain with a novel member of the Armadillo family. Neuroreport 8:2085–2090

Zhou J, Qu J, Yi X, Graber K, Huber L, Wang X, Gerdes A, Li F (2007) Upregulation of gamma-catenin compensates for the loss of beta-catenin in adult cardiomyocytes. Am J Physiol Heart Circ Physiol 292:270–276

Part V
How AJs Affect Tissue Homeostasis and Disease

Chapter 15
Adherens Junctions and Stem Cells

Terry Lechler

Abstract The specification, maintenance, division and differentiation of stem cells are integral to the development and homeostasis of many tissues. These stem cells often live in specialized anatomical areas, called niches. While niches can be complex, most involve cell-cell interactions that are mediated by adherens junctions. A diverse array of functions have been attributed to adherens junctions in stem cell biology. These include physical anchoring to the niche, control of proliferation and division orientation, regulation of signaling cascades and of differentiation. In this review, a number of model stem cell systems that highlight various functions of adherens junctions are discussed. In addition, a summary of the current understanding of adherens junction function in mammalian tissues and embryonic and induced pluripotent stem cells is provided. This analysis demonstrates that the roles of adherens junctions are surprisingly varied and integrated with both the anatomy and the physiology of the tissue.

15.1 Introduction

Both the development and the homeostasis of many tissues in our body relies on specialized stem cells. In adults, these cells are functionally defined by their ability to self-renew, to exist for extended periods of time and to give rise to daughter cells that contribute to the differentiated cell pool of a tissue. Because of their importance in maintaining tissues, both proliferation and differentiation decisions are highly regulated in these cells. Much of that regulation comes from specialized environments in which many stem cells are found, called niches. Niches may be easily identifiable anatomical structures, like the bulge of the hair follicle or the crypts of the small intestine, but in many cases are less well defined. The complexity of niches is also quite variable. While they can be seemingly simple in some contexts, there may also be contributions from extracellular matrix components, homophilic and heterophilic cell-cell interactions, association with vasculature or neurons and soluble factors. This environment is thought to provide protection for the stem cell,

T. Lechler (✉)
Departments of Cell Biology and Dermatology, Duke University Medical Center,
310 Nanaline Duke Bldg, 3709, 27710 Durham, NC USA
e-mail: terry.lechler@duke.edu

T. Harris (ed.), *Adherens Junctions: From Molecular Mechanisms to Tissue Development and Disease,* Subcellular Biochemistry 60,
DOI 10.1007/978-94-007-4186-7_15, © Springer Science+Business Media Dordrecht 2012

possibly providing a level of insulation from stressors. It also often maintains stem cells in a quiescent state, but this can rapidly change as these cells receive inputs from their environment (Li and Xie 2005; Ohlstein et al. 2004).

There is complex regulation of stem cell behavior by many pathways including cell-cell adhesion. AJs (AJs) play profound roles in controlling cell adhesion, proliferation, polarity and cytoskeletal organization in many cell types, including stem cells.

Below we discuss several roles that AJs play in stem cell and their niches. These include direct anchoring of the stem cell to niche cells, organizing the niche, controlling cell division orientation, regulating signaling pathways and affecting the mechanics of the cells. While β-catenin also plays essential roles in Wnt-signaling pathways, which regulate many facets of stem cell behavior, these non-adhesive functions will not be discussed in this review (Nusse et al. 2008).

Because cell adhesion is integrated into almost all aspects of cell physiology, analysis of adherens junction function in diverse stem cell systems has yielded a great diversity of roles making it difficult to predict the phenotype due to their loss. We begin with a discussion of the roles of AJs in *Drosophila* stem cell systems where relatively simple niches exist and considerable studies in the role of cell adhesion have been performed. We also discuss the various roles that AJs play in the development, growth and maintenance of embryonic stem cells and induced pluripotent stem cells. In addition, the roles of AJs in the less defined stem cell niches associated with a number of adult mammalian stem cells is considered. Finally, we discuss potential mechanical functions of AJs in regulating cell fate decisions.

15.2 AJs and Niche Association—*Drosophila* Ovary Germ Stem Cells

The simplest and the first characterized function for AJs in stem cells is association with the niche. As discussed above, the niche can be relatively simple or it can be a complex structure composed of basement membrane, vasculature, nerves and a number of different cell types that can interact directly or indirectly with the stem cell. The anchoring of the stem cell to surrounding cells is important in a number of tissues for both stem cell/niche association as well as maintenance of stemness.

The most elegant example of this can be found in germline stem cells of the *Drosophila melanogaster* ovary (Gonzalez-Reyes 2003; Xi 2009). The *Drosophila* ovary is responsible for the continued production of eggs throughout the lifetime of the female fly. It is composed of a number of tubes, the anterior ends of which are termed the germaria (Fig. 15.1). These are the structures that house the stem cells and their supporting cells. The supporting cells are called cap cells and they make direct cell-cell contacts with the two or three germ stem cells (GSCs) that reside in each germarium. These cell contacts include AJs as both the *Drosophila* homologs of E-cadherin and β-catenin accumulate at the interface between cap and germ cells, and AJ-like structures are visible by electron microscopic analysis (Song et al. 2002). This cadherin-rich interface develops just after specification of the cap cells.

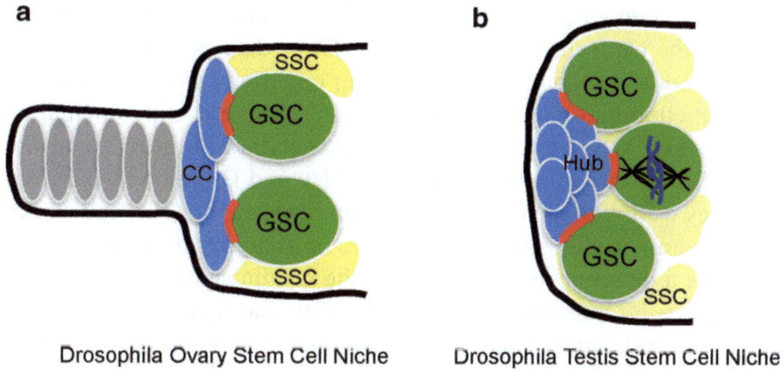

Fig. 15.1 *Drosophila* germline stem cell niches. **a** Ovary stem cell niche. *CC* cap cell, *GSC* germline stem cell, *SSC* somatic stem cell. Adherens junctions between the *GSC* and *CC*s are highlighted in red. **b** Testis stem cell niche. *GSC*s contact hub cells with adherens junctions and align their mitotic spindles with this site

At the early stages of germ cell association, only low levels of adherens junction components are seen between the cells. Functionally, cadherin in the germ cells is required both for their efficient initial association with the cap cells as well as their continued attachment to them (Song et al. 2002). Loss of cadherin in primordial germ cells decreases their ability to stably integrate into the germarium. Loss of cadherin in established germ stem cells causes an increased rate of their loss from the niche (Song et al. 2002). In addition to the GSCs, ovaries contain somatic epithelial stem cells that also require cadherin for their maintenance (Song and Xie 2002).

Additional evidence for cadherin-mediated attachment of GSCs came from analysis of Rab11 GTP-ase mutants (Bogard et al. 2007). Rab11 localizes to and is required for the function of recycling endosomes (Hsu and Prekeris 2010). Cadherins normally undergo recycling in these endosomal compartments (Desclozeaux et al. 2008; Lock and Stow 2005). In Rab11 mutants, cadherin and β-catenin levels are decreased at the GSC-cap cell interface and mutant GSCs are lost from the niche (Bogard et al. 2007).

The association of GSCs with cap cells is required for efficient BMP signal transduction between these two cell types (Chen et al. 2011a). In the ovary, BMP signaling occurs very locally and cells that are not in immediate contact with cap cells do not show active BMP signaling. It is presently not clear whether AJs simply provide close apposition of cell membranes for efficient signaling and communication, or whether there are additional or distinct functions for them. For example, in *Drosophila* testis, evidence suggests that the AJs may act as scaffolda for signaling molecules as well (see below) (Michel et al. 2011).

A well defined cellular niche like the cap cells have a limited surface area for interaction with stem cells. Examination of GSCs has led to the appreciation that stem cells can compete for position at the niche. Differentiation defective GSCs outcompete wild-type cells for niche occupancy and this occurs in a cadherin-dependent manner (Jin et al. 2008). In fact, simply increasing the cadherin levels in GSCs makes them able to outcompete wild type cells by increasing their area of attachment to the cap cells (Jin et al. 2008). These data suggest that intrinsic and extrinsic cues could regulate stem cell number or activity simply by mediating the physical interaction between cap cells and GSCs. There is experimental evidence for this occurring. When cap cells are mutant for the insulin receptor, there is a decrease in the number of associated GSCs (Hsu and Drummond-Barbosa 2009). This correlates well with the levels of cadherin between the GSC and cap cell. It is not yet known how insulin signaling impinges on adhesion status in these cells and whether this is a direct or indirect effect of the insulin signaling pathway. It does, however, provide an important precedent for extrinsic regulation of cell adhesion controlling stem cell maintenance. This also implies that mutations in stem cells can be positively selected for if they result in more robust niche association.

As flies age, the number of GSCs decreases (Pan et al. 2007; Wallenfang et al. 2006). While this is a complex and poorly understood process, aging is associated with a decrease in the levels of cadherin complexes formed between cap cells and GSCs (Pan et al. 2007). Further, experimentally decreasing cadherin levels resulted in an increased rate of GSC loss with age while over expression of cadherin resulted in fewer lost GSCs. Both the age related control of cadherin levels and their interaction with the insulin signaling pathway point to complex extrinsic regulation of stem-niche association. How well this translates to the function of more complex niches in mammals is still unclear.

15.3 Stem Cells in the *Drosophila* Testis—Anchoring and Spindle Orientation

Stem cells of the *Drosophila* testis are similarly dependent upon cadherin for their maintenance (Inaba et al. 2010; Voog et al. 2008). Like the ovary, the testis contains both germline and somatic stem cells termed cyst stem cells (CysSCs). In addition, there is a group of somatic cells, called the hub, that directly interact with the GSCs (Fig. 15.1). The hub cells therefore form part of the physical niche for the GSCs in the testis, with cadherin complexes concentrating at the interface between these cells (Inaba et al. 2010; Voog et al. 2008). The role of cadherin in hub cells and germ cells is distinct. Surprisingly, loss of cadherin by RNAi in the hub alone does not result in niche defects, yet loss of cadherin in the GSCs results in their loss (Voog et al. 2008). The hub cells express both fasciclin II and DN-cadherin, which likely preserves their attachment to one another when DE-cadherin is lost. Although not tested, DN-DE heterodimers or other cell adhesion systems could mediate the association of germ cells with hub cells when DE-cadherin is lost in the hub. When cadherin is lost in the CysSCs, these cells cannot be maintained and are lost (Voog et al. 2008).

Cadherin likely plays a role beyond simple adhesion in the male GSCs. When these cells divide, they orient their mitotic spindle perpendicular to the hub cell (Yamashita et al. 2003). After division, this results in one cell that remains a GSC attached to the hub, and a second cell that has been displaced from the hub and begins to differentiate down the sperm pathway. This process requires capture of one of the centrosomes/spindle poles by the region of the GSC cortex associated with the hub. Because cadherin is localized at this site, it was an excellent candidate for mediating the spindle orientation. However, due to the loss of GSCs from the niche upon loss of DE-cadherin, it has not been possible to look directly at the effect of loss of function mutants on spindle orientation. Instead, expression of a mutant cadherin construct lacking the extracellular domain, and therefore defective in adhesion, has been performed (Inaba et al. 2010). The mutant cadherin is localized around the entire cell cortex rather than being highly enriched at the GSC-hub junction, and results in misoriented mitotic spindles. Proper spindle orientation in GSCs requires the microtubule binding protein, APC2, and mutant cadherin expression causes loss of the normal polarization of APC2 to the GSC-hub junction (Inaba et al. 2010; Yamashita et al. 2003). Therefore, in the GSCs the cortical AJ patch acts as a polarity marker for the cell and allows for spindle orientation. A similar role for AJs has been proposed in the planar polarized divisions of sensory organ precursor cells in the fly (Le Borgne et al. 2002). However, beyond a role for APC2, little is known about how a cortical patch of AJs orients the mitotic spindle. While the assumption is that cadherin is acting in a canonical AJ-like manner, it has not yet been demonstrated that loss of AJ complex components in the hub cell results in spindle orientation defects in the GSCs. DE-cadherin lacking its extracellular domain can exert a dominant influence on spindle orientation in GSCs. Similar experiments in hub cells should determine whether it is the cadherin directly or the AJ complex that mediates the spindle orientation effect. Because AJs have been implicated in controlling spindle orientation in epithelial cells, this is likely a more generalizable phenomenon (den Elzen et al. 2009).

Like in the ovary, there is a decrease in DE-cadherin expression in the hub cells during aging (Boyle et al. 2007). While this is consistent with impaired adhesion affecting the loss of germ cells with age, this has not been directly tested.

In addition to adhesion and spindle orientation, new data suggests that the cortical patch containing AJs that forms between hub cells and GSCs is a signaling platform. Active BMP receptor is found highly enriched at this site and this localized activation correlates with levels of cadherin at the interface (Michel et al. 2011). Signaling complexes are known to associate with AJs in a number of contexts (McLachlan and Yap 2007). In this case, the polarity resulting from hub cell-GSC interaction is transduced into a highly localized activation of the BMP receptor (Michel et al. 2011). What role the AJs play in the localization and activation of BMP receptor signaling requires further analysis.

15.4 *Drosophila* Intestinal Stem Cells—AJs Role in Signaling in an Epithelial Tissue

While the *Drosophila* ovary and testis have provided insight into how cadherin anchors stem cells in their niche, the intestine offers insights into cadherin functions in stem cells in an epithelial tissue. The *Drosophila* gut is a simple epithelium with stem cells that divide to regenerate themselves and give rise to enteroblasts

(Ohlstein and Spradling 2006). Differentiation into enteroblasts requires Notch signaling—high levels of the Notch ligand Delta are present in the stem cell and lead to activation of Notch signaling and differentiation of the enteroblast (Ohlstein and Spradling 2006, 2007). In the absence of DE-cadherin, the number of cells with active Notch signaling decreased (Maeda et al. 2008) and normal differentiation was perturbed. This is consistent with either robust cell-cell adhesion being required for Notch signaling, with the cadherin complex playing a more direct role in the signaling process, and/or with DE-cadherin maintaining the stem cells.

15.5 Roles of AJs in Mammalian Embryonic Stem Cells and Induced Pluripotent Cells

E-cadherin plays profound roles not only in intact tissues, but also in cultured cells. A very specialized and important example of this is in embryonic stem cells (ESCs). First cultured from preimplantation mouse blastocysts, these cells are pluripotent and can differentiate *in vitro* into many cell lineages (Evans and Kaufman 1981; Martin 1981). They can incorporate into blastocysts resulting in the formation of chimeric animals with ESC contribution to all tissue types. Human embryonic stem cells also exist, though they are somewhat distinct from mESCs. While culture of mouse cells relies on leukemia inhibitory factor (LIF) and bone morphogenetic proteins (BMPs), hESCs are often grown with fibroblast growth factor-2 and TGF-β family members activin and nodal. These differences are significant and likely reflect more than just species differences.

There are also epiblast stem cells (EpiSCs) that can be derived from either pre- or post-implantation mouse embryos (mESCs are only derived from the inner cell mass of preimplantation embryos) (Najm et al. 2011; Brons et al. 2007; Tesar et al. 2007). EpiSCs are thought to be a more committed form of the mESCs as they cannot contribute to chimerism. Despite this difference, they still express many markers that are often used to define ESCs—such as Oct-4 and Nanog and can be easily converted back into ESCs. While it has not been possible to test the *in vivo* pluripotency of human ESCs and determine whether they can contribute to chimerism (as mESCs do), or whether they are more similar to EpiSCs in this assay, it is clear that human ESCs are pluripotent in culture and can be differentiated into many lineages.

E-cadherin is required very early in development. Genetic ablation of E-cadherin results in defects before implantation of the embryo, but compaction, an adhesion-dependent event, does occur (Riethmacher et al. 1995). This is likely due to the presence of a maternal pool of E-cadherin because inhibition of E-cadherin homodimerization with inhibitory antibodies results in failure in compaction, as does antisense RNA inhibition of E-cadherin (Ao and Erickson 1992; Hyafil et al. 1980; Vestweber et al. 1985). These data demonstrate that E-cadherin plays important adhesion roles at very early stages of development. However, it does not offer full insight into the roles that E-cadherin plays in ESCs.

Important roles for AJs/E-cadherin have been reported in three aspects of ESC biology—establishment of ES/iPS cells, maintenance of pluripotency and differential potential. Both human and mouse ESCs are characterized by their formation of tight colonies. Differentiation of these cells is associated with their spreading. E-cadherin is expressed in ESCs, and plays significant roles in their maintenance and differentiation. As this is a young field, there are conflicting reports in the literature making it difficult to fully understand E-cadherin function. Here we discuss some of the findings and suggest possible interpretations and areas that require further clarification.

Cells have been isolated from the inner cell mass of E-cadherin null mouse embryos and used to establish ESC lines (Soncin et al. 2009). These behave as ESCs in that they can be propagated in LIF-containing ESC media and express pluripotent markers like Oct-4. However, distinct from wild-type mESCs, E-cadherin null ESCs do not differentiate when LIF is removed from the media. Similar results have been found with RNAi and peptide-inhibitor mediated loss of E-cadherin function (Soncin et al. 2009). These data suggest that E-cadherin promotes differentiation under some conditions. Although the E-cadherin null cells resemble EpiSCs (not ESCs) in their growth capabilities, they display distinct transcriptional profiles (Soncin et al. 2011). The fact that E-cadherin null cells can switch from a LIF/BMP mode of renewal to an activin/Nodal type suggests that E-cadherin functions in pathways more complex than simple adhesion. This supports a more active role for E-cadherin complexes in regulating signaling in these cells. However, while the E-cadherin null ESC/EpiSCs did not differentiate fully, it is not clear that they are *bona fide* ESCs either.

In contrast to these findings, other groups have reported an essential function for E-cadherin in maintaining the pluripotency of ESCs (Chen et al. 2011b; Li et al. 2010a; Redmer et al. 2011; Li et al. 2010b). Knockdown of E-cadherin in both hESCs and mESCs was shown to result in decreased expression of the pluripotency marker Oct-4. Whether these cells are actually differentiating or becoming epiblast-like SCs will require further investigation. It may be that their status does not resemble either ESCs or EpiSCs in entirety. Unfortunately, the most relevant assay, the ability to contribute to chimerism cannot be performed because E-cadherin is required during differentiation for cell adhesion. That said, the fact that E-cadherin null ES-like cells could be established and propagated demonstrates that these cells either have alternative mechanisms or adaptive methods of retaining some level of pluripotency even upon loss of E-cadherin.

Similar discrepancies exist for stem cells devoid of β-catenin, which is complicated by β-catenin's dual roles in cell adhesion and the Wnt pathway. Several lines have been generated with phenotypes more similar to EpiSCs, and others have reported the generation of β-catenin null ESCs (Lyashenko et al. 2011; Wray et al. 2011).

While the detailed effects of loss of E-cadherin in ESCs requires further examination, it is clear that E-cadherin ligation can have significant effects on these cells. When mESCs are plated on substrates of E-cadherin, they no longer grow as colonies (Nagaoka et al. 2006). Despite this, they maintain their growth and their pluripotent ability in contributing to all three germ layers both in culture and when injected into blastocysts. Similarly, mESCs were better maintained on fibroblast feeders that were engineered to express E-cadherin (Horie et al. 2010). These data suggest that ligation of E-cadherin, rather than tight cell-cell association is needed to maintain ESCs. In addition, they provide an alternative way to culture cells that makes them more amenable to manipulation. Similar results have been found in hESCs (Nagaoka et al. 2010). hES cells grew in defined media in the presence of E-cadherin-coated substrates as well as they did on matrigel and maintained their ability to form multiple cell lineages in teratomas. Therefore, in both hESCs and mESCs, exogenous E-cadherin can function to maintain stem cell identity.

E-cadherin ligation can also improve the efficiency of mESC and iPS derivation. A short exposure of early stage blastomeres to E-cadherin resulted in a significant increase in the production of ESC lines, which correlated well with early proliferation of the cells (Gonzalez et al. 2011). Induced pluripotent stem cells are generated by a number of methods—but classically through the transduction of the Yamanaka factors—Oct4, Sox2, Klf4 and myc (Takahashi and Yamanaka 2006). This is an inefficient process with only a fraction of cells becoming iPS cells. E-cadherin and epithelial cell adhesion molecule (EpCAM) are both upregulated in iPS cells, but not in incompletely reprogrammed cells—allowing for an enrichment process. Not only is E-cadherin expressed in iPS cells, but its upregulation promotes the process. Exogenous expression of E-cadherin is able to increase the rate of iPS formation by about 4 fold (Chen et al. 2010). In contrast, loss of E-cadherin (and loss of cell-cell contact) inhibits iPS generation (Chen et al. 2010).

Finally, E-cadherin can also regulate the differentiation of ESCs into defined lineages. Differentiation into neuronal cell types and endoderm and hepatocyte cells is promoted by growth on feeders or matrix with the E-cadherin extracellular domain (Haque et al. 2011; Moore et al. 2011). These data make the situation more complex as it suggests that E-cadherin can have divergent functions depending upon the cells used and the growth environment to which they are exposed.

15.6 Roles of AJs in Tissue Specific Stem Cells

15.6.1 Intestinal Stem Cells

The intestinal epithelium is organized into finger-like projections called villi (which are lined with differentiated enterocytes, goblet cells and enteroendocrine cells) and invaginations into the mesenchyme called crypts of Lieberkühn (Fig. 15.2). The crypts contain not only the stem cells, but also a rapidly proliferating transit amplifying cell population and, in the small intestine, the terminally differentiated Paneth cells. There is still debate about the organization and hierarchy within the stem cell compartment (Snippert et al. 2010; Tian et al. 2011). There are cells at the base of the crypt, interspersed among Paneth cells, that express Lgr5, are proliferative and

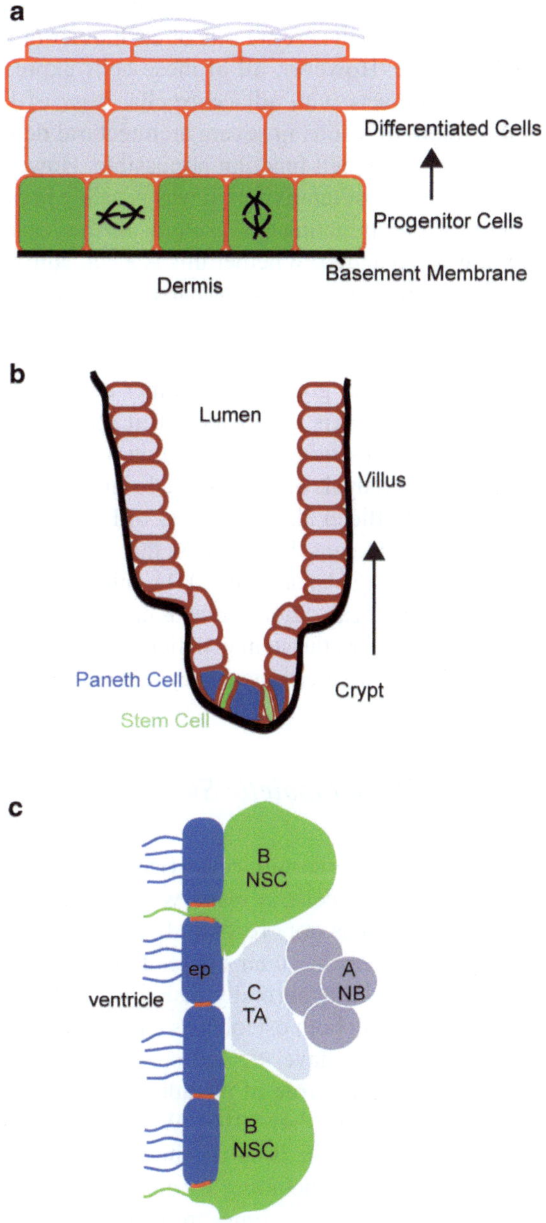

Fig. 15.2 Mammalian stem cell niches. **a** Interfollicular epidermis. *Progenitor* cells (*green*) reside in the basal layer, attached to the *basement membrane*. **b** Intestinal epithelium. *Stem* cell reside in the *crypt* compartment and *Paneth* cells serve as part of their physical niche. **c** Neural stem cell niche. Neural stem cells, or Type B cells (*green*), are in contact with ependymal cells (*blue*) that line the lateral ventricles. They give rise to transit-amplifying (*TA*) type C cells which give rise to neurblasts (type A cells). Adherens junctions are highlighted in red between neural stem cells and the ependymal cells

self renew. Lineage tracing has shown that they can contribute to all intestinal lineages over extended periods of time (Barker et al. 2007). However, there is also a population of cells, sometimes marked by Bmi1 or mTert expression, that are found at a higher position, often termed +4 (indicating its cell number counting from the base of the crypt) (Montgomery et al. 2011; Sangiorgi and Capecchi 2008). These cells are less proliferative, but also contribute to all lineages over extended periods. Whether these cells are subsets of each other or are hierarchically organized is still under debate. However, all of these cells express E-cadherin and make extensive cell-cell contacts with adjacent cells. Loss of E-cadherin throughout the intestinal epithelium results in severe architectural defects resulting in lethality—making analysis of stem cell function impossible. However, by modulating recombination-inducing doses of tamoxifen, it was possible to perturb E-cadherin function without completely eliminating the protein. In this case, the zone of proliferation in the crypt was expanded. Whether this is a cell autonomous effect of loss of E-cadherin or a tissue response to loss of barrier activity is not yet known. It is also not clear whether the stem cells as well as the transit amplifying cells become more proliferative. In addition, cell lineage specification and organization was altered. There was a decrease in goblet cell number and Paneth cells were localized throughout the villus-crypt axis (Schneider et al. 2011). The mislocalization of Paneth cells suggested that E-cadherin may play a role in cell sorting. Further evidence for this comes from analysis of EphB-based signaling, which is known to pattern the villus-crypt axis (Batlle et al. 2002). Loss of EphB3 results in a similar mislocalization of Paneth cells. Recent data suggest that EphB regulates ADAM10-mediated E-cadherin cleavage (Solanas et al. 2011). Inhibition of ADAM10 also leads to disruption of Paneth cell localization (Solanas et al. 2011). Therefore, E-cadherin may function to properly pattern the stem cell niche and maintain Paneth cells (an important niche component) in an area where they make direct contact with stem cells.

15.6.2 Hematopoietic Stem Cells

While hematopoietic stem cells (HSCs) have served as the best characterized adult mammalian stem cells in many ways, both visualizing them *in vivo* and identifying their niche has only recently been achieved. Because of this, it has been more difficult to determine their cell-cell contacts in the intact tissue (Singbrant et al. 2011li). These cells normally reside in the bone marrow and are thought to make associations with both osteoblasts and vasculature. Studies on the roles of AJs components in HSCs have yielded some opposing results. Much of the controversy has surrounded the role of N-cadherin. For a full discussion of this, the reader is directed to (Li and Zon 2010). While these studies suggest that N-cadherin does not play a major role in anchoring or proliferation of HSCs, there are likely additional cadherins, including E-cadherin, which could functionally substitute. Evidence against this idea, comes from analysis of mice in which both β-catenin and its homolog γ-catenin (plakoglobin) were eliminated (Jeannet et al. 2008; Koch et al. 2008). In these mice, which should lack functional AJs, the long-term maintenance of the HSCs was normal. However, expression of a dominant-negative N-cadherin in the HSCs resulted in decreased anchoring and long-term engraftment (Hosokawa et al. 2010). Conversely, expression of wild-type N-cadherin in HSCs decreased cell cycling and increased their lodging in the bone marrow. Therefore, while data supporting a strong role for AJs in controlling HSC physiology or niche architecture is under debate, clearly more research is needed to clarify the issue.

15.6.3 Epidermal Stem Cells

The epidermis is a stratified squamous epithelium that acts as a barrier between us and the outside world. Because of its multi-layered architecture the use of cell-cell adhesion structures are somewhat different than in simple epithelia. All the living layers of the epidermis have AJs but there are no clear zonula adherens in these cells (Fig. 15.2). In proliferative basal cells of the epidermis, AJs are not restricted to lateral surfaces, but also cover the apical surface, which is a cell-cell interface in this tissue.

There are a number of stem and progenitor cells in the epidermis. Most well-studied are the stem cells that reside within the bulge region of the hair follicle (Fuchs 2009; Gambardella and Barrandon 2003; Jaks et al. 2010). These cells give rise to all the cell lineages in the hair follicles and can also transiently contribute to the interfollicular epidermis upon wounding. Additional distinct stem cells reside in the region around the bulge and have been reported to have unique properties and perhaps different abilities to contribute long-term to different lineages. In the interfollicular epidermis, progenitor cells lie within the innermost basal layer. There is still debate whether the cells in the basal layer are an essentially homogenous population of progenitors or whether there are distinct groups of stem cells and their progeny (Clayton et al. 2007; Kaur and Potten 2011).

Most loss of function studies on adherens junction components have relied on targeted gene ablation using the keratin 5 or 14 promoter. This promoter is active in all basal cells of the interfollicular epidermis, as well as the bulge and outer root sheath cells of the hair follicle. Because essentially all epidermal cells are derived from the keratin 5/14 population, this complicates analysis as not only stem cells, but also their progeny, which includes some niche cells, have also lost the protein. It is therefore often difficult to unravel direct effects on stem cells verses secondary effects due to changes in tissue architecture or physiology. We present below the effects of ablation of each of the core AJ components which yield surprisingly different phenotypes—highlighting the diverse roles of proteins within this complex.

E/P-cadherin E-cadherin has been ablated from both adult epidermis (Krox20-Cre) and embryonic epidermis (K14-Cre) (Tinkle et al. 2004; Tunggal et al. 2005; Young et al. 2003). Phenotypes vary somewhat depending on the timing of loss of the protein and genetic background. Early loss in some backgrounds results in perinatal lethality with barrier defects (Tunggal et al. 2005). This is due to an inability to form tight junctions within the differentiated layers of the epidermis, and is unlikely related to defects in progenitor cells. The epidermal cells do not show significant changes in cell shape or adhesion ultrastructurally, suggesting that desmosomes are responsible for the majority of cell adhesion in this tissue. In other strain backgrounds P-cadherin can partially substitute for E-cadherin and the loss of both of these proteins results in a more severe phenotype—including perinatal lethality with barrier defects (Tinkle et al. 2008). In this case, more significant cell architecture and cell-cell adhesion defects were noted, as well as a loss of cell polarity. Therefore, it is likely that P-cadherin (which is upregulated in basal cells upon E-cadherin ablation) can perform these functions in progenitor cells. While no significant changes in proliferation were noted, cadherins are important to protect cells from apoptosis in the progenitor pool (Tinkle et al. 2008). Therefore, within the epidermis, cadherins do not perform an essential anchoring function, and are not primary regulators of proliferation, differentiation or potency.

α-catenin Loss of α-catenin results in dramatic changes in epidermal architecture, differentiation and proliferation (Vasioukhin et al. 2001). Most notably, the epidermis becomes hyperproliferative resulting in a phenotype closely resembling squamous cell carcinoma. Several explanations have been reported to explain the hyperproliferative phenotype. These include an increase in insulin receptor substrate driven MAP kinase activation, loss of polarity and division orientation and loss of cytoplasmic sequestration of the transcription factor YAP1 (Lechler and Fuchs 2005; Schlegelmilch et al. 2011; Silvis et al. 2011; Vasioukhin et al. 2001). Of these, the influence on YAP1 is perhaps most compelling. α-catenin can directly interact with YAP1, protecting it from dephosphorylation and nuclear accumulation (Schlegelmilch et al. 2011; Silvis et al. 2011). In addition, knockdown of YAP1 prevents some of the transformed phenotypes of α-catenin null cells. Thus, in epidermal progenitors, α-catenin plays profound roles in controlling cell proliferation through affecting signaling pathways. Whether any of the other pleiotropic effects due to α-catenin loss also contribute to the phenotype is still under investigation.

β-catenin While β-catenin has essential roles in the Wnt pathway in epidermal development and homeostasis (see review, Watt and Collins 2008), it is not clear that its role in AJs is relevant. This is likely due, at least in part, to functional compensation of β-catenin by plakoglobin, a structurally similar protein that is usually a constituent of the desmosome.

p120-catenin Similar to loss of α-catenin, loss of p120 results in hyperproliferation and tumor formation in mice (Perez-Moreno et al. 2006, 2008). The mechanism underlying this phenotype is quite distinct, however. While adherens junction components are decreased in the tissue, they appear to be sufficient to mediate basic cell adhesion, polarity and tight junction formation. Instead these mice activate the NF-kB pathway through elevated levels of the small GTPase Rho (Perez-Moreno et al. 2006, 2008). The hyperproliferation phenotype of the progenitor cells occurs in response to immune cell stimulation that is caused by NF-kB signaling. In this way, an adherens junction component has a non-cell autonomous effect by regulating epidermal signaling to the immune system.

15.6.4 Neural Stem Cells

The best characterized adult neuronal stem cells are those of the subventricular zone of the lateral ventricles (Fig. 15.2). The ventricle is lined by ependymal cells which form part of the niche for the NSCs. Rosettes are formed by a centrally located NSC surrounded by ependymal cells, with which they make direct contact (Mirzadeh et al. 2008). In addition, NSCs contact astrocytes, neuroblasts and are closely associated with the vasculature through cellular processes. Both *in vivo* and cultured cell approaches have been used to address roles of cell adhesion molecules in these cells.

Loss of E-cadherin by Cre-mediated recombination in nestin-postive cells results in an increase in proliferation in neuroblasts, but not NSCs (Karpowicz et al. 2009). In young mice, there was no change in DNA label-retaining cell number (marking putative NSCs that are slowly proliferating), but their number decreased as the mice aged. Whether this effect is due to slight changes in proliferation, in anchoring of stem cells, or in symmetric/asymmetric cell divisions is not yet known. These phenotypes were recapitulated in culture—isolated E-cadherin null NSC produced fewer colonies upon repeated passaging (Karpowicz et al. 2009). While affecting stem cell behavior, loss of E-cadherin did not substantially perturb niche architecture, possibly due to compensation by N-cadherin which is also expressed in the niche. In culture, blocking N-cadherin resulted in decreased neurosphere formation (cellular aggregates derived from stem cells) and increased production of glial progeny (Yagita et al. 2009). While additional experiments are required to test the role of N-cadherin *in vivo*, analysis of β-catenin ablation in the nestin-positive cells suggests that AJs play an important role in stem cell biology. These mice displayed disorganized brains and in neurosphere assays for NSC activity, the cells were unable to adhere to one another or form colonies (Holowacz et al. 2011). Plating cells in collagen matrices prevented their dissociation and under these conditions, β-catenin null cells outperformed their wild-type counterparts. In addition, β-catenin was required for survival of neural progenitors, with increased apoptosis seen in knockout cultures (Holowacz et al. 2011). Additional supporting evidence for cadherin function in the neural stem cell niche comes from analysis of Numb/Numblike and Ankyrin3 mutants. Both of these result in perturbations in lateral membranes and the ability of cadherins to be stably maintained at cell junctions (Kuo et al. 2006; Paez-Gonzalez et al. 2011; Rasin et al. 2007). In both cases, this results in disturbances of the niche architecture. Ablating Ankyrin3, which is expressed in the ependymal cells, results in defects in N-cadherin at the lateral membranes, and in the loss of niche organization and production of neurons. Therefore, misregulation of cadherin impacts the niche and thus affects stem cell activity.

Analysis of α-catenin ablation leads to even more severe phenotypes in the developing brain (Klezovitch et al. 2004). While this has not been performed in the adult brain, developmentally it bears some resemblance to the hyperproliferative phenotypes seen in the developing epidermis. In this case, however, at least some of the phenotype has been ascribed to activation of the Hedgehog signaling pathway

(Klezovitch et al. 2004). Whether Yap1 signaling also plays a role in this hyperproliferation has not yet been addressed. Similar hyperproliferative responses were noted after ablation of RhoA (a small GTPase that localizes to AJs) in the developing brain (Katayama et al. 2011).

15.7 AJs and Mechanosensing/Signaling

An area of emerging interest is how mechanical forces are sensed and responded to in tissues. AJs are known mechanosensors and thus serve as ideal candidates for both generating and responding to tissue forces (le Duc et al. 2010; Liu et al. 2010; Yonemura et al. 2010). While there is significant literature demonstrating that mechanical properties of extracellular matrix can control cell fate decisions in stem cells, there is less evidence that this occurs with cell-cell adhesion (Assoian and Klein 2008; Cohen and Chen 2008). Likely, this is due, in part, to the difficulty in teasing apart mechanical aspects of AJs from adhesion and signaling roles. However, our emerging knowledge of how AJs affect cytoskeleton architecture, myosin-dependent contraction, signaling pathways that can independently contribute to cortical tension, and cross-talk between cell-cell and cell-ECM adhesion suggests that mechanical control of stem cell activity by AJs will be important.

15.8 Summary and Future Perspectives

AJs have varied and complex roles in stem cell biology. An important lesson to be taken away from completed studies is that the roles of AJs vary between cell types and the roles of individual AJs components often vary within a single type of stem cell. This suggests an enlightening future for adherens junction research in stem cells. One of the biggest challenges for future researchers will be unraveling the contributions of primary effects and secondary effects of adherens junction perturbation. More thorough descriptions of niches and the ability to specifically perturb AJs in defined cell populations will help in this endeavor. However, equally important is parsing out the effects of AJs on adhesion, polarity, patterning, signaling, proliferation, spindle orientation, and cortical tension—thus discovering the underlying mechanisms by which AJs regulate stem cell behavior.

Acknowledgements Work on cell adhesion and cytoskeleton organization is supported by a grant to TL from NIH/NIAMS (R01AR055926).

References

Ao A, Erickson RP (1992) Injection of Antisense RNA specific for E-cadherin demonstrates that E-cadherin facilitates compaction, the first differentiative step of the mammalian embryo. Antisense Res Dev 2:153–163

Assoian RK, Klein EA (2008) Growth control by intracellular tension and extracellular stiffness. Trends Cell Biol 18:347–352

Barker N, van Es JH, Kuipers J, P. Kujala, van den Born M, Cozijnsen M, Haegebarth A, Korving J, Begthel H, Peters PJ, Clevers H (2007) Identification of stem cells in small intestine and colon by marker gene Lgr5. Nature 449:1003–1007

Batlle E, Henderson JT, Beghtel H, Van Den Born MM, Sancho E, Huls G, Meeldijk J, Robertson J, van de Wetering M, Pawson T, Clevers H (2002) Beta-catenin and TCF mediate cell positioning in the intestinal epithelium by controlling the expression of EphB/ephrinB. Cell 111:251–263

Bogard N, Lan L, Xu J, Cohen RS (2007) Rab11 maintains connections between germline stem cells and niche cells in the Drosophila ovary. Development 134:3413–3418

Boyle M, Wong C, Rocha M, Jones DL (2007) Decline in self-renewal factors contributes to aging of the stem cell niche in the Drosophila testis. Cell Stem Cell 1:470–478

Brons IG, Smithers LE, Trotter MW, Rugg-Gunn P, Sun B, Chuva de Sousa Lopes SM, Howlett SK, Clarkson A, Ahrlund-Richter L, Pedersen RA, Vallier L (2007) Derivation of pluripotent epiblast stem cells from mammalian embryos. Nature 448:191–195

Chen HF, Chuang CY, Lee WC, Huang HP, Wu HC, Ho HN, Chen YJ, Kuo HC (2011a) Surface Marker epithelial cell adhesion molecule and E-cadherin facilitate the identification and selection of induced pluripotent stem cells. Stem Cell Rev 7:722–735

Chen, S, Wang S, Xie T (2011b) Restricting self-renewal signals within the stem cell niche: multiple levels of control. Curr Opin Genet Dev 6:684–689

Chen T, Yuan D, Wei B, Jiang J, Kang J, Ling K, Gu Y, Li J, Xiao L, Pei G (2010) E-cadherin-mediated cell-cell contact is critical for induced pluripotent stem cell generation. Stem Cells 28:1315–1325

Clayton E, Doupe DP, Klein AM, Winton DJ, Simons BD, Jones PH (2007) A single type of progenitor cell maintains normal epidermis. Nature 446:185–189

Cohen DM, Chen CS (2008) Mechanical control of stem cell differentiation. StemBook

den Elzen N, Buttery CV, Maddugoda MP, Ren G, Yap AS (2009) Cadherin adhesion receptors orient the mitotic spindle during symmetric cell division in mammalian epithelia. Mol Biol Cell 20:3740–3750

Desclozeaux M, Venturato J, Wylie FG, Kay JG, Joseph SR, Le HT, Stow JL (2008) Active Rab11 and functional recycling endosome are required for E-cadherin trafficking and lumen formation during epithelial morphogenesis. Am J Physiol Cell Physiol 295:C545--C556

Evans MJ, Kaufman MH (1981) Establishment in culture of pluripotential cells from mouse embryos. Nature 292:154–156

Fuchs E (2009) The tortoise and the hair: slow-cycling cells in the stem cell race. Cell 137:811–819

Gambardella L, Barrandon Y (2003) The multifaceted adult epidermal stem cell. Curr Opin Cell Biol 15:771–777

Gonzalez S, Ibanez E, Santalo J (2011) Influence of e-cadherin-mediated cell adhesion on mouse embryonic stem cells derivation from isolated blastomeres. Stem Cell Rev 7:494–505

Gonzalez-Reyes A (2003) Stem cells, niches and cadherins: a view from drosophila. J Cell Sci 116:949–954

Haque A, Hexig B, Meng Q, Hossain S, Nagaoka M, Akaike T (2011) The effect of recombinant E-cadherin substratum on the differentiation of endoderm-derived hepatocyte-like cells from embryonic stem cells. Biomaterials 32:2032–2042

Holowacz T, Huelsken J, Dufort D, Van Der Kooy D (2011) Neural stem cells are increased after loss of beta-catenin, but neural progenitors undergo cell death. Eur J Neurosci 33:1366–1375

Horie M, Ito A, Kiyohara T, Kawabe Y, Kamihira M (2010) E-cadherin gene-engineered feeder systems for supporting undifferentiated growth of mouse embryonic stem cells. J Biosci Bioeng 110:582–587

Hosokawa K, Arai F, Yoshihara H, Iwasaki H, Hembree M, Yin T, Nakamura Y, Gomei Y, Takubo K, Shiama H, Matsuoka S, Li L, Suda T (2010) Cadherin-based adhesion is a potential target for niche manipulation to protect hematopoietic stem cells in adult bone marrow. Cell Stem Cell 6:194–198

Hsu HJ, Drummond-Barbosa D (2009) Insulin levels control female germline stem cell maintenance via the niche in drosophila. Proc Natl Acad Sci U S A 106:1117–1121

Hsu VW, Prekeris R (2010) Transport at the recycling endosome. Curr Opin Cell Biol 22:528–534

Hyafil F, Morello D, Babinet C, Jacob F (1980) A cell surface glycoprotein involved in the compaction of embryonal carcinoma cells and cleavage stage embryos. Cell 21:927–934

Inaba M, Yuan H, Salzmann V, Fuller MT, Yamashita YM (2010) E-cadherin is required for centrosome and spindle orientation in drosophila male germline stem cells. PLoS One 5:e12473

Jaks V, Kasper M, Toftgard R (2010) The hair follicle-a stem cell zoo. Exp Cell Res 316:1422–1428

Jeannet G, Scheller M, Scarpellino L, Duboux S, Gardiol N, Back J, Kuttler F, Malanchi I, Birchmeier W, Leutz A, Huelsken J, Held W (2008) Long-term, multilineage hematopoiesis occurs in the combined absence of beta-catenin and gamma-catenin. Blood 111:142–149

Jin Z, Kirilly D, Weng C, Kawase E, Song X, Smith S, Schwartz J, Xie T (2008) Differentiation-defective stem cells outcompete normal stem cells for niche occupancy in the drosophila ovary. Cell Stem Cell 2:39–49

Karpowicz P, Willaime-Morawek S, Balenci L, DeVeale B, Inoue T, Van Der Kooy D (2009) E-Cadherin regulates neural stem cell self-renewal. J Neurosci 29:3885–3896

Katayama K, Melendez J, Baumann JM, Leslie JR, Chauhan BK, Nemkul N, Lang RA, Kuan CY, Zheng Y, Yoshida Y (2011) Loss of RhoA in neural progenitor cells causes the disruption of adherens junctions and hyperproliferation. Proc Natl Acad Sci U S A 108:7607–7612

Kaur P, Potten CS (2011) The interfollicular epidermal stem cell saga: sensationalism versus reality check. Exp Dermatol 20:697–702

Klezovitch O, Fernandez TE, Tapscott SJ, Vasioukhin V (2004) Loss of cell polarity causes severe brain dysplasia in Lgl1 knockout mice. Genes Dev 18:559–571

Koch U, Wilson A, Cobas M, Kemler R, Macdonald HR, Radtke F (2008) Simultaneous loss of beta- and gamma-catenin does not perturb hematopoiesis or lymphopoiesis. Blood 111:160–164

Kuo CT, Mirzadeh Z, Soriano-Navarro M, Rasin M, Wang D, Shen J, Sestan N, Garcia-Verdugo J, Alvarez-Buylla A, Jan LY, Jan YN (2006) Postnatal deletion of Numb/Numblike reveals repair and remodeling capacity in the subventricular neurogenic niche. Cell 127:1253–1264

Le Borgne R, Bellaiche Y, Schweisguth F (2002) Drosophila E-cadherin regulates the orientation of asymmetric cell division in the sensory organ lineage. Curr Biol 12:95–104

le Duc Q, Shi Q, Blonk I, Sonnenberg A, Wang N, Leckband D, de Rooij J (2010) Vinculin potentiates E-cadherin mechanosensing and is recruited to actin-anchored sites within adherens junctions in a myosin II-dependent manner. J Cell Biol 189:1107–1115

Lechler T, Fuchs E (2005) Asymmetric cell divisions promote stratification and differentiation of mammalian skin. Nature 437:275–280

Li L, Xie T (2005) Stem cell niche: structure and function. Annu Rev Cell Dev Biol 21:605–631

Li P, Zon LI (2010) Resolving the controversy about N-cadherin and hematopoietic stem cells. Cell Stem Cell 6:199–202

Li D, Zhou J, Wang L, Shin ME, Su P, Lei X, Kuang H, Guo W, Yang H, Cheng L, Tanaka TS, Leckband DE, Reynolds AB, Duan E, Wang F (2010a) Integrated biochemical and mechanical signals regulate multifaceted human embryonic stem cell functions. J Cell Biol 191:631–644

Li L, Wang BH, Wang S, Moalim-Nour L, Mohib K, Lohnes D, Wang L (2010b) Individual cell movement, asymmetric colony expansion, rho-associated kinase, and E-cadherin impact the clonogenicity of human embryonic stem cells. Biophys J 98:2442–2451

Liu Z, Tan JL, Cohen DM, Yang MT, Sniadecki NJ, Ruiz SA, Nelson CM, Chen CS (2010c) Mechanical tugging force regulates the size of cell-cell junctions. Proc Natl Acad Sci U S A 107:9944–9949

Lock JG, Stow JL (2005) Rab11 in recycling endosomes regulates the sorting and basolateral transport of E-cadherin. Mol Biol Cell 16:1744–1755

Lyashenko N, Winter M, Migliorini D, Biechele T, Moon RT, Hartmann C (2011) Differential requirement for the dual functions of beta-catenin in embryonic stem cell self-renewal and germ layer formation. Nat Cell Biol 13:753–761

Maeda K, Takemura M, Umemori M, Adachi-Yamada T (2008) E-cadherin prolongs the moment for interaction between intestinal stem cell and its progenitor cell to ensure notch signaling in adult drosophila midgut. Genes Cells 13:1219–1227

Martin GR (1981) Isolation of a pluripotent cell line from early mouse embryos cultured in medium conditioned by teratocarcinoma stem cells. Proc Natl Acad Sci U S A 78:7634–7638

McLachlan RW, Yap AS (2007) Not so simple: the complexity of phosphotyrosine signaling at cadherin adhesive contacts. J Mol Med (Berl) 85:545–554

Michel M, Raabe I, Kupinski AP, Perez-Palencia R, Bokel C (2011) Local BMP receptor activation at adherens junctions in the drosophila germline stem cell niche. Nat Commun 2:415

Mirzadeh Z, Merkle FT, Soriano-Navarro M, Garcia-Verdugo JM, Alvarez-Buylla A (2008) Neural stem cells confer unique pinwheel architecture to the ventricular surface in neurogenic regions of the adult brain. Cell Stem Cell 3:265–278

Montgomery RK, Carlone DL, Richmond CA, Farilla L, Kranendonk ME, Henderson DE, Baffour-Awuah NY, Ambruzs DM, Fogli LK, Algra S, Breault DT (2011) Mouse telomerase reverse transcriptase (mTert) expression marks slowly cycling intestinal stem cells. Proc Natl Acad Sci U S A 108:179–184

Moore RN, Cherry JF, Mathur V, Cohen R, Grumet M, Moghe PV (2011) E-Cadherin-expressing feeder cells promote neural lineage restriction of human embryonic stem cells. Stem Cells Dev 21:30-41

Nagaoka M, Koshimizu U, Yuasa S, Hattori F, Chen H, Tanaka T, Okabe M, Fukuda K, Akaike T (2006) E-cadherin-coated plates maintain pluripotent ES cells without colony formation. PLoS One 1:e15

Nagaoka M, Si-Tayeb K, Akaike T, Duncan SA (2010) Culture of human pluripotent stem cells using completely defined conditions on a recombinant E-cadherin substratum. BMC Dev Biol 10:60

Najm FJ, Chenoweth JG, Anderson PD, Nadeau JH, Redline RW, McKay RD, Tesar PJ (2011) Isolation of epiblast stem cells from preimplantation mouse embryos. Cell Stem Cell 8:318–325

Nusse R, Fuerer C, Ching W, Harnish K, Logan C, Zeng A, ten Berge D, Kalani Y (2008) Wnt signaling and stem cell control. Cold Spring Harb Symp Quant Biol 73:59–66

Ohlstein B, Kai T, Decotto E, Spradling A (2004) The stem cell niche: theme and variations. Curr Opin Cell Biol 16:693–699

Ohlstein B, Spradling A (2006) The adult drosophila posterior midgut is maintained by pluripotent stem cells. Nature 439:470–474

Ohlstein B, Spradling A (2007) Multipotent drosophila intestinal stem cells specify daughter cell fates by differential notch signaling. Science 315:988–992

Paez-Gonzalez P, Abdi K, Luciano D, Liu Y, Soriano-Navarro M, Rawlins E, Bennett V, Garcia-Verdugo JM, Kuo CT (2011) Ank3-dependent SVZ niche assembly is required for the continued production of new neurons. Neuron 71:61–75

Pan L, Chen S, Weng C, Call G, Zhu D, Tang H, Zhang N, Xie T (2007) Stem cell aging is controlled both intrinsically and extrinsically in the drosophila ovary. Cell Stem Cell 1:458–469

Perez-Moreno M, Davis MA, Wong E, Pasolli HA, Reynolds AB, Fuchs E (2006) p120-catenin mediates inflammatory responses in the skin. Cell 124:631–644

Perez-Moreno M, Song W, Pasolli HA, Williams SE, Fuchs E (2008) Loss of p120 catenin and links to mitotic alterations, inflammation, and skin cancer. Proc Natl Acad Sci U S A 105:15399–15404

Rasin MR, Gazula VR, Breunig JJ, Kwan KY, Johnson MB, Liu-Chen S, Li HS, Jan LY, Jan YN, Rakic P, Sestan N (2007) Numb and Numbl are required for maintenance of cadherin-based adhesion and polarity of neural progenitors. Nat Neurosci 10:819–827

Redmer T, Diecke S, Grigoryan T, Quiroga-Negreira A, Birchmeier W, Besser D (2011) E-cadherin is crucial for embryonic stem cell pluripotency and can replace OCT4 during somatic cell reprogramming. EMBO Rep 12:720–726

Riethmacher D, Brinkmann V, Birchmeier C (1995) A targeted mutation in the mouse E-cadherin gene results in defective preimplantation development. Proc Natl Acad Sci U S A 92:855–859

Sangiorgi E, Capecchi MR (2008) Bmi1 is expressed in vivo in intestinal stem cells. Nat Genet 40:915–920

Schlegelmilch K, Mohseni M, Kirak O, Pruszak J, Rodriguez JR, Zhou D, Kreger BT, Vasioukhin V, Avruch J, Brummelkamp TR, Camargo FD (2011) Yap1 acts downstream of alpha-catenin to control epidermal proliferation. Cell 144:782–795

Schneider MR, Dahlhoff M, Horst D, Hirschi B, Trulzsch K, Muller-Hocker J, Vogelmann R, Allgauer M, Gerhard M, Steininger S, Wolf E, Kolligs FT (2011) A key role for E-cadherin in intestinal homeostasis and paneth cell maturation. PLoS One 5:e14325

Silvis MR, Kreger BT, Lien WH, Klezovitch O, Rudakova GM, Camargo FD, Lantz DM, Seykora JT, Vasioukhin V (2011) alpha-catenin is a tumor suppressor that controls cell accumulation by regulating the localization and activity of the transcriptional coactivator Yap1. Sci Signal 4:ra33

Singbrant S, Askmyr M, Purton LE, Walkley CR (2011) Defining the hematopoietic stem cell niche: the chicken and the egg conundrum. J Cell Biochem 112:1486–1490

Snippert HJ, Van Der Flier LG, Sato T, van Es JH, Van Den Born M, Kroon-Veenboer C, Barker N, Klein AM, van Rheenen J, Simons BD, Clevers H (2010) Intestinal crypt homeostasis results from neutral competition between symmetrically dividing Lgr5 stem cells. Cell 143:134–144

Solanas G, Cortina C, Sevillano M, Batlle E (2011) Cleavage of E-cadherin by ADAM10 mediates epithelial cell sorting downstream of EphB signalling. Nat Cell Biol 13:1100–1107

Soncin F, Mohamet L, Eckardt D, Ritson S, Eastham AM, Bobola N, Russell A, Davies S, Kemler R, Merry CL, Ward CM (2009) Abrogation of E-cadherin-mediated cell-cell contact in mouse embryonic stem cells results in reversible LIF-independent self-renewal. Stem Cells 27:2069–2080

Soncin F, Mohamet L, Ritson S, Hawkins K, Bobola N, Zeef L, Merry CL, Ward CM (2011) E-cadherin acts as a regulator of transcripts associated with a wide range of cellular processes in mouse embryonic stem cells. PLoS One 6:e21463

Song X, Xie T (2002) DE-cadherin-mediated cell adhesion is essential for maintaining somatic stem cells in the drosophila ovary. Proc Natl Acad Sci U S A 99:14813–14818

Song X, Zhu CH, Doan C, Xie T (2002) Germline stem cells anchored by adherens junctions in the drosophila ovary niches. Science 296:1855–1857

Takahashi K, Yamanaka S (2006) Induction of pluripotent stem cells from mouse embryonic and adult fibroblast cultures by defined factors. Cell 126:663–676

Tesar PJ, Chenoweth JG, Brook FA, Davies TJ, Evans EP, Mack DL, Gardner RL, McKay RD (2007) New cell lines from mouse epiblast share defining features with human embryonic stem cells. Nature 448:196–199

Tian H, Biehs B, Warming S, Leong KG, Rangell L, Klein OD, de Sauvage FJ (2011) A reserve stem cell population in small intestine renders Lgr5-positive cells dispensable. Nature 478:255–259

Tinkle CL, Lechler T, Pasolli HA, Fuchs E (2004) Conditional targeting of E-cadherin in skin: insights into hyperproliferative and degenerative responses. Proc Natl Acad Sci U S A 101:552–557

Tinkle CL, Pasolli HA, Stokes N, Fuchs E (2008) New insights into cadherin function in epidermal sheet formation and maintenance of tissue integrity. Proc Natl Acad Sci U S A 105:15405–15410

Tunggal JA, Helfrich I, Schmitz A, Schwarz H, Gunzel D, Fromm M, Kemler R, Krieg T, Niessen CM (2005) E-cadherin is essential for in vivo epidermal barrier function by regulating tight junctions. EMBO J 24:1146–1156

Vasioukhin V, Bauer C, Degenstein L, Wise B, Fuchs E (2001) Hyperproliferation and defects in epithelial polarity upon conditional ablation of alpha-catenin in skin. Cell 104:605–617

Vestweber D, Ocklind C, Gossler A, Odin P, Obrink B, Kemler R (1985) Comparison of two cell-adhesion molecules, uvomorulin and cell-CAM 105. Exp Cell Res 157:451–461

Voog J, D'Alterio C, Jones DL (2008) Multipotent somatic stem cells contribute to the stem cell niche in the drosophila testis. Nature 454:1132–1136

Wallenfang MR, Nayak R, DiNardo S (2006) Dynamics of the male germline stem cell population during aging of drosophila melanogaster. Aging Cell 5:297–304

Watt FM, Collins CA (2008) Role of beta-catenin in epidermal stem cell expansion, lineage selection, and cancer. Cold Spring Harb Symp Quant Biol 73:503–512

Wray J, Kalkan T, Gomez-Lopez S, Eckardt D, Cook A, Kemler R, Smith A (2011) Inhibition of glycogen synthase kinase-3 alleviates Tcf3 repression of the pluripotency network and increases embryonic stem cell resistance to differentiation. Nat Cell Biol 13:838–845

Xi R (2009) Anchoring stem cells in the niche by cell adhesion molecules. Cell Adh Migr 3:396–401

Yagita Y, Sakurai T, Tanaka H, Kitagawa K, Colman DR, Shan W (2009) N-cadherin mediates interaction between precursor cells in the subventricular zone and regulates further differentiation. J Neurosci Res 87:3331–3342

Yamashita YM, Jones DL, Fuller MT (2003) Orientation of asymmetric stem cell division by the APC tumor suppressor and centrosome. Science 301:1547–1550

Yonemura S, Wada Y, Watanabe T, Nagafuchi A, Shibata M (2010) alpha-Catenin as a tension transducer that induces adherens junction development. Nat Cell Biol 12:533–542

Young P, Boussadia O, Halfter H, Grose R, Berger P, Leone DP, Robenek H, Charnay P, Kemler R, Suter U (2003) E-cadherin controls adherens junctions in the epidermis and the renewal of hair follicles. EMBO J 22:5723–5733

Chapter 16
Adherens Junctions and Cancer

Valeri Vasioukhin

Abstract Cadherins and catenins are the central cell–cell adhesion molecules in adherens junctions (AJs). This chapter reviews the knowledge concerning the role of cadherins and catenins in epithelial cancer and examines the published literature demonstrating the changes in the expression and function of these proteins in human cancer and the association of these changes with patient outcomes. The chapter also covers the mechanistic studies aiming at uncovering the significance of changes in cadherin and catenin expression in cancer and potential molecular mechanisms responsible for the causal role of AJs in cancer initiation and progression.

16.1 Introduction

The ability to adhere to each other is one of the most fundamental cellular functions necessary for the formation of metazoan organisms. Cell–cell adhesion of properly polarized cells is directly responsible for complex architecture of all organs and tissues. The mechanisms of intercellular adhesion are very complex and differ significantly between cells in different organs and tissues. The cell–cell adhesion is especially strong in epithelial tissues, where the cells can tightly seal their membranes to prevent the formation of gaps between neighboring cells. This is important, because the primary function of epithelia is formation of a barrier that separates organism from both external and internal environments (epidermis, intestines, epithelial tubes in various glandular organs) and also separates different tissues from each other (blood vessels).

Tumors originating in epithelial tissues are known as carcinomas and these cancers account for the majority of human malignancies. The most significant manifestation of carcinoma is the focal loss of normal tissue architecture and aberrant accumulation of epithelial cells. Since maintenance of normal tissue organization is one of the primary functions of cell–cell adhesion, it is perhaps not surprising that cell–cell ad-

V. Vasioukhin (✉)
Division of Human Biology, Fred Hutchinson Cancer Research Center, Seattle, WA 98109, USA
e-mail: vvasiouk@fhcrc.org

Department of Pathology and Institute for Stem Cell and Regenerative Medicine,
University of Washington, Seattle,WA 98195, USA,

T. Harris (ed.), *Adherens Junctions: From Molecular Mechanisms to Tissue Development and Disease,* Subcellular Biochemistry 60,
DOI 10.1007/978-94-007-4186-7_16, © Springer Science+Business Media Dordrecht 2012

hesion structures are often abnormal in epithelial tumors. Interestingly, weakening of intercellular adhesion in carcinomas has been noted by pathologists many decades ago (Coman 1944). While gaps and separations between the neighboring cells are often visible on histological sections of low-grade early tumors, they are especially noticeable in more lethal high-grade carcinomas, which are more likely to spread to distant organs and kill the patient. Only recently, with the advancements of biochemistry and molecular biology it became possible to decipher the molecular mechanisms responsible for cell–cell adhesion in normal tissues and the loss or weakening of cell–cell adhesion in epithelial tumors. Interestingly, while cells can form different types of cell–cell adhesion structures, the adherens junctions (AJs) generated by classical cadherins are especially important for normal epithelial self organization and are also frequently affected in epithelial tumors. This chapter will summarize and review the literature concerning the role and mechanisms of AJs in tumor initiation and progression.

16.2 Core Proteins Involved in the Formation and Maintenance of the AJs

Many proteins localize to the AJs, however, only few groups of proteins are essential for AJ formation. These proteins belong to the cadherin and catenin families (Fig. 16.1). Cadherins are a large family of proteins that contains several calcium-binding cadherin domains in the extracellular portion of the molecule. Cadherin domains are directly responsible for the homophilic and calcium-dependent adhesive interactions between cadherin molecules on the neighboring cells. According to the overall domain structure and the number of cadherin domains, the cadherin family can be divided into classical, atypical and protocadherin cadherins. The classical cadherins include the best-studied E(epithelial)-, N(neural)-, P(placental)-, and VE(vasculo-endothelial)-cadherins. The cytoplasmic portion of these transmembrane proteins can bind to catenins and this confers strong adhesive interactions between the membranes of adjacent cells (Cavey and Lecuit 2009; Meng and Takeichi 2009). Since the role of nonclassical and atypical cadherins in mammalian cancer is still poorly understood, this chapter will focus on the role and significance of classical cadherins in cancer initiation and progression.

Catenins are the proteins that interact with cytoplasmic domain of classical cadherins and are required for strengthening of cell–cell contacts and proper AJ formation (Fig. 16.1). p120-catenin directly interacts with cadherins and plays an important role in delivery and stabilization of cadherins at the plasma membrane (Chen et al. 2003b; Davis et al. 2003; Ishiyama et al. 2010). In addition, p120-catenin is also a prominent negative regulator of RhoA and a positive regulator of Rac1 small GTPases, which orchestrate the dynamics of the actin cytoskeleton at the AJs (Anastasiadis and Reynolds 2001; Noren et al. 2000). β-catenin or a similar protein γ-catenin (plakoglobin) also directly interacts with the cytoplasmic domain of cadherin. Importantly, both β-catenin and plakoglobin bind to α-catenin and the principal function of these proteins in the AJs is to bring α-catenin to the cadherin–catenin

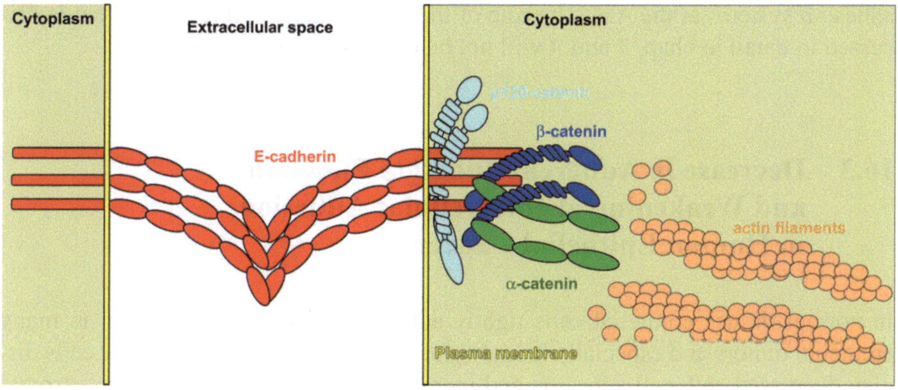

Fig. 16.1 Major proteins involved in the formation of the AJs. Classical cadherins are the transmembrane proteins containing multiple extracellular cadherin domains, which are directly involved in *cis*- and *trans-adhesive* interactions. Cytoplasmic domains of cadherins interact with p120- and β-catenins. β-catenin binds to α-catenin, which provides a functional link to the actin cytoskeleton. Actin filaments help to cluster the cadherin–catenin complexes at the membrane and strengthen the adhesive interactions between the neighboring cells

complex at the cell membrane. In addition to its critical role in the AJs, β-catenin is also a crucial transcriptional co-activator in the canonical Wnt signal transduction pathway, which orchestrates normal development and, when abnormally activated, can cause cancer. β-catenin translocates to the nucleus, binds to TCF family of transcription factors and activates transcription of genes that regulate morphogenetic, as well as, self-renewal and differentiation programs (Clevers 2006; Wend et al. 2010).

At the AJs, β-catenin links cadherin with α-catenin. α-catenin is the only catenin containing a well-defined actin-binding domain, and it is believed, that it functionally links cadherin–catenin proteins at the membrane with the actin cytoskeleton. The mechanisms of this linkage are poorly understood because purified cadherin-β-catenin-α-catenin protein complexes cannot bind to actin filaments (Drees et al. 2005; Yamada et al. 2005). α-catenin may be involved in regulation of actin polymerization at the AJs (Kobielak et al. 2004; Kovacs et al. 2002; Vasioukhin et al. 2000) and, additionally, it may use other actin-binding proteins Eplin or Vinculin to link to the actin filaments indirectly (Abe and Takeichi 2008; Yonemura et al. 2010). Mammalian genomes contain three α-catenin genes, αE(epithelial), αN-(neural) and αT(testicular), which differ primarily by their tissue-specific pattern of expression (Janssens et al. 2001). In this chapter, we will concentrate primarily on epithelial αE-catenin.

In addition to cadherin–catenin protein complexes, the Nectin-afadin cell–cell adhesion system is also present at the AJs (Ogita et al. 2010). Nectins and Nectin-like molecules are transmembrane adhesion proteins containing immunoglobulin-like domains, which interact in *trans* with each other and are responsible for weak calcium-independent cell–cell adhesion. Cytoplasmic domains of Nectins and Nectin-like proteins bind to afadin, which contains actin-binding domain and links Nectin–afadin molecules at the membrane to the actin cytoskeleton. Interestingly, afadin also binds to α-catenin and this interaction can link Nectin and cadherin adhesion systems at the AJs. The role of the Nectin–afadin adhesion system is discussed in detail in chap. 7 and it will not be covered in this chapter.

16.3 Decrease in Adherens Junction Function and Weakening of Intercellular Adhesion in Human Epithelial Cancer

In normal tissues epithelial cells tightly adhere to each other; however, in many epithelial tumors and especially in high-grade advanced cancers the tumor cells display decreased cell–cell adhesion and increased invasion into surrounding stromal tissues. Generally, tumors displaying invasion and local dissemination of epithelial tumor cells are considered to be more likely to result in metastatic progression and death of the patient. Therefore, knowledge about the molecular mechanisms responsible for the differences between well-circumscribed tumors containing tightly adhering cells and tumors containing loosely adhering cells is very important for understanding the biology of cancer progression and metastasis. Since AJs play an important role in normal cell–cell adhesion, many investigators analyzed the AJs in human tumors. Immunostainings of various tumor types with antibodies recognizing specific proteins of the AJs resulted in numerous papers describing the changes in these proteins expression and the correlations between these changes and patient outcomes.

16.3.1 Changes in Cadherin Expression. Decrease in E-Cadherin and Upregulation of Mesenchymal Cadherins in High-Grade Invasive Carcinoma

E-cadherin is the most prominently expressed classical cadherin in normal epithelial cells. Immunostaining with anti-E-cadherin antibodies revealed significant decreases in the expression levels of this protein in the variety of human epithelial tumors (Berx and van Roy 2009). Interestingly, not all the tumors are exactly the same, as there are significant differences between different organs and even different tumor types within the organ. In breast carcinoma, loss of E-cadherin expression is an early and highly penetrant event (up to 85%) in lobular tumors (Berx et al. 1995; Berx and Van Roy 2001; Cowin et al. 2005). Conversely, E-cadherin expression in ductal breast carcinoma is decreased only in advanced high-grade tumors (Jeschke et al. 2007; Nagae et al. 2002; Oka et al. 1993; Park et al. 2007; Pedersen et al. 2002; Rakha et al. 2005).

Similar to lobular breast carcinoma, the diffuse-type gastric cancer also displays the early onset loss of E-cadherin expression (92%) (Mayer et al. 1993). In contrast, intestinal-type gastric cancer shows loss of E-cadherin expression primarily in high-grade invasive tumors (Shun et al. 1998). In squamous cell carcinoma, prostate cancer, non-small cell lung carcinoma and colon cancer the expression of E-cadherin is usually maintained in early low-grade tumors and is downregulated in a subset of late high-grade tumors (Bohm et al. 1994; Schipper et al. 1991; Schuhmacher et al. 1999; Umbas et al. 1992). Loss or decrease in expression of E-cadherin usually correlates with tumor invasiveness, distant metastasis and unfavorable patient outcome (Asgeirsson et al. 2000; Dolled-Filhart et al. 2006; Heimann et al. 2000; Nakopoulou et al. 2002; Park et al. 2007; Pedersen et al. 2002; Rakha et al. 2005).

E-cadherin is not the only classical cadherin expressed in normal epithelial tissues. In addition to E-cadherin, epithelial cells often express other cadherins. For example, skin keratinocytes express both E-cadherin and P-cadherin and loss of expression of both of these genes is required for disruption of the AJs (Tinkle et al. 2008b). In contrast to E-cadherin, which is expressed at relatively similar levels by all epithelial cells within the tissue, P-cadherin often displays much more differential levels of expression. For example, in skin epidermis only basal and hair germ cells express P-cadherin. Similarly, in breast epithelium, P-cadherin is expressed primarily by myoepithelial cells, Thus, perhaps not surprisingly, the results of the analyses of P-cadherin expression in epithelial tumors were very different from what was found for E-cadherin. In breast cancer, P-cadherin is often overexpressed and its expression correlates with poor patient prognosis (Paredes et al. 2007; Turashvili et al. 2011). P-cadherin is not expressed in normal adult colon, but its expression is activated in colon cancer (Milicic et al. 2008). In contrast, in oral squamous cell carcinoma loss of P-cadherin expression is a hallmark of aggressive tumors (Lo Muzio et al. 2004; Munoz-Guerra et al. 2005).

While expression of E-cadherin in high-grade epithelial tumors is downregulated, tumor cells often show increased expression of N-cadherin, which is normally present only in mesenchymal cells and neurons (Cavallaro et al. 2002; Wheelock et al. 2008). This phenomenon is called "cadherin switching" and it is considered to be one of the hallmarks of epithelial to mesenchymal transition (EMT). EMT is a normal process of a dramatic change in cellular morphology which occurs during specific stages of normal embryonic development, when highly adhesive epithelial cells downregulate expression of epithelial cadherins, upregulate expression of mesenchymal cadherins and become highly motile and invasive (Shook and Keller 2003). Notable examples of these developmental events include gastrulation, formation of neural crest cells, heart valve formation and delamination of muscle precursor cells from the dermomyotome during muscle morphogenesis (Micalizzi et al. 2010). This ability of epithelial cells to drastically change their phenotype and become highly invasive mesenchymal cells is maintained in epithelial tumors and EMT is likely to play an important role during tumor dissemination and metastasis (Hugo et al. 2007). While downregulation of epithelial cadherins is clearly the causal event responsible for the transition from epithelial to mesenchymal phenotype, the role of upregulation of N-cadherin is not completely clear. Early experiments with established cancer cell lines indicated that upregulation of N-cadherin in epithelial cells can cause a prominent increase in cell migration and invasion (Nieman et al. 1999; Zahir and Weaver 2004). N-cadherin can activate cell migration in tissue culture by interacting with fibroblast growth factor receptor (FGFR) and stimulating FGF signaling, as well as by activating Rac1 and Cdc42 small GTPases, which regulate the actin cytoskeleton (Wheelock et al. 2008). *In vivo* experiments revealed a more complex picture. Overexpression of N-cadherin in breast tumors, which were induced by upregulation of ErbB2, did not produce a significant phenotype; however, overexpression of N-cadherin in breast carcinoma, which was induced by expression of polyoma virus middle T antigen, resulted in increased metastasis (Hulit et al. 2007; Knudsen et al. 2005). Thus, it is likely that the role of N-cadherin upregulation during EMT in epithelial tumors depends on the nature of the signaling pathways responsible for tumor initiation and progression.

16.3.2 Changes in Expression of Catenins

In addition to E-cadherin, AJs in epithelial cells also require expression of catenins, which were also analyzed in a range of epithelial tumors. In general, immunostaining with anti-α-catenin antibodies revealed changes that are very similar to the changes in expression of E-cadherin. Tumors showing loss or decrease in expression of E-cadherin, often also show the downregulation of α-catenin (Aaltomaa et al. 1999; Kimura et al. 2000; Richmond et al. 1997; Toyoyama et al. 1999; Umbas et al. 1992; Zhou et al. 2005). 81% of breast carcinomas display loss or decrease in staining for α-catenin (Rimm et al. 1995). α-catenin is absent in 50% of squamous cell carcinomas of esophagus (Setoyama et al. 2007) and in 41% of gastric carcinomas (Zhou et al. 2005). Decrease or loss of expression of α-catenin often correlate with invasive phenotype, lymph node metastasis, recurrence of the disease and poor patient outcome (Aaltomaa et al. 1999, 2005; Kadowaki et al. 1994; Nakanishi et al. 1997; Richmond et al. 1997; Setoyama et al. 2007; van Oort et al. 2007).

In addition to its important role in AJs, β-catenin is also a transcriptional co-activator, which can cause tumor formation, when it is constitutively stabilized and localized to the nucleus. The important role of β-catenin in cell transformation is the primary reason for the extensive analyses of this protein in variety of human malignancies. The amount of information about β-catenin in human cancer is massive and it has been extensively covered in excellent recent reviews (Clevers 2006; Heuberger and Birchmeier 2010; Klaus and Birchmeier 2008; Wend et al. 2010). Overall, β-catenin is prominently upregulated in colorectal cancer, in subsets of medulloblastomas and basal cell carcinomas, and type of hair follicle cancer called pilomatricoma (Clevers 2006; El-Bahrawy et al. 2003; Gat et al. 1998; Gibson et al. 2010; Saldanha et al. 2004; Yamazaki et al. 2001). While there are many studies that report the presence of nuclear β-catenin in various additional tumor types, these findings are not always corroborated by other laboratories. For example, the presence of significant levels of nuclear β-catenin was reported in breast cancer tumors in one report (Lin et al. 2000), but this finding was not confirmed in the follow up papers (Dolled-Filhart et al. 2006; Gillett et al. 2001; Pedersen et al. 2002; Wong et al. 2002). In many epithelial tumors, the levels of junctional β-catenin parallels the expression of E-cadherin and it is downregulated in the advanced high-grade tumors (Aaltomaa et al. 2005; Carico et al. 2001; Chung et al. 2007; Dolled-Filhart et al. 2006; Fukumaru et al. 2007; Ishizaki et al. 2004; Jaggi et al. 2005; Nakanishi et al. 1997).

16.4 Molecular Mechanisms Responsible for the Decrease of Cadherin-Catenin-Mediated Adhesion in Human Epithelial Tumors

As described earlier, extensive literature indicates that epithelial cadherins and associated catenins are frequently downregulated in advanced human epithelial tumors. The molecular mechanisms responsible for this downregulation are very diverse and include regulation at the level of genes, gene transcription, and multiple post-transcriptional stages.

16.4.1 Mutation and Deletion of Cadherin and Catenin Genes in Human Cancer

The most direct effects on gene function are gene mutations, deletions or amplifications. While "next generation" sequencing approaches currently make it possible to analyze tumor cell mutations in all human genes, earlier efforts concentrated on sequencing of specific genes in DNA isolated from human tumors. Cancer genome re-sequencing of *E-cadherin (CDH1)* revealed frequent inactivating mutations in *E-cadherin* in lobular breast and diffuse-type gastric cancers, where up to 50% of primary tumors contain mutations in *E-cadherin* (Becker et al. 1993, 1994;. Berx et al. 1995, 1996). In breast cancer, the wild-type allele of *E-cadherin* is often lost and the overall loss of heterozygosity of *E-cadherin* is a frequent event in breast cancer (Berx et al. 1996; Cleton-Jansen et al. 2001). Besides the lobular breast and diffuse-type gastric cancer, mutations in *E-cadherin* are not frequent in other carcinomas (Endo et al. 2001; Risinger et al. 1994; Soares et al. 1997; Taddei et al. 2000; Wijnhoven et al. 1999).

The homozygous loss-of-function mutations in *E-cadherin, β-catenin(Ctnnb1)* and *αE-catenin (Ctnna1)* genes are embryonic lethal in mice and it is likely that they result in lethality in humans (Haegel et al. 1995; Huelsken et al. 2000; Larue et al. 1994; Torres et al. 1997). While mice with heterozygous mutations in *E-cadherin, β-catenin* and *αE-catenin* genes appear normal, germ line mutation in one allele of *E-cadherin* in humans strongly predisposes to development of diffuse-type gastric cancer (Guilford et al. 1998). Interestingly, the probability of development of lobular breast cancer in females carrying the germ line mutation in *E-cadherin* is only slightly higher than in general population (Pharoah et al. 2001).

Loss-of-function mutations in *α- and β-catenin genes* have been described in cell lines derived from various human epithelial tumors; however, there is no knowledge about the prevalence of these mutations in primary tumors. Cancer genome re-sequencing projects identified mutations and deletions in αE- and *αN-catenin genes*; however, since only few tumors were sequenced, it is impossible to conclude the overall prevalence of these mutations (Ding et al. 2010; Wood et al. 2007).

In contrast to the loss-of-function mutations, protein stabilizing mutations in *β-catenin* are frequently found in primary epithelial tumors (Polakis 2000). These mutations concentrate in the amino-terminal part of the protein and often affect serine and threonine amino acids, which need to be phosphorylated for protein degradation (see below). Thus, the functional outcome of these mutations is stabilization of the protein and constitutive activation of the canonical Wnt signaling pathway, which is causally involved in cancer initiation. Mutations in *β-catenin* are found in 27% of intestinal type gastric cancers (Park et al. 1999), in up to 20% of hepatocellular carcinomas (de La Coste et al. 1998; Miyoshi et al. 1998), in 15% of pediatric kidney cancers (Koesters et al. 1999), in 61% of anaplastic thyroid carcinoma (Garcia-Rostan et al. 1999), and in 75% of hair follicle tumors called pilomatricoma (Chan et al. 1999).

16.4.2 Transcriptional Downregulation of Cadherin and Catenin Gene Expression

While decreases in the expression of epithelial cadherins and loss of cell–cell adhesion occur in the majority of advanced epithelial tumors, *E-cadherin* is frequently mutated or lost in only a few specific tumor types. Indeed, the most frequent mechanisms responsible for the loss of E-cadherin function is transcriptional downregulation of its expression (Berx and van Roy 2009). Expression of *E-cadherin* is downregulated by either promoter hypermethylation or an increase in expression of transcriptional factors that down-regulate *E-cadherin* promoter activity.

Promoter hypermethylation is likely the most frequent event responsible for the loss of *E-cadherin* expression in advanced tumors. Even in gastric diffuse-type tumors containing an inactivating mutation of *E-cadherin*, the second nonmutant allele of *E-cadherin* is silenced via methylation of a CpG island in the promoter of the *E-cadherin* gene (Grady et al. 2000). Abnormally increased methylation of the *E-cadherin* promoter is found in primary gastric, breast, prostate, non-small cell lung, thyroid, bladder, cervical, esophageal, renal, colorectal, hepatocellular and other types of cancer (Chen et al. 2003a; Garinis et al. 2002; Graff et al. 1995, 1998; Kim et al. 2007; Matsumura et al. 2001; Nojima et al. 2001; Ribeiro-Filho et al. 2002; Si et al. 2001; Tamura et al. 2000; Yoshiura et al. 1995). Increased methylation of the *E-cadherin* promoter enhances the binding of methyl-CpG-interacting proteins MeCP2 and MBP2, which recruit HDACs that cause histone 3 (H3) deacetylation and shut down transcription (Koizume et al. 2002). Moreover, proteins that bind to methylated H3 at the *E-cadherin* promoter also play an important role in regulation of *E-cadherin* expression. Methyl-H3K9-binding protein MPP8 localizes to the *E-cadherin* promoter and mediates *E-cadherin* silencing by directing DNA methylation via interaction with DNA methyltransferase 3A (Kokura et al. 2010). Interestingly, in some cell types transcription factors critical for proper epithelial differentiation may neutralize the effect of E-cadherin promoter hypermethylation. For example, the members of the forkhead transcription factor family FOXA1/2 are downregulated in pancreatic ductal adenocarcinoma resulting in the loss of *E-cadherin* expression. Re-expression of FOXA1/2 in pancreatic cancer cells with extensive *E-cadherin* promoter hypermethylation results in re-activation of *E-cadherin* expression (Song et al. 2010).

In addition to promoter methylation, several transcription repressors can mediate downregulation of *E-cadherin* expression through direct interaction with *E-cadherin* promoter. Earlier somatic cell hybrid experiments indicated the presence of factors that utilize E-box sequences within the *E-cadherin* promoter to downregulate its activity (Giroldi et al. 1997). Later studies identified multiple E-box-binding transcriptional repressors that can bind to the *E-cadherin* promoter, downregulate its transcriptional activity and promote EMT, cell migration and invasion. Specifically, the Zinc-finger protein SNAIL represses *E-cadherin* promoter by the recruitment of the H3K27 histone methyltransferases polycomb repressive complex 2 (PRC2) and Sin3 A/HDAC1/2 (Batlle et al. 2000; Cano et al. 2000; Herranz et al. 2008; Peinado et al. 2004). E-box binding proteins ZEB1/2 downregulate *E-cadherin* expression by interacting with the transcription co-repressor CtBP-1 (Comijn et al. 2001; Grooteclaes and Frisch 2000; Shi et al. 2003). In addition, ZEB1 can regulate *E-cadherin* independently of CtBP, by interacting with the SWI/SNF chromatin-remodeling protein BRG1 (Sanchez-Tillo et al. 2010). In addition to Snail and ZEB1/2, basic helix-loop-helix proteins E12/E47 and Twist repress the transcription of *E-cadherin* through direct interaction with its promoter (Perez-Moreno et al. 2001; Yang et al. 2004). The forkhead domain transcription factor FOXQ1 is overexpressed in high-grade basal-type breast cancers. It downregulates expression of *E-cadherin* and promotes EMT, gain of stem cell-like properties, and acquisition of resistance to chemotherapy-induced apoptosis (Qiao et al. 2011). EZH2 is a histone-lysine-methylase and the member of the polycomb group of proteins involved in transcriptional repression of target genes. EZH2 is upregulated in prostate, breast and bladder cancers and it mediates tumor aggressiveness by transcriptional silencing of E-cadherin (Cao et al. 2008). Bmi1 is another polycomb-group protein frequently overexpressed in human cancers. Bmi1 is expressed in stem cells, it maintains self-renewal, and it cooperates with Twist1 to repress *E-cadherin* and *p16INK4a* and to promote epithelial-mesenchymal transition and tumor initiating capacity (Yang et al. 2010). *Zeppo1* (zinc finger elbow-related proline domain protein 1) is amplified and overexpressed in breast cancer. It binds to Groucho, represses *E-cadherin* expression and promotes breast cancer progression in mouse models of metastatic breast cancer (Slorach et al. 2011). Overall, various transcriptional repressors are utilized during normal development to quickly downregulate *E-cadherin* expression and promote EMT. During progression of epithelial tumors, these factors are also upregulated and this causes tumor cell EMT and promotes tumor invasion and metastasis (Peinado et al. 2007; Yang et al. 2004).

16.4.3 Posttranscriptional Inactivation of Cadherin–Catenin Adhesion System by MicroRNAs

While transcriptional regulation of *E-cadherin* has been extensively studied, it is clear that cells can employ multiple posttranscriptional mechanisms to control the levels of epithelial cadherins. A number of microRNAs exercise a potent control over E-cadherin protein production at the level of translation. MicroRNAs usually recognize the 3' UTR region of their target mRNAs and negatively regulate protein translation. The Weinberg laboratory recently demonstrated that a microRNA upregulated in breast cancer cells, miR-9, directly targets E-cadherin mRNA (Ma et al. 2010). Overexpression of miR-9 in a nonmetastatic cell line downregulated E-cadherin and enabled the cells to form metastases. Conversely, downregulation of miR-9 levels in highly malignant cells inhibited metastasis. Interestingly, transcription of miR-9 itself was controlled by MYC and MYCN proteins, which are strongly implicated in cancer. Similar to miR-9, a microRNA overexpressed in esophageal squamous cell carcinoma, miR-92a, also directly targets E-cadherin 3'UTR and promotes cell migration and invasion (Chen et al. 2011). Highly expressed in breast cancer stem cells miR-495 also directly targets E-cadherin and promotes tumorigenesis in mice. Interestingly, the transcription factor E12/E47, previously implicated in the direct regulation of *E-cadherin*, was also shown to control the expression of miR-495 (Hwang-Verslues et al. 2011).

While the majority of microRNAs target the 3'UTR of mRNAs and negatively regulate protein translation, some microRNAs target promoter regions and activate transcription. For example, miR-373 has a binding site in the *E-cadherin* promoter and it activates *E-cadherin* transcription (Place et al. 2008). Besides this noncanonical regulation, miR-373 can target LATS2 and CD44 mRNAs and promote tumor invasion and metastasis (Huang et al. 2008; Voorhoeve et al. 2006).

While miR-9, mir-92a and miR-495 directly regulate E-cadherin mRNA, other microRNAs act on transcriptional factors that can control *E-cadherin* expression. For example, miR-141, mir-200 and mir-205 miRNAs directly repress ZEB1 and ZEB2 (Burk et al. 2008; Gregory et al. 2008; Korpal et al. 2008; Olson et al. 2009; Park et al. 2008). In addition, the miR-200 family directly targets Suz12, a subunit of a polycomb repressor complex (PRC2), which negatively regulates *E-cadherin* expression and promotes formation of breast cancer stem cells (Iliopoulos et al. 2010). miR-708 is downregulated in human renal cell carcinoma, and in normal cells, this microRNA targets *E-cadherin* regulators ZEB2 and BMI1 (Saini et al. 2011). In some cases, the microRNA regulation of epithelial cadherin expression is even more complex. miR-221 and miR-222 are expressed in basal-type breast cancer, where they target the 3'UTR of TRPS1 (trichorhinophalangeal syndrome type 1), which is a member of the GATA family of transcriptional repressors directly repressing expression of ZEB2 (Stinson et al. 2011).

Overall, microRNAs elicit a powerful control of E-cadherin expression in normal epithelial cells and deregulation of these microRNAs can have profound consequences for the loss of epithelial phenotype and increased invasion and metastasis in high-grade epithelial tumors.

16.4.4 Posttranslational Effects on the Cadherin–Catenin Adhesion System

During development epithelial cells constantly remodel their cell–cell adhesion structures to allow for normal tissue morphogenesis. Moreover, in the majority of adult organs, homeostasis is maintained through constant tissue renewal via a well-orchestrated process of stem cell-mediated self-renewal and differentiation. These processes also involve constant epithelial cell movements within the tissue, which implies the need for quick cell–cell junction breakdown and re-formation that often takes place without the overall loss of the epithelial phenotype. To accomplish this task, cells evolved several mechanisms eliciting posttranslational control of AJ function. Cadherin–catenin complexes can be quickly regulated by phosphorylation-mediated disruption of cadherin–catenin protein complexes that is often coupled with endocytosis of cadherin and its degradation.

Protein phosphorylation is a powerful mechanism that can be rapidly employed to change protein function. Stability of the AJs depends on the efficient formation of cadherin–catenin protein complexes. Phosphorylation of cadherins and catenins influences their binding affinities and overall AJ complex stability. E-cadherin at the membrane associates with multiple receptor-type tyrosine kinases (RTKs) including IGF1R, ErbB2, Met and EGFR (Canonici et al. 2008; Hiscox and Jiang 1999; Ochiai et al. 1994; Reshetnikova et al. 2007). Activation of RTKs results in decreases in the affinity between cadherin and catenins and disassembly of the cadherin–catenin complexes (Behrens et al. 1993; Hamaguchi et al. 1993). RTKs and/or Src-family kinases phosphorylate E-cadherin and this creates a binding site for Hakai, which mono-ubiquitinates E-cadherin and promotes its interaction with mono-ubiquitin-binding protein HRS, internalization and subsequent lysosome-mediated degradation (Fujita et al. 2002; Palacios et al. 2005; Shen et al. 2008; Toyoshima et al. 2007). The AJ protein Shrew-1 plays an important role in EGF-induced endocytosis of E-cadherin; however, the mechanism of Shrew-1 function is not well understood (Gross et al. 2009). Endocytosed E-cadherin can also be ubiquitinated by MDM 2, which is overexpressed in breast cancer, and this is followed by E-cadherin degradation (Yang et al. 2006). Similar to E-cadherin, β-catenin is a prominent substrate for tyrosine phosphorylation by Src-family, BCL-Abl, MET and RET tyrosine kinases and this phosphorylation decreases the affinities between β-catenin and both E-cadherin and α-catenin, and in addition, causes an activation of β-catenin-mediated transcription (Brembeck et al. 2004; Coluccia et al. 2006, 2007; Gujral et al. 2008; Lilien and Balsamo 2005; Roura et al. 1999; Zeng et al. 2006).

p120-catenin was first discovered as a prominent substrate for Src-family tyrosine kinases (Reynolds et al. 1994). Tyrosine phosphorylation of p120-catenin modulates its binding to and inhibition of RhoA and this can have a significant impact on the stability of the AJs (Castano et al. 2007). Src can also disrupt AJs and promote EMT by phosphorylation and targeted degradation of the Rac1 activator Tiam1, which is required for AJ formation and maintenance (Woodcock et al. 2009).

In addition to tyrosine kinases, the stability of the AJs is also regulated by serine-threonine phosphorylation. The cadherin cytoplasmic tail contains a serine-rich domain and its phosphorylation by casein kinase I results in the disruption of cadherin–catenin complexes (Ochiai et al. 1994). Serine-threonine phosphorylation of β-catenin by GSK3β plays a critical role in the destruction of cytoplasmic β-catenin and the negative regulation of its transcriptional activity (Aberle et al. 1997; Clevers 2006). In contrast, phosphorylation by AKT and cyclic AMP-dependent protein kinases causes activation of β-catenin transcriptional activity (Fang et al. 2007; Hino et al. 2005).

AJs link to the actin cytoskeleton and changes in the actin cytoskeleton play crucial roles in both AJs formation and disruption. Cadherin–catenin clustering is regulated by the actin cytoskeleton (Angres et al. 1996; Hirano et al. 1992). Rho-family GTPases are the general regulators of the actin cytoskeleton, and they play a crucial role in the regulation of cell–cell adhesion, cell migration and invasion (Ellenbroek and Collard 2007). While α-catenin is the only catenin with a well-defined f-actin binding domain, p120-catenin is the principal regulator of Rho family GTPases. P120-catenin binds and inactivates RhoA and activates Rac1, and this promotes AJ stabilization (Reynolds 2007). While some dynamics of the actin cytoskeleton are necessary for AJ formation, sustained activation of either RhoA or Rac1, which is frequently induced by oncogenes, is likely to destabilize cadherin–catenin contacts and stimulate cell migration and invasion (Gimond et al. 1999; Lozano et al. 2008; Zhong et al. 1997). For example, Abl and Arg intracellular tyrosine kinases impact AJs assembly through regulation of Rho-ROCK-myosin signaling pathway (Zandy et al. 2007). Inhibition of these kinases results in activation of Rho-ROCK pathway and disruption of the AJs, which is mediated by actomyosin contraction.

Cell surface levels of cadherin can be rapidly altered by activation of cadherin endocytosis. For example, during normal embryonic development a dramatic EMT takes place during gastrulation. In gastrulation, activin/nodal members of the TGF-β superfamily induce expression of Fibronectin Leucine-rich Repeat Transmembrane 3 (FLRT3), a transmembrane protein containing extracellular leucine-rich repeats, and the small GTPase Rnd1 (Ogata et al. 2007). These two proteins interact physically and decrease cell adhesion by sequestering cadherin through a dynamin-dependent endocytosis pathway.

We previously discussed the important role of p120-catenin in cadherin stabilization at the plasma membrane (Kowalczyk and Reynolds 2004). Interestingly, the interaction site between p120-catenin and cadherin extends over the cadherin residues required for clathrin-mediated endocytosis and Hakai-dependent ubiquitination of this protein (Ishiyama et al. 2010). Thus, interaction between cadherin and p120-catenin should inhibit cadherin internalization and stabilize it at the cell surface. Upon endocytosis, E-cadherin can associate with the MDM 2 E3 ubiquitin ligase, which ubiquitinates cadherin and targets it for degradation (Yang et al. 2006). Interestingly, expression levels of MDM2 negatively correlate with the levels of E-cadherin in clinical samples of breast cancer, and overexpression of MDM 2 promotes E-cadherin internalization and degradation in breast cancer cell lines (Yang et al. 2006).

The TGF-β signal transduction pathway is a critical regulator of EMT in various epithelial tumors. Activated by TGF-β treatment SMAD3/4 transcription factors interact with Snail1 at the promoters of *E-cadherin* and tight-junctional genes and co-repress their transcription to induce EMT (Vincent et al. 2009). Another mechanism responsible for TGF-β-induced EMT involves phosphorylation of Ser 43 of hnRNP E1 by protein kinase Akt2, which results in the release of the hnRNP from the 3' UTR of Disabled-2 (Dab2) and interleukin-like EMT inducer (ILEI) transcripts and translational activation of Dab2 and ILEI (Chaudhury et al. 2010; Hussey et al. 2011). Dab2 mediates directional trafficking and polarized distribution of several cell surface proteins including E-cadherin (Yang et al. 2007). Moreover, Dab2 is a critical protein responsible for EMT in TGF-β-treated cells (Prunier and Howe 2005).

Progression of epithelial tumors is frequently associated with activation of cell surface proteases, which promote tumor invasion and metastasis. Interestingly, many of these enzymes directly cleave cadherin molecules. Cleavage of cadherins by metalloproteinases (MMPs), disintegrin and metalloproteinases (ADAM) or γ-secretase results in the disruption of cadherin-mediated adhesion (Ferber et al. 2008; Lochter et al. 1997; Marambaud et al. 2002; Maretzky et al. 2005; Solanas et al. 2011). MMPs can be responsible for EMT induced by potent oncogenes. For example, in lung epithelial cells, oncogenic K-Ras can promote disruption of the AJs and EMT via ERK-mediated induction of MMP-9 and cleavage of E-cadherin at two sites (Wang et al. 2009). Interestingly, both extracellular and intracellular fragments of cadherin generated after cleavage can have an important function in regulation of cell adhesion and signaling. The extracellular fragments have dominant negative activity, binding to cadherin on the cell surface and interfering with its function in cell–cell adhesion (Damsky et al. 1983). The cytoplasmic domain of E-cadherin can translocate to the nucleus and regulate p120-catenin-Kaiso transcription activity (Ferber et al. 2008). Similarly, upon translocation to the nucleus, the cytoplasmic domain of N-cadherin enhances transcription activity of β-catenin and represses CBP/CREB-mediated transcription (Marambaud et al. 2003; Shoval et al. 2007).

To summarize, the advanced epithelial tumors may employ a variety of mechanisms to down-regulate the function of epithelial cadherins and decrease epithelial cell adhesion. We will now discuss the potential clinical significance of these events.

16.5 Causal Role of AJ Proteins in Cancer Initiation and Progression

Analyses of human tumors clearly demonstrated a decrease of the epithelial cadherin–catenin cell–cell adhesion system in advanced malignancies. This, however, does not necessarily indicate that loss of epithelial-type AJs is causally involved in tumor initiation or progression. One of the potential causal proofs frequently used in the clinical literature is the evidence by association. Indeed, primary tumors demonstrating loss of E-cadherin expression statistically are more likely to present with metastasis and result in poor patient outcome. However, one has to be careful in making conclusions based on associations, because there is a potential caveat in this simple explanation. Since decrease in E-cadherin expression is more frequently seen in histologically advanced tumors, the association between decreased E-cadherin expression and tumor metastasis may be circumstantial and not causal. An experimental approach is usually necessary to establish causality. We will now discuss the potential causal connection between the cadherin–catenin adhesion system and cancer initiation and progression.

Many cell lines generated from human tumors display decreased expression of E-cadherin and/or catenins. Re-expression of missing cell–cell adhesion molecules and the analysis of resulting cell lines provided a powerful tool for the investigation of the role of AJ proteins in human cancer. Many groups also used a complementary loss-of-function approach to decrease the expression of cadherins or catenins in normal epithelial cells or in cancer cell lines with the cadherin–catenin adhesion system. Overall, studies with E-cadherin demonstrated that it plays a critical role in the establishment and maintenance of epithelial cell phenotypes and the attenuation of EMT, cell migration, invasion and metastasis (Behrens et al. 1989; Chen and Obrink 1991; Frixen et al. 1991; Vleminckx et al. 1991). Re-expression of α-catenin in cell lines missing α-catenin resulted in even more dramatic phenotypes and caused not only reversion of EMT and attenuation of cell invasion, but also decreased rates of cell proliferation and attenuated primary tumor formation in immunocompromised mice (Bullions et al. 1997; Ewing et al. 1995; Watabe et al. 1994). The role of p120-catenin in human cancer cell lines is even more complex and it appears to depend on the expression of E-cadherin. In cells expressing E-cadherin, p120-catenin inhibits Ras and attenuates cell proliferation; however, in cell lines missing E-cadherin, p120-catenin does not attenuate Ras and, instead, it activates Rac1-MAPK signaling and promotes cell proliferation (Dohn et al. 2009; Soto et al. 2008).

Cell lines are a powerful experimental model, which can quickly assess the tumorigenic potential and significance of specific genetic and epigenetic changes that take place in human tumors. However, experiments with cell lines are unable to capture the complexity of tumor initiation and progression in human patients, because of the important role of three-dimensional tissue organization and overall cellular diversity that is missing in these models (Bissell and Hines 2011; Lee and Vasioukhin 2008). Genetic experiments with mice are usually necessary to capture all the details of normal tissue microenvironment and three-dimentional tissue organization. The Christofori laboratory generated the first genetic evidence of an important role of E-cadherin in tumor progression. The progression from well-differentiated adenoma to invasive carcinoma in the mouse model of pancreatic β-cell carcinoma (Rip1Tag2 mice) is marked by the decrease in E-cadherin expression (Perl et al. 1998). Re-expression of E-cadherin in these tumors *in vivo* resulted in the arrest of tumor progression at the adenoma stage and, conversely, expression of a dominant-negative E-cadherin caused enhanced tumor invasion and metastasis (Perl et al. 1998). In a different genetic cancer model, the growth and metastatic progression of non-small-cell lung cancer driven by C-Raf overexpression was enhanced by simultaneous deletion of E-cadherin (Ceteci et al. 2007).

Unlike the majority of human epithelial tumors, E-cadherin is lost early in lobular breast carcinoma. These tumors also display frequent inactivation of the p53 tumor suppressor gene. Tissue-specific deletion of tumor suppressor p53 in mice is sufficient to cause development of breast cancer. While deletion of E-cadherin in the same tissues was not sufficient to cause cancer, simultaneous deletion of E-cadherin and p53 resulted in earlier development of invasive and metastatic lobular breast carcinoma (Derksen et al. 2006). Interestingly, deletion of E-cadherin in breast epithelial cells in this model caused resistance to anoikis, an apoptotic cell death caused by the loss of attachment to the extracellular matrix.

While the experiments with genetic inactivation of E-cadherin clearly demonstrated that the loss of E-cadherin expression is playing a causal role in tumor progression, it was also very clear that E-cadherin is not a canonical tumor suppressor, because deletion of E-cadherin was not sufficient to cause tumor formation (Derksen et al. 2006). Somewhat different results were obtained in the experiments with genetic ablation of p120- and α-catenins. Deletion of α-catenin in hair follicle stem and progenitor cells caused the formation of prominent inflammatory skin lesions and development of skin squamous cell carcinoma tumors (Silvis et al. 2011). Moreover, transplantation of *α-catenin*$^{-/-}$keratinocytes on the skin of nude mice resulted in wound-like microenvironment and formation of skin lesions resembling squamous cell carcinoma (Kobielak and Fuchs 2006). Similarly, transplantation of *p120-catenin*$^{-/-}$ keratinocytes on the skin of nude mice caused formation of prominent inflammatory lesions and tumor-like growth (Perez-Moreno et al. 2008). Conditional deletion of p120-catenin in the mouse oral cavity, esophagus, and forestomach results in inflammation and invasive squamous cell cancer (Stairs et al. 2011). Similarly, ablation of p120-catenin in intestinal epithelial cells results in chronic inflammation and formation of tumors (Smalley-Freed et al. 2011). These genetic *in vivo* experiments demonstrate that both p120- and α-catenins can function as tumor suppressors.

Unlike p120 and α-catenins, β-catenin is not critical for AJ formation in epithelial cells, because loss of β-catenin can be compensated by plakoglobin, which can provide a link between cadherins and α-catenin. In contrast to p120- and α-catenins, β-catenin is a very potent proto-oncogene playing a central role in the canonical Wnt signal transduction pathway (Clevers 2006). Constitutive activation of Wnt signaling is oncogenic in many organs and tissues. β-catenin levels are negatively regulated by a destruction complex containing APC, Axin, GSK3β and Casein kinase I (CKI). GSK3β and CKI phosphorylate β-catenin and target it for degradation. Deletion or mutation of the phosphorylation sites stabilizes β-catenin and prominently increases its transcriptional activity. Expression of a stabilized form of β-catenin in the variety of mouse organs and tissues resulted in development of cancer, essentially proving that stabilization of β-catenin is sufficient for tumor development (Clevers 2006). Inactivation of β-catenin destruction complex also results in increase in β-catenin levels and signaling activities. Mutation of one allele of APC in mice results in upregulation of β-catenin and development of multiple polyps and eventually intestinal cancer (Clevers 2004). Similarly, mutation of APC in humans results in Familial adenomatous polyposis (FAP), which is an inherited disorder characterized by development of multiple intestinal polyps and predisposition to cancer of the large intestine, as well as fibromas, osteomas and medulloblastomas. In general, human and mouse phenotypes vary significantly, depending on the nature of the APC gene mutation. Interestingly, while loss of one allele of APC should be theoretically sufficient to activate β-catenin in all cell types, the most prevalent phenotype in APC mutants is the development of intestinal tumors (Clevers 2004). The reasons for such specificity are not known, but it may indicate an exquisite sensitivity of the intestinal stem cells to changes in β-catenin signaling. Since transcriptional activity of β-catenin is required for the maintenance of stem and progenitor cells in many organs and tissues, conditional deletion of *β-catenin* usually results in the depletion of these cell populations and failure of normal development or adult organ homeostasis (Fevr et al. 2007; Machon et al. 2003; Zechner et al. 2003). This provides a significant opportunity for therapeutic intervention, as tumors are likely to be very sensitive to the loss of β-catenin signaling activity (Malanchi et al. 2008).

Overall, loss-of-function and gain-of-function experiments using cell lines and genetically engineered mice demonstrated that AJ proteins play an important role in epithelial tumor initiation and progression. E-cadherin downregulation in epithelial tumors is not sufficient for tumor initiation, but it promotes tumor invasion and metastasis. P120- and α-catenins can function as tumor suppressors and cause tumor development in some organs and tissues. Finally, β-catenin is a proto-oncogene, since upregulation of β-catenin protein levels is sufficient for tumor development. We will now discuss potential molecular mechanisms responsible for tumor- and metastasis suppression by AJs.

16.6 Mechanisms of AJ Proteins in Preventing Cancer Initiation and Progression

16.6.1 Maintenance of the Epithelial Phenotype and Adhesion-Mediated Attenuation of Tumor Invasion

One of the most important functions of the AJs is mediation of strong intercellular adhesion between epithelial cells, which is necessary for the three-dimensional architecture of organs and tissues. This function is mediated by the connection of AJs to the actin cytoskeleton, which helps to generate the forces necessary to drive the membranes of neighboring cells together and promote cell–cell adhesion (Cavey and Lecuit 2009; Meng and Takeichi 2009). During epithelial tumor development, properly functioning AJs continue to mediate strong cell–cell adhesion between the tumor cells and this should attenuate primary tumor invasion, which is necessary for tumor progression and dissemination. Thus, the purely adhesive function of AJ proteins should be able to suppress tumor progression and metastasis by simply preventing the tumor cells from separating from the bulk of the tumor and invading the stromal cell compartment (Fig. 16.2a). It has been assumed that the decrease in expression of E-cadherin and the increase in expression of N-cadherin in tumor cells would reduce the affinity of tumor cells for each other and promote their affinity for mesenchymal cells, expressing N-cadherin. Theoretically, this should promote tumor cell invasion. The recent development of novel experimental techniques that include conditional gene knockouts and gene replacements technologies made it possible to test this hypothesis experimentally in live animals. The experimental results demonstrated that this explanation is probably too simplistic. Ablation of classical cadherins (both E- and P-cadherin) or α-catenin in epidermal keratinocytes in mice causes prominent cell–cell adhesion defects, but does not result in the loss of epithelial cell phenotypes and intermingling of mutant epithelial cells with stromal fibroblasts (Tinkle et al. 2008b; Vasioukhin et al. 2001). Moreover, genetic replacement of E-cadherin with N-cadherin in intestinal epithelial cells does not result in disruption of tissue morphology and intermingling of N-cadherin-expressing epithelial cells with the stromal cells (Libusova et al. 2010). It appears that simple ablation of the AJs in normal epithelial cells *in vivo* does not result in a complete loss of the epithelial cell phenotype. Since this does happen in many tumor cells, it is likely that AJs do play an important role in the maintenance of epithelial phenotype, but this function is reinforced by some unknown additional mechanisms in the normal cells, and these mechanisms may be lost in epithelial tumors.

16.6.2 Modulation of Growth Factor Receptor Signaling

Cell surface growth factor receptors play a critical role in regulation of cell proliferation and many of these proteins function as proto-oncogenes or tumor-suppressors. Cadherin–catenin protein complexes at the membrane interact with multiple growth factor receptors and this profoundly influences their signaling outputs (Fig. 16.2b). For example, E-cadherin interacts with and negatively regulates signaling of multiple receptor type tyrosine kinases (Qian et al. 2004; Takahashi and Suzuki 1996). Furthermore, some of the mutations of E-cadherin that are found in primary tumors result in decreased binding between E-cadherin and epithelial growth factor receptor (EGFR) and this enhances EGFR activity (Bremm et al. 2008). Interestingly, E-cadherin-mediated adhesion impacts only some, but not all EGFR downstream signaling pathways. For example, clustering of E-cadherin at the cell surface with

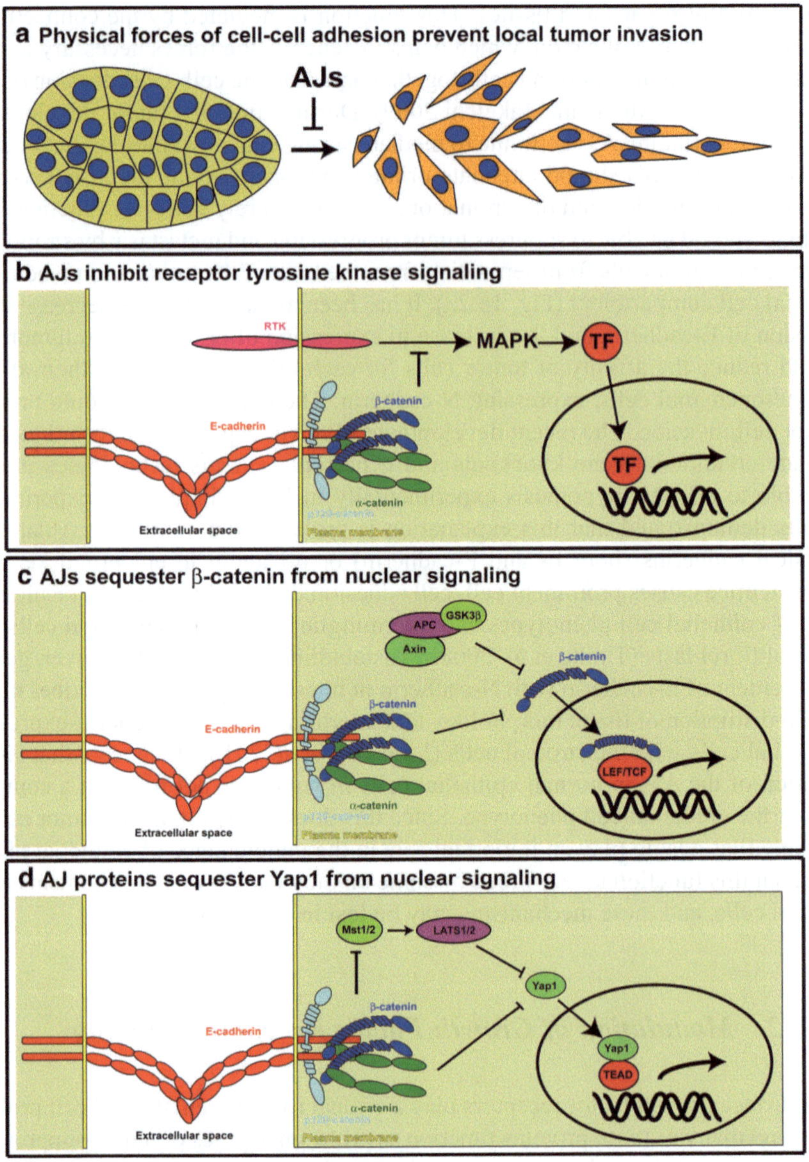

Fig. 16.2 Mechanisms of Adherens Junction function in cancer. **a** AJs in epithelial cells prevent tumor cell dispersion and local tissue invasion, which attenuates tumor progression and metastasis. **b** Epithelial cadherins bind to the variety of tyrosine kinase receptor proteins and negatively regulate their signaling outputs. **c** In the tumor cells with attenuated cytoplasmic β-catenin destruction machinery, cadherin sequesters β-catenin from the nucleus and negatively regulates β-catenin signaling activity. **d** AJ proteins support Hippo pathway signaling by negatively regulating nuclear localization of Yap1

E-cadherin-coated beads specifically impacts EGFR-mediated activation of STAT5 (Perrais et al. 2007). E-cadherin can also regulate signaling of growth factor receptors indirectly through neuronal cell adhesion molecule (NCAM) (Lehembre et al. 2008). Loss of E-cadherin expression results in upregulation of NCAM, its translocation to the lipid rafts, where it activates non-receptor tyrosine kinase Fyn, leading to the phosphorylation and activation of the focal adhesion kinase and the assembly of integrin-mediated focal adhesions, cell spreading and EMT (Lehembre et al. 2008).

In contrast to the relationship between EGFR and E-cadherin, fibroblast growth factor receptor (FGFR) interaction with N-cadherin potentiates FGFR signaling and causes increased activation of MAPK pathway and upregulation of MMP9, which promotes cellular invasion (Suyama et al. 2002). Similarly, interaction between N-cadherin and platelet-derived growth factor-receptor β (PDGFRβ) promotes PDGF signaling-induced cell migration (Theisen et al. 2007).

Cadherin-mediated adhesion can also potentiate specific signaling branches downstream from the cell surface receptor-type tyrosine kinases. For example, VE-cadherin is necessary for proper vascular endothelial growth factor receptor-mediated activation of phosphoinositide 3 (PI3)-kinase (Carmeliet et al. 1999; Kang et al. 2007). Similarly, when Ewing sarcoma cells are grown in anchorage-independent conditions in soft agar, they upregulate E-cadherin, which increases ErbB4-phosphatidylinositol 3-kinase signaling and promotes cell survival (Kang et al. 2007). This PI3-kinase pathway-associated pro-survival function of AJs may play an important role during tumor cell dissemination and formation of metastatic lesions. Indeed, while primary epithelial tumors frequently show the decrease in expression of E-cadherin, the metastatic lesions in the same patients often display the reemergence of E-cadherin expression (Bukholm et al. 2000; Hung et al. 2006; Imai et al. 2004). During later stages of metastasis, cancer cells have to survive in foreign microenvironments, which are often lacking the proper extracellular matrix proteins, that otherwise facilitate formation of integrin-based adhesion structures and associated activation of PI3-kinase signaling. In these conditions, AJs may promote activation of PI3-kinase signaling necessary for cell survival and, thus, promote metastasis.

In addition to cadherins, catenins can also profoundly change the growth factor receptor signaling pathways. p120-catenin-mediated inhibition of RhoA-ROCK pathway is necessary for anchorage-independent growth of MDCK tumor cells overexpressing activated Src or Rac1 proteins (Dohn et al. 2009). Quite surprisingly, in lobular breast cancer developing upon ablation of *E-cadherin* and *p53*, p120-catenin promotes anoikis resistance by indirect upregulation of Rho-Rock signaling (Schackmann et al. 2011). Although it is confusing, p120-catenin is known to have different and some times opposite functions depending on whether the cells express E-cadherin (Soto et al. 2008). Similar to p120-catenin, α-catenin also impacts the signaling by the growth factor receptors. While the mechanisms are still not well understood, the ablation of α-catenin in skin keratinocytes results in increased insulin-like growth factor (IGFR)–Ras-MAPK signaling (Vasioukhin et al. 2001). IGF signaling can activate both Akt and MAPK kinase pathways, but only the MAPK pathway was affected in *αE-catenin*$^{-/-}$ keratinocytes.

16.6.3 Negative Regulation of Oncogenic β-Catenin Signaling

β-catenin is a potent proto-oncogene and upregulation of its levels *in vivo* results in the development of cancer in a variety of organs and tissues (Clevers 2006; Polakis 2000). Since AJs also utilize β-catenin, upregulation of the levels of epithelial cadherins may sequester β-catenin to the AJs and attenuate its transcriptional activity in the nucleus (Fig. 16.2c). This type of relationship has been demonstrated in variety of model systems. Overexpression of E-cadherin and α-catenin in Xenopus and Drosophila embryos and cancer cell lines results in attenuation of β-catenin-mediated signaling (Giannini et al. 2000; Gottardi et al. 2001; Heasman et al. 1994; Merdek et al. 2004; Onder et al. 2008; Orsulic et al. 1999; Sanson et al. 1996; Sehgal et al. 1997; Simcha et al. 1998). Similarly, inactivation of epithelial cadherins in many cancer cell lines with an activated β-catenin signaling pathway results in increased β-catenin signaling (Kuphal and Behrens 2006; Onder et al. 2008). β-catenin is also hyperactive in *E-cadherin*$^{-/-}$ embryonic stem cells (Orsulic et al. 1999). In contrast, inactivation of cadherins in other cell lines does not cause increase in β-catenin-mediated signaling (Kuphal and Behrens 2006; van de Wetering et al. 2001). Similarly, loss-of-function experiments involving epithelial cadherins and α-catenin in live organisms for the most part do not show impacts on β-catenin signaling. Tissue-specific inactivation of *E-cadherin* and *α-catenin* in keratinocytes, as well as the knockout of *α-catenin* in the developing brain do not cause increases in β-catenin signaling (Lien et al. 2006; Vasioukhin et al. 2001; Young et al. 2003). Moreover, genetic inactivation of epithelial cadherins in mouse models of epithelial cancer often promotes tumor progression, but rarely causes concomitant increase in β-catenin signaling. For example, loss of E-cadherin promotes pancreatic cancer progression, but it causes no changes in β-catenin signaling (Herzig et al. 2007). Similarly, deletion of E-cadherin in the mouse model of breast and skin cancer accelerates tumor development and promotes metastasis, but it does not impact β-catenin signaling (Derksen et al. 2006). Conditional deletion of *αE-catenin* in skin hair follicle stem cells results in the development of inflammatory lesions and squamous cell carcinoma, but β-catenin signaling is not activated in *αE-catenin*$^{-/-}$ cells (Silvis et al. 2011). In contrast, inactivation of cadherin function by overexpression of dominant-negative E-cadherin in a mouse model of Raf-driven lung cancer results in increased β-catenin signaling (Ceteci et al. 2007).

Overall, it appears that sequestration of β-catenin to the AJs can attenuate β-catenin nuclear signaling, but this is especially evident in cells where the β-catenin destruction machinery is inactivated and the excess of β-catenin is not efficiently cleared by degradation (Heuberger and Birchmeier 2010).

16.6.4 Positive Regulation of the Tumor-Suppressive Hippo Signaling

Initially discovered in Drosophila, the Hippo pathway in mammalian organisms regulates the size of the organs, and protects them from tumor development (Pan 2010; Zhao et al. 2011). The canonical Hippo pathway consists of a kinase cascade that culminates in phosphorylation of transcriptional co-activators Yap1 and Taz (*WWTR1*) (Fig. 16.2d). Activation of Hippo signaling results in phosphorylation and degradation of Yap and Taz. Decreased Hippo pathway activity results in nuclear translocation of Yap and Taz, interaction with TEADs, as well as several other transcription factors, and transcriptional regulation of genes involved in proliferation, differentiation and apoptotic cell death. Constitutive activation of Yap1 or Taz signaling in mammary epithelial cells results in cell transformation (Chan et al. 2008; Dong et al. 2007; Overholtzer et al. 2006). Moreover, tissue specific activation of Yap1 in liver and skin progenitors results in development of liver cancer and skin squamous cell carcinoma (Dong et al. 2007; Schlegelmilch et al. 2011). Interestingly, Hippo signaling is upregulated by increased cell density and it appears to be a critical pathway regulating contact inhibition of cell proliferation (Zhao et al. 2007). This is intriguing because contact inhibition of cell proliferation is regulated by cadherins and catenins (Takahashi and Suzuki 1996; Vasioukhin et al. 2001).

Several recent studies discovered a functional connection between the Hippo pathway and the AJ protein α-catenin (Schlegelmilch et al. 2011; Silvis et al. 2011). Mass Spectrometry analysis identified α-catenin as a prominent Yap1 interacting partner in keratinocytes and the loss of α-catenin expression resulted in constitutive nuclear localization and activation of Yap1 (Schlegelmilch et al. 2011). Moreover, constitutively nuclear Yap1 was necessary for tumor formation associated with loss of α-catenin in keratinocytes (Silvis et al. 2011). Interestingly, in keratinocytes α-catenin did not regulate Mst1/2 and Lats1/2, the canonical kinases of the Hippo pathway. Instead, it interacted with Yap1 and this interaction attenuated its nuclear translocation (Schlegelmilch et al. 2011; Silvis et al. 2011). In addition, this functional impact on Yap1 localization was specific for α-catenin, because the knockdowns of epithelial cadherins did not affect Yap1 localization (Schlegelmilch et al. 2011). This is consistant with the prominent phenotypic differences between α-catenin-null and E-/P-cadherin-null epidermises. While both show disruption of the AJs, only ablation of *α-catenin* results in epidermal hyperplasia (Tinkle et al. 2008a; Vasioukhin et al. 2001). In contrast to keratinocytes, disruption of AJs is sufficient for nuclear translocation of Yap1 in breast epithelial MCF10 A cells (Kim et al. 2011). In these cells, AJ activation of the canonical Hippo kinase cascade is responsible for the contact inhibition of cell proliferation. The mechanisms responsible for the connection between AJs and the Hippo kinases in MCF10 A cells are not well understood. It is quite intriguing that the tumor suppressor NF2 is implicated in regulation of the Hippo signaling (Zhang et al. 2010), and it is also a direct interactor of α-catenin (Gladden et al. 2010). Future research will help to determine whether NF2-α-catenin interaction is an important mechanistic link between the AJs and the Hippo signaling pathway.

16.7 Summary and Future Perspectives

AJs play a critical role in human epithelial tumors. Loss of epithelial cadherins and disruption of AJs is an early and causal event in lobular breast carcinoma and diffuse type gastric cancer. In other tumor types, disruption of the AJs usually happens during the transition from low-grade well-differentiated tumors to high-grade poorly differentiated invasive cancer. Loss of epithelial cadherin usually correlates with poor patient outcome, and mechanistic experiments in cell lines and model organisms demonstrated a causal role of AJs in tumor initiation and progression. While significant progress has been achieved in elucidation of the signaling events responsible for the tumor and metastasis suppressor function of cadherins and catenins, the available knowledge is still fragmented and quite rudimentary. This direction represents an exciting avenue for the future research.

References

Aaltomaa S, Lipponen P, Ala-Opas M, Eskelinen M, Kosma VM (1999) Alpha-catenin expression has prognostic value in local and locally advanced prostate cancer. Br J Cancer 80:477–482

Aaltomaa S, Karja V, Lipponen P, Isotalo T, Kankkunen JP, Talja M, Mokka R (2005) Reduced alpha- and beta-catenin expression predicts shortened survival in local prostate cancer. Anticancer Res 25:4707–4712

Abe K, Takeichi M (2008) EPLIN mediates linkage of the cadherin catenin complex to F-actin and stabilizes the circumferential actin belt. Proc Natl Acad Sci U S A 105:13–19

Aberle H, Bauer A, Stappert J, Kispert A, Kemler R (1997) Beta-catenin is a target for the ubiquitin-proteasome pathway. Embo J 16:3797–3804

Anastasiadis PZ, Reynolds AB (2001) Regulation of Rho GTPases by p120-catenin. Curr Opin Cell Biol 13:604–610

Angres B, Barth A, Nelson WJ (1996) Mechanism for transition from initial to stable cell-cell adhesion: kinetic analysis of E-cadherin-mediated adhesion using a quantitative adhesion assay. J Cell Biol 134:549–557

Asgeirsson KS, Jonasson JG, Tryggvadottir L, Olafsdottir K, Sigurgeirsdottir JR, Ingvarsson S, Ogmundsdottir HM (2000) Altered expression of E-cadherin in breast cancer. patterns, mechanisms and clinical significance. Eur J Cancer 36:1098–1106

Batlle E, Sancho E, Franci C Dominguez D, Monfar M, Baulida J, Garcia De Herreros A (2000) The transcription factor snail is a repressor of E-cadherin gene expression in epithelial tumour cells. Nat Cell Biol 2:84–89

Becker KF, Atkinson MJ, Reich U Huang HH, Nekarda H, Siewert JR, Hofler H (1993) Exon skipping in the E-cadherin gene transcript in metastatic human gastric carcinomas. Hum Mol Genet 2:803–804

Becker KF, Atkinson MJ, Reich U, Becker I, Nekarda H, Siewert JR, Hofler H (1994) E-cadherin gene mutations provide clues to diffuse type gastric carcinomas. Cancer Res 54:3845–3852

Behrens J, Mareel MM, Van Roy FM, Birchmeier W (1989) Dissecting tumor cell invasion: epithelial cells acquire invasive properties after the loss of uvomorulin-mediated cell-cell adhesion. J Cell Biol 108:2435–2447

Behrens J, Vakaet L, Friis R, Winterhager E, Van Roy F, Mareel MM, Birchmeier W (1993) Loss of epithelial differentiation and gain of invasiveness correlates with tyrosine phosphorylation of the E-cadherin/beta-catenin complex in cells transformed with a temperature-sensitive v-SRC gene. J Cell Biol 120:757–766

Berx G, Van Roy F (2001) The E-cadherin/catenin complex: an important gatekeeper in breast cancer tumorigenesis and malignant progression. Breast Cancer Res 3:289–293

Berx G, van Roy F (2009) Involvement of members of the cadherin superfamily in cancer. Cold Spring Harb Perspect Biol 1:a003129

Berx G, Cleton-Jansen AM, Nollet F, de Leeuw WJ, van de Vijver M, Cornelisse C, van Roy F (1995) E-cadherin is a tumour/invasion suppressor gene mutated in human lobular breast cancers. Embo J 14:6107–6115

Berx G, Cleton-Jansen AM, Strumane K, de Leeuw WJ, Nollet F,van Roy F, Cornelisse C(1996) E-cadherin is inactivated in a majority of invasive human lobular breast cancers by truncation mutations throughout its extracellular domain. Oncogene 13:1919–1925

Bissell MJ, Hines WC (2011) Why don't we get more cancer? A proposed role of the microenvironment in restraining cancer progression. Nat Med 17:320–329

Bohm M, Totzeck B, Birchmeier W, Wieland I (1994) Differences of E-cadherin expression levels and patterns in primary and metastatic human lung cancer. Clin Exp Metastasis 12:55–62

Brembeck FH, Schwarz-Romond T, Bakkers J, Wilhelm S, Hammerschmidt M, Birchmeier W (2004) Essential role of BCL9–2 in the switch between beta-catenin's adhesive and transcriptional functions. Genes Dev 18:2225–2230

Bremm A, Walch A, Fuchs M, Mages J, Duyster J, Keller G, Hermannstadter C, Becker KF, Rauser S, Langer R, von Weyhern CH, Hofler H, Luber B (2008) Enhanced activation of epidermal growth factor receptor caused by tumor-derived E-cadherin mutations. Cancer Res 68:707–714

Bukholm IK, Nesland JM, Borresen-Dale AL (2000) Re-expression of E-cadherin, alpha-catenin and beta-catenin, but not of gamma-catenin, in metastatic tissue from breast cancer patients (seecomments). J Pathol 190:15–19

Bullions LC, Notterman DA, Chung LS, Levine AJ (1997) Expression of wild-type alpha-catenin protein in cells with a mutant alpha-catenin gene restores both growth regulation and tumor suppressor activities. Mol Cell Biol 17:4501–4508

Burk U, Schubert J, Wellner U, Schmalhofer O, Vincan E, Spaderna S, Brabletz T (2008) A reciprocal repression between ZEB1 and members of the miR-200 family promotes EMT and invasion in cancer cells. EMBO Rep 9:582–589

Cano A, Perez-Moreno MA, Rodrigo I, Locascio A, Blanco MJ, del Barrio MG, Portillo F, Nieto MA (2000) The transcription factor snail controls epithelial-mesenchymal transitions by repressing E-cadherin expression. Nat Cell Biol 2:76–83

Canonici A, Steelant W, Rigot V, Khomitch-Baud A, Boutaghou-Cherid H, Bruyneel E, Van Roy F, Garrouste F, Pommier G, Andre F (2008) Insulin-like growth factor-I receptor, E-cadherin and alpha v integrin form a dynamic complex under the control of alpha-catenin. Int J Canc 122:572–582

Cao Q, Yu J, Dhanasekaran SM, Kim JH, Mani RS, Tomlins SA, Mehra R, Laxman B, Cao X, Kleer CG, Varambally S, Chinnaiyan AM (2008) Repression of E-cadherin by the polycomb group protein EZH2 in cancer. Oncogene 27:7274–7284

Carico E, Atlante M, Bucci B, Nofroni I, Vecchione A (2001) E-cadherin and alpha-catenin expression during tumor progression of cervical carcinoma. Gynecol Oncol 80:156–161

Carmeliet P, Lampugnani MG, Moons L, Breviario F, Compernolle V, Bono F, Balconi G, Spagnuolo R, Oostuyse B, Dewerchin M, Zanetti A, Angellilo A, Mattot V, Nuyens D, Lutgens E, Clotman F, de Ruiter MC, Gittenberger-de Groot A, Poelmann R, Lupu F, Herbert JM, Collen D, Dejana E (1999) Targeted deficiency or cytosolic truncation of the VE-cadherin gene in mice impairs VEGF-mediated endothelial survival and angiogenesis. Cell 98:147–157

Castano J, Solanas G, Casagolda D, Raurell I, Villagrasa P, Bustelo XR, Garcia de Herreros A, Dunach M (2007) Specific phosphorylation of p120-catenin regulatory domain differently modulates its binding to RhoA. Mol Cell Biol 27:1745–1757

Cavallaro U, Schaffhauser B, Christofori G (2002) Cadherins and the tumour progression: is it all in a switch? Cancer Lett 176:123–128

Cavey M, Lecuit T (2009) Molecular bases of cell-cell junctions stability and dynamics. Cold Spring Harb Perspect Biol 1:a002998

Ceteci F, Ceteci S, Karreman C, Kramer BW, Asan E, Gotz R, Rapp UR (2007) Disruption of tumor cell adhesion promotes angiogenic switch and progression to micrometastasis in RAF-driven murine lung cancer. Cancer Cell 12:145–159

Chan EF, Gat U, McNiff JM, Fuchs E (1999) A common human skin tumour is caused by activating mutations in beta-catenin. Nat Genet 21:410–413

Chan SW, Lim CJ, Guo K, Ng CP, Lee I, Hunziker W, Zeng Q, Hong W (2008) A role for TAZ in migration, invasion, and tumorigenesis of breast cancer cells. Cancer Res 68:2592–2598

Chaudhury A, Hussey GS, Ray PS, Jin G, Fox PL, Howe PH (2010) TGF-beta-mediated phosphorylation of hnRNP E1 induces EMT via transcript-selective translational induction of Dab2 and ILEI. Nat Cell Biol 12:286–293

Chen CL, Liu SS, Ip SM, Wong LC, Ng TY, Ngan HY (2003a) E-cadherin expression is silenced by DNA methylation in cervical cancer cell lines and tumours. Eur J Cancer 39:517–523

Chen WC, Obrink B (1991) Cell-cell contacts mediated by E-cadherin (uvomorulin) restrict invasive behavior of L-cells. J Cell Biol 114:319–327

Chen X, Kojima S, Borisy GG, Green KJ (2003b) p120 catenin associates with kinesin and facilitates the transport of cadherin-catenin complexes to intercellular junctions. J Cell Biol 163:547–557

Chen ZL, Zhao XH, Wang JW, Li BZ, Wang Z, Sun J, Tan FW, Ding DP, Xu XH, Zhou F, Tan XG, Hang J, Shi SS, Feng XL, He J (2011) microRNA-92a promotes lymph node metastasis of human esophageal squamous cell carcinoma via E-cadherin. J Biol Chem 286:10725–10734

Chung Y, Lam AK, Luk JM, Law S, Chan KW, Lee PY, Wong J (2007) Altered E-cadherin expression and p120 catenin localization in esophageal squamous cell carcinoma. Ann Surg Oncol 14:3260–3267

Cleton-Jansen AM, Callen DF, Seshadri R, Goldup S, McCallum B, Crawford J, Powell JA, Settasatian C, van Beerendonk H, Moerland EW, Smit VT, Harris WH, Millis R, Morgan NV, Barnes D, Mathew CG, Cornelisse CJ (2001) Loss of heterozygosity mapping at chromosome arm 16q in 712 breast tumors reveals factors that influence delineation of candidate regions. Cancer Res 61:1171–1177

Clevers H (2004) Wnt breakers in colon cancer. Cancer Cell 5:5–6

Clevers H (2006) Wnt/beta-catenin signaling in development and disease. Cell 127:469–480

Coluccia AM, Benati D, Dekhil H, De Filippo A, Lan C, Gambacorti-Passerini C (2006) SKI-606 decreases growth and motility of colorectal cancer cells by preventing pp60(c-Src)-dependent tyrosine phosphorylation of beta-catenin and its nuclear signaling. Cancer Res 66:2279–2286

Coluccia AM, Vacca A, Dunach M, Mologni L, Redaelli S, Bustos VH, Benati D, Pinna LA, Gambacorti-Passerini C (2007) Bcr-Abl stabilizes beta-catenin in chronic myeloid leukemia through its tyrosine phosphorylation. Embo J 26:1456–1466

Coman D (1944) Decreased mutual adhesiveness, a property of cells from squamous cell carcinomas. Cancer Res 4:625–629

Comijn J, Berx G, Vermassen P, Verschueren K, van Grunsven L, Bruyneel E, Mareel M, Huylebroeck D, van Roy F (2001) The two-handed E box binding zinc finger protein SIP1 downregulates E-cadherin and induces invasion. Mol Cell 7:1267–1278

Cowin P, Rowlands TM, Hatsell SJ (2005) Cadherins and catenins in breast cancer. Curr Opin Cell Biol 17:499–508

Damsky CH, Richa J, Solter D, Knudsen K, Buck CA (1983) Identification and purification of a cell surface glycoprotein mediating intercellular adhesion in embryonic and adult tissue. Cell 34:455–466

Davis MA, Ireton RC, Reynolds AB (2003) A core function for p120-catenin in cadherin turnover. J Cell Biol 163:525–534

de La Coste A, Romagnolo B Billuart P, Renard CA, Buendia MA, Soubrane O, Fabre M, Chelly J, Beldjord C, Kahn A, Perret C (1998) Somatic mutations of the beta-catenin gene are frequent in mouse and human hepatocellular carcinomas. Proc Natl Acad Sci U S A 95:8847–8851

Derksen PW, Liu X, Saridin F, Van Der Gulden H, Zevenhoven J, Evers B, van Beijnum JR, Griffioen AW, Vink J, Krimpenfort P, Peterse JL, Cardiff RD, Berns A, Jonkers J (2006) Somatic inactivation of E-cadherin and p53 in mice leads to metastatic lobular mammary carcinoma through induction of anoikis resistance and angiogenesis. Cancer Cell 10:437–449

Ding L, Ellis MJ, Li S, Larson DE, Chen K, Wallis JW, Harris CC, McLellan MD, Fulton RS, Fulton LL, Abbott RM, Hoog J, Dooling DJ, Koboldt DC, Schmidt H, Kalicki J, Zhang Q, Chen L, Lin L, Wendl MC, McMichael JF, Magrini VJ, Cook L, McGrath SD, Vickery TL, Appelbaum E, Deschryver K, Davies S, Guintoli T, Crowder R, Tao Y, Snider JE, Smith SM, Dukes AF, Sanderson GE, Pohl CS, Delehaunty KD, Fronick CC, Pape KA, Reed JS, Robinson JS, Hodges JS, Schierding W, Dees ND, Shen D, Locke DP, Wiechert ME, Eldred JM, Peck JB, Oberkfell BJ, Lolofie JT, Du F, Hawkins AE, O'Laughlin MD, Bernard KE, Cunningham M, Elliott G, Mason MD, Thompson DM Jr, Ivanovich JL, Goodfellow PJ, Perou CM, Weinstock GM, Aft R, Watson M, Ley TJ, Wilson RK, Mardis ER (2010) Genome remodelling in a basal-like breast cancer metastasis and xenograft. Nature 464:999–1005

Dohn MR, Brown MV, Reynolds AB (2009) An essential role for p120-catenin in Src- and Rac1-mediated anchorage-independent cell growth. J Cell Biol 184:437–450

Dolled-Filhart M, McCabe A, Giltnane J, Cregger M, Camp RL, Rimm DL (2006) Quantitative in situ analysis of beta-catenin expression in breast cancer shows decreased expression is associated with poor outcome. Cancer Res 66:5487–5494

Dong J, Feldmann G, Huang J, Wu S, Zhang N, Comerford SA, Gayyed MF, Anders RA, Maitra A, Pan D (2007) Elucidation of a universal size-control mechanism in Drosophila and mammals. Cell 130:1120–1133

Drees F, Pokutta S, Yamada S, Nelson WJ, Weis WI (2005) Alpha-catenin is a molecular switch that binds E-cadherin-beta-catenin and regulates actin-filament assembly. Cell 123:903–915

El-Bahrawy M, El-Masry N, Alison M, Poulsom R,Fallowfield M (2003) Expression of beta-catenin in basal cell carcinoma. Br J Dermatol Syph 148:964–970

Ellenbroek SI, Collard JG (2007) Rho GTPases: functions and association with cancer. Clin Exp Metastasis 24:657–672

Endo K, Ashida K, Miyake N, Terada T (2001) E-cadherin gene mutations in human intrahepatic cholangiocarcinoma. J Pathol 193:310–317

Ewing CM, Ru N, Morton RA, Robinson JC, Wheelock MJ, Johnson KR, Barrett JC, Isaacs WB (1995) Chromosome 5 suppresses tumorigenicity of PC3 prostate cancer cells: correlation with re-expression of alpha-catenin and restoration of E-cadherin function. Cancer Res 55:4813–4817

Fang D, Hawke D, Zheng Y, Xia Y, Meisenhelder J, Nika H, Mills GB, Kobayashi R, Hunter T, Lu Z (2007) Phosphorylation of beta-catenin by AKT promotes beta-catenin transcriptional activity. J Biol Chem 282:11221–11229

Ferber EC, Kajita M, Wadlow A, Tobiansky L, Niessen C, Ariga H, Daniel J, Fujita Y (2008) A role for the cleaved cytoplasmic domain of e-cadherin in the nucleus. J Biol Chem 283:12691–12700

Fevr T, Robine S, Louvard D, Huelsken J (2007) Wnt/beta-catenin is essential for intestinal homeostasis and maintenance of intestinal stem cells. Mol Cell Biol 27:7551–7559

Frixen UH, Behrens J, Sachs M, Eberle G, Voss B, Warda A, Lochner D, Birchmeier W (1991) E-cadherin-mediated cell-cell adhesion prevents invasiveness of human carcinoma cells. J Cell Biol 113:173–185

Fujita Y, Krause G, Scheffner M, Zechner D, Leddy HE, Behrens J, Sommer T, Birchmeier W (2002) Hakai, a c-Cbl-like protein, ubiquitinates and induces endocytosis of the E-cadherin complex. Nat Cell Biol 4:222–231

Fukumaru K, Yoshii N, Kanzaki T, Kanekura T (2007) Immunohistochemical comparison of beta-catenin expression by human normal epidermis and epidermal tumors. J Dermatol 34:746–753

Garcia-Rostan G, Tallini G, Herrero A, D'Aquila TG, Carcangiu ML, Rimm DL (1999) Frequent mutation and nuclear localization of beta-catenin in anaplastic thyroid carcinoma. Cancer Res 59:1811–1815

Garinis GA, Menounos PG, Spanakis NE, Papadopoulos K, Karavitis G, Parassi I, Christeli E, Patrinos GP, Manolis EN, Peros G (2002) Hypermethylation-associated transcriptional silencing of E-cadherin in primary sporadic colorectal carcinomas. J Pathol 198:442–449

Gat U, DasGupta R, Degenstein L, Fuchs E (1998) De Novo hair follicle morphogenesis and hair tumors in mice expressing a truncated beta-catenin in skin. Cell 95:605–614

Giannini AL, Vivanco M, Kypta RM (2000) alpha-catenin inhibits beta-catenin signaling by preventing formation of a beta-catenin*T-cell factor*DNA complex. J Biol Chem 275:21883–21888

Gibson P, Tong Y, Robinson G, Thompson MC, Currle DS, Eden C, Kranenburg TA, Hogg T, Poppleton H, Martin J, Finkelstein D, Pounds S, Weiss A, Patay Z, Scoggins M, Ogg R, Pei Y, Yang ZJ, Brun S, Lee Y, Zindy F, Lindsey JC, Taketo MM, Boop FA, Sanford RA, Gajjar A, Clifford SC, Roussel MF, McKinnon PJ, Gutmann DH, Ellison DW, Wechsler-Reya R, Gilbertson RJ (2010) Subtypes of medulloblastoma have distinct developmental origins. Nature 468:1095–1099

Gillett CE, Miles DW, Ryder K, Skilton D, Liebman RD, Springall RJ, Barnes DM, Hanby AM (2001) Retention of the expression of E-cadherin and catenins is associated with shorter survival in grade III ductal carcinoma of the breast. J Pathol Bacteriol 193:433–441

Gimond C, Van Der Flier A, van Delft S, Brakebusch C, Kuikman I, Collard JG, Fassler R, Sonnenberg A (1999) Induction of cell scattering by expression of beta1 integrins in beta1-deficient epithelial cells requires activation of members of the rho family of GTPases and downregulation of cadherin and catenin function. J Cell Biol 147:1325–1340

Giroldi LA, Bringuier PP, de Weijert M, Jansen C, van Bokhoven A, Schalken JA (1997) Role of E boxes in the repression of E-cadherin expression. Biochem Biophys Res Commun 241:453–458

Gladden AB, Hebert AM, Schneeberger EE, McClatchey AI (2010) The NF2 tumor suppressor, Merlin, regulates epidermal development through the establishment of a junctional polarity complex. Dev Cell 19:727–739

Gottardi CJ, Wong E, Gumbiner BM (2001) E-cadherin suppresses cellular transformation by inhibiting beta-catenin signaling in an adhesion-independent manner. J Cell Biol 153:1049–1060

Grady WM, Willis J, Guilford PJ, Dunbier AK, Toro TT, Lynch H, Wiesner G, Ferguson K, Eng C, Park JG, Kim SJ, Markowitz S (2000) Methylation of the CDH1 promoter as the second genetic hit in hereditary diffuse gastric cancer. Nat Genet 26:16–17

Graff JR, Herman JG, Lapidus RG, Chopra H, Xu R, Jarrard DF, Isaacs WB, Pitha PM, Davidson NE, Baylin SB (1995) E-cadherin expression is silenced by DNA hypermethylation in human breast and prostate carcinomas. Cancer Res 55:5195–5199

Graff JR, Greenberg VE, Herman JG, Westra WH, Boghaert ER, Ain KB, Saji M, Zeiger MA, Zimmer SG, Baylin SB (1998) Distinct patterns of E-cadherin CpG island methylation in papillary, follicular, Hurthle's cell, and poorly differentiated human thyroid carcinoma. Cancer Res 58:2063–2066

Gregory PA, Bert AG, Paterson EL, Barry SC, Tsykin A, Farshid G, Vadas MA, Khew-Goodall Y, Goodall GJ (2008) The miR-200 family and miR-205 regulate epithelial to mesenchymal transition by targeting ZEB1 and SIP1. Nat Cell Biol 10:593–601

Grooteclaes ML, Frisch SM (2000) Evidence for a function of CtBP in epithelial gene regulation and anoikis. Oncogene 19:3823–3828

Gross JC, Schreiner A, Engels K, Starzinski-Powitz A (2009) E-cadherin surface levels in epithelial growth factor-stimulated cells depend on adherens junction protein shrew-1. Mol Biol Cell 20:3598–3607

Guilford P, Hopkins J, Harraway J, McLeod M, McLeod N, Harawira P, Taite H, Scoular R, Miller A, Reeve AE (1998) E-cadherin germline mutations in familial gastric cancer. Nature 392:402–405

Gujral TS, van Veelen W, Richardson DS, Myers SM, Meens JA, Acton DS, Dunach M, Elliott BE, Hoppener JW, Mulligan LM (2008) A novel RET kinase-beta-catenin signaling pathway contributes to tumorigenesis in thyroid carcinoma. Cancer Res 68:1338–1346

Haegel H, Larue L, Ohsugi M, Fedorov L, Herrenknecht K, Kemler R (1995) Lack of beta-catenin affects mouse development at gastrulation. Development 121:3529–3537

Hamaguchi M, Matsuyoshi N, Ohnishi Y, Gotoh B, Takeichi M, Nagai Y (1993) p60v-src causes tyrosine phosphorylation and inactivation of the N-cadherin-catenin cell adhesion system. Embo J 12:307–314

Heasman J, Crawford A, Goldstone K, Garner-Hamrick P, Gumbiner B, McCrea P, Kintner C, Noro CY, Wylie C (1994) Overexpression of cadherins and underexpression of beta-catenin inhibit dorsal mesoderm induction in early Xenopus embryos. Cell 79:791–803

Heimann R, Lan F, McBride R, Hellman S (2000) Separating favorable from unfavorable prognostic markers in breast cancer: the role of E-cadherin. Cancer Res 60:298–304

Herranz N, Pasini D, Diaz VM, Franci C, Gutierrez A, Dave N, Escriva M, Hernandez-Munoz I, Di Croce L, Helin K, Garcia de Herreros A, Peiro S (2008) Polycomb complex 2 is required for E-cadherin repression by the Snail1 transcription factor. Mol Cell Biol 28:4772–4781

Herzig M, Savarese F, Novatchkova M, Semb H, Christofori G (2007) Tumor progression induced by the loss of E-cadherin independent of beta-catenin/Tcf-mediated Wnt signaling. Oncogene 26:2290–2298

Heuberger J, Birchmeier W (2010) Interplay of cadherin-mediated cell adhesion and canonical Wnt signaling. Cold Spring Harb Perspect Biol 2:a002915

Hino S, Tanji C, Nakayama KI, Kikuchi A (2005) Phosphorylation of beta-catenin by cyclic AMP-dependent protein kinase stabilizes beta-catenin through inhibition of its ubiquitination. Mol Cell Biol 25:9063–9072

Hirano S, Kimoto N, Shimoyama Y, Hirohashi S, Takeichi M (1992) Identification of a neural alpha-catenin as a key regulator of cadherin function and multicellular organization. Cell 70:293–301

Hiscox S, Jiang WG (1999) Association of the HGF/SF receptor, c-met, with the cell-surface adhesion molecule, E-cadherin, and catenins in human tumor cells. Biochem Biophys Res Commun 261:406–411

Huang Q, Gumireddy K, Schrier M, le Sage C, Nagel R, Nair S, Egan DA, Li A, Huang G, Klein-Szanto AJ, Gimotty PA, Katsaros D, Coukos G, Zhang L, Pure E, Agami R (2008) The microRNAs miR-373 and miR-520c promote tumour invasion and metastasis. Nat Cell Biol 10:202–210

Huelsken J, Vogel R, Brinkmann V, Erdmann B, Birchmeier C, Birchmeier W (2000) Requirement for beta-catenin in anterior-posterior axis formation in mice. J Cell Biol 148:567–578

Hugo H, Ackland ML, Blick T, Lawrence MG, Clements JA, Williams ED, Thompson EW (2007) Epithelial–mesenchymal and mesenchymal--epithelial transitions in carcinoma progression. J Cell Physiol 213:374–383

Hulit J, Suyama K, Chung S, Keren R, Agiostratidou G, Shan W, Dong X, Williams TM, Lisanti MP, Knudsen K, Hazan RB (2007) N-cadherin signaling potentiates mammary tumor metastasis via enhanced extracellular signal-regulated kinase activation. Cancer Res 67:3106–3116

Hung KF, Chang CS, Liu CJ, Lui MT, Cheng CY, Kao SY (2006) Differential expression of E-cadherin in metastatic lesions comparing to primary oral squamous cell carcinoma. J Oral Pathol Med 35:589–594

Hussey GS, Chaudhury A, Dawson AE, Lindner DJ, Knudsen CR, Wilce MC, Merrick WC, Howe PH (2011) Identification of an mRNP complex regulating tumorigenesis at the translational elongation step. Mol Cell 41:419–431

Hwang-Verslues WW, Chang PH, Wei PC, Yang CY, Huang CK, Kuo WH, Shew JY, Chang KJ, Lee EY, Lee WH (2011) miR-495 is upregulated by E12/E47 in breast cancer stem cells, and promotes oncogenesis and hypoxia resistance via downregulation of E-cadherin and REDD1. Oncogene 30:2463–2474

Iliopoulos D, Lindahl-Allen M, Polytarchou C, Hirsch HA, Tsichlis PN, Struhl K (2010) Loss of miR-200 inhibition of Suz12 leads to polycomb-mediated repression required for the formation and maintenance of cancer stem cells. Mol Cell 39:761–772

Imai T, Horiuchi A, Shiozawa T, Osada R, Kikuchi N, Ohira S, Oka K, Konishi I (2004) Elevated expression of E-cadherin and alpha-, beta-, and gamma-catenins in metastatic lesions compared with primary epithelial ovarian carcinomas. Hum Pathol 35:1469–1476

Ishiyama N, Lee SH, Liu S, Li GY, Smith MJ, Reichardt LF, Ikura M (2010) Dynamic and static interactions between p120 catenin and E-cadherin regulate the stability of cell-cell adhesion. Cell 141:117–128

Ishizaki Y, Omori Y, Momiyama M, Nishikawa Y, Tokairin T, Manabe M, Enomoto K (2004) Reduced expression and aberrant localization of p120catenin in human squamous cell carcinoma of the skin. J Dermatol Sci. 34:99–108

Jaggi M, Johansson SL, Baker JJ, Smith LM, Galich A, Balaji KC (2005) Aberrant expression of E-cadherin and beta-catenin in human prostate cancer. Urol Oncol 23:402–406

Janssens B, Goossens S, Staes K, Gilbert B, van Hengel J, Colpaert C, Bruyneel E, Mareel M, van Roy F (2001) alphaT-catenin: a novel tissue-specific beta-catenin-binding protein mediating strong cell-cell adhesion. J Cell Sci 114:3177–3188

Jeschke U, Mylonas I, Kuhn C, Shabani N, Kunert-Keil C, Schindlbeck C, Gerber B, Friese K (2007) Expression of E-cadherin in human ductal breast cancer carcinoma in situ, invasive carcinomas, their lymph node metastases, their distant metastases, carcinomas with recurrence and in recurrence. Anticancer Res 27:1969–1974

Kadowaki T, Shiozaki H, Inoue M, Tamura S, Oka H, Doki Y, Iihara K, Matsui S, Iwazawa T, Nagafuchi A, et al (1994) E-cadherin and alpha-catenin expression in human esophageal cancer. Cancer Res 54:291–296

Kang HG, Jenabi JM, Zhang J, Keshelava N, Shimada H, May WA, Ng T, Reynolds CP, Triche TJ, Sorensen PH (2007) E-cadherin cell-cell adhesion in ewing tumor cells mediates suppression of anoikis through activation of the ErbB4 tyrosine kinase. Cancer Res 67:3094–3105

Kim DS, Kim MJ, Lee JY, Kim YZ, Kim EJ, Park JY (2007) Aberrant methylation of E-cadherin and H-cadherin genes in nonsmall cell lung cancer and its relation to clinicopathologic features. Cancer 110:2785–2792

Kim NG, Koh E, Chen X, Gumbiner BM (2011) E-cadherin mediates contact inhibition of proliferation through Hippo signaling-pathway components. Proc Natl Acad Sci U S A 108:11930–11935

Kimura K, Endo Y, Yonemura Y, Heizmann CW, Schafer BW, Watanabe Y, Sasaki T (2000) Clinical significance of S100A4 and E-cadherin-related adhesion molecules in non-small cell lung cancer. Int J Oncol 16:1125–1131

Klaus A, Birchmeier W (2008) Wnt signalling and its impact on development and cancer. Nature reviews. Cancer 8:387–398

Knudsen KA, Sauer C, Johnson KR, Wheelock MJ (2005) Effect of N-cadherin misexpression by the mammary epithelium in mice. J Cell Biochem 95:1093–1107

Kobielak A, Fuchs E (2006) Links between alpha-catenin, NF-kappaB, and squamous cell carcinoma in skin. Proc Natl Acad Sci U S A 103:2322–2327

Kobielak A, Pasolli HA, Fuchs E (2004) Mammalian formin-1 participates in adherens junctions and polymerization of linear actin cables. Nat Cell Biol 6:21–30

Koesters R, Ridder R, Kopp-Schneider A, Betts D, Adams V, Niggli F, Briner J, von Knebel Doeberitz M (1999) Mutational activation of the beta-catenin proto-oncogene is a common event in the development of Wilms' tumors. Cancer Res 59:3880–3882

Koizume S, Tachibana K, Sekiya T, Hirohashi S, Shiraishi M (2002) Heterogeneity in the modification and involvement of chromatin components of the CpG island of the silenced human CDH1 gene in cancer cells. Nucleic Acids Res 30:4770–4780

Kokura K, Sun L, Bedford MT, Fang J (2010) Methyl-H3K9-binding protein MPP8 mediates E-cadherin gene silencing and promotes tumour cell motility and invasion. EMBO J 29:3673–3687

Korpal M, Lee ES, Hu G, Kang Y (2008) The miR-200 family inhibits epithelial-mesenchymal transition and cancer cell migration by direct targeting of E-cadherin transcriptional repressors ZEB1 and ZEB2. J Biol Chem 283:14910–14914

Kovacs EM, Goodwin M, Ali RG, Paterson AD, Yap AS (2002) Cadherin-directed actin assembly: E-cadherin physically associates with the Arp2/3 complex to direct actin assembly in nascent adhesive contacts. Curr Biol 12:379–382

Kowalczyk AP, Reynolds AB (2004) Protecting your tail: regulation of cadherin degradation by p120-catenin. Curr Opin Cell Biol 16:522–527

Kuphal F, Behrens J (2006) E-cadherin modulates Wnt-dependent transcription in colorectal cancer cells but does not alter Wnt-independent gene expression in fibroblasts. Exp Cell Res 312:457–467

Larue L, Ohsugi M, Hirchenhain J, Kemler R (1994) E-cadherin null mutant embryos fail to form a trophectoderm epithelium. Proc Natl Acad Sci U S A 91:8263–8267

Lee M, Vasioukhin V (2008) Cell polarity and cancer--cell and tissue polarity as a non-canonical tumor suppressor. J Cell Sci 121:1141–1150

Lehembre F, Yilmaz M, Wicki A, Schomber T, Strittmatter K, Ziegler D, Kren A, Went P, Derksen PW, Berns A, Jonkers J, Christofori G (2008) NCAM-induced focal adhesion assembly: a functional switch upon loss of E-cadherin. EMBO J 27:2603–2615

Libusova L, Stemmler MP, Hierholzer A, Schwarz H, Kemler R (2010) N-cadherin can structurally substitute for E-cadherin during intestinal development but leads to polyp formation. Development 137:2297–2305

Lien WH, Klezovitch O, Fernandez TE, Delrow J, Vasioukhin V (2006) alphaE-catenin controls cerebral cortical size by regulating the hedgehog signaling pathway. Science 311:1609–1612

Lilien J, Balsamo J (2005) The regulation of cadherin-mediated adhesion by tyrosine phosphorylation/dephosphorylation of beta-catenin. Curr Opin Cell Biol 17:459–465

Lin SY, Xia W, Wang JC, Kwong KY, Spohn B, Wen Y, Pestell RG, Hung MC (2000) Beta-catenin, a novel prognostic marker for breast cancer: its roles in cyclin D1 expression and cancer progression. Proc Natl Acad Sci U S A 97:4262–4266

Lo Muzio L, Pannone G, Mignogna MD, Staibano S, Mariggio MA, Rubini C, Procaccini M, Dolci M, Bufo P, De Rosa G, Piattelli A (2004) P-cadherin expression predicts clinical outcome in oral squamous cell carcinomas. Histol Histopathol 19:1089–1099

Lochter A, Galosy S, Muschler J, Freedman N, Werb Z, Bissell MJ (1997) Matrix metalloproteinase stromelysin-1 triggers a cascade of molecular alterations that leads to stable epithelial-to-mesenchymal conversion and a premalignant phenotype in mammary epithelial cells. J Cell Biol 139:1861–1872

Lozano E, Frasa MA, Smolarczyk K, Knaus UG, Braga VM (2008) PAK is required for the disruption of E-cadherin adhesion by the small GTPase Rac. J Cell Sci 121:933–938

Ma L, Young J, Prabhala H, Pan E, Mestdagh P, Muth D, Teruya-Feldstein J, Reinhardt F, Onder TT, Valastyan S, Westermann F, Speleman F, Vandesompele J, Weinberg RA (2010) miR-9, a MYC/MYCN-activated microRNA, regulates E-cadherin and cancer metastasis. Nat Cell Biol 12:247–256

Machon O, Van Den Bout CJ, Backman M, Kemler R, Krauss S (2003) Role of beta-catenin in the developing cortical and hippocampal neuroepithelium. Neuroscience 122:129–143

Malanchi I, Peinado H, Kassen D, Hussenet T, Metzger D, Chambon P, Huber M, Hohl D, Cano A, Birchmeier W, Huelsken J (2008) Cutaneous cancer stem cell maintenance is dependent on beta-catenin signalling. Nature 452:650–653

Marambaud P, Shioi J Serban G, Georgakopoulos A, Sarner S, Nagy V, Baki L, Wen P, Efthimiopoulos S, Shao Z, Wisniewski T, Robakis NK (2002) A presenilin-1/gamma-secretase cleavage releases the E-cadherin intracellular domain and regulates disassembly of adherens junctions. Embo J 21:1948–1956

Marambaud P, Wen PH Dutt A, Shioi J, Takashima A, Siman R, Robakis NK (2003) A CBP binding transcriptional repressor produced by the PS1/epsilon-cleavage of N-cadherin is inhibited by PS1 FAD mutations. Cell 114:635–645

Maretzky T, Reiss K, Ludwig A, Buchholz J, Scholz F, Proksch E, de Strooper B, Hartmann D, Saftig P (2005) ADAM10 mediates E-cadherin shedding and regulates epithelial cell-cell adhesion, migration, and beta-catenin translocation. Proc Natl Acad Sci U S A 102:9182–9187

Matsumura T, Makino R, Mitamura K (2001) Frequent down-regulation of E-cadherin by genetic and epigenetic changes in the malignant progression of hepatocellular carcinomas. Clin Cancer Res 7:594–599

Mayer B, Johnson JP, Leitl F, Jauch KW, Heiss MM, Schildberg FW, Birchmeier W, Funke I (1993) E-cadherin expression in primary and metastatic gastric cancer: down-regulation correlates with cellular dedifferentiation and glandular disintegration. Cancer Res 53:1690–1695

Meng W, Takeichi M (2009) Adherens junction: molecular architecture and regulation. Cold Spring Harb Perspect Biol 1:a002899

Merdek KD, Nguyen NT, Toksoz D (2004) Distinct activities of the alpha-catenin family, alpha-catulin and alpha-catenin, on beta-catenin-mediated signaling. Mol Cell Biol 24:2410–2422

Micalizzi DS, Farabaugh SM, Ford HL (2010) Epithelial-mesenchymal transition in cancer: parallels between normal development and tumor progression. J Mammary Gland Biol Neoplasia 15:117–134

Milicic A, Harrison LA, Goodlad RA, Hardy RG, Nicholson AM, Presz M, Sieber O, Santander S, Pringle JH, Mandir N, East P, Obszynska J, Sanders S, Piazuelo E, Shaw J, Harrison R, Tomlinson IP, McDonald SA, Wright NA, Jankowski JA (2008) Ectopic expression of P-cadherin correlates with promoter hypomethylation early in colorectal carcinogenesis and enhanced intestinal crypt fission in vivo. Cancer Res 68:7760–7768

Miyoshi Y, Iwao K, Nagasawa Y, Aihara T, Sasaki Y, Imaoka S, Murata M, Shimano T, Nakamura Y (1998) Activation of the beta-catenin gene in primary hepatocellular carcinomas by somatic alterations involving exon 3. Cancer Res 58:2524–2527

Munoz-Guerra MF, Marazuela EG, Fernandez-Contreras ME, Gamallo C (2005) P-cadherin expression reduced in squamous cell carcinoma of the oral cavity: an indicatior of poor prognosis. Cancer 103:960–969

Nagae Y, Kameyama K, Yokoyama M, Naito Z, Yamada N, Maeda S, Asano G, Sugisaki Y, Tanaka S (2002) Expression of E-cadherin catenin and C-erbB-2 gene products in invasive ductal-type breast carcinomas. Journal of Nippon Medical School = Nihon Ika Daigaku zasshi 69:165–171

Nakanishi Y, Ochiai A, Akimoto S, Kato H, Watanabe H, Tachimori Y, Yamamoto S, Hirohashi S (1997) Expression of E-cadherin, alpha-catenin, beta-catenin and plakoglobin in esophageal carcinomas and its prognostic significance: immunohistochemical analysis of 96 lesions. Oncology 54:158–165

Nakopoulou L, Gakiopoulou-Givalou H, Karayiannakis AJ, Giannopoulou I, Keramopoulos A, Davaris P, Pignatelli M (2002) Abnormal alpha-catenin expression in invasive breast cancer correlates with poor patient survival. Histopathology 40:536–546

Nieman MT, Prudoff RS, Johnson KR, Wheelock MJ (1999) N-cadherin promotes motility in human breast cancer cells regardless of their E-cadherin expression. J Cell Biol 147:631–644

Nojima D, Nakajima K, Li LC, Franks J, Ribeiro-Filho L, Ishii N, Dahiya R (2001) CpG methylation of promoter region inactivates E-cadherin gene in renal cell carcinoma. Mol Carcinog 32:19–27

Noren NK, Liu BP, Burridge K, Kreft B (2000) p120 catenin regulates the actin cytoskeleton via Rho family GTPases. J Cell Biol 150:567–580

Ochiai A, Akimoto S, Kanai Y, Shibata T, Oyama T, Hirohashi S (1994) c-erbB-2 gene product associates with catenins in human cancer cells. Biochem Biophys Res Commun 205:73–78

Ogata S, Morokuma J, Hayata T, Kolle G, Niehrs C, Ueno N, Cho KW (2007) TGF-beta signaling-mediated morphogenesis: modulation of cell adhesion via cadherin endocytosis. Genes Dev 21:1817–1831

Ogita H, Rikitake Y, Miyoshi J, Takai Y (2010) Cell adhesion molecules nectins and associating proteins: Implications for physiology and pathology. Proc Jpn Acad Ser B Phys Biol Sci 86:621–629

Oka H, Shiozaki H, Kobayashi K, Inoue M, Tahara H, Kobayashi T, Takatsuka Y, Matsuyoshi N, Hirano S, Takeichi M et al (1993) Expression of E-cadherin cell adhesion molecules in human breast cancer tissues and its relationship to metastasis. Cancer Res 53:1696–1701

Olson P, Lu J, Zhang H, Shai A, Chun MG, Wang Y, Libutti SK, Nakakura EK, Golub TR, Hanahan D (2009) MicroRNA dynamics in the stages of tumorigenesis correlate with hallmark capabilities of cancer. Genes Dev. 23:2152–2165

Onder TT, Gupta PB, Mani SA, Yang J, Lander ES, Weinberg RA (2008) Loss of E-cadherin promotes metastasis via multiple downstream transcriptional pathways. Cancer Res 68:3645–3654

Orsulic S, Huber O, Aberle H, Arnold S, Kemler R (1999) E-cadherin binding prevents beta-catenin nuclear localization and beta-catenin/LEF-1-mediated transactivation. J Cell Sci 112(Pt 8):1237–1245

Overholtzer M, Zhang J, Smolen GA, Muir B, Li W, Sgroi DC, Deng CX, Brugge JS, Haber DA (2006) Transforming properties of YAP, a candidate oncogene on the chromosome 11q22 amplicon. Proc Natl Acad Sci U S A 103:12405–12410

Palacios F, Tushir JS, Fujita Y, D'Souza-Schorey C (2005) Lysosomal targeting of E-cadherin: a unique mechanism for the down-regulation of cell-cell adhesion during epithelial to mesenchymal transitions. Mol Cell Biol 25:389–402

Pan D (2010) The hippo signaling pathway in development and cancer. Dev Cell 19:491–505

Paredes J, Correia AL, Ribeiro AS, Albergaria A, Milanezi F, Schmitt FC (2007) P-cadherin expression in breast cancer: a review. Breast Cancer Res 9:214

Park D, Karesen R, Axcrona U, Noren T, Sauer T (2007) Expression pattern of adhesion molecules (E-cadherin, alpha-, beta-, gamma-catenin and claudin-7), their influence on survival in primary breast carcinoma, and their corresponding axillary lymph node metastasis. Apmis 115:52–65

Park SM, Gaur AB, Lengyel E, Peter ME (2008) The miR-200 family determines the epithelial phenotype of cancer cells by targeting the E-cadherin repressors ZEB1 and ZEB2. Genes Dev 22:894–907

Park WS, Oh RR, Park JY, Lee SH, Shin MS, Kim YS, Kim SY, Lee HK, Kim PJ, Oh ST, Yoo NJ, Lee JY (1999) Frequent somatic mutations of the beta-catenin gene in intestinal-type gastric cancer. Cancer Res 59:4257–4260

Pedersen KB, Nesland JM, Fodstad O, Maelandsmo GM (2002) Expression of S100A4, E-cadherin, alpha- and beta-catenin in breast cancer biopsies. Br J Cancer 87:1281–1286

Peinado H, Ballestar E, Esteller M, Cano A (2004) Snail mediates E-cadherin repression by the recruitment of the Sin3 A/histone deacetylase 1 (HDAC1)/HDAC2 complex. Mol Cell Biol 24:306–319

Peinado H, Olmeda D, Cano A (2007) Snail, Zeb and bHLH factors in tumour progression: an alliance against the epithelial phenotype? Nat Rev 7:415–428

Perez-Moreno M, Song W, Pasolli HA, Williams SE, Fuchs E (2008) Loss of p120 catenin and links to mitotic alterations, inflammation, and skin cancer. Proc Natl Acad Sci U S A 105:15399–15404

Perez-Moreno MA, Locascio A, Rodrigo I, Dhondt G, Portillo F, Nieto MA, Cano A (2001) A new role for E12/E47 in the repression of E-cadherin expression and epithelial-mesenchymal transitions. J Biol Chem 276:27424–27431

Perl AK, Wilgenbus P, Dahl U, Semb H, Christofori G (1998) A causal role for E-cadherin in the transition from adenoma to carcinoma. Nature 392:190–193

Perrais M, Chen X, Perez-Moreno M, Gumbiner BM (2007) E-Cadherin Homophilic Ligation Inhibits Cell Growth and Epidermal Growth Factor Receptor Signaling Independently of Other Cell Interactions. Mol Biol Cell 18:2013–2025

Pharoah PD, Guilford P, Caldas C (2001) Incidence of gastric cancer and breast cancer in CDH1 (E-cadherin) mutation carriers from hereditary diffuse gastric cancer families. Gastroenterology 121:1348–1353

Place RF, Li LC, Pookot D, Noonan EJ, Dahiya R (2008) MicroRNA-373 induces expression of genes with complementary promoter sequences. Proc Natl Acad Sci U S A 105:1608–1613

Polakis P (2000) Wnt signaling and cancer. Genes Dev 14:1837–1851

Prunier C, Howe PH (2005) Disabled-2 (Dab2) is required for transforming growth factor beta-induced epithelial to mesenchymal transition (EMT). J Biol Chem 280:17540–17548
Qian X, Karpova T, Sheppard AM, McNally J, Lowy DR (2004) E-cadherin-mediated adhesion inhibits ligand-dependent activation of diverse receptor tyrosine kinases. Embo J 23:1739–1748
Qiao Y, Jiang X, Lee ST, Karuturi RK, Hooi SC, Yu Q (2011) FOXQ1 regulates epithelial-mesenchymal transition in human cancers. Cancer Res 71:3076–3086
Rakha EA, Abd El Rehim D, Pinder SE, Lewis SA, Ellis IO (2005) E-cadherin expression in invasive non-lobular carcinoma of the breast and its prognostic significance. Histopathology 46:685–693
Reshetnikova G, Troyanovsky S, Rimm DL (2007) Definition of a direct extracellular interaction between Met and E-cadherin. Cell Biol Int 31:366–373
Reynolds AB (2007) p120-catenin: Past and present. Biochim Biophys Acta 1773:2–7
Reynolds AB, Daniel J, McCrea PD, Wheelock MJ, Wu J, Zhang Z (1994) Identification of a new catenin: the tyrosine kinase substrate p120cas associates with E-cadherin complexes. Mol Cell Biol 14:8333–8342
Ribeiro-Filho LA, Franks J, Sasaki M, Shiina H, Li LC, Nojima D, Arap S, Carroll P, Enokida H, Nakagawa M, Yonezawa S, Dahiya R (2002) CpG hypermethylation of promoter region and inactivation of E-cadherin gene in human bladder cancer. Mol Carcinog 34:187–198
Richmond PJ, Karayiannakis AJ, Nagafuchi A, Kaisary AV, Pignatelli M (1997) Aberrant E-cadherin and alpha-catenin expression in prostate cancer: correlation with patient survival. Cancer Res 57:3189–3193
Rimm DL, Koslov ER, Kebriaei P, Cianci CD, Morrow JS (1995) Alpha 1(E)-catenin is an actin-binding and -bundling protein mediating the attachment of F-actin to the membrane adhesion complex. Proc Natl Acad Sci U S A 92:8813–8817
Risinger JI, Berchuck A, Kohler MF, Boyd J (1994) Mutations of the E-cadherin gene in human gynecologic cancers. Nat Genet 7:98–102
Roura S, Miravet S, Piedra J, Garcia de Herreros A, Dunach M (1999) Regulation of E-cadherin/Catenin association by tyrosine phosphorylation. J Biol Chem 274:36734–36740
Saini S, Yamamura S, Majid S, Shahryari V, Hirata H, Tanaka Y, Dahiya R (2011) MicroRNA-708 induces apoptosis and suppresses tumorigenicity in renal cancer cells. Cancer Res
Saldanha G, Ghura V, Potter L, Fletcher A (2004) Nuclear beta-catenin in basal cell carcinoma correlates with increased proliferation. Br J Dermatol 151:157–164
Sanchez-Tillo E, Lazaro A, Torrent R, Cuatrecasas M, Vaquero EC, Castells A, Engel P, Postigo A (2010) ZEB1 represses E-cadherin and induces an EMT by recruiting the SWI/SNF chromatin-remodeling protein BRG1. Oncogene 29:3490–3500
Sanson B, White P, Vincent JP (1996) Uncoupling cadherin-based adhesion from wingless signalling in Drosophila. Nature 383:627–630
Schackmann RC, van Amersfoort M, Haarhuis JH, Vlug EJ, Halim VA, Roodhart JM, Vermaat JS, Voest EE, Van Der Groep P, van Diest PJ, Jonkers J, Derksen PW (2011) Cytosolic p120-catenin regulates growth of metastatic lobular carcinoma through Rock1-mediated anoikis resistance. J Clin Invest 121:3176–3188
Schipper JH, Frixen UH, Behrens J, Unger A, Jahnke K, Birchmeier W (1991) E-cadherin expression in squamous cell carcinomas of head and neck: inverse correlation with tumor dedifferentiation and lymph node metastasis. Cancer Res 51:6328–6337
Schlegelmilch K, Mohseni M, Kirak O, Pruszak J, Rodriguez JR, Zhou D, Kreger BT, Vasioukhin V, Avruch J, Brummelkamp TR, Camargo FD (2011) Yap1 Acts Downstream of alpha-Catenin to Control Epidermal Proliferation. Cell 144:782–795
Schuhmacher C, Becker I, Oswald S, Atkinson MJ, Nekarda H, Becker KF, Mueller J, Siewert JR, Hofler H (1999) Loss of immunohistochemical E-cadherin expression in colon cancer is not due to structural gene alterations. Virchows Arch 434:489–495
Sehgal RN, Gumbiner BM, Reichardt LF (1997) Antagonism of cell adhesion by an alpha-catenin mutant, and of the Wnt-signaling pathway by alpha-catenin in Xenopus embryos. J Cell Biol 139:1033–1046

Setoyama T, Natsugoe S, Okumura H, Matsumoto M, Uchikado Y, Yokomakura N, Ishigami S, Aikou T (2007) alpha-catenin is a significant prognostic factor than E-cadherin in esophageal squamous cell carcinoma. J Surg Oncol 95:148–155

Shen Y, Hirsch DS, Sasiela CA, Wu WJ (2008) Cdc42 regulates E-cadherin ubiquitination and degradation through an epidermal growth factor receptor to Src-mediated pathway. J Biol Chem 283:5127–5137

Shi Y, Sawada J, Sui G, Affarel B, Whetstine JR, Lan F, Ogawa H, Luke MP, Nakatani Y (2003) Coordinated histone modifications mediated by a CtBP co-repressor complex. Nature 422:735–738

Shook D, Keller R (2003) Mechanisms, mechanics and function of epithelial-mesenchymal transitions in early development. Mech Dev 120:1351–1383

Shoval I, Ludwig A, Kalcheim C (2007) Antagonistic roles of full-length N-cadherin and its soluble BMP cleavage product in neural crest delamination. Development 134:491–501

Shun CT, Wu MS, Lin JT, Wang HP, Houng RL, Lee WJ, Wang TH, Chuang SM (1998) An immunohistochemical study of E-cadherin expression with correlations to clinicopathological features in gastric cancer. Hepatogastroenterology 45:944–949

Si HX, Tsao SW, Lam KY, Srivastava G, Liu Y, Wong YC, Shen ZY, Cheung AL (2001) E-cadherin expression is commonly downregulated by CpG island hypermethylation in esophageal carcinoma cells. Cancer Lett 173:71–78

Silvis MR, Kreger BT, Lien WH, Klezovitch O, Rudakova GM, Camargo FD, Lantz DM, Seykora JT, Vasioukhin V (2011) {alpha}-Catenin is a tumor suppressor that controls cell accumulation by regulating the localization and activity of the transcriptional coactivator Yap1. Sci Signal 4:ra33

Simcha I, Shtutman M, Salomon D, Zhurinsky J, Sadot E, Geiger B, Ben-Ze'ev A (1998) Differential nuclear translocation and transactivation potential of beta-catenin and plakoglobin. J Cell Biol 141:1433–1448

Slorach EM, Chou J, Werb Z (2011) Zeppo1 is a novel metastasis promoter that represses E-cadherin expression and regulates p120-catenin isoform expression and localization. Genes Dev 25:471–484

Smalley-Freed WG, Efimov A, Short SP, Jia P, Zhao Z, Washington MK, Robine S, Coffey RJ, Reynolds AB (2011) Adenoma formation following limited ablation of p120-catenin in the mouse intestine. PLoS ONE 6(5):e19880. doi:10.1371/journal.pone.0019880

Soares P, Berx G, van Roy F, Sobrinho-Simoes M (1997) E-cadherin gene alterations are rare events in thyroid tumors. International journal of cancer. Int J Cancer 70:32–38

Solanas G, Cortina C, Sevillano M, Batlle E (2011) Cleavage of E-cadherin by ADAM10 mediates epithelial cell sorting downstream of EphB signalling. Nat Cell Biol 13:1100–1107

Song Y, Washington MK, Crawford HC (2010) Loss of FOXA1/2 is essential for the epithelial-to-mesenchymal transition in pancreatic cancer. Cancer Res 70:2115–2125

Soto E, Yanagisawa M, Marlow LA, Copland JA, Perez EA, Anastasiadis PZ (2008) p120 catenin induces opposing effects on tumor cell growth depending on E-cadherin expression. J Cell Biol 183:737–749

Stairs DB, Bayne LJ, Rhoades B, Vega ME, Waldron TJ, Kalabis J, Klein-Szanto A, Lee JS, Katz JP, Diehl JA, Reynolds AB, Vonderheide RH, Rustgi AK (2011) Deletion of p120-catenin results in a tumor microenvironment with inflammation and cancer that establishes it as a tumor suppressor gene. Cancer Cell 19:470–483

Stinson S, Lackner MR, Adai AT, Yu N, Kim HJ, O'Brien C, Spoerke J, Jhunjhunwala S, Boyd Z, Januario T, Newman RJ, Yue P, Bourgon R, Modrusan Z, Stern HM, Warming S, de Sauvage FJ, Amler L, Yeh RF, Dornan D (2011) miR-221/222 Targeting of Trichorhinophalangeal 1 (TRPS1) promotes epithelial-to-mesenchymal transition in breast cancer. Sci Signal 4:pt5

Suyama K, Shapiro I, Guttman M, Hazan RB (2002) A signaling pathway leading to metastasis is controlled by N-cadherin and the FGF receptor. Cancer Cell 2:301–314

Taddei I, Piazzini M, Bartoletti R, Dal Canto M, Sardi I (2000) Molecular alterations of E-cadherin gene: possible role in human bladder carcinogenesis. Int J Mol Med 6:201–208

Takahashi K, Suzuki K (1996) Density-dependent inhibition of growth involves prevention of EGF receptor activation by E-cadherin-mediated cell-cell adhesion. Exp Cell Res 226:214–222

Tamura G, Yin J, Wang S, Fleisher AS, Zou T, Abraham JM, Kong D, Smolinski KN, Wilson KT, James SP, Silverberg SG, Nishizuka S, Terashima M, Motoyama T, Meltzer SJ (2000) E-Cadherin gene promoter hypermethylation in primary human gastric carcinomas. J Natl Cancer Inst 92:569–573

Theisen CS, Wahl JK, 3rd, Johnson KR, Wheelock MJ (2007) NHERF links the N-cadherin/catenin complex to the platelet-derived growth factor receptor to modulate the actin cytoskeleton and regulate cell motility. Mol Biol Cell 18:1220–1232

Tinkle CL, Pasoli HA, Stokes N, Fuchs E (2008a) New insights into cadherin function in epidermal sheet formation and maintenance of tissue integrity. Proc Natl Acad Sci U S A 105(40):15225-15226, doi:10.1073/pnas.0808458105

Tinkle CL, Pasolli HA, Stokes N, Fuchs E (2008b) New insights into cadherin function in epidermal sheet formation and maintenance of tissue integrity. Proc Natl Acad Sci U S A 105:15405–15410

Torres M, Stoykova A, Huber O, Chowdhury K, Bonaldo P, Mansouri A, Butz S, Kemler R, Gruss P (1997) An alpha-E-catenin gene trap mutation defines its function in preimplantation development. Proc Natl Acad Sci U S A 94:901–906

Toyoshima M, Tanaka N, Aoki J, Tanaka Y, Murata K, Kyuuma M, Kobayashi H, Ishii N, Yaegashi N, Sugamura K (2007) Inhibition of tumor growth and metastasis by depletion of vesicular sorting protein Hrs: its regulatory role on E-cadherin and beta-catenin. Cancer Res 67:5162–5171

Toyoyama H, Nuruki K, Ogawa H, Yanagi M, Matsumoto H, Nishijima H, Shimotakahara T, Aikou T, Ozawa M (1999) The reduced expression of e-cadherin, alpha-catenin and gamma-catenin but not beta-catenin in human lung cancer. Oncol Rep 6:81–85

Turashvili G, McKinney SE, Goktepe O, Leung SC, Huntsman DG, Gelmon KA, Los G, Rejto PA, Aparicio SA (2011) P-cadherin expression as a prognostic biomarker in a 3992 case tissue microarray series of breast cancer. Mod Pathol 24:64–81

Umbas R, Schalken JA, Aalders TW, Carter BS, Karthaus HF, Schaafsma HE, Debruyne FM, Isaacs WB (1992) Expression of the cellular adhesion molecule E-cadherin is reduced or absent in high-grade prostate cancer. Cancer Res 52:5104–5109

van de Wetering M, Barker N, Harkes IC, Van Der Heyden M, Dijk NJ, Hollestelle A, Klijn JG, Clevers H, Schutte M (2001) Mutant E-cadherin breast cancer cells do not display constitutive Wnt signaling. Cancer Res 61:278–284

van Oort IM, Tomita K, van Bokhoven A, Bussemakers MJ, Kiemeney LA, Karthaus HF, Witjes JA, Schalken JA (2007) The prognostic value of E-cadherin and the cadherin-associated molecules alpha-, beta-, gamma-catenin and p120ctn in prostate cancer specific survival: a long-term follow-up study. Prostate 67:1432–1438

Vasioukhin V, Bauer C, Yin M, Fuchs E (2000) Directed actin polymerization is the driving force for epithelial cell-cell adhesion. Cell 100:209–219

Vasioukhin V, Bauer C, Degenstein L, Wise B, Fuchs E (2001) Hyperproliferation and defects in epithelial polarity upon conditional ablation of alpha-catenin in skin. Cell 104:605–617

Vincent T, Neve EP, Johnson JR, Kukalev A, Rojo F, Albanell J, Pietras K, Virtanen I, Philipson L, Leopold PL, Crystal RG, de Herreros AG, Moustakas A, Pettersson RF, Fuxe J (2009) A SNAIL1-SMAD3/4 transcriptional repressor complex promotes TGF-beta mediated epithelial-mesenchymal transition. Nat Cell Biol 11:943–950

Vleminckx K, Vakaet L Jr, Mareel M, Fiers W, van Roy F (1991) Genetic manipulation of E-cadherin expression by epithelial tumor cells reveals an invasion suppressor role. Cell 66:107–119

Voorhoeve PM, le Sage C, Schrier M, Gillis AJ, Stoop H, Nagel R, Liu YP, van Duijse J, Drost J, Griekspoor A, Zlotorynski E, Yabuta N, De Vita G, Nojima H, Looijenga LH, Agami R (2006) A genetic screen implicates miRNA-372 and miRNA-373 as oncogenes in testicular germ cell tumors. Cell 124:1169–1181

Wang XQ, Li H, Van Putten V, Winn RA, Heasley LE, Nemenoff RA (2009) Oncogenic K-Ras regulates proliferation and cell junctions in lung epithelial cells through induction of cyclooxygenase-2 and activation of metalloproteinase-9. Mol Biol Cell 20:791–800

Watabe M, Nagafuchi A, Tsukita S, Takeichi M (1994) Induction of polarized cell-cell association and retardation of growth by activation of the E-cadherin-catenin adhesion system in a dispersed carcinoma line. J Cell Biol 127:247–256
Wend P, Holland JD, Ziebold U, Birchmeier W (2010) Wnt signaling in stem and cancer stem cells. Semin Cell Dev Biol 21:855–863
Wheelock MJ, Shintani Y, Maeda M, Fukumoto Y, Johnson KR (2008) Cadherin switching. J Cell Sci 121:727–735
Wijnhoven BP, de Both NJ, van Dekken H, Tilanus HW, Dinjens WN (1999) E-cadherin gene mutations are rare in adenocarcinomas of the oesophagus. Br J Cancer 80:1652–1657
Wong SC, Lo SF, Lee KC, Yam JW, Chan JK, Wendy Hsiao WL (2002) Expression of frizzled-related protein and Wnt-signalling molecules in invasive human breast tumours. J Pathol 196:145–153
Wood LD, Parsons DW, Jones S, Lin J, Sjöblom T, Leary RJ, Shen D, Boca SM, Barber T, Ptak J, Silliman N, Szabo S, Dezso Z, Ustyanksky V, Nikolskaya T, Nikolsky Y, Karchin R, Wilson PA, Kaminker JS, Zhang Z, Croshaw R, Willis J, Dawson D, Shipitsin M, Willson JK, Sukumar S, Polyak K, Park BH, Pethiyagoda CL, Pant PV, Ballinger DG, Sparks AB, Hartigan J, Smith DR, Suh E, Papadopoulos N, Buckhaults P, Markowitz SD, Parmigiani G, Kinzler KW, Velculescu VE, Vogelstein B (2007) The genomic landscapes of human breast and colorectal cancers. Science 318 (5853):1108–1113
Woodcock SA, Rooney C, Liontos M, Connolly Y, Zoumpourlis V, Whetton AD, Gorgoulis VG, Malliri A (2009) SRC-induced disassembly of adherens junctions requires localized phosphorylation and degradation of the rac activator tiam1. Mol Cell 33:639–653
Yamada S, Pokutta S, Drees F, Weis WI, Nelson WJ (2005) Deconstructing the cadherin-catenin-actin complex. Cell 123:889–901
Yamazaki F, Aragane Y, Kawada A, Tezuka T (2001) Immunohistochemical detection for nuclear beta-catenin in sporadic basal cell carcinoma. Br J Dermatol 145:771–777
Yang DH, Cai KQ, Roland IH, Smith ER, Xu XX (2007) Disabled-2 is an epithelial surface positioning gene. J Biol Chem 282:13114–13122
Yang J, Mani SA, Donaher JL, Ramaswamy S, Itzykson RA, Come C, Savagner P, Gitelman I, Richardson A, Weinberg RA (2004) Twist, a master regulator of morphogenesis, plays an essential role in tumor metastasis. Cell 117:927–939
Yang JY, Zong CS, Xia W, Wei Y, Ali-Seyed M, Li Z, Broglio K, Berry DA, Hung MC (2006) MDM 2 promotes cell motility and invasiveness by regulating E-cadherin degradation. Mol Cell Biol 26:7269–7282
Yang MH, Hsu DS, Wang HW, Wang HJ, Lan HY, Yang WH, Huang CH, Kao SY, Tzeng CH, Tai SK, Chang SY, Lee OK, Wu KJ (2010) Bmi1 is essential in Twist1-induced epithelial-mesenchymal transition. Nat Cell Biol 12:982–992
Yonemura S, Wada Y, Watanabe T, Nagafuchi A, Shibata M (2010) Alpha-Catenin as a tension transducer that induces adherens junction development. Nat Cell Biol 12:533–542
Yoshiura K, Kanai Y, Ochiai A, Shimoyama Y, Sugimura T, Hirohashi S (1995) Silencing of the E-cadherin invasion-suppressor gene by CpG methylation in human carcinomas. Proc Natl Acad,Sci U S A 92:7416–7419
Young P, Boussadia O, Halfter H, Grose R, Berger P, Leone DP, Robenek H, Charnay P, Kemler R, Suter U (2003) E-cadherin controls adherens junctions in the epidermis and the renewal of hair follicles. Embo J 22:5723–5733
Zahir N, Weaver VM (2004) Death in the third dimension: apoptosis regulation and tissue architecture. Curr Opin Genet Dev 14:71–80
Zandy NL, Playford M, Pendergast AM (2007) Abl tyrosine kinases regulate cell-cell adhesion through Rho GTPases. Proc Natl Acad Sci U S A 104:17686–17691
Zechner D, Fujita Y, Hulsken J, Muller T, Walther I, Taketo MM, Crenshaw EB 3rd, Birchmeier W, Birchmeier C (2003) Beta-Catenin signals regulate cell growth and the balance between progenitor cell expansion and differentiation in the nervous system. Dev Biol 258:406–418
Zeng G, Apte U, Micsenyi A, Bell A, Monga SP (2006) Tyrosine residues 654 and 670 in beta-catenin are crucial in regulation of Met-beta-catenin interactions. Exp Cell Res 312:3620–3630

Zhang N, Bai H, David KK, Dong J, Zheng Y, Cai J, Giovannini M, Liu P, Anders RA, Pan D (2010) The Merlin/NF2 tumor suppressor functions through the YAP oncoprotein to regulate tissue homeostasis in mammals. Dev Cell 19:27–38

Zhao B, Wei X, Li W, Udan RS, Yang Q, Kim J, Xie J, Ikenoue T, Yu J, Li L, Zheng P, Ye K, Chinnaiyan A, Halder G, Lai ZC, Guan KL (2007) Inactivation of YAP oncoprotein by the Hippo pathway is involved in cell contact inhibition and tissue growth control. Genes Dev 21:2747–2761

Zhao B, Tumaneng K, Guan KL (2011) The Hippo pathway in organ size control, tissue regeneration and stem cell self-renewal. Nat Cell Biol 13:877–883

Zhong C, Kinch MS, Burridge K (1997) Rho-stimulated contractility contributes to the fibroblastic phenotype of Ras-transformed epithelial cells. Mol Biol Cell 8:2329–2344

Zhou YN, Xu CP, Chen Y, Han B, Yang SM, Fang DC (2005) Alpha-catenin expression is decreased in patients with gastric carcinoma. World J Gastroenterol 11:3468–3472

Chapter 17
Adherens Junctions and Pathogen Entry

Georgios Nikitas and Pascale Cossart

Abstract Epithelia are highly organised structures protecting underlying tissues against microbial pathogens. Epithelial morphogenesis and maintenance is mediated by cell-cell adhesion molecules organised in junctional complexes, such as the *adherens junctions*. The tight organisation of these complexes and their interactions with cellular factors render the epithelia impermeable to potential invaders. Nevertheless, pathogens have developed strategies to target, interact and manipulate junctional complexes, in order to disrupt or cross the epithelial barriers and cause infection. Bacteria, viruses and parasites access the junctional molecular components either directly, often taking advantage of physiological alterations in epithelial polarity, or indirectly, by delivering into cells molecular factors that destabilise junctional integrity. Importantly, microbial interactions with junctional components are instrumental not only to elucidate mechanisms of invasion, but also to unravel fundamental physiological properties of the epithelial barriers, at the cellular and tissular level.

17.1 Introduction

Cell-cell junctions and cell-matrix interactions are functionally involved in sealing cells together to form a cellular sheet, attaching them to their neighbours and to the extracellular matrix, and facilitating passage of signaling molecules from one cell to the next. The sealing of the epithelial sheet and the formation of an impermeable barrier are mediated mostly by the tight junctions. Cell-cell adhesion is established and maintained by adherens junctions (AJs)—which are located below tight junctions in polarized epithelial cells—together with desmosomes and hemidesmosomes. Gap junctions mediate intercellular passage of small molecules. Thus, AJs

P. Cossart (✉)
Institut Pasteur, Unité des Interactions Bactéries-Cellules, 75015 Paris, France
e-mail: pcossart@pasteur.fr

Inserm, U604, 75015 Paris, France

INRA, USC2020, 75015 Paris, France

G. Nikitas
Biomedical Research Foundation Academy of Athens, 11527 Athens Greece

T. Harris (ed.), *Adherens Junctions: From Molecular Mechanisms to Tissue Development and Disease,* Subcellular Biochemistry 60,
DOI 10.1007/978-94-007-4186-7_17, © Springer Science+Business Media Dordrecht 2012

are key molecular players for barrier functionality. In polarized cells, the AJs link cells into a continuous sheet while separating the apical and basolateral membranes. Three classes of proteins are required for AJ formation: (i) adhesion receptors that span the intercellular space, such as cadherins and nectins, (ii) adaptor proteins, such as catenins and afadin, that link the adhesive components to the cytoskeleton and (iii) cytoskeletal proteins, such as actin that anchors the adhesive components to the intracellular space (Niessen and Gottardi 2008).

The main adhesive components of AJs are the cadherin/catenin and nectin/afadin complexes. Cadherins are single-pass transmembrane glycoproteins that mediate Ca^{2+}-dependent intercellular adhesion. Cadherins are key molecules in the morphogenesis from the morula stage to the building of a variety of organs and in the maintenance of epithelial polarity and tissue architecture. The "classical" cadherins are the first reported family of adhesion molecules found in the AJs and include E-cadherin (epithelial), N-cadherin (neuronal), P-cadherin (placental) and R-cadherin (retinal) (Hulpiau and van Roy 2009). The cadherin cytoplasmic domain associates with α-, β-, and p120-catenin forming the cadherin/catenin complex (CCC). α-catenin is crucial for actin rearrangements at AJs. Recent studies show that α-catenin is not the direct link to the underlying cytoskeleton but the driving force of changes in actin dynamics (Drees et al. 2005). β-catenin is crucial for the adhesion properties of the complex and provides the binding link for α-catenin. p120-catenin controls steady-state levels of cadherins and regulates actin dynamics as well. Overall, the CCC contains factors involved in its stabilization at the cell surface (p120-catenin and β-catenin), intercellular adhesion (cadherins), association to the cytoskeleton (α-catenin) and regulation of cytoskeleton dynamics (p120-catenin and α-catenin) (van Roy and Berx 2008).

The CCC associates with the nectin/afadin complex, and α-catenin was shown to be critical for this interaction (Takai and Nakanishi 2003). Nectins are transmembrane proteins that form a family of Ca^{2+}-independent immunoglobulin (Ig)-like adhesion molecules. In epithelial cells nectins are present in AJs and interact with the actin cytoskeleton via afadin. Nectins form homodimers, and are linked to the CCC via α-catenins. However, the way nectins bind to α-catenins remains unknown. Of note, nectins mediate a Ca^{2+}-independent adhesion, while Ca^{+2}-dependent cadherin adhesion is essential for epithelial cohesion. Moreover, afadin null mice show loss of the junctional complex and epithelial structure, without significant alteration of the localization of cadherins. Thus, nectin/afadin-driven AJs formation and CCC-based adhesion seem uncoupled. Overall, the CCC and the nectin/afadin complexes are the crucial structural components of AJs that mediate cell adhesion and confer tissue homeostasis in health and disease.

17.2 Adherens Junctions and Pathogens

Despite the strong adhesive properties of AJs and their localization in the epithelial sheet below tight junctions, microbes have developed strategies to target AJs in order to disrupt and/or cross epithelial barriers and cause infection.

17.2.1 Adherens Junctions and Bacteria

***Listeria monocytogenes* targets directly E-cadherin** *Listeria monocytogenes* (*Lm*) is a food-borne pathogen able to cross three host barriers, that is the intestinal, the blood-brain and the feto-placental barrier, causing gastroenteritis, encephalitis, meningitis, and materno-fetal infections, respectively. The bacterial proteins InlA and InlB are involved in *Lm* entry into non-phagocytic cells. The receptor for InlA is E-cadherin. The receptor for InlB is Met, together with gC1qR and proteoglycans. *In vitro* studies have shown that upon InlA-E-cadherin interaction AJ components are recruited, E-cadherin clusters, Src tyrosine kinase is activated, E-cadherin is modified and clathrin is assembled triggering actin recruitment and bacterial internalization. InlB-Met interaction results in PI3 kinase activation, membrane rearrangements, actin remodeling, and bacteria internalization (Bonazzi and Cossart 2011). Both InlA-Ecad and InlB-Met interactions are species specific, that is InlA interacts with human, guinea pig and gerbil E-cadherin and InlB with human, mouse and gerbil Met. Thus to study listeriosis *in vivo* genetically engineered mice models have been developed. In hosts permissive to InlA, such as transgenic mice expressing human E-cadherin at the level of enterocytes and knock-in mice expressing "humanized" mouse E-cadherin, *Lm* invades intestinal epithelial cells directly and crosses the intestinal epithelium to disseminate into deeper organs (Disson et al. 2008; Lecuit et al. 2001; Nikitas et al. 2011). InlB is not involved in the process.

Nevertheless, E-cadherin localization below tight junctions raised serious concerns regarding its luminal accessibility. It is now well documented that *Lm* invades intestinal villi at sites of luminally accessible E-cadherin, that is (i) junctions between mucus-secreting goblet cells and adjacent enterocytes, (ii) extruding enterocytes at the tip and lateral sides of intestinal villi, and (iii) villus epithelial folds. Moreover, the intestinal goblet cells are identified as *Lm* preferential cell targets (Nikitas et al. 2011; Pentecost et al. 2006, 2010). Of note, in the intestinal environment and prior to intestinal invasion, *Lm* expresses virulence factors that mediate *Lm* intestinal survival and barrier crossing (Toledo-Arana et al. 2009). Thus, *Lm* takes advantage of an intestinal defence mechanism, namely mucus secretion, to target the epithelium, uncovering an unsuspected Achilles' heel of the innate immune system that *Lm* takes advantage of.

Importantly, InlA-E-cadherin interaction mediates *Lm* intestinal crossing by transcytosis, translocation into the underlying *lamina propria* (Nikitas et al. 2011) (Fig. 17.1), and dissemination into deeper tissues and organs, such as the liver and spleen (Lecuit et al. 2001; Nikitas et al. 2011). Unexpectedly, some of the well characterized *Lm* virulence factors such as ActA or LLO are dispensable for barrier crossing, clearly demonstrating that *InlA* interaction with E-cad is the critical factor for the onset of the disease. Nevertheless, at early stages of infection, the InlA-dependent *Lm* tissue invasion at the level of intestinal villi does not trigger a host response, as determined by host gene expression microarrays (Lecuit et al. 2007). Thus, *Lm* transcytosis through the epithelial cells of intestinal villi provides the opportunity for bacterial translocation into the underlying *lamina propria* undetected. This is in sharp contrast with enteropathogens such as *Shigella flexneri* that are contained at the intestinal level with strong and destructive inflammatory responses.

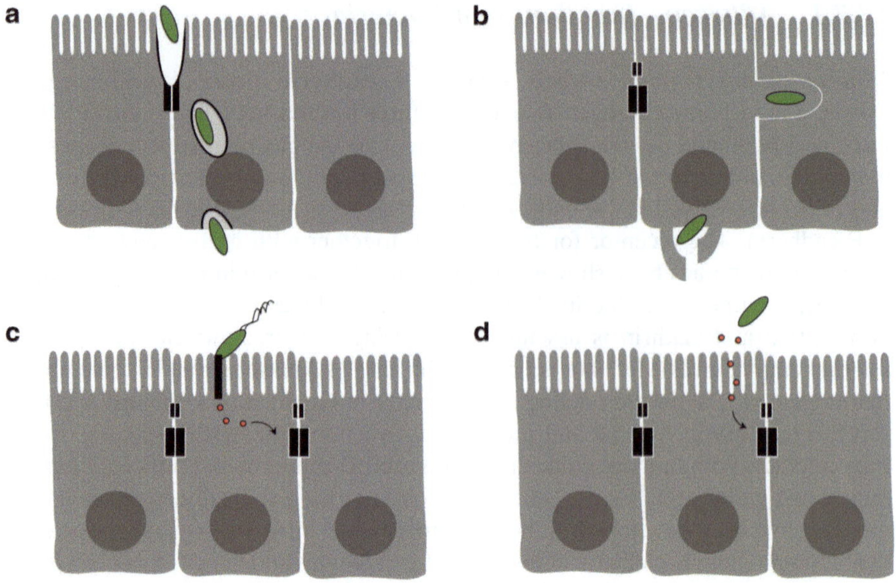

Fig. 17.1 Schematic representation of bacterial strategies to target intestinal adherens junctions. **a** *Listeria monocytogenes* (*green*) interacts with luminally accessible E-cadherin (*black*) and is internalized. Upon entry the *Listeria monocytogenes*-containing vacuole is rapidly translocated in a microtubule-dependent manner to the basolateral pole of the cell, where it is exocytosed. **b** *Shigella flexneri* that has crossed the epithelium enters polarized cells via the basal pole. The cell-to-cell spread depends on E-cadherin. **c** *Helicobacter pylori* adheres to epithelial cell surface and delivers CagA into the host cytoplasm. CagA activates the hepatocyte growth factor receptor (HGF-R) inducing the internalization and redistribution of E-cadherin and disturbing the integrity of AJs. **d** *Bacteroides fragilis* secretes BFT, which cleaves the extracellular domain of E-cadherin, disrupting adherens junctions. (adapted from Sousa et al. 2005b)

Finally, *Lm* crossing of the feto-placental barrier is dependent on both InlA- and InlB (Disson et al. 2008). Understanding how InlA and InlB act in concert to mediate placental invasion and why this contrasts the InlA-dependent intestinal translocation is critical for a complete understanding of the disease.

Overall, *Lm* surface proteins interactions with their cellular receptors are instrumental to understand how the bacterium manipulates cellular junctions to cause infection, and render *Lm* a unique model for cellular and molecular microbiology and immunology (Stavru et al. 2011).

***Shigella flexneri* may target β-catenin** *Shigella flexneri* (*Sf*) invades and multiplies within the human intestinal epithelium leading to severe inflammatory colitis, called bacillary dysentery (or shigellosis). *In vitro* studies have shown that *Sf* delivers a subset of virulence associated proteins via a type III secretion system (TTSS) (protein injectisome) into host cells subverting cellular functions and promoting invasion. Upon entry, *Sf* escapes from its vacuolar compartment, multiplies intracellularly and by using actin-based motility spreads from an infected cell to the neighboring cell after formation of a protrusion, and of a two membrane vacuole which is rapidly lyzed. *In vivo Sf* initially crosses the epithelium by transcytosis through M cells and invades resident macrophages and dendritic cells. Bacterial multiplication within macrophages results in a massive inflammatory response and cell death. *Sf* released from dying macrophages subsequently enters the surrounding epithelium via the basolateral surface (Fig. 17.1).

Bacterial genes involved in *Sf* cell entry, intra- and intercellular spread are encoded by an invasion plasmid that contains the *ipa* operon coding for the IpaA-D effectors of the TTSS. IpaB and IpaC, as well as, the outer-membrane protein IcsA are crucial for cell-cell spread. IpaC interacts directly with β-catenin. Interestingly, IpaC-associated β-catenin is tyrosine-phosphorylated and the CCC is destabilized (Shaikh et al. 2003). Moreover, cadherin expression is required for cell-to-cell spread (Sansonetti et al. 1994). Thus, *Sf*-induced cell adhesion perturbation seems to facilitate bacterial entry at the basolateral surface of epithelial cells and mediates cell-to-cell spread (Fig. 17.1).

***Helicobacter pylori* opens the junctions via intracellular signaling** *Helicobacter pylori* (*Hp*) infects gastric mucosa causing chronic atrophic gastritis, peptic ulcers, gastric adenocarcinoma and/or mucosa-associated lymphoid tissue lymphomas. *In vitro* data show that upon *Hp* attachment to the epithelium the bacterial effector CagA is injected via the type-IV secretion system (T4SS) into the host cytoplasm. CagA activates the hepatocyte growth factor receptor (HGF-R) inducing the internalization and redistribution of E-cadherin and disturbing the integrity of AJs (Fig. 17.1). Interestingly, a CagA-independent opening of AJs has also been recently shown. Specifically, *Hp* lipopolysaccharide acts as a Toll-like receptor 2 (TLR2) agonist on the apical surface of gastric mucosa in *Hp* patients and TLR2 activation was shown to induce calcium fluxes that mediate the activation of Ca^{2+}-dependent cysteine proteases the calpains that cleave proteolytically E-cadherin and β-catenin leading to AJs disassembly (O'Connor et al. 2011; Oliveira et al. 2009). In addition, despite the fact that *Hp* is an extracellular pathogen that colonizes the gastric epithelium, it can be detected in an intracellular location *in vitro* as well as *in vivo* (Oh et al. 2005). Nevertheless, the contribution of the *Hp* intracellular life to pathology remains undetermined.

***Neisseria meningitis* opens endothelial AJs via signaling events** *Neisseria meningitis* (*Nm*) is a commensal of the human nasopharyngeal mucosa that is able to cross the blood-brain barrier, causing severe sepsis and cerebrospinal meningitis. *In vitro Nm* adheres on brain endothelial cells, divides on their apical surfaces and forms microcolonies. These structures trigger the formation of membrane rearrangements and bacterial transcytosis and/or loosening of the epithelium and crossing of the barrier. This process is type-IV pili-dependent. In fact, these structures are critical for bacterial adhesion and triggering of signaling events that lead to the depletion of junctional components at the sites of bacterial-cell interactions and the opening of the junctions. *Nm* induces the recruitment of polarity proteins as well as of AJs and tight junctions, altering the integrity of the endothelial junctions in the brain (Coureuil et al. 2009). Specifically, apical polarity proteins, that is Par3, Par6 and PKCζ, which are typically recruited at sites of initial cell-cell contacts, are enriched underneath *Nm* microcolonies, mediating the recruitment of AJ and TJ components, that is VE-cadherin, p120, β-catenin, ZO-1 and claudin, and forming ectopic spot-like junctions that are believed to facilitate the opening of the paracelullar route. Of note, these signaling events are similar to leukocyte extravasation suggesting that *Nm* hijacks this machinery to promote its crossing through the blood-brain barrier (Coureuil et al. 2009; Lemichez et al. 2010).

Streptococcus pneumoniae **directly interacts with E-cadherin** *Streptococcus pneumoniae* (*Sp*) colonizes the nasopharynx and is able to disseminate into the lungs, causing pneumonia and sepsis. In addition, *Sp* is the causative agent of non-invasive diseases, such as otitis, sinusitis and pneumonia. *Sp* interacts with pulmonary epithelial cells and vascular endothelial cells of the alveolar capillaries to traverse the barrier and reach the blood circulation. The pneumococcal surface adhesin A (PsaA) is critical for adherence and colonization. PsaA binds to nasopharyngeal epithelial cells through interaction with its human cellular receptor, E-cadherin (Anderton et al. 2007). Alternatively, pneumococci adherence to pulmonary epithelial cells and vascular endothelial cells *in vitro* is increased by plasminogen, the inactive form of plasmin, which is a serine protease involved in dissolving blood clots. Interestingly, pneumococci coated with active plasmin cleave VE-cadherin, the main component of endothelial AJs, promoting pneumococcal migration through cell barriers. Thus, PsaA-cadherin dependent *Sp* adherence and plasmin-mediated junctional degradation promote pneumococci paracellular crossing of host barriers and dissemination in the host (Attali et al. 2008).

Bacteroides fragilis **enterotoxin mediates E-cadherin cleavage** *Bacteroides fragilis* (*Bf*) is a gut commensal, which corresponds to approximately 0.5% of the bacteria present in the stools. Nevertheless, *Bf* is the most commonly isolated species of Bacteriodaceae in anaerobic infections that originate from the gastrointestinal flora. Strains of *Bf* associated with diarrhea in children produce a zinc-dependent metalloprotease BFT (*B. fragilis* enterotoxin). BFT alters junctional function in polarized epithelial cells by cleaving the extracellular domain of E-cadherin and thus gaining access to the paracellular space (Wu et al. 1998) (Fig. 17.1). Moreover, BFT is involved in the activation of β-catenin signaling in intestinal epithelial cells, rendering this organism able to contribute to oncogenic transformation in the colon (Wu et al. 2003).

17.2.2 Adherens Junctions and Viruses

The human herpes simplex viruses 1 and 2 (HSV-1 and HSV-2), bovine herpes virus-1 (BHV-1), and pseudorabies virus (PRV), are neurotropic with a short replicative cycle. They display a broad host range and associate with AJ components to cause infection (Gonzalez-Mariscal et al. 2009).

HSV-1, HSV-2, BHV-1 and PRV interact directly with nectins HSV-1 produces fever blisters on the lips, and infects the cornea of the eye causing keratitis. HSV-2 causes genital herpes, a sexually transmitted disease. Nectin-1 is the cellular receptor for HSV-1, while both nectin-1 and nectin-2 are the receptors for HSV-2 (Geraghty et al. 1998; Warner et al. 1998). The HSV glycoprotein D binds to the amino terminal V like domain of nectins 1 and 2. Accessibility to nectins results in apical infection of human epithelial cells by HSV (Galen et al. 2006). Nevertheless, nectin needs to be rendered accessible to HSV in order to serve efficiently as a receptor (Yoon and Spear 2002). However, the mechanism enabling nectin accessibility to viral infection remains unknown. BHV-1, that causes rhinotracheitis, conjunctivitis, genital and upper respiratory infections in young calves, uses nectin-1 for cell entry during infection. However, the nectin-1 binding sites for BHV-1 glycoprotein D are not identical to those for HSV-1 and HSV-2 (Struyf et al. 2002). PRV is transmitted via nose-to-nose or fecal-oral contact. Indirect transmission occurs via inhalation. The virus replicates in the respiratory tract, especially in the nasal and pharyngeal mucosa and spreads along cranial nerves to the brain, while via the lymph and blood disseminates to internal organs. Viral entry to its target cells is initiated by binding of the envelope glycoprotein C to the cellular heparin sulphate proteoglycans at the plasma membrane. Next, the glycoprotein D interacts with nectin-1, nectin-2 and nectin-like protein-5 (ncl-5) leading to viral fusion, entry and microtubule-dependent viral transportation to the nucleus (Gonzalez-Mariscal et al. 2009; Yoon and Spear 2002).

Poliovirus interacts directly with nectin-like protein-5 PV is the causative agent of poliomyelitis. PV drains from the gastrointestinal tract into cervical and mesenteric lymph nodes and then to the blood circulation causing viremia and occasionally CNS infection. Viral replication in neurons of the spinal cord, brain stem and motor cortex leads to muscle paralysis and severe cases respiratory arrest and death. The cellular receptor for PV is nectin-like protein-5 (ncl-5). The PV viral protein 1 (VP1) interacts with the amino terminal variable Ig like domain of ncl-5. Ncl-5 is located in the follicle associated epithelium and germinal centres of tonsils and Peyer's patches and on enterocytes of the small intestine and colon. Upon receptor docking VP1 and VP4 are inserted into the cell membrane leading to pore formation, through which virion RNA is injected into the cytoplasm. Thus the empty capsid does not enter. Alternatively PV can also be uptaken by a clathrin and caveoli-independent process (Hogle 2002). Interestingly, not all tissues that express ncl-5 are sites of poliovirus replication. This is due to the presence of soluble isoforms of ncl-5 that compete for virion binding, the absence of host factors essential for translation initiation, post-translational modifications of viral proteins, as well as the anti-viral effect of a mounted IFNα/β response (Gonzalez-Mariscal et al. 2009). Nevertheless, PV crossing of the blood-brain barrier is a rather effective process since viral accumulation in the CNS is a hundred times higher than that of albumin (Gonzalez-Mariscal et al. 2009). However, the exact mechanisms of viral translocation remain to be elucidated.

Rhinoviruses alter E-cadherin expression Rhinovirus (RV) is the causative agent of common cold, and a risk factor for several diseases, such as asthma, rhinosinusitis, and respiratory tract bacterial infections. *In vitro* studies have shown that RV infection of nasal and airway epithelial cells decreases the expression of TJ and AJ components, notably E-cadherin, altering the permeability of the barrier and leading to paracellular entry (Gavala et al. 2011). Once infection is established, virus-induced as well as pro-inflammatory response-mediated damage to the airway tissue results in barrier destruction, epithelial edema, cell shedding and mucus production, which altogether cause airway obstruction and respiratory symptoms (Gavala et al. 2011).

17.2.3 Adherens Junctions and Parasites

***Candida albicans* interacts directly with N- and E-cadherin** *Candida albicans* (*Ca*) is a fungal pathogen, causing oropharyngeal, vulvovaginal and haematogenously disseminated candidiasis. Hyphae internalization into host cells is the first step for tissue invasion. *In vitro*, *Ca* interacts with N-cadherin and E-cadherin to enter endothelial and epithelial cells, respectively. These interactions are dependent on two invasins, Als1 and Als3, and lead to the recruitment of clathrin, dynamin and cortactin at the internalization sites, cytoskeletal rearrangements, internalization and host cell damage. Interestingly, Als3 shares some structural similarity with *Lm* InlA, and it has been reported that *Ca*, like *Lm*, hijacks the clathrin-dependent endocytic machinery to invade host cells (Moreno-Ruiz et al. 2009).

***Trypanosoma cruzi* alters cadherin and β-catenin expression** The protozoan parasite *Trypanosoma cruzi* (*Tc*) is the causative agent of Chagas disease, with clinical manifestations ranging from dilatation of enteric viscera to inflammatory cardiomyopathy. *Tc* is capable of infecting endothelial cells. Nevertheless, it is not yet clear whether the parasite infects and lyses the vascular endothelium first, or breaches the barrier without establishing infection, in order to spread in the surrounding tissue. In cardiomyocytes *Tc* is shown to disrupt AJs by reducing the expression of both cadherin and β-catenin. This causes disorganization of the CCC, potentially disturbing the anchorage of myofibrils to the plasma membrane of the muscle cells and leading to consequent loss of cell tension and to cardiomyopathy (de Melo et al. 2008).

***Trichomonas vaginalis* causes alterations in E-cadherin expression patterns** *Trichomonas vaginalis* (*Tv*) is a protozoan parasite that causes trichomoniasis, a common, non-viral, sexually transmitted disease, associated with preterm delivery, infertility, cervical cancer and increased transmission of HIV. *Tv* adheres to epithelial cells causing alterations in the expression patterns of AJ components, enlargement of the spaces between epithelial cells and loss of barrier integrity. These effects are dependent on AP65 adhesin (da Costa et al. 2005).

17.3 Conclusion

Microbial interactions with AJ components are increasingly recognized as critical events during infectious processes, providing invaluable information on the mechanisms of infection and host immune responses, as well as host physiology, at both the cellular and tissular levels. The following points are worth highlighting:

Pathogens Interact Directly or Indirectly With Cellular Components of AJs *Lm* targets directly luminally accessible E-cadherin, taking advantage of the junctional remodelling occurring at the level of goblet cells, extruding cells and epithelial invaginations. On the other hand *Hp* disturbs AJ integrity indirectly via signalling events mainly promoted by the CagA effector injected into the host cells.

Pathogen-AJ Interactions are Key Events to Initiate Infection During PV entry for example, the VP-1-Ncl-5 interaction leads to membrane docking and viral entry upon pore formation, as a prelude for replication. It is also the case for bacterial entry. *Lm* internalization, mediated by InlA-E-cadherin interaction, is the initial event in the whole process of infection.

Pathogen Species Specificity and Cellular Tropism Often Depend on Species Specific or Tissue Specific Interactions with AJs In the case of PV, wild type mice are resistant to infection since they do not express Ncl-5. Thus, transgenic mice expressing Ncl-5 at the level of the intestine showed susceptibility in CNS infection. Of note, these mice lack the interferon α-/β- receptor, clearly demonstrating that Ncl-5 as well as IFN-α/β control tissue tropism and pathogenicity (Ida-Hosonuma et al. 2005). In the case of bacterial pathogens, the *Lm* tropism for the intestinal barrier and its species specificity relies on the nature of the InlA-E-cadherin interaction (Lecuit et al. 2001).

AJ-Pathogen Interactions are Instrumental to Understand Host Physiology In the case of bacterial pathogens, *Lm* InlA-E-cadherin-dependent infection has led to the identification of novel molecular players and cellular mechanisms involved also in the formation of AJs (Bonazzi et al. 2011; Sousa et al. 2005a). Similarly, viruses have been instrumental to further comprehend endocytic mechanisms, such as clathrin-mediated endocytosis, macropinocytosis and caveolar/lipid raft-mediated endocytosis (Schelhaas 2010).

Overall, pathogen-AJs interactions set the basis for the development of study models to answer fundamental questions on human physiology and to develop novel therapeutic strategies against infectious agents.

Acknowledgements Georgios Nikitas was a recipient of the Marie Curie Actions Early Stage Researcher INTRAPATH programme which he received while preparing his PhD thesis in the Laboratory of Microbes and Host Barriers, under the direction of Marc Lecuit in the Pasteur Institute.

References

Anderton JM, Rajam G, Romero-Steiner S, Summer S, Kowalczyk AP, Carlone GM, Sampson JS, Ades EW (2007) E-cadherin is a receptor for the common protein pneumococcal surface adhesin A (PsaA) of streptococcus pneumoniae. Microb Pathog 42:225–236

Attali C, Durmort C, Vernet T, Di Guilmi AM (2008) The interaction of Streptococcus prneumoniae with plasmin mediates transmigration across endothelial and epithelial monolayers by intercellular junction cleavage Infect Immun 76:5350–5356

Bonazzi M, Cossart P (2011) Host-pathogen interactions: impenetrable barriers or entry portals? The role of cell-cell adhesion during infection. J Cell Biol 195:349–358

Bonazzi M, Vasudevan L, Mallet A, Sachse M, Sartori A, Prevost MC, Roberts A, Taner SB, Wilbur JD, Brodsky FM et al (2011) Clathrin phosphorylation is required for actin recruitment at sites of bacterial adhesion and internalization. J Cell Biol 195:525–536
Coureuil M, Mikaty G, Miller F, Lecuyer H, Bernard C, Bourdoulous S, Dumenil G, Mege RM, Weksler BB, Romero IA et al (2009) Meningococcal type IV pili recruit the polarity complex to cross the brain endothelium. Science 325:83–87
da Costa RF, de Souza W, Benchimol M, Alderete JF, Morgado-Diaz JA (2005) Trichomonas vaginalis perturbs the junctional complex in epithelial cells. Cell Res 15:704–716
de Melo TG, Meirelles Mde N, Pereira MC (2008) Trypanosoma cruzi alters adherens junctions in cardiomyocytes. Microbes Infect 10:1405–1410
Disson O, Grayo S, Huillet E, Nikitas G, Langa-Vives F, Dussurget O, Ragon M, Le Monnier A, Babinet C, Cossart P et al (2008) Conjugated action of two species-specific invasion proteins for fetoplacental listeriosis. Nature 455:1114–1118
Drees F, Pokutta S, Yamada S, Nelson WJ, Weis WI (2005) Alpha-catenin is a molecular switch that binds E-cadherin-beta-catenin and regulates actin-filament assembly. Cell 123:903–915
Galen B, Cheshenko N, Tuyama A, Ramratnam B, Herold BC (2006) Access to nectin favors herpes simplex virus infection at the apical surface of polarized human epithelial cells. J Virol 80:12209–12218
Gavala ML, Bertics PJ, Gern JE (2011) Rhinoviruses, allergic inflammation, and asthma. Immunol Rev 242:69–90
Geraghty RJ, Krummenacher C, Cohen GH, Eisenberg RJ, Spear PG (1998) Entry of alphaherpesviruses mediated by poliovirus receptor-related protein 1 and poliovirus receptor. Science 280:1618–1620
Gonzalez-Mariscal L, Garay E, Lechuga S (2009) Virus interaction with the apical junctional complex. Front Biosci 14:731–768
Hogle JM (2002) Poliovirus cell entry: common structural themes in viral cell entry pathways. Annu Rev Microbiol 56:677–702
Hulpiau P, van Roy F (2009) Molecular evolution of the cadherin superfamily. Int J Biochem Cell Biol 41:349–369
Ida-Hosonuma M, Iwasaki T, Yoshikawa T, Nagata N, Sato Y, Sata T, Yoneyama M, Fujita T, Taya C, Yonekawa H et al (2005) The alpha/beta interferon response controls tissue tropism and pathogenicity of poliovirus. J Virol 79:4460–4469
Lecuit M, Sonnenburg JL, Cossart P, Gordon JI (2007) Functional genomic studies of the intestinal response to a foodborne enteropathogen in a humanized gnotobiotic mouse model. J Biol Chem 282:15065–15072
Lecuit M, Vandormael-Pournin S, Lefort J, Huerre M, Gounon P, Dupuy C, Babinet C, Cossart P (2001) A transgenic model for listeriosis: role of internalin in crossing the intestinal barrier. Science 292:1722–1725
Lemichez E, Lecuit M, Nassif X, Bourdoulous S (2010) Breaking the wall: targeting of the endothelium by pathogenic bacteria. Nat Rev Microbiol 8:93–104
Moreno-Ruiz E, Galan-Diez M, Zhu W, Fernandez-Ruiz E, d'Enfert C, Filler SG, Cossart P, Veiga E (2009) Candida albicans internalization by host cells is mediated by a clathrin-dependent mechanism. Cell Microbiol 11:1179–1189
Niessen CM, Gottardi CJ (2008) Molecular components of the adherens junction. Biochim Biophys Acta 1778:562–571
Nikitas G, Deschamps C, Disson O, Niault T, Cossart P, Lecuit M (2011) Transcytosis of Listeria monocytogenes across the intestinal barrier upon specific targeting of goblet cell accessible E-cadherin. J Exp Med 208:2263–2277
O'Connor PM, Lapointe TK, Jackson S, Beck PL, Jones NL, Buret AG (2011) Helicobacter pylori activates calpain via toll-like receptor 2 to disrupt adherens junctions in human gastric epithelial cells. Infect Immun 79:3887–3894
Oh JD, Karam SM, Gordon JI (2005) Intracellular Helicobacter pylori in gastric epithelial progenitors. Proc Natl Acad Sci U S A 102:5186–5191

Oliveira MJ, Costa AM, Costa AC, Ferreira RM, Sampaio P, Machado JC, Seruca R, Mareel M, Figueiredo C (2009) CagA associates with c-Met, E-cadherin, and p120-catenin in a multiproteic complex that suppresses Helicobacter pylori-induced cell-invasive phenotype. J Infect Dis 200:745–755

Pentecost M, Kumaran J, Ghosh P, Amieva MR (2010) Listeria monocytogenes internalin B activates junctional endocytosis to accelerate intestinal invasion. PLoS Pathog 6(5): e1000900. doi:10.1371/journal.ppat.1000900

Pentecost M, Otto G, Theriot JA, Amieva MR (2006) Listeria monocytogenes invades the epithelial junctions at sites of cell extrusion. PLoS Pathog 2(1): e3. doi:10.1371/journal.ppat.0020003

Sansonetti PJ, Mounier J, Prevost MC, Mege RM (1994) Cadherin expression is required for the spread of Shigella flexneri between epithelial cells. Cell 76:829–839

Schelhaas M (2010) Come in and take your coat off – how host cells provide endocytosis for virus entry. Cell Microbiol 12:1378–1388

Shaikh N, Terajima J, Watanabe H (2003) IpaC of Shigella binds to the C-terminal domain of beta-catenin. Microb Pathog 35:107–117

Sousa S, Cabanes D, Archambaud C, Colland F, Lemichez E, Popoff M, Boisson-Dupuis S, Gouin E, Lecuit M, Legrain P et al (2005a) ARHGAP10 is necessary for alpha-catenin recruitment at adherens junctions and for Listeria invasion. Nat Cell Biol 7:954–960

Sousa S, Lecuit M, Cossart P (2005b) Microbial strategies to target, cross or disrupt epithelia. Curr Opin Cell Biol 17:489–498

Stavru F, Archambaud C, Cossart P (2011) Cell biology and immunology of Listeria monocytogenes infections: novel insights. Immunol Rev 240:160–184

Struyf F, Martinez WM, Spear PG (2002) Mutations in the N-terminal domains of nectin-1 and nectin-2 reveal differences in requirements for entry of various alphaherpesviruses and for nectin-nectin interactions. J Virol 76:12940–12950

Takai Y, Nakanishi H (2003) Nectin and afadin: novel organizers of intercellular junctions. J Cell Sci 116:17–27

Toledo-Arana A, Dussurget O, Nikitas G, Sesto N, Guet-Revillet H, Balestrino D, Loh E, Gripenland J, Tiensuu T, Vaitkevicius K et al (2009) The Listeria transcriptional landscape from saprophytism to virulence. Nature 459:950–956

van Roy F, Berx G (2008) The cell-cell adhesion molecule E-cadherin. Cell Mol Life Sci 65:3756–3788

Warner MS, Geraghty RJ, Martinez WM, Montgomery RI, Whitbeck JC, Xu R, Eisenberg RJ, Cohen GH, Spear PG (1998) A cell surface protein with herpesvirus entry activity (HveB) confers susceptibility to infection by mutants of herpes simplex virus type 1, herpes simplex virus type 2, and pseudorabies virus. Virology 246:179–189.

Wu S, Lim KC, Huang J, Saidi RF, Sears CL (1998) Bacteroides fragilis enterotoxin cleaves the zonula adherens protein, E-cadherin. Proc Natl Acad Sci USA 95:14979–14984

Wu S, Morin PJ, Maouyo D, Sears CL (2003) Bacteroides fragilis enterotoxin induces c-Myc expression and cellular proliferation. Gastroenterology 124:392–400

Yoon M, Spear PG (2002) Disruption of adherens junctions liberates nectin-1 to serve as receptor for herpes simplex virus and pseudorabies virus entry. J Virol 76:7203–7208

Index

T. Harris (ed.), *Adherens Junctions: From Molecular Mechanisms to Tissue Development and Disease,* Subcellular Biochemistry 60,
DOI 10.1007/978-94-007-4186-7, © Springer Science+Business Media Dordrecht 2012

MIX
Papier aus verantwortungsvollen Quellen
Paper from responsible sources
FSC® C105338

If you have any concerns about our products,
you can contact us on
ProductSafety@springernature.com

In case Publisher is established outside the EU,
the EU authorized representative is:
Springer Nature Customer Service Center GmbH
Europaplatz 3, 69115 Heidelberg, Germany

Printed by Libri Plureos GmbH
in Hamburg, Germany